NATIONAL GEOGRAPHIC SCIENCE

TEACHER'S EDITION

LIFE SCIENCE

NATIONAL GEOGRAPHIC School Publishing

PROGRAM AUTHORS

Randy Bell, Ph.D.

Malcolm B. Butler, Ph.D.

Kathy Cabe Trundle, Ph.D.

Judith S. Lederman, Ph.D.

David W. Moore, Ph.D.

NATIONAL GEOGRAPHIC SCIENCE

Built to target your state standards, ***National Geographic Science*** is a research-based program that brings science learning to life through the lens of National Geographic.

Promote Science success as you share

Grade 3

The National Geographic Experience . . .

- **Immerses Students in the Nature of Science and Inquiry.**
- **Unlocks the Big Ideas in Science for All Learners.**
- **Builds Scientific and Content Literacy.**

Grade 4

Grade 5

Immerse Students in . . . The Nature of Science.

In ***National Geographic Science*** process skills build at each grade level to ensure a complete understanding of the Nature of Science. This chart shows how process skills and the Nature of Science work together to help students think and act like scientists.

	Grade K	Grades 1 & 2
PROCESS SKILLS	**OBSERVE**	**OBSERVE & INFER**
Nature of Science	• **Science knowledge is based on evidence.** • **Science knowledge can change based on new evidence.**	• **Science conclusions are based on observation and inference.** • **Science theories are based partly on things that cannot be observed.**

Immerse Students in . . . Leveled, Hands-on Inquiry.

National Geographic Science provides students with abundant and relevant hands-on explorations to facilitate a thorough understanding of key Science concepts. The four levels of inquiry in ***National Geographic Science*** are designed to help students build confidence and competence in scientific thought and inquiry.

Explore Activity

The *Explore Activity* builds background for the unit. This activity *Engages* students as they *Explore*.

Explore Activity

Investigate Star Positions

Question How do some stars appear to move across the sky?

Science Process Vocabulary

observe verb

Scientists often use tools to take a close look at, or **observe**, objects and events.

Materials

safety goggles

Star Finder

What to Do

1. Put on your safety goggles. Cut out both parts of the Star Finde[r]. Cut out the viewing window in the top piece of the Star Finder.

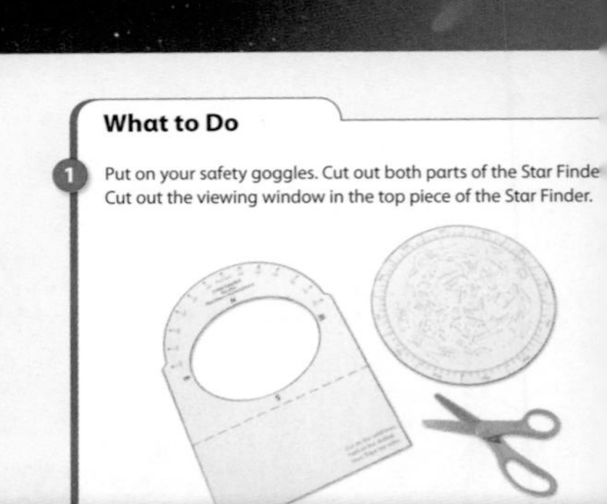

Directed Inquiry

In *Directed Inquiry,* the teacher gives direct instruction throughout the activity. Students are given opportunities to *Explain* what they have done, *Elaborate* by asking further questions, and *Evaluate* by using a self-reflection rubric.

Directed Inquiry

Investigate Weathering

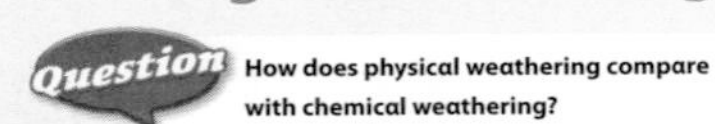

Question How does physical weathering compare with chemical weathering?

Science Process Vocabulary

model noun

A **model** can show how a process, such as weathering, works.

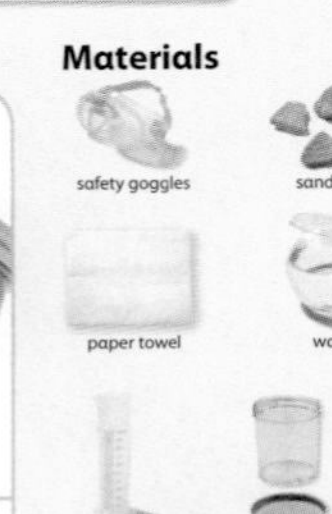

predict verb

When you want to explain what

Materials

safety goggles, sandstone, paper towel, water, graduated cylinder, 2 jars with lids

What to Do

1. Put on your safety goggles. **Predict** what will happen if you rub 2 pieces of sandstone together. Hold the pieces of sandstone over the paper towel and rub them together for a few seconds. **Observe** what happens. Record your observations in your science notebook.

2. **Measure** 250 mL of water into a jar. Place 5 pieces of sandstone in the jar. Put the lid on the jar securely.

3. Use the hand lens to observe the sandstone and the bottom of the jar. Record your

Grade 3

CLASSIFY

- **There is often no single "right" answer in science.**

Grade 4

PREDICT/HYPOTHESIZE

- **Scientific theories provide the base upon which predictions and hypotheses are built.**

Grade 5

DESIGN EXPERIMENTS

- **There is no single, scientific method that all scientists follow.**
- **There are a number of ways to do science.**

Also includes

Science in a Snap!

Offers quick investigations to activate understanding of science concepts

Guided Inquiry

In *Guided Inquiry*, students become independent learners with guidance from the teacher. Students manipulate variables, provide *Explanations, Elaborate* by asking further questions, and *Evaluate* by using a self-reflection rubric.

Guided Inquiry

Investigate Erosion

How does the way water moves on soil affect the way the soil moves?

Science Process Vocabulary

variable noun

A **variable** is a part of an experiment that you can change.

You change only one **variable** while you keep all the other parts the same. You control the parts that do not change.

Materials

safety goggles
3 plastic containers
masking tape
small paper cup

Do an Experiment

Write your plan in your science notebook.

Make a Hypothesis

In this investigation, you will pour water through holes in a cup onto soil. Water moves slowly through small holes and quickly through large holes. How will this affect the amount of erosion you observe? Write your hypothes

Identify, Manipulate, and Control Variables

Which variable will you change?
Which variable will you observe or measure?
Which variables will you keep the same?

What to Do

1. Put on your safety goggles. Label the plastic containers 1, 2, and 3. Put one paper

Open Inquiry

In *Open Inquiry,* students choose their own questions, create their own plans, carry out their plans, collect and record their own data, look for patterns, and share that data. Students *Explain* their results, *Elaborate* by asking further questions, and *Evaluate* by using a self-reflection rubric.

Open Inquiry

Do Your Own Investigation

Choose one of these questions, or make up one of your own to do your investigation.

- How can you use shadows caused by the sun to tell time?
- If I cool half an alum solution at room temperature and half in a cold temperature, will the crystals that form be the same?
- What happens to sand particles of different sizes when they are blown by the wind?
- How does gravity affect soil on a slope?
- What happens when pure water and tap water evaporate?
- How is air temperature different over land and water?

Science Process Vocabulary

hypothesis noun

Open Inquiry Checklist

Here is a checklist you can use when you investigate.

- ☐ Choose a **question** or make up one of your own.
- ☐ Gather the materials you will use.
- ☐ If needed, make a **hypothesis** or a **prediction.**
- ☐ If needed, identify, manipulate, and control **variables.**
- ☐ Make a **plan** for your **investigation.**
- ☐ Carry out your **plan.**
- ☐ Collect and record **data. Analyze** your data.
- ☐ Explain and **share** your results.
- ☐ Tell what you **conclude.**
- ☐ Think of another question.

Cave of Crystals, Naica, Chihuahua, Mexico

The Big Ideas in Science Unlocked and In Depth

The Big Ideas in Science should be targeted and focused for the teacher and the student. Therefore, ***National Geographic Science*** was created to be targeted and focused on these Big Ideas in Science as well.

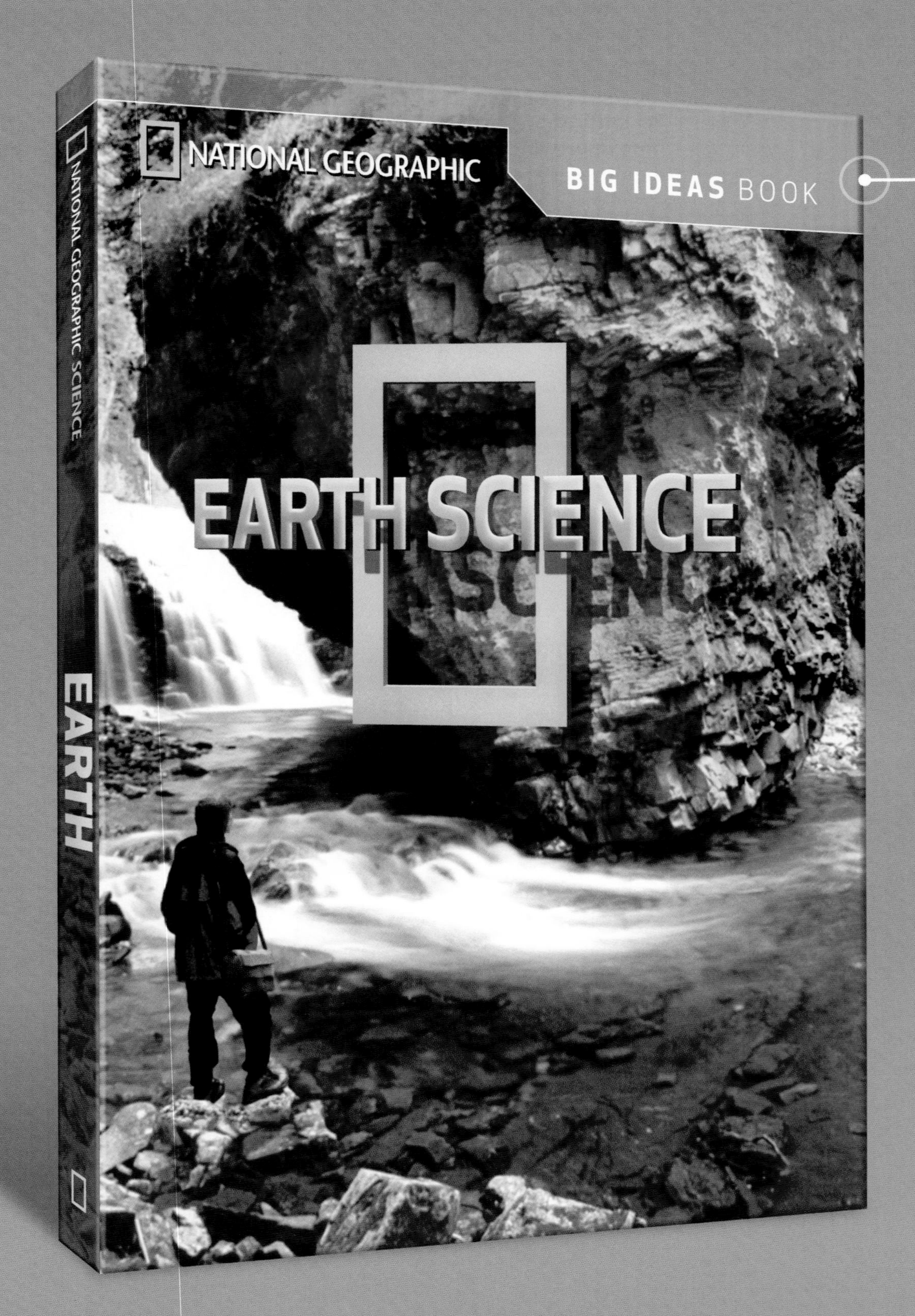

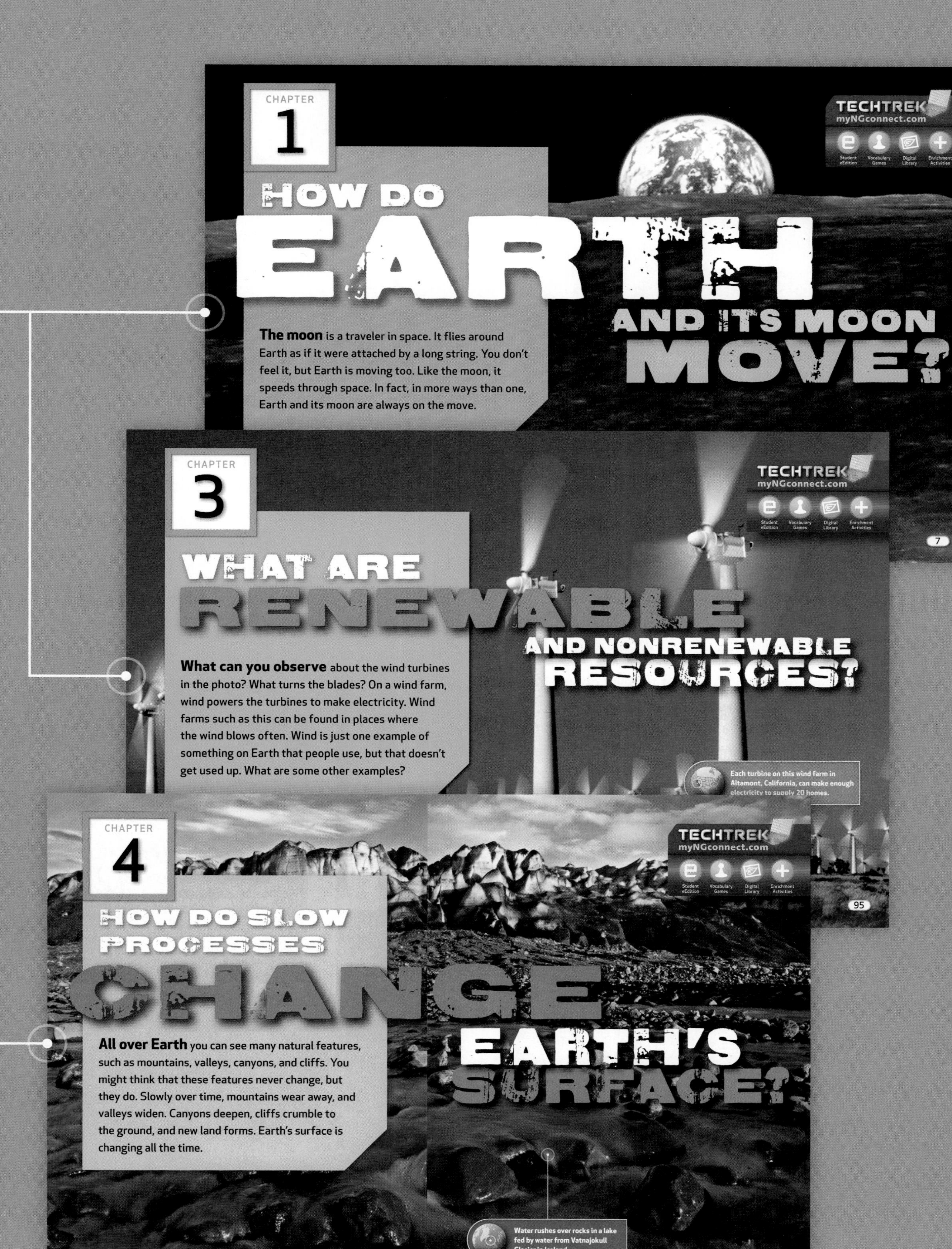

CHAPTER 1

HOW DO EARTH AND ITS MOON MOVE?

The moon is a traveler in space. It flies around Earth as if it were attached by a long string. You don't feel it, but Earth is moving too. Like the moon, it speeds through space. In fact, in more ways than one, Earth and its moon are always on the move.

TECHTREK myNGconnect.com
Student eEdition | Vocabulary Games | Digital Library | Enrichment Activities

7

CHAPTER 3

WHAT ARE RENEWABLE AND NONRENEWABLE RESOURCES?

What can you observe about the wind turbines in the photo? What turns the blades? On a wind farm, wind powers the turbines to make electricity. Wind farms such as this can be found in places where the wind blows often. Wind is just one example of something on Earth that people use, but that doesn't get used up. What are some other examples?

TECHTREK myNGconnect.com
Student eEdition | Vocabulary Games | Digital Library | Enrichment Activities

Each turbine on this wind farm in Altamont, California, can make enough electricity to supply 20 homes.

95

CHAPTER 4

HOW DO SLOW PROCESSES CHANGE EARTH'S SURFACE?

All over Earth you can see many natural features, such as mountains, valleys, canyons, and cliffs. You might think that these features never change, but they do. Slowly over time, mountains wear away, and valleys widen. Canyons deepen, cliffs crumble to the ground, and new land forms. Earth's surface is changing all the time.

TECHTREK myNGconnect.com
Student eEdition | Vocabulary Games | Digital Library | Enrichment Activities

Water rushes over rocks in a lake fed by water from Vatnajokull Glacier in Iceland.

NATIONAL GEOGRAPHIC

Build Scientific Literacy for Your Future Scientists

Students will build scientific literacy and gain an understanding of the Nature of Science from real-life National Geographic Explorers. Your students will learn that Science is:

- A way of knowing
- Empirically based and consistent with evidence
- Subject to change when new evidence presents itself
- A creative process

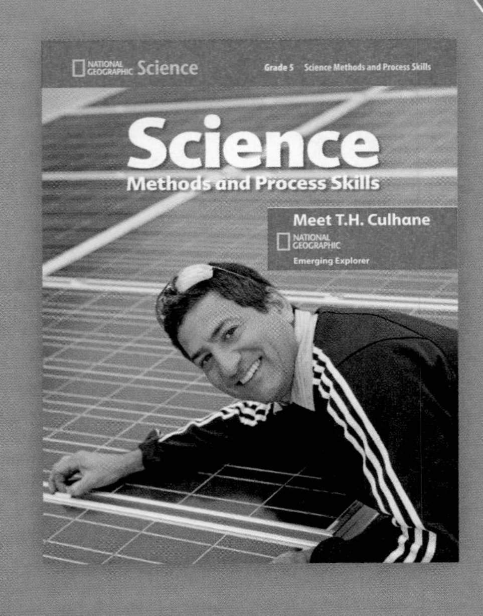

Science
Methods and Process Skills
Meet T.H. Culhane
Emerging Explorer

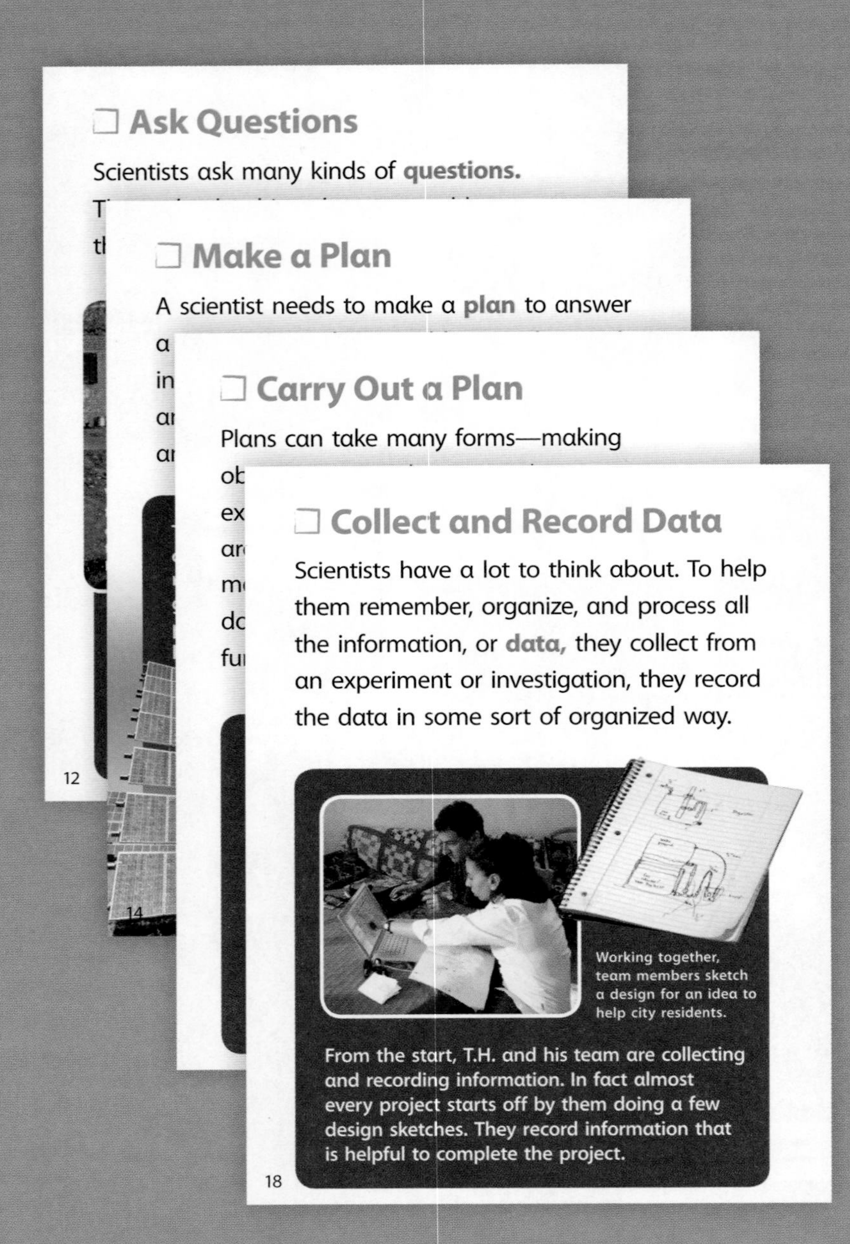

Ask Questions
Scientists ask many kinds of questions.
Make a Plan
A scientist needs to make a plan to answer
Carry Out a Plan
Plans can take many forms—making
Collect and Record Data
Scientists have a lot to think about. To help them remember, organize, and process all the information, or data, they collect from an experiment or investigation, they record the data in some sort of organized way.
Working together, team members sketch a design for an idea to help city residents.
From the start, T.H. and his team are collecting and recording information. In fact almost every project starts off by them doing a few design sketches. They record information that is helpful to complete the project.
18

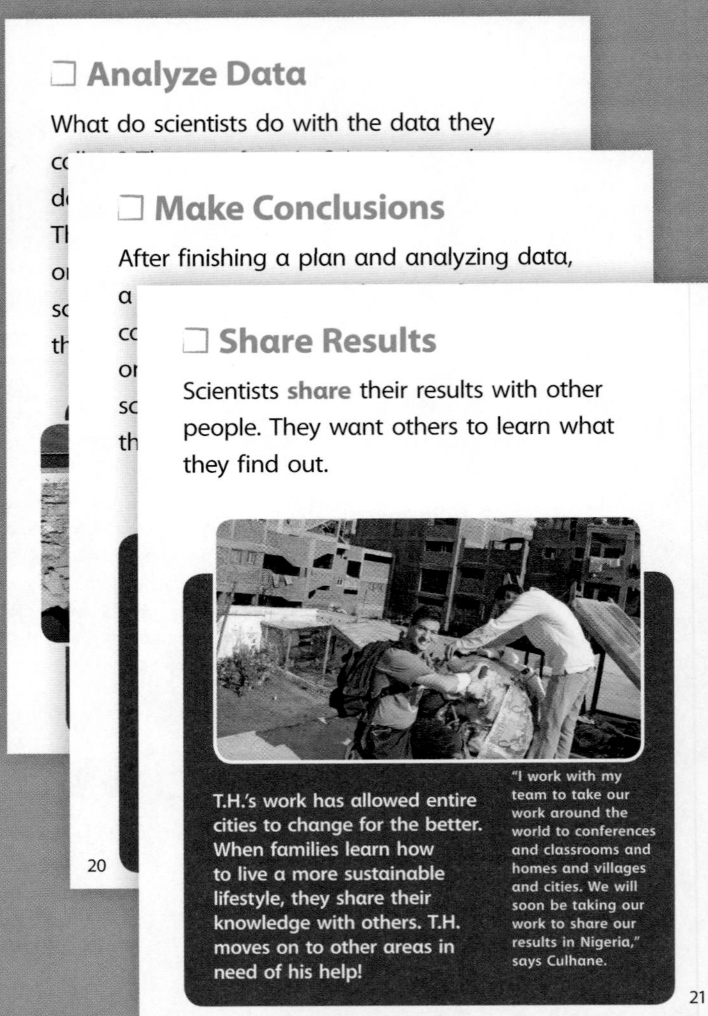

Analyze Data
What do scientists do with the data they
Make Conclusions
After finishing a plan and analyzing data,
Share Results
Scientists share their results with other people. They want others to learn what they find out.
T.H.'s work has allowed entire cities to change for the better. When families learn how to live a more sustainable lifestyle, they share their knowledge with others. T.H. moves on to other areas in need of his help!
"I work with my team to take our work around the world to conferences and classrooms and homes and villages and cities. We will soon be taking our work to share our results in Nigeria," says Culhane.
20
21

Access Science in Our World

Become An Expert

Students, through each chapter's Big Idea, learn concepts specifically tied to your state standards.

In the ***Become An Expert*** section of each chapter, students apply what they learned through concrete examples found throughout our world.

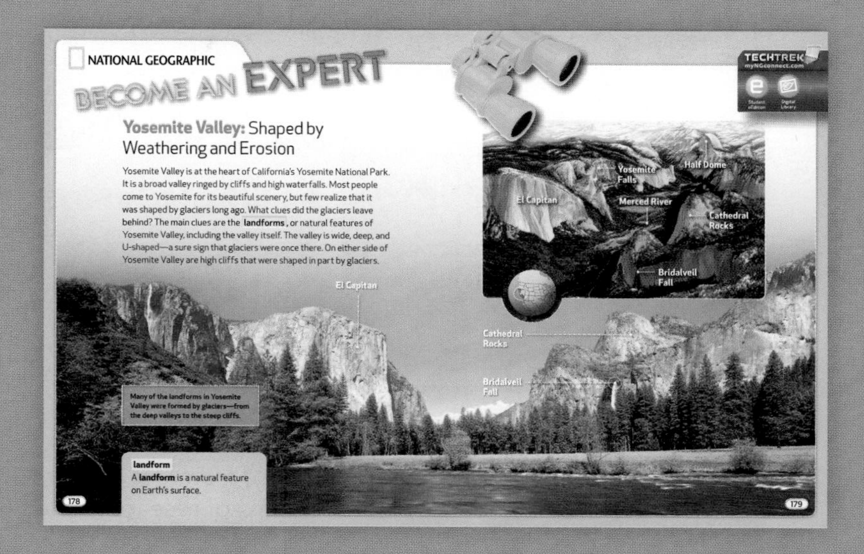

NATIONAL GEOGRAPHIC

BECOME AN EXPERT

Yosemite Valley: Shaped by Weathering and Erosion

Yosemite Valley is at the heart of California's Yosemite National Park. It is a broad valley ringed by cliffs and high waterfalls. Most people come to Yosemite for its beautiful scenery, but few realize that it was shaped by glaciers long ago. What clues did the glaciers leave behind? The main clues are the **landforms**, or natural features of Yosemite Valley, including the valley itself. The valley is wide, deep, and U-shaped—a sure sign that glaciers were once there. On either side of Yosemite Valley are high cliffs that were shaped in part by glaciers.

Many of the landforms in Yosemite Valley were formed by glaciers—from the deep valleys to the steep cliffs.

landform
A **landform** is a natural feature on Earth's surface.

178 179

Reading Comprehension Strategies

In the Teacher's Edition, teachers are offered four comprehension strategies to ensure that content learning is deep and lasting.

- Preview and Predict
- Monitor and Fix Up
- Make Inferences
- Sum Up

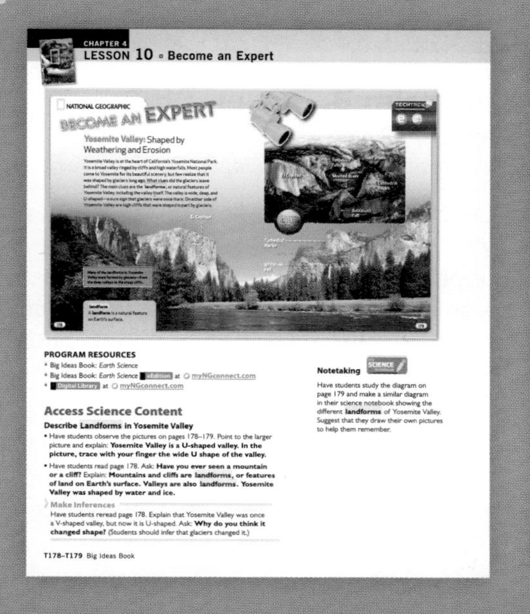

CHAPTER 4
LESSON 10 ■ Become an Expert

PROGRAM RESOURCES

- Big Ideas Book: *Earth Science*
- Big Ideas Book: *Earth Science* eEdition at myNGconnect.com
- Digital Library at myNGconnect.com

Access Science Content

Describe Landforms in Yosemite Valley

- Have students observe the pictures on pages 178–179. Point to the larger picture and explain: **Yosemite Valley is a U-shaped valley. In the picture, trace with your finger the wide U shape of the valley.**
- Have students read page 178. Ask: **Have you ever seen a mountain or a cliff?** Explain: **Mountains and cliffs are landforms, or features of land on Earth's surface. Valleys are also landforms. Yosemite Valley was shaped by water and ice.**

Make Inferences

Have students reread page 178. Explain that Yosemite Valley was once a V-shaped valley, but now it is U-shaped. Ask: **Why do you think it changed shape?** (Students should infer that glaciers changed it.)

Notetaking

Have students study the diagram on page 179 and make a similar diagram in their science notebook showing the different **landforms** of Yosemite Valley. Suggest that they draw their own pictures to help them remember.

T178–T179 Big Ideas Book

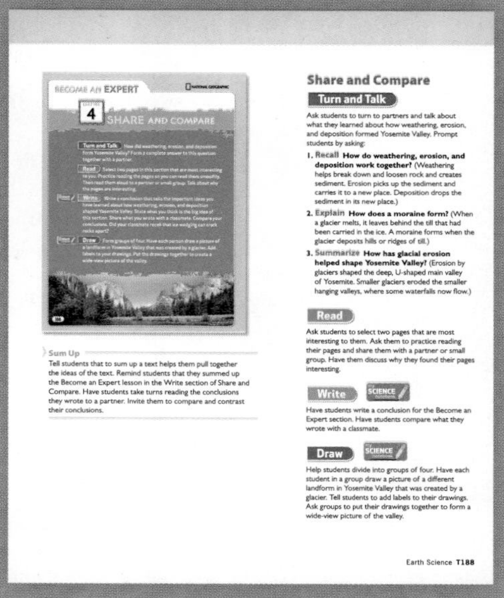

Sum Up

Tell students that to sum up a text helps them pull together the ideas of the text. Remind students that they summed up the Become an Expert lesson in the Write section of Share and Compare. Have students take turns reading the conclusions they wrote to a partner. Invite them to compare and contrast their conclusions.

Share and Compare

Turn and Talk

Ask students to turn to partners and talk about what they learned about how weathering, erosion, and deposition formed Yosemite Valley. Prompt students by asking:

1. Recall **How do weathering, erosion, and deposition work together?** (Weathering helps break down and loosen rock and creates sediment. Erosion picks up the sediment and carries it to a new place. Deposition drops the sediment in its new place.)
2. Explain **How does a moraine form?** (When a glacier melts, it leaves behind the till that had been carried in the ice. A moraine forms when the glacier deposits hills or ridges of till.)
3. Summarize **How has glacial erosion helped shape Yosemite Valley?** (Erosion by glaciers shaped the deep, U-shaped main valley of Yosemite. Smaller glaciers eroded the smaller hanging valleys, where some waterfalls now flow.)

Read

Ask students to select two pages that are most interesting to them. Ask them to practice reading their pages and share them with a partner or small group. Have them discuss why they found their pages interesting.

Write

Have students write a conclusion for the Become an Expert section. Have students compare what they wrote with a classmate.

Draw

Help students divide into groups of four. Have each student in a group draw a picture of a different landform in Yosemite Valley that was created by a glacier. Tell students to add labels to their drawings. Ask groups to put their drawings together to form a wide-view picture of the valley.

Earth Science T188

Explore on Your Own Leveled Books

Explore on Your Own books provide an opportunity for group or independent reading. These books are leveled for your low-mid achieving students and for your mid-high achieving students. These books use the same concept at different reading levels to motivate all students to read independently.

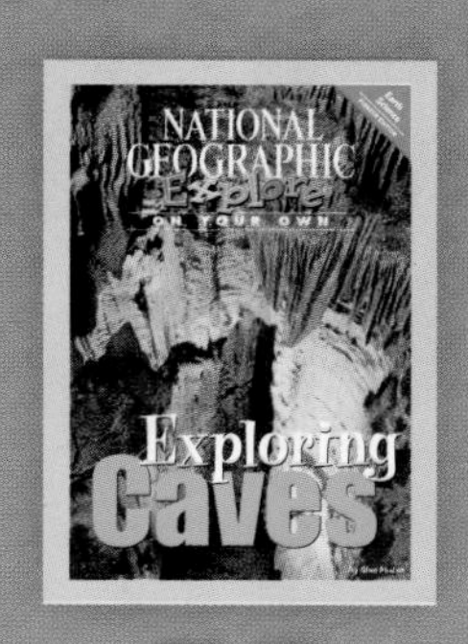

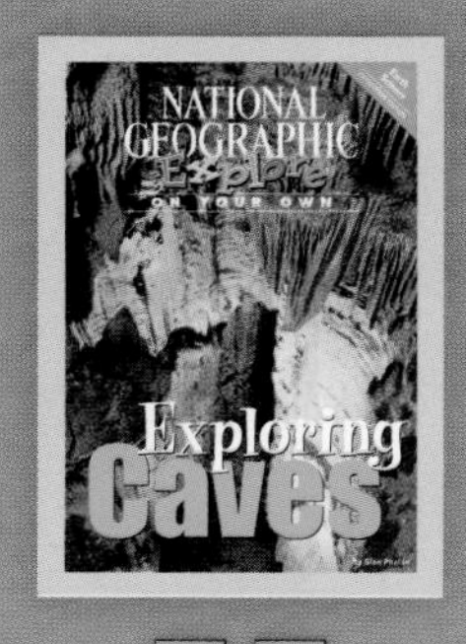

NATIONAL GEOGRAPHIC SCIENCE

Program Authors

Randy Bell, Ph.D.

Associate Professor of Science Education,
University of Virginia, Charlottesville, Virginia

SCIENCE

Dr. Randy Bell began his career as a forest researcher in the Pacific Northwest. His interest in sharing science with others led him to pursue a teaching license, and he taught science for six years in rural eastern Oregon. Currently, Dr. Bell teaches pre-service teachers, provides professional development for practicing teachers, and researches and develops curricular materials. He has received numerous teaching awards as well as the Early Career Research Award from the National Association for Research in Science Teaching. Dr. Bell has authored dozens of research articles and books, and his two primary areas of research focus on teaching and learning about the nature of science and assessing the impact of educational technology.

Malcolm B. Butler, Ph.D.

Associate Professor of Science Education,
University of South Florida, St. Petersburg, Florida

SCIENCE

Dr. Malcolm B. Butler's teaching and research address multicultural issues in the classroom. With a specialization in physics, he has worked to support typically underserved student populations and has interests in the areas of writing to learn in science, science content for elementary teachers, and coastal and environmental education professional development for teachers. He has written and contributed to several academic journals, including *The Journal of Research in Science Teaching, The Journal of Science Teacher Education, The Journal of Multicultural Education,* and *Science Activities.*

Kathy Cabe Trundle, Ph.D.

Associate Professor of Early Childhood Science Education,
The School of Teaching and Learning,
The Ohio State University, Columbus, Ohio

SCIENCE

Dr. Kathy Cabe Trundle has enjoyed her work in science education programs for more than 25 years, including 10 years as a public school teacher. Her research focuses on children and adults' understandings of Earth and space science concepts, including misconceptions and instructional interventions that promote scientific ideas and understandings. Her publications include research reports, professional articles, and curriculum materials. She is also very active in the development of science teacher education and professional development programs. Dr. Cabe Trundle was the recipient of the 2008 Outstanding Teacher Educator of the Year, presented by the Association for Science Teacher Education, an international organization.

Judith Sweeney Lederman, Ph.D.

SCIENCE

Director of Teacher Education, Associate Professor of Science Education,
Department of Mathematics and Science Education,
Illinois Institute of Technology, Chicago, Illinois

Dr. Judith S. Lederman's teaching experiences include pre-K to 12th grade science, masters and doctoral level science education courses, and post-graduate professional development for in-service teachers. She is known nationally and internationally for her work in the teaching and learning of Scientific Inquiry and Nature of Science in both formal and informal settings. In 2008 she was awarded a Fulbright Fellowship to work with South African university and museum educators and K–12 science teachers. Dr. Lederman served on the Board of Directors of the National Science Teachers Association (NSTA) and as President of the Council for Elementary Science International (CESI).

David W. Moore, Ph.D.

LITERACY

Professor of Education, College of Teacher Education and Leadership,
Arizona State University, Tempe, Arizona

Dr. David W. Moore taught reading in Arizona public schools before entering college teaching. He currently is in the Division of Educational Leadership and Innovation at Arizona State University where he researches and teaches courses in literacy across the content areas. He is actively involved with several professional associations. His thirty-year publication record balances research reports, professional articles, book chapters, and books. Noteworthy publications include a history of content area literacy instruction and a quantitative and qualitative review of graphic organizer research. He currently is preparing the sixth edition of *Developing Readers and Writers in the Content Areas: K–12.*

NATIONAL GEOGRAPHIC SCIENCE

Program Reviewers

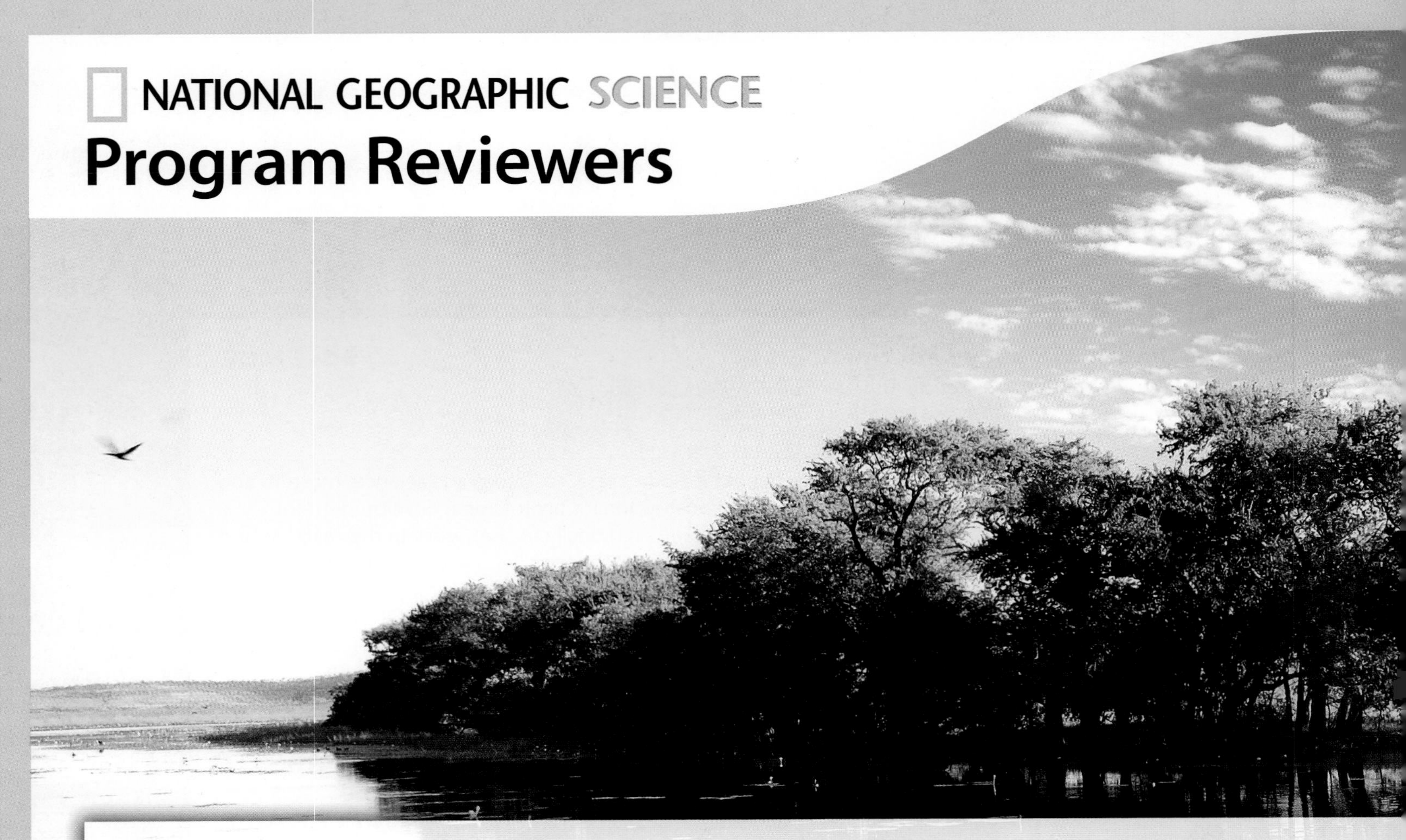

Amani Abuhabsah
Teacher, Dawes Elementary
Chicago, IL

Maria Aida Alanis, Ph.D.
Elementary Science Instructional Coordinator, Austin Independent School District
Austin, TX

Jamillah Bakr
Science Mentor Teacher, Cambridge Public Schools
Cambridge, MA

Gwendolyn Battle-Lavert
Assistant Professor of Education, Indiana Wesleyan University
Marion, IN

Carmen Beadles
Retired Science Instructional Coach, Dallas Independent School District
Dallas, TX

Andrea Blake-Garrett, Ed.D.
Science Educational Consultant,
Newark, NJ

Lori Bowen
Science Specialist, Fayette County Schools
Lexington, KY

Pamela Breitberg
Lead Science Teacher, Zapata Academy
Chicago, IL

Carol Brueggeman
K–5 Science/Math Resource Teacher, District 11
Colorado Springs, CO

Miranda Carpenter
Teacher, MS Academy Leader, Imagine School
Bradenton, FL

Samuel Carpenter
Teacher, Coonley Elementary
Chicago, IL

Diane E. Comstock
Science Resource Teacher, Cheyenne Mountain School District
Colorado Springs, CO

Kelly Culbert
K–5 Science Lab Teacher, Princeton Elementary
Orange County, FL

Karri Dawes
K–5 Science Instructional Support Teacher, Garland Independent School District
Garland, TX

Richard Day
Science Curriculum Specialist, Union Public Schools
Tulsa, OK

Michele DeMuro
Teacher/Educational Consultant,
Monroe, NY

Richard Ellenburg
Science Lab Teacher, Camelot Elementary
Orlando, FL

Beth Faulkner
Brevard Public Schools Elementary Training Cadre, Science Point of Contact, Teacher, NBCT, Apollo Elementary
Titusville, FL

Kim Feltre
Science Supervisor, Hillsborough School District
Newark, NJ

Judy Fisher
Elementary Curriculum Coordinator, Virginia Beach Schools
Virginia Beach, VA

Anne Z. Fleming
Teacher, Coonley Elementary
Chicago, IL

Becky Gill, Ed.D.
Principal/Elementary Science Coordinator, Hough Street Elementary
Barrington, IL

Rebecca Gorinac
Elementary Curriculum Director, Port Huron Area Schools
Port Huron, MI

Anne Grall Reichel, Ed.D.
Educational Leadership/ Curriculum and Instruction Consultant,
Barrington, IL

Mary Haskins, Ph.D.
Professor of Biology, Rockhurst University
Kansas City, MO

Arlene Hayman
Teacher, Paradise Public School District
Las Vegas, NV

DeLene Hoffner
Science Specialist, Science Methods Professor, Regis University, Academy 20 School District
Colorado Springs, CO

Cindy Holman
District Science Resource Teacher, Jefferson County Public Schools
Louisville, KY

Sarah E. Jesse
Instructional Specialist for Hands-on Science, Rutherford County Schools
Murfreesboro, TN

Dianne Johnson
Science Curriculum Specialist, Buffalo City School District
Buffalo, NY

Kathleen Jordan
Teacher, Wolf Lake Elementary
Orlando, FL

Renee Kumiega
Teacher, Frontier Central School District
Hamburg, NY

Edel Maeder
K–12 Science Curriculum Coordinator, Greece Central School District
North Greece, NY

Trish Meegan
Lead Teacher, Coonley Elementary
Chicago, IL

Donna Melpolder
Science Resource Teacher, Chatham County Schools
Chatham, NC

Melissa Mishovsky
Science Lab Teacher, Palmetto Elementary
Orlando, FL

Nancy Moore
Educational Consultant,
Port Stanley, Ontario, Canada

Melissa Ray
Teacher, Tyler Run Elementary
Powell, OH

Shelley Reinacher
Science Coach, Auburndale Central Elementary
Auburndale, FL

Kevin J. Richard
Science Education Consultant, Office of School Improvement, Michigan Department of Education
Lansing, MI

Cathe Ritz
Teacher, Louis Agassiz Elementary
Cleveland, OH

Rose Sedely
Science Teacher, Eustis Heights Elementary
Eustis, FL

Robert Sotak, Ed.D.
Science Program Director, Curriculum and Instruction, Everett Public Schools
Everett, WA

Karen Steele
Teacher, Salt Lake City School District
Salt Lake City, UT

Deborah S. Teuscher
Science Coach and Planetarium Director, Metropolitan School District of Pike Township
Indianapolis, IN

Michelle Thrift
Science Instructor, Durrance Elementary
Orlando, FL

Cathy Trent
Teacher, Ft. Myers Beach Elementary
Ft. Myers Beach, FL

Jennifer Turner
Teacher, PS 146
New York, NY

Flavia Valente
Teacher, Oak Hammock Elementary
Port St. Lucie, FL

Deborah Vannatter
District Coach, Science Specialist, Evansville Vanderburgh School Corporation
Evansville, IN

Katherine White
Science Coordinator, Milton Hershey School
Hershey, PA

Sandy Yellenberg
Science Coordinator, Santa Clara County Office of Education
Santa Clara, CA

Hillary Zeune de Soto
Science Strategist, Lunt Elementary
Las Vegas, NV

NATIONAL GEOGRAPHIC SCIENCE

Explorers and Scientists

These explorers and scientists help students understand real-world science, the nature of science, and science inquiry.

CONSTANCE ADAMS

National Geographic Emerging Explorer
Space Architect

"When you have a brand new problem, you need as many tools as you can get. Who knows? An approach from a very different field might give you the insight you need. For example, I'm working to forge communication between advanced engineering and consumer product design to bring more user-centered designs to aerospace."

STEPHON ALEXANDER, Ph.D.

National Geographic Emerging Explorer
Theoretical Physicist

"My childhood was full of surprises. It taught me the idea of embracing the unknown. I cope with unexpected events by making up theories about why they may be happening."

THOMAS TAHA RASSAM "T.H." CULHANE

National Geographic Emerging Explorer
Urban Planner

"During rain forest ecology fieldwork with the Dayak of Borneo and the Maya Itza of Guatemala's jungle villages, I witnessed a culture that used every part of the environment to survive and thrive. This inspired me to rethink urban living along those same ecological principles. Now I want to bring that message to the rest of the world."

LUKE DOLLAR, Ph.D.

National Geographic Emerging Explorer
Conservation Scientist

"When you're in the field, it's muddy, sweaty, stinky, gritty—there's no glamour to it at all. But it's great fun. I wake up every morning knowing I'm one of the luckiest guys on Earth because I'm doing exactly what I want to do, and it's going to make a difference."

MARIANNE DYSON

Science Writer and Former NASA Flight Controller

"Children who learn to observe, describe, and predict forces in the world will develop the skills to tackle the challenges of a future increasingly dependent on technology."

MARIA FADIMAN, Ph.D.

National Geographic Emerging Explorer
Ethnobotanist

"I was born with a passion for conservation and a fascination with indigenous cultures. Ethnobotany lets me bring it all together. On my first trip to the rain forest, I met a woman who was in terrible pain because no one in her village could remember which plant would cure her. I saw that knowledge was truly being lost, and in that moment, I knew this was what I wanted to do with my life."

BEVERLY GOODMAN, Ph.D.

National Geographic Emerging Explorer
Geo-Archaeologist

"We can learn about our past, present, and future by studying the sea. Coastal regions present a major challenge. I hope the clues I am collecting help avert future catastrophe."

MADHULIKA GUHATHAKURTA, Ph.D.

NASA Astrophysicist

"Earth is inside the atmosphere of the sun. The sun is so big that if you dropped the entire Earth onto the sun's surface, it would barely make a decent sunspot."

AINISSA RAMIREZ, Ph.D.

Physicist

"In a technologically-driven society, we need citizens who can comprehend science. Knowledge of science is necessary in order to fully appreciate the world's beauty."

ALBERT YU-MIN LIN, Ph.D.

National Geographic Emerging Explorer
Material Scientist, Digital Explorer

"I spent much of my young adulthood dreaming of exploration and have realized that goals can only be reached by many different approaches."

TIM SAMARAS

National Geographic Emerging Explorer
Severe-Storms Researcher

"I started to love science when I was a young child. Now I am studying tornadoes to help unlock their secrets for a better understanding of how and why they form."

GREG MARSHALL

National Geographic Filmmaker
Marine Biologist, Conservationist, Inventor

"Science is a great adventure in exploration, discovery, and learning new things. We want to expand our knowledge and inspire others to help conserve and protect animals and their habitats."

TIERNEY THYS, Ph.D.

National Geographic Emerging Explorer
Marine Biologist, Filmmaker

"I think what it takes to be an explorer is simply to never lose your curiosity, never lose your desire to learn more, to ask questions, and to keep pushing deeper and deeper into what you're studying. It's a vital time to be an explorer; there's never been a better, more important time."

MIREYA MAYOR, Ph.D.

National Geographic Emerging Explorer
Primatologist, Conservationist

"The more questions I asked, the more it became clear to me that much about our natural world still remained a mystery."

KATEY WALTER ANTHONY, Ph.D.

National Geographic Emerging Explorer
Aquatic Ecologist, Biogeochemist

"We are researching the greenhouse gas that could have the most powerful effect of all on global warming. It's a worldwide responsibility to reduce our carbon footprint and its effects on the atmosphere."

NATIONAL GEOGRAPHIC
SCIENCE
K-5

Take your students from the classroom to the world!

Grade K

Life Science
- Plants
- Animals

Earth Science
- Day and Night
- Weather and Seasons

Physical Science
- Observing Objects
- How Things Move

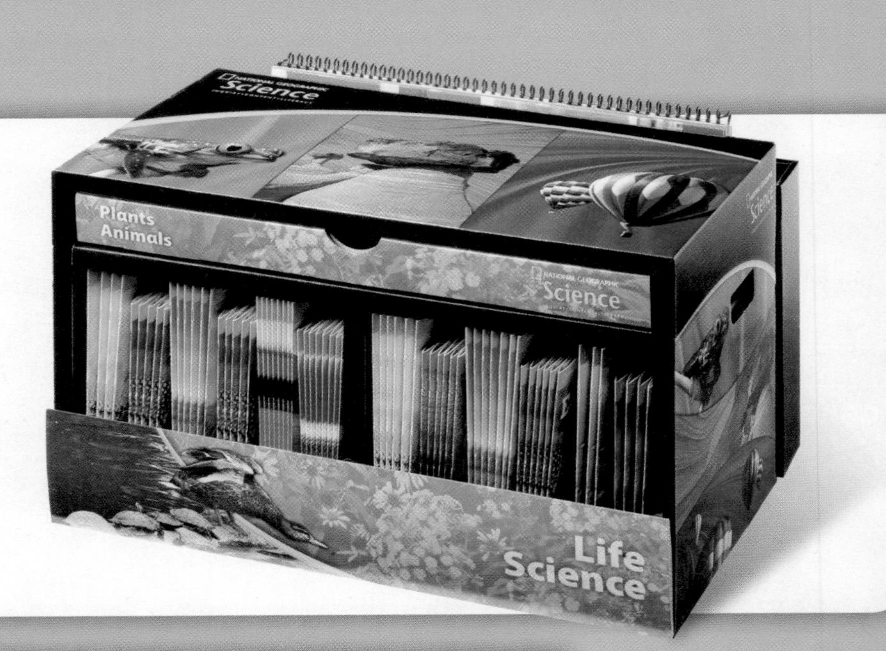

Grades 1-2
- Living Things
- Plants and Animals
- Sun, Moon, and Stars
- Land and Water
- Pushes and Pulls
- Properties
- Habitats
- Life Cycles
- Weather
- Rocks and Soil
- Solids, Liquids, and Gases
- Forces and Motion

Grades K–5

myNGconnect.com

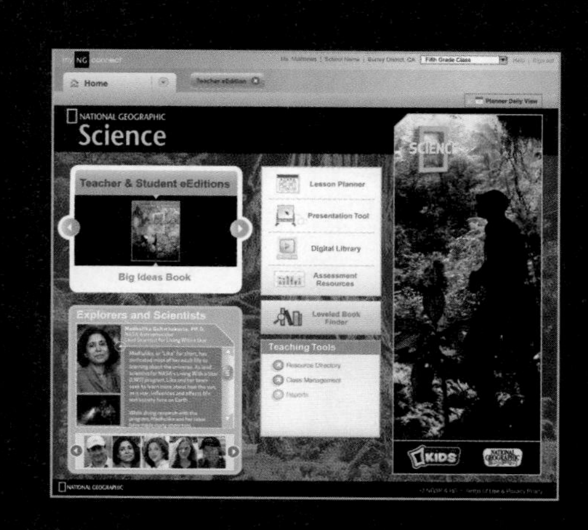

Grade 3

Big Ideas Books

Science Inquiry and Writing Book

Explore on Your Own Books

Grade 4

Big Ideas Books

Science Inquiry and Writing Book

Explore on Your Own Books

Grade 5

Big Ideas Books

Science Inquiry and Writing Book

Explore on Your Own Books

NATIONAL GEOGRAPHIC
SCIENCE

STUDENT
Unit Components

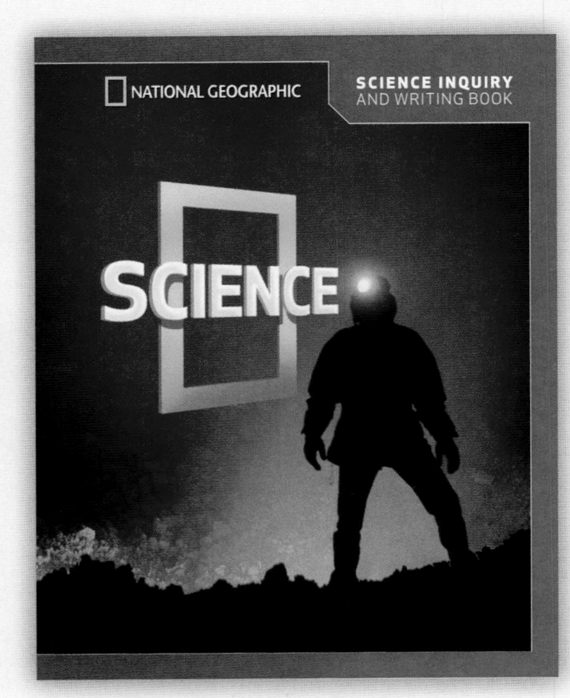

Science Inquiry and Writing Book

Big Ideas Books

TECHTREK
myNGconnect.com

Unit Launch Videos

Explore on Your Own Books

National Geographic Digital Library

Enrichment Activities

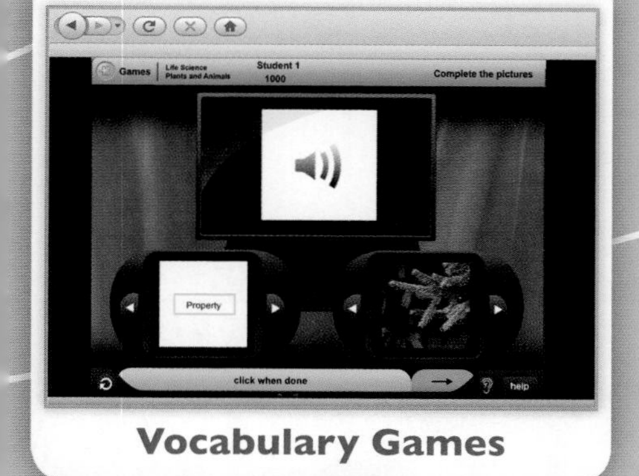

Vocabulary Games

National Geographic Kids Web Site

National Geographic Explorer! Web Site

NATIONAL GEOGRAPHIC
SCIENCE

TEACHER
Unit Components

Teacher's Editions

Grade 4 Science Methods and Process Skills
Teacher's Guide

NATIONAL GEOGRAPHIC Science
Grade 4 Science Methods and Process Skills
Science
Methods and Process Skills
Meet Luke Dollar
NATIONAL GEOGRAPHIC
Emerging Explorer

Science Methods and Process Skills Book / Teacher's Guide

TECHTREK
myNGconnect.com

www.myNGconnect.com

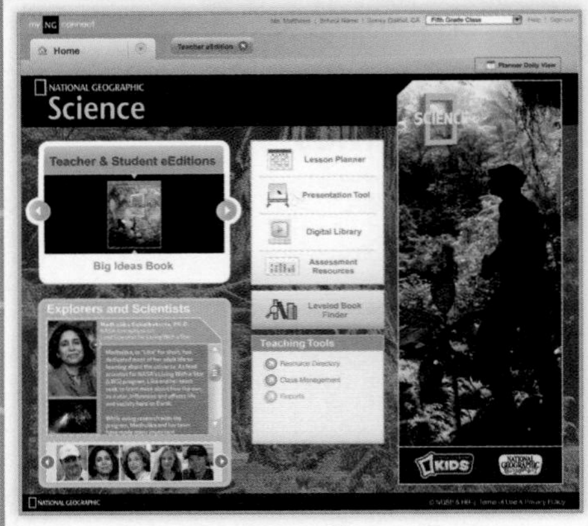

Integrated Online System

Science Inquiry Kit

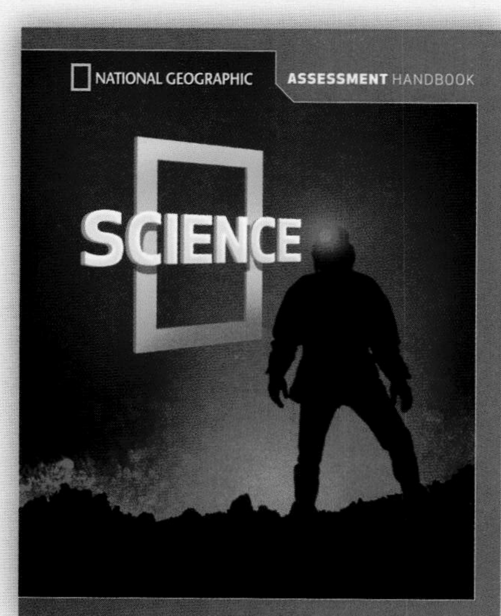

Assessment Handbook

Learning Masters

National Geographic ExamView® CD-ROM

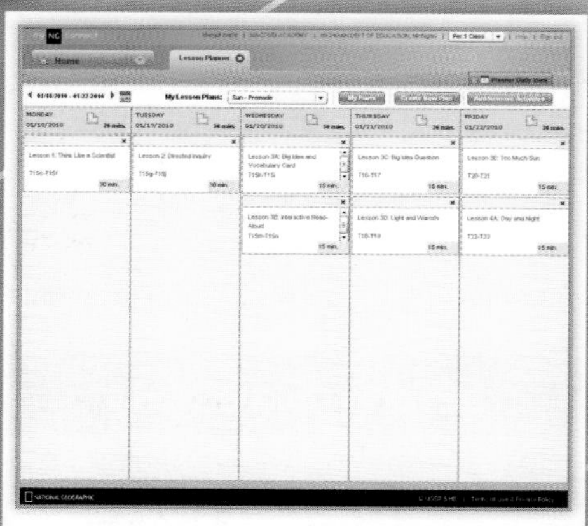

Online Lesson Planner

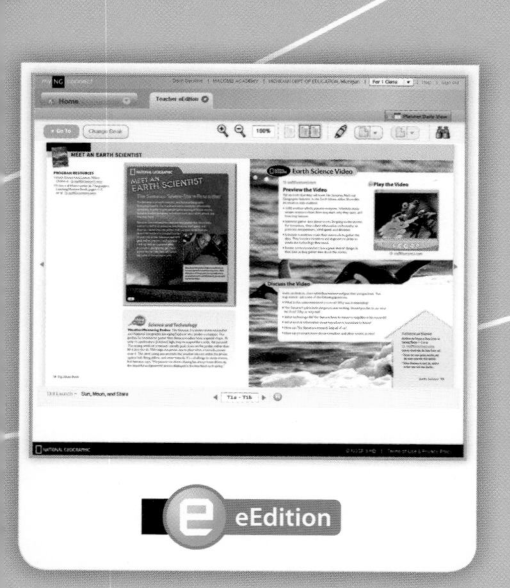

eEdition

National Geographic

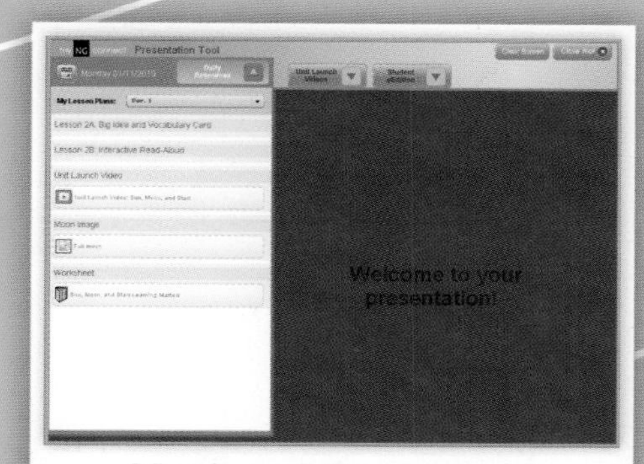

National Geographic

Assessment

Assessment Design

National Geographic Science assessments have been designed so that frequent, varied assessment informs instruction every step of the way. Chapter and Benchmark Tests provide a window into student thinking about scientific concepts throughout the instructional cycle.

Instruct

Develop an understanding of and provide explicit and systematic instruction in:

- The Big Ideas in Science
- Scientific Inquiry
- Science Academic Vocabulary

Assess to Monitor Progress

Chapter Tests provide opportunities for students to apply their understanding of key concepts. Use **Chapter Tests** for timely information about student progress as you deliver instruction.

Show Success

The **Benchmark Test** measures a student's overall progress in understanding the Life, Earth, and Physical units' Big Ideas.

Assessment Tools

National Geographic Science offers an array of assessments in a variety of formats.

ASSESSMENT TOOLS	Assessment Handbook	PDFs myNGConnect.com	Exam*View*®
CHAPTER TESTS • Chapter Tests provide immediate feedback on students' understanding of standards-based scientific concepts. • The tests are administered at the end of each chapter to provide an early indicator of the students' progress.	✓	✓	✓
CHAPTER SELF-ASSESSMENT • The Chapter Self-Assessment empowers students to rate their own understanding of the chapter's concepts and share their opinions about future interests. • The self-assessment is completed by students before they take the Chapter Test.	✓	✓	
BENCHMARK TEST • The Benchmark Test measures students' progress toward understanding the unit's Big Ideas. • The test is administered at the end of the unit to measure the students' understanding of the science standards taught in the unit.	✓	✓	✓
INQUIRY RUBRICS • Inquiry Rubrics assess students' performance of skills in Inquiry activities. • The rubrics are completed after students have finished each Inquiry activity.	✓	✓	
INQUIRY SELF-REFLECTIONS • Inquiry Self-Reflections engage students in evaluating their own performance of inquiry skills and in sharing their opinions about the activity process. • The self-reflections are completed by students after they have finished each Inquiry activity.	✓	✓	

Unit Overview and Pacing Guide

Nature of Science / Science Notebook

	DAY 1	20 minutes	**Setting Up a Science Notebook** • *Big Ideas Book*	*page SN3*

Life Science Big Ideas Book and Inquiry

eEdition

Unit Launch	DAY 1	30 minutes	**What Is Life Science?**	*page T2*
			Meet a Scientist/Unit Launch Video • *Big Ideas Book*	*page T4*
Chapter 1	DAY 2	30 minutes	**Explore Activity** *Investigate a Plant Life Cycle* • *Science Inquiry and Writing Book, page 16*	*page T5e*
	DAYS 3–6	35 minutes each	**How Do Plants Grow and Reproduce?** • *Big Ideas Book*	*page T6–T7*
	DAY 7	50 minutes	**Think Like a Scientist** Math in Science: Recording Observations • *Science Inquiry and Writing Book, page 20*	*page T33a*
			Directed Inquiry *Investigate Flowers* • *Science Inquiry and Writing Book, page 24*	*page T33e*
	DAY 8	50 minutes	**Conclusion and Review**	*page T34*
			NATIONAL GEOGRAPHIC LIFE SCIENCE EXPERT *Plant Biologist*	*page T36*
			NATIONAL GEOGRAPHIC BECOME AN EXPERT *Bristlecone Pines: The Old Ones of the Mountains* • *Big Ideas Book*	*page T38–T39*
			Explore ON YOUR OWN *Serious Survivors* PIONEER PATHFINDER	*page T44a*
Chapter 2	DAY 9	30 minutes	**Directed Inquiry** *Investigate Earthworm Behavior* • *Science Inquiry and Writing Book, page 28*	*page T45e*
	DAYS 10–13	35 minutes each	**How Do Animals Grow and Change?** • *Big Ideas Book*	*page T46–T47*
	DAY 14	20 minutes	**Guided Inquiry** *Investigate Insect Life Cycles* • *Science Inquiry and Writing Book, page 32*	*page T69a*
	DAY 15	50 minutes	**Conclusion and Review**	*page T70*
			NATIONAL GEOGRAPHIC LIFE SCIENCE EXPERT *Lepidopterist*	*page T72*
			NATIONAL GEOGRAPHIC BECOME AN EXPERT *Nature's Transformers* • *Big Ideas Book*	*page T74–T75*
			Explore ON YOUR OWN *Web Wizards* PIONEER PATHFINDER	*page T84a*

Chapter 3	DAY 16	40 minutes	Directed Inquiry *Investigate Owl Pellets* • *Science Inquiry and Writing Book, page 36*	*page T85e*
	DAYS 17–21	35 minutes each	**How Do Living Things Depend on Their Environment?** • *Big Ideas Book*	*page T86–T87*
	DAY 22	40 minutes	Guided Inquiry *Investigate Food Chains and Webs* • *Science Inquiry and Writing Book, page 40*	*page T119a*
	DAY 23	50 minutes	**Conclusion and Review**	*page T120*
			NATIONAL GEOGRAPHIC LIFE SCIENCE EXPERT *Behavorial Ecologist*	*page T122*
			NATIONAL GEOGRAPHIC BECOME AN EXPERT *Bats: Winged Wonders of the Night* • *Big Ideas Book*	*page T124–T125*
			Explore ON YOUR OWN *Wild Ponies* PIONEER PATHFINDER	*page T132a*
Chapter 4	DAY 24	40 minutes	Directed Inquiry *Investigate Plant Fossils* • *Science Inquiry and Writing Book, page 44*	*page T133g*
	DAYS 25–29	35 minutes each	**How Do Adaptations Help Living Things Survive?** • *Big Ideas Book*	*page T134–T135*
	DAY 30	30 minutes	Guided Inquiry *Investigate Animal Survival* • *Science Inquiry and Writing Book, page 48*	*page T169a*
	DAY 31	50 minutes	**Conclusion and Review**	*page T170*
			NATIONAL GEOGRAPHIC LIFE SCIENCE EXPERT *Herpetologist*	*page T172*
			NATIONAL GEOGRAPHIC BECOME AN EXPERT *Snakes: Adaptations for Many Environments* • *Big Ideas Book*	*page T174–T175*

NATIONAL GEOGRAPHIC

Unit Overview and Pacing Guide, *continued*

	Day	Time	Activity	Page
Chapter 5	**DAY 32**	**20** minutes	Directed Inquiry *Investigate Plants and Water* • *Science Inquiry and Writing Book, page 52*	*page T181e*
	DAYS 33–38	**35** minutes each	**How Do Living Things Interact with Their Environment?** • *Big Ideas Book*	*page T182–T183*
	DAY 39	**40** minutes	Guided Inquiry *Investigate Colors in Green Leaves* • *Science Inquiry and Writing Book, page 56*	*page T213a*
	DAY 40	**50** minutes	**Conclusion and Review**	*page T214*
			NATIONAL GEOGRAPHIC LIFE SCIENCE EXPERT *Conservationist*	*page T216*
			NATIONAL GEOGRAPHIC BECOME AN EXPERT *Green Movements Around the World* • *Big Ideas Book*	*page T218–T219*
Chapter 6	**DAY 41**	**30** minutes	Directed Inquiry *Investigate How Your Brain Can Work* • *Science Inquiry and Writing Book, page 60*	*page T229g*
	DAYS 42–45	**30** minutes each	**How Do the Parts of an Organism Work Together?** • *Big Ideas Book*	*page T230–T231*
	DAY 46	**30** minutes	Guided Inquiry *Investigate Exercise and Heart Rate* • *Science Inquiry and Writing Book, page 64*	*page T255a*
	DAY 47	**50** minutes	**Conclusion and Review**	*page T256*
			NATIONAL GEOGRAPHIC LIFE SCIENCE EXPERT *Pharmacologist*	*page T258*
			NATIONAL GEOGRAPHIC BECOME AN EXPERT *Technology and Body Systems* • *Big Ideas Book*	*page T260–T261*
Unit Wrap Up	**DAY 48**	**50** minutes	Open Inquiry *Do Your Own Investigation* • *Science Inquiry and Writing Book, page 68*	*page T268b*
			Write Like a Scientist Write About an Investigation: How Animals Get Energy • *Science Inquiry and Writing Book, page 70*	*page T268d*
	DAY 49	**30** minutes	**Think Like a Scientist** How Scientists Work: Designing Investigations and Using Evidence • *Science Inquiry and Writing Book, page 78*	*page T268j*
	DAY 50	**50** minutes *(Select one unit project.)*	**Make a Life Cycle Poster**	*page T268n*
			Make a Web Page About Living Things and Their Environment	*page T268o*
			Act Out an Interview with a Conservationist	*page T268p*
			Make a Book About Body Systems • *Big Ideas Book*	*page T268q*

Nature of Science/ Science Notebook

Understanding the Nature of Science

In *National Geographic Science* process skills build at each grade level to ensure a complete understanding of the Nature of Science. This chart shows how process skills and the Nature of Science work together to help students think and act like scientists. The Nature of Science is embedded in all parts of the program, not just in Inquiry.

	Grade K	Grades 1 & 2
PROCESS SKILLS	**OBSERVE**	**OBSERVE & INFER**
Nature of Science	• Science knowledge is based on evidence. • Science knowledge can change based on new evidence.	• Science conclusions are based on observation and inference. • Science theories are based partly on things that cannot be observed.

Contents

Grade 3

CLASSIFY

- There is often no single "right" answer in science.

Grade 4

PREDICT/HYPOTHESIZE

- Scientific theories provide the base upon which predictions and hypotheses are built.

Grade 5

DESIGN EXPERIMENTS

- There is no single, scientific method that all scientists follow.
- There are a number of ways to do science.

Why Use a Science Notebook?

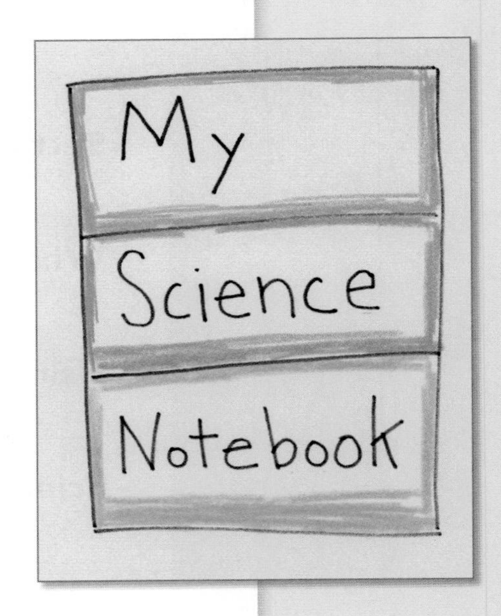

RECORD PREDICTIONS AND HYPOTHESES

A science notebook is a place for students to record predictions and hypotheses, much like scientists record their predictions and hypotheses. Science notebooks make science more meaningful for students because they are able to predict, hypothesize, record, and reflect on their work. Students will learn through making predictions and hypotheses that scientific theories provide the basis upon which predictions and hypotheses are built, an important tenet of the Nature of Science.

DEVELOP INQUIRY SKILLS

Using a science notebook can help students develop inquiry skills and practice science process skills such as observe, sort, infer, predict, collect data, and conclude. Predict and hypothesize are two key process skills that will help students understand the Nature of Science in *National Geographic Science.*

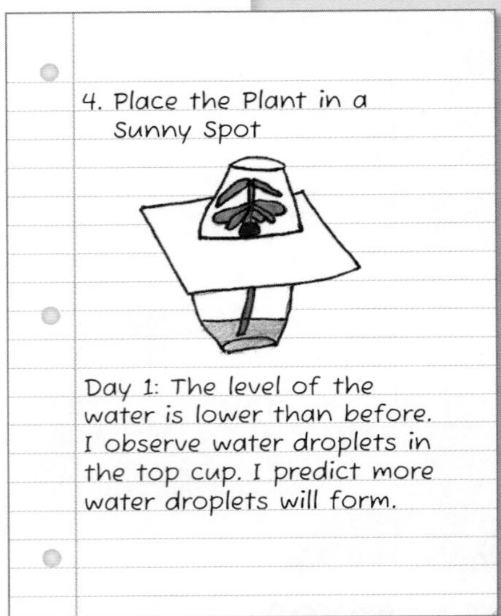

4. Place the Plant in a Sunny Spot
Day 1: The level of the water is lower than before. I observe water droplets in the top cup. I predict more water droplets will form.

ENHANCE UNDERSTANDING OF SCIENCE CONTENT

A science notebook can help students understand science content. Writing about what they learn helps students understand science concepts and improves organizational skills. Encourage students to use science notebooks to present findings to share with others.

DIFFERENTIATE INSTRUCTION

A science notebook provides a way to support differentiated instruction. You can design science notebook use in your classroom so that students move at their own pace. This may include flexible time frames for completing work or small groups to support student learning.

Cows are herbivores. They eat grass and other plants.

EVALUATE

You can use the science notebooks for evaluation. Review the notebooks and see what understandings students have reached and what material you need to teach again. You may have students use their notebooks for self-reflection by marking places where they think they have successfully thought about and accomplished a section.

Setting Up a Science Notebook

Guide your students to set up a structure that will support their learning and your evaluation of science understanding.

TEACHING TIPS

- Have students use the science notebook daily or on a regular basis. Assign a student to pass them out at the beginning of each science period. Collecting the notebooks each day and keeping them in a central place may help them last longer.
- Students may want to tape or staple a ribbon into the notebook and use it to mark their place.
- Tell students to use labels and/or captions for all drawings.
- Students can include graphic organizers to help them understand and organize information. A variety of graphic organizer options are found in the Learning Masters Book.
- You may want to use sticky notes to indicate corrections or additions that students need to make in their science notebook.

1 Introduce the science notebook to students.

Say: **Like scientists, you will use this science notebook for your science work. It's your own notebook to use every time you do science. Like a real scientist, you will write notes in it, draw pictures, write about experiments, and write vocabulary words.**

2 Have students choose an organization style.

DAILY RECORD: Students can make entries, in order, as they move through the unit. They can take notes about the Nature of Science, content lessons, and inquiry activities.

SEPARATE SECTIONS: Students can create separate sections for predictions, hypotheses, inquiry activities, vocabulary, notetaking, questions, reflections, and so on.

3 Have students choose the notebook.

Some students may want to put looseleaf paper in a binder.

Other students may choose a spiral notebook, a two-pocket folder, or a composition book. Still other students might choose a combination of these.

4 Have students make and/or decorate the cover.

Instruct students to include their names and grade on the cover.

5 Have students write "Contents" on the first right-hand page.

Suggest that students save several pages for the Table of Contents and fill in the information as they continue through the unit.

6 Tell students that they should number the pages as they go along.

Tell students to add the appropriate page numbers to the Table of Contents as they proceed through the unit.

What's In a Science Notebook?

Science notebook references are included throughout the Teacher's Edition. A science notebook is a place to record predictions and hypotheses. Students can record their predictions and hypotheses in a variety of ways:

STUDENT DRAWINGS

- Have students draw pictures to illustrate the Nature of Science (science is based on predictions and hypotheses) and their understanding of science concepts.

TABLES, CHARTS, AND GRAPHS

- Have students draw tables, charts, and graphs to record information or data.

NOTES

- Encourage students to jot down notes from each lesson in their science notebook. They can include graphic organizers, charts, lists, questions, and sketches. Suggestions for notetaking appear in the Teacher's Edition at point of use.

Life Cycle of a Praying Mantis

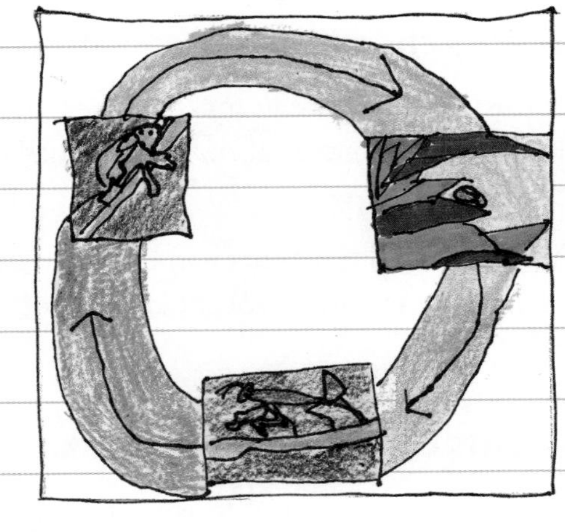

Adult	Eggs	Nymph
large has wings	hundreds laid at once hatch after 3 weeks	looks like adult no wings

A praying mantis passes through incomplete metamorphosis. The life cycle has three stages.

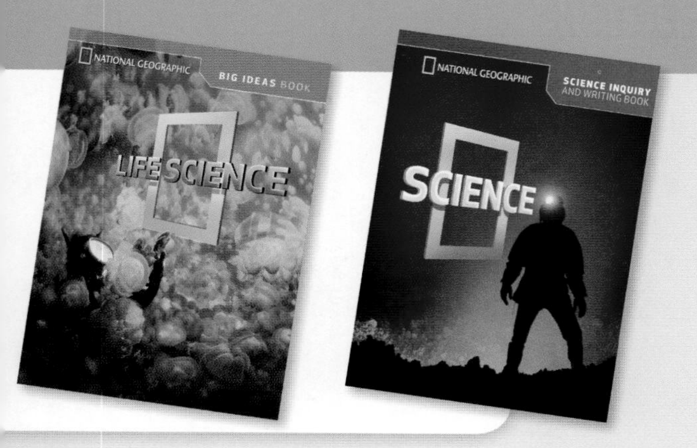

Praying Mantis

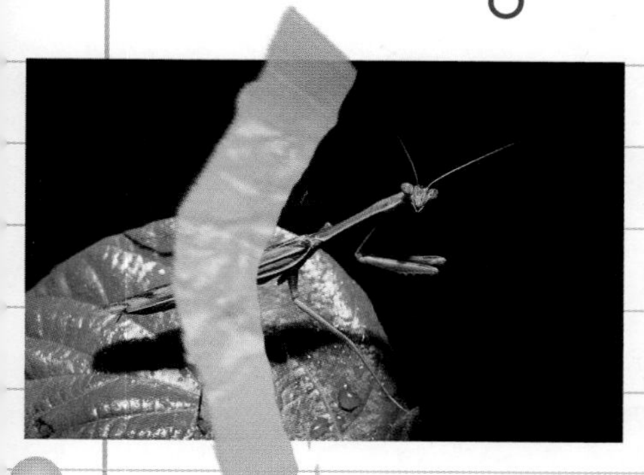

I want to know more about what a praying mantis eats.

In which parts of the world do these insects live?

COLLECTED OBJECTS

- photographs of insects or other animals
- seeds, fallen leaves
- small animal shells or other hard parts

REFLECTIVE AND ANALYTICAL ENTRIES

- You might want to give students prompts or frames to guide them as they write in their science notebook. For example:

  I want to find out ________.

  If ________, then ________.

  What would happen if I changed ________?

  I think ________ because ________.

  The most important thing I learned in this chapter was ________.

  I was surprised to learn ________.

OTHER QUESTIONS STUDENTS HAVE

- Students may have a variety of questions. Have them record questions and help them research the answers.

Integrated Technology

- **Digital Camera** Suggest that students use digital cameras to take photos. The photos can be included in their science notebook.
- **Computer Presentation** Encourage students to share their ideas. They can share their notebook with each other, present their ideas to the class, or talk about their ideas in small groups. They can also make computer presentations as appropriate.

What's In a Science Notebook? continued

Animal Life Cycles

I learned that all animals have life cycles. They become adults and show the traits they inherited. Animals also have other traits because of their environment.

STUDENT REPORTS

- Students can answer one or two reflective questions at the end of each chapter. Or you can assign special projects or reports for them to write in their science notebook.

Name ______________________ Date ____________ my SCIENCE notebook

Chapter 1 Science Vocabulary

Choose from the words in the box to complete each sentence.

conifer
fertilization
inherited
photosynthesis
pollination
seed dispersal

1. A longleaf pine is a(n) conifer.
2. The shape of a plant's leaves is a(n) inherited characteristic.
3. Plants use the process of photosynthesis to make food using the energy of sunlight.
4. The shape and color of a flower may attract insects for pollination.
5. A dandelion uses the wind for seed dispersal.
6. The joining of an egg and sperm cell is called fertilization.

Write a caption for the drawing. Use one or more vocabulary words.

Sample answer: Pollination is the movement of pollen between flower parts. Bees help pollinate sunflowers.

Learning Master 12 Life Science

© NGSP & HB

SCIENCE ACADEMIC VOCABULARY

- You can have students include the Vocabulary Learning Masters in their science notebook. See page SN11 for other suggestions for using the science notebook with Science Academic Vocabulary.

Name ______________________ Date ____________ my SCIENCE notebook

Directed Inquiry

Investigate Flowers

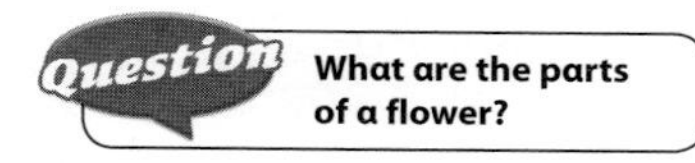

Record

Draw the whole flower in the box below. Use the Parts of a Flower Diagram to help you label each part. Be sure to draw the parts to scale.

Whole Flower

Drawings will vary. Students should draw and label the petals, anther, stamen, ovary, and pistil.

Learning Master 18 Life Science

© NGSP & HB

INQUIRY ACTIVITIES

- Use Learning Masters or have students write notes about the activities in their notebook. See pages SN8–SN10 for suggestions for using a science notebook with inquiry activities.

Using a Science Notebook for Inquiry Activities

The inquiry activities in *National Geographic Science* provide an opportunity for students to ask questions and do investigations much like scientists do. Writing what they learn will help students understand why they are doing the activity and what it teaches them.

ASK A QUESTION

- Every inquiry activity begins with a question that shows the purpose of the activity. Have students write the question in their science notebook.

How can a plant affect the water in its environment?

BUILD VOCABULARY

- Have students write the Science Process Vocabulary words and their definitions in their science notebook.

Observe:
When you observe, you use your senses to learn about an object or event.
Infer:
When you infer, you use what you know and what you observe to draw a conclusion.

MAKE A PREDICTION

- Have students write a statement predicting what will happen in the activity. Encourage students to use their prior knowledge and experience to make the prediction.

I know that plants need water to live. So I predict that the level of the water in the cup will go down.

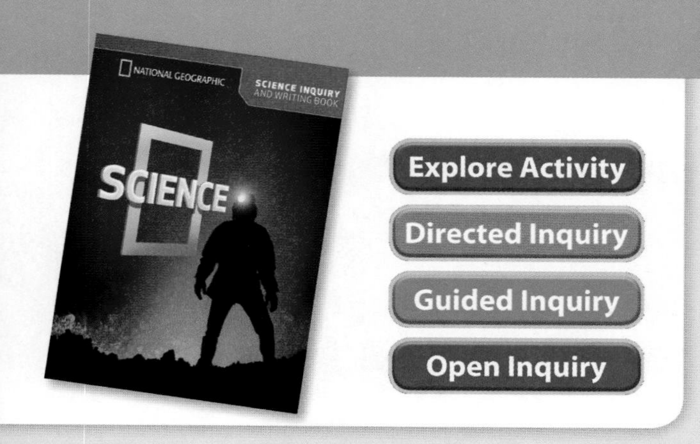

4. Place the Plant in a Sunny Spot

Day 1: The level of the water is lower than before. I observe water droplets in the top cup. I predict more water droplets will form.

WHAT TO DO

- Have students write or draw the steps of the activity in their notebook.

MAKE AND RECORD PREDICTIONS AND HYPOTHESES

- Have students record their predictions and hypotheses in the table on the Learning Master or have them draw and fill in their own table or graph. Students can write or make drawings to record their predictions and hypotheses.

Using a Science Notebook for Inquiry Activities continued

EXPLAIN AND CONCLUDE

- Have students examine data, or evidence, and use this evidence to make predictions and hypotheses, develop explanations, and draw conclusions about their results. New conclusions can then be made based on further predictions and hypotheses. In doing these activities, students are ready to learn that scientific conclusions are based on evidence and can change with new evidence.

Plants take in water through their stems as they live and grow.

THINK OF ANOTHER QUESTION

- Have students reflect about what they have learned. Then have them use their observations to think of other questions that they could study through an investigation. Have them write their questions in their science notebook.

Do larger plants take in more water than smaller plants?

Science Academic Vocabulary

Students can use their science notebook as their own vocabulary resource.

Cows are herbivores. They eat grass and other plants.

Have students designate a special section of the science notebook for Science Academic Vocabulary.

- Write the vocabulary words and their definitions on a Science Word Wall for students to copy.
- Encourage students to use the graphic organizers from the Learning Masters Book or draw their own for recording vocabulary.
  - Students can write the word, draw a picture, write a definition in their own words, and write a sentence using the word.
  - Students can fill in a Word Web.
  - Students can make their own vocabulary cards or use the Vocabulary Cards Graphic Organizer, writing the word on one side and a picture and/or a definition on the other side.
- Encourage students to write ideas about vocabulary in their science notebook.

Using a Science Notebook for Unit Activities

Students can summarize and synthesize the unit's content with culminating unit entries.

STUDENT REFLECTIONS

At the end of each chapter, help students think about what they learned. Writing will help them understand the Nature of Science, science content, and science inquiry. This helps students relate the three parts of science to their daily lives. Ask them to write answers to questions such as these in their science notebook:

- What is the main idea of this chapter?
- What is the most surprising thing I learned?
- What is the most important thing I learned?
- How does the life cycle of a seed plant compare to the life cycle of a conifer?
- What happens to an animal as it goes through metamorphosis?
- What are the parts of a food chain?
- What are some examples of adaptations, and how do they help a living thing survive in its environment?
- How can human actions affect the lives of plants and animals?
- What are some examples of systems of the human body, and how do they work together to keep the body alive?
- What is a hypothesis? Give an example.

USING THE SCIENCE NOTEBOOK WITH THE UNIT WRAP UP

- Use the science notebook as a place to respond to the Unit Wrap Up projects found on pages T268n–T268q.

What food chains do animals and plants form in my neighborhood? How can I find out?

I would like to learn more about the different animals that I see outdoors near my house. I see many birds, squirrels, insects, and a few frogs. I want to know how they get the food they need to survive.

Big Ideas About Life Science

Student Big Ideas Book
Science Inquiry and Writing Book

LIFE
SCIENCE
1

CONTENTS

Chapter 3

Chapter 4

CONTENTS

Chapter 5

Chapter 6

Unit Inquiry

❶ Introduce

Tap Prior Knowledge

- Have students preview the photos on pages 2–3. Ask: **What do all the photos have in common?** (All show examples of living things.) **How can you tell that something is alive?** (Possible answers: It grows and changes; it moves on its own; it responds to its environment.)

Set a Purpose and Read

- Read the heading. Tell students that this is a preview of what they'll learn in the *Life Science* unit.
- Have students read pages 2–3.

❷ Teach

How Do Plants Grow and Reproduce?

- Read aloud the title of the chapter. Ask: **Do you think flowers help a plant grow or reproduce?** (reproduce) Explain that flowers make fruits and seeds, which is how many plants reproduce, or make more of their own kind.

How Do Animals Grow and Change?

- Point to the photo of elephants on page 2. Say: **The young elephants are smaller versions of the adult elephant but will grow to look like the adults.**
- Ask: **What other animals begin their lives as smaller versions of adults?** (many animals, including nearly all fishes, reptiles, birds, and mammals) Say: **In this chapter, we'll learn how animals grow and change.**

How Do Living Things Depend on Their Environment?

- Point to the photo of the caribou and mountain on page 2. Ask: **What are some of the living parts of the caribou's environment?** (plants, such as grasses) **How does the caribou depend on plants?** (for food) **What are some of the nonliving parts of the environment?** (air, water, rocks) **How does the caribou depend on air and water?** (It needs oxygen in air to breathe and water to drink.)
- Say: **In this chapter, we'll learn how living things depend on their environment to survive.**

What Is Life Science?

Life science is the study of all the living things around you and how they interact with one other and with the environment. This type of science investigates how living things are similar to and different from one another, how they live and reproduce, and how they function in the environment. Life science includes the study of humans, as well as all the other kinds of living things on Earth. People who study living things and the environment are called life scientists.

You will learn about these aspects of life science in this unit:

HOW DO PLANTS GROW AND REPRODUCE?

Plants go through life cycles that result in more plants of the same kind. Plants inherit characteristics from the parent plants, although some characteristics may be shaped by the environment.

HOW DO ANIMALS GROW AND CHANGE?

Animals go through life cycles in which many changes occur. Some of these changes are shaped by characteristics inherited from the parents, while others are a result of the environment in which animals are growing and reproducing.

HOW DO LIVING THINGS DEPEND ON THEIR ENVIRONMENT?

The environment includes both living and nonliving parts. The nonliving parts of the environment determine what kinds of living things are there. But the living things in all environments interact with one another, moving energy from the sun through all living things. Life scientists study how living things depend on the environment.

2

HOW DO ADAPTATIONS HELP LIVING THINGS SURVIVE?

Plants and animals have adaptations that help them move, get food, and even reproduce. Adaptations can be body parts or plant parts. Some adaptations are behaviors, such as communication.

HOW DO LIVING THINGS INTERACT WITH THEIR ENVIRONMENT?

Seasonal changes in the environment cause changes in plants and animals. Plants and animals cause changes in the environment too. The actions of humans and all animals have effects—large and small—on their surroundings.

HOW DO THE PARTS OF AN ORGANISM WORK TOGETHER?

Humans have many body systems. Each body system does a job in the body. Many different organs make up the different systems. Life scientists study how body systems work together.

3

Differentiated Instruction

ELL Language Support for Describing the Study of Life Science

BEGINNING

Have students copy the titles of the six chapters. Then distribute dictionaries and have partners look up and discuss potentially unfamiliar words in the titles, such as *reproduce*, *environment*, *adaptations*, and *interact*.

INTERMEDIATE

Help students use Academic Language Frames to describe life science.

- *Life science is the study of the ____ around you.*
- *Living things ____ when they make more of their own kind.*
- *Living things interact with their ____.*

ADVANCED

Have students choose one of the photos on pages 2–3 and write a new caption for it. Remind students that the caption should relate to a topic in life science.

Teach, continued

How Do Adaptations Help Living Things Survive?

- Point to the photo of the bird on page 3. Ask: **Do you think the tufted feathers on this bird's head help it move, get food, or reproduce? How so?** (Possible answer: Reproduce. They help the bird attract a mate.)
- Say: **In this chapter, you will learn about some adaptations in plants and animals that help them to get what they need to survive.**

How Do Living Things Interact with Their Environment?

- Point to the photo of the caterpillars on page 3. Ask: **What are these caterpillars doing?** (eating leaves) **How might eating all of the leaves of the tree affect the tree?** (Leaves make food for the tree, so the tree might die if it lost all its leaves.)
- Say: **Living things change an environment when they use it to meet their needs.** Ask: **How might human actions affect living things?** Discuss how cities, industries, and farms have changed habitats all over Earth.

How Do the Parts of an Organism Work Together?

- Point to the photo of the girl eating on page 3. Ask: **What parts of the girl's body are used to eat and digest food?** (Answers include teeth, tongue, stomach, intestines, and other parts.)
- Say: **In this chapter, we will learn that the body has many organs and body systems. Each body system works together with other systems to complete complex tasks.**

3 Wrap Up

- Ask: **In summary, what do life scientists study?** (how plants and animals grow and reproduce; how living things depend on each other and their environment; how adaptations help living things get what they need to survive; how living things impact their environment; how body systems work together)
- Ask: **Which of these chapters are you most curious to read? Why?** (Answers will vary.)

MEET A SCIENTIST / UNIT LAUNCH VIDEO

PROGRAM RESOURCES

- *Life Science* Unit Launch Video: Online at myNGconnect.com
- Science at Home Letter (in 7 languages): Learning Masters Book, pages 1–7, or at myNGconnect.com

NATIONAL GEOGRAPHIC

MEET A SCIENTIST

Luke Dollar: Conservation Scientist

Luke Dollar is a conservation scientist and National Geographic Emerging Explorer. Luke is devoted to ensuring that every minute and dollar is spent on conservation efforts that really count. "I take the money I raise or borrow straight to the ground level and get more bang for my buck," reports Luke.

Luke's recent focus has been on researching and studying the elusive predator found only on the island of Madagascar—the fosa. As one of the most important species on the island, the fosa plays a crucial role in maintaining the equilibrium of Madagascar's entire food chain.

Luke's research has two parts: tracking the hard-to-find fosa and monitoring forests. To carry out his research, Luke and his team observe and count fosa.

4

Think Like a Scientist: How Scientists Work

Luke Dollar spends much of his time in the remote, dense forests of Madagascar. Here he tracks, observes, and measures trapped fosas. He also tries to convince others to preserve Madagascar forests. "When I arrived, my work was on the ground and I could see the habitat disappearing," Luke said. "I needed a way to quantify the qualitative claims being made by conservation management. I used satellite images, and pictures don't lie. Comparing an image of an area taken at one point in time, with an image of the same area later, provides an irrefutable measure of the success or failure of conservation programs, no sugarcoating."

NATIONAL GEOGRAPHIC

Life Science Video

myNGconnect.com

Preview the Video

Tell students that they will meet Luke Dollar, National Geographic Emerging Explorer, in the *Life Science* video. Share this information with students:

- Conservation scientists advise governments about the wise use and management of forests and other natural resources.
- Madagascar is an island off Africa. Many types of plants and animals live in Madagascar and nowhere else.
- The fosa lives in the forests of Madagascar. It is a carnivore.
- Since people began living on the island 2,000 years ago, nearly 90 percent of Madagascar's original forest has been destroyed. Many of the island's plants and animals are now endangered, meaning they soon could become extinct.

Play the Video

myNGconnect.com

Discuss the Video

Invite students to share what they learned and give their perspectives. You may want to ask some of the following questions.

- What in the video interested you most? Why was it interesting?
- Luke Dollar's job is both difficult and important. Would you like to do what he does? Why or why not?
- What kind of information about the fosa did Luke Dollar find out?
- How is Luke Dollar's research important to people everywhere, not just on Madagascar?
- How can you learn more about fosas and the forests of Madagascar?

Science at Home

Distribute the Science at Home Letter on Learning Masters 1–7, or at myNGconnect.com. Students should take the letter home and

- Discuss plants, animals, and how they depend on and interact with their environment.
- Follow directions with their families to write a skit about the life cycle of a familiar plant or animal.

Science Through Literacy

Teaching students reading comprehension strategies can help them access science content and support their scientific inquiries. National Geographic Science focuses on four key reading comprehension strategies.

Each chapter throughout the unit includes notes to support your teaching of the four key reading comprehension strategies.

Reading Comprehension Strategies

Preview and Predict

- Look over the text.
- Form ideas about how the text is organized and what it says.
- Confirm ideas about how the text is organized and what it says.

Monitor and Fix Up

- Think about whether the text is making sense and how it relates to what you know.
- Identify comprehension problems and clear up the problems.

Make Inferences

- Use what you know to figure out what is not said or shown directly.

Sum Up

- Pull together the text's big ideas.

How Do Plants Grow and Reproduce?

LIFE SCIENCE

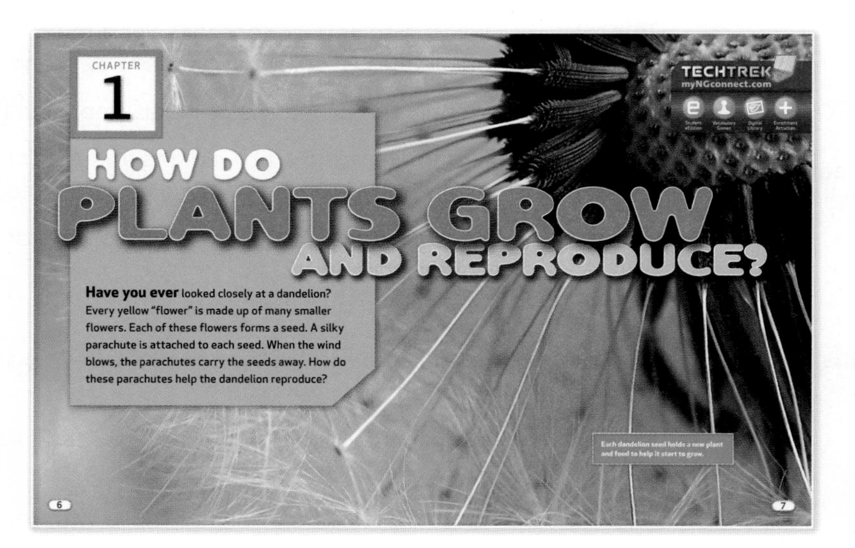
CHAPTER
1
HOW DO
PLANTS GROW
AND REPRODUCE?
Have you ever looked closely at a dandelion? Every yellow "flower" is made up of many smaller flowers. Each of these flowers forms a seed. A silky parachute is attached to each seed. When the wind blows, the parachutes carry the seeds away. How do these parachutes help the dandelion reproduce?
TECHTREK
myNGconnect.com

After reading Chapter 1, you will be able to:

- Identify the major structures of a seed plant and relate them to their functions. **HOW PLANTS GROW**
- Classify plants according to their characteristics. **HOW PLANTS GROW, LIFE CYCLES OF SEED PLANTS**
- Describe the main processes involved in the reproduction of flowering plants. **FLOWERS, FRUITS AND SEEDS**
- Compare and contrast the life cycles of flowering plants and conifers. **LIFE CYCLES OF SEED PLANTS**
- Identify individual differences in organisms of the same kind. **CHARACTERISTICS OF PLANTS**
- Explain that although most characteristics of plants are inherited, some characteristics can be affected by factors in the environment. **CHARACTERISTICS OF PLANTS**
- Science in a Snap! Describe the main processes involved in the reproduction of flowering plants. **FRUITS AND SEEDS**

5

Chapter 1 Contents

Preview and Predict

Tell students that they can preview a text by looking over it to get an idea of what the text is about and how it is organized. Tell students that predicting is forming ideas about what they will read. Students should confirm their ideas with others.

Have students preview and make predictions about the chapter. Remind them to use section heads, vocabulary words, pictures, and their background knowledge to make predictions.

CHAPTER 1 ▫ How Do Plants Grow and Reproduce?

LESSON	PACING	OBJECTIVES
1 **Explore Activity** *Investigate a Plant Life Cycle* pages T5e–T5h **Science Inquiry and Writing Book** pages 16–19	**30** minutes	Investigate and Explore (answer a question; make and compare observations; collect and record data and observations; share findings). Identify the processes of sexual reproduction in flowering plants. Communicate the procedures and results of investigations and explanations through drawings, data tables, and writings. Recognize and comprehend the orders of magnitude associated with small physical quantities.
2 **Big Idea Question and Vocabulary** pages T6–T7—T8–T9 **How Plants Grow** pages T10–T13	**35** minutes	Identify the major structures of a seed plant and relate them to their functions. Identify the major structures of a seed plant and relate them to their functions, including photosynthesis. Classify plants according to those with seeds and those without seeds.
3 **Flowers** pages T14–T17	**25** minutes	Describe the parts of a flower. Describe the processes of pollination and fertilization in flowering plants.
4 **Fruits and Seeds** pages T18–T23	**30** minutes	Describe the main processes of seed and fruit formation in flowering plants. Describe how the seeds of flowering plants are dispersed. Describe the germination and growth of flowering plants.
5 NATIONAL GEOGRAPHIC **We Need Honeybees!** pages T24–T25	**20** minutes	Describe the importance of honeybees in the pollination of flowering plants.
6 **Life Cycles of Seed Plants** pages T26–T31	**30** minutes	Compare and contrast the life cycles of flowering plants and conifers.

VOCABULARY	RESOURCES	ASSESSMENT
observe **compare**	Science Inquiry and Writing Book: *Life Science* Science Inquiry Kit: *Life Science* Explore Activity: Learning Masters 8–11	Inquiry Rubric: Assessment Handbook, page 182 Inquiry Self-Reflection: Assessment Handbook, page 189 Reflect and Assess, page T5h
photosynthesis **conifer**	Vocabulary: Learning Master 12 *Life Science* Big Ideas Book	Assess, page T13
pollination **fertilization**	Extend Learning: Learning Masters 13–14	Assess, page T17
seed dispersal	Science Inquiry and Writing Book: *Life Science*	Assess, page T23
		Assess, page T25
		Assess, page T31

TECHNOLOGY RESOURCES

STUDENT RESOURCES

myNGconnect.com

- **Student eEdition**
- Big Ideas Book
- Science Inquiry and Writing Book
- Explore on Your Own Books
- **Read with Me**
- **Vocabulary Games**
- **Enrichment Activities**
- **Digital Library**

National Geographic Kids

National Geographic Explorer!

TEACHER RESOURCES

myNGconnect.com

- **Teacher eEdition**
- Teacher's Edition
- Science Inquiry and Writing Book
- Explore on Your Own Books
- Online Lesson Planner
- National Geographic Unit Launch Videos
- Assessment Handbook
- **Presentation Tool**
- **Digital Library**

NGSP ExamView CD-ROM

IF TIME IS SHORT...

FAST FORWARD.

CHAPTER 1 ▫ How Do Plants Grow and Reproduce?

LESSON	PACING	OBJECTIVES
7 **Characteristics of Plants** pages T32–T33	**20** minutes	Identify individual differences in organisms of the same kind. Explain that although most characteristics of plants are inherited, some characteristics can be affected by factors in a plant's environment.
8 **Think Like a Scientist** *Math in Science: Recording Observations* pages T33a–T33d **Science Inquiry and Writing Book** pages 20–23	**20** minutes	Summarize information from charts and graphs to answer scientific questions. Compare and contrast the stages in the life cycles of plants. Interpret organized observations and measurements, recognizing simple patterns, sequences, and relationships. Explain that science focuses solely on the natural world.
9 Directed Inquiry *Investigate Flowers* pages T33e–T33h **Science Inquiry and Writing Book** pages 24–27	**30** minutes	Investigate through Directed Inquiry (answer a question; make and compare observations; collect and record data and observations; generate explanations and conclusions based on evidence; share findings; ask questions based on observations to increase understanding). Identify processes of sexual reproduction in flowering plants, including pollination. Make observations using simple tools and use relevant safety procedures to measure and describe organisms by basic characteristics and their uses.
10 **Conclusion and Review** pages T34–T35	**15** minutes	
11 NATIONAL GEOGRAPHIC LIFE SCIENCE EXPERT *Plant Biologist* pages T36–T37 NATIONAL GEOGRAPHIC BECOME AN EXPERT *Bristlecone Pines: The Old Ones of the Mountains* pages T38–T39—T44	**35** minutes	Describe how plant biologists investigate the characteristics of plants. Describe the life cycle of bristlecone pines.

FAST FORWARD
ACCELERATED PACING GUIDE

DAY 1 30 minutes

Explore Activity

Investigate a Plant Life Cycle, page T5e

DAY 2 35 minutes

NATIONAL GEOGRAPHIC LIFE SCIENCE EXPERT *Plant Biologist*, page T36

NATIONAL GEOGRAPHIC BECOME AN EXPERT *Bristlecone Pines: The Old Ones of the Mountains*, page T38

VOCABULARY	RESOURCES	ASSESSMENT
inherited	Share and Compare: Learning Master 15	Assess, page T33
	Science Inquiry and Writing Book: *Life Science* Think Like a Scientist: Learning Masters 16–17	Assess, page T33d
compare **infer**	Science Inquiry and Writing Book: *Life Science* Science Inquiry Kit: *Life Science* Directed Inquiry: Learning Masters 18–23	Inquiry Rubric: Assessment Handbook, page 182 Inquiry Self-Reflection: Assessment Handbook, page 190 Reflect and Assess, page T33h
		Assess the Big Idea, page T35 Chapter 1 Test: Assessment Handbook, pages 4–6 NGSP ExamView CD-ROM
		Assess, pages T37, T40–T41, T42–T43

TECHNOLOGY RESOURCES

STUDENT RESOURCES

myNGconnect.com

- **Student eEdition**
- Big Ideas Book
- Science Inquiry and Writing Book
- Explore on Your Own Books
- **Read with Me**
- **Enrichment Activities**
- **Digital Library**

National Geographic Kids

National Geographic Explorer!

TEACHER RESOURCES

myNGconnect.com

- **Teacher eEdition**
- Teacher's Edition
- Science Inquiry and Writing Book
- Explore on Your Own Books
- Online Lesson Planner
- National Geographic Unit Launch Videos
- Assessment Handbook
- **Presentation Tool**
- **Digital Library**

NGSP ExamView CD-ROM

DAY 3 30 minutes

Directed Inquiry

Investigate Flowers, page T33e

Objectives

Students will be able to:

- Investigate and Explore (answer a question; make and compare observations; collect and record data and observations; share findings).
- Identify the processes of sexual reproduction in flowering plants.
- Communicate the procedures and results of investigations and explanations through drawings, data tables, and writings.
- Recognize and comprehend the orders of magnitude associated with small physical quantities.

Science Process Vocabulary

observe, compare

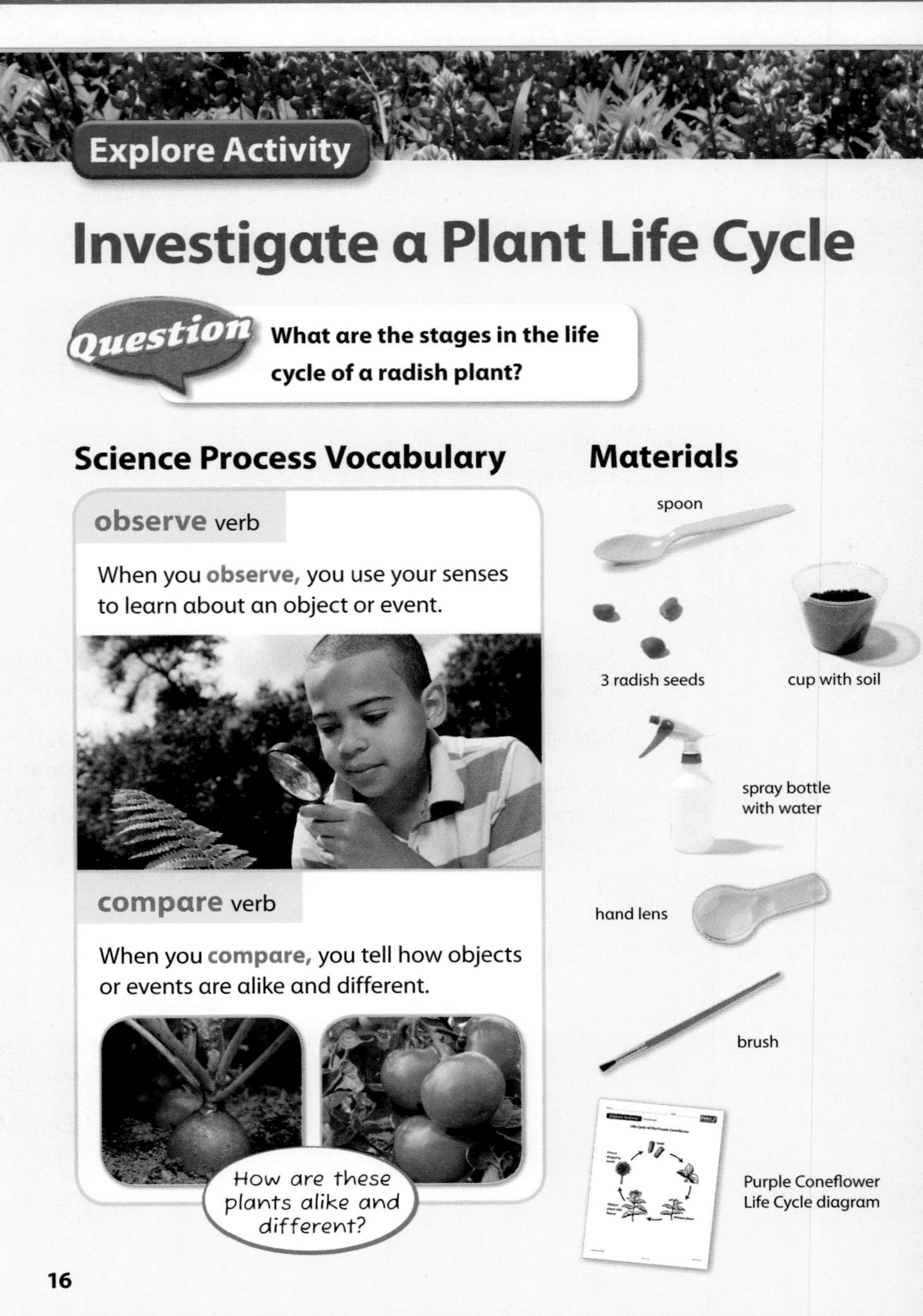

Explore Activity

Investigate a Plant Life Cycle

Question What are the stages in the life cycle of a radish plant?

Science Process Vocabulary

observe verb

When you **observe,** you use your senses to learn about an object or event.

compare verb

When you **compare,** you tell how objects or events are alike and different.

How are these plants alike and different?

Materials

spoon
3 radish seeds
cup with soil
spray bottle with water
hand lens
brush
Purple Coneflower Life Cycle diagram

16

PROGRAM RESOURCES

- Science Inquiry and Writing Book: *Life Science*
- Science Inquiry and Writing Book eEdition at myNGconnect.com
- Inquiry eHelp at myNGconnect.com
- Science Inquiry Kit: *Life Science*
- Learning Masters Book, pages 8–11, or at myNGconnect.com
- Inquiry Rubric: Assessment Handbook, page 182, or at myNGconnect.com
- Inquiry Self-Reflection: Assessment Handbook, page 189, or at myNGconnect.com

MATERIALS

Kit materials are listed in italics.

spoon; *3 radish seeds*; *plastic cup (9 oz)*; *potting soil*; *spray bottle*; water; *hand lens*; *brush*; Purple Coneflower Life Cycle Learning Master

❶ Introduce

Tap Prior Knowledge

- Have students identify the stem, leaves, flowers, and seeds of a plant. Lead students in a discussion about parts of a plant and some of the functions of the parts.

MANAGING THE INVESTIGATION

Time

30 minutes for setup, then 10 minutes a day over a few weeks for observation

Groups

Small groups of 4

Advance Preparation

- Make copies of the Purple Coneflower Life Cycle Learning Master.
- Supply each group with a 9-oz plastic cup about ¾ full of potting soil.
- Fill the spray bottles with water.

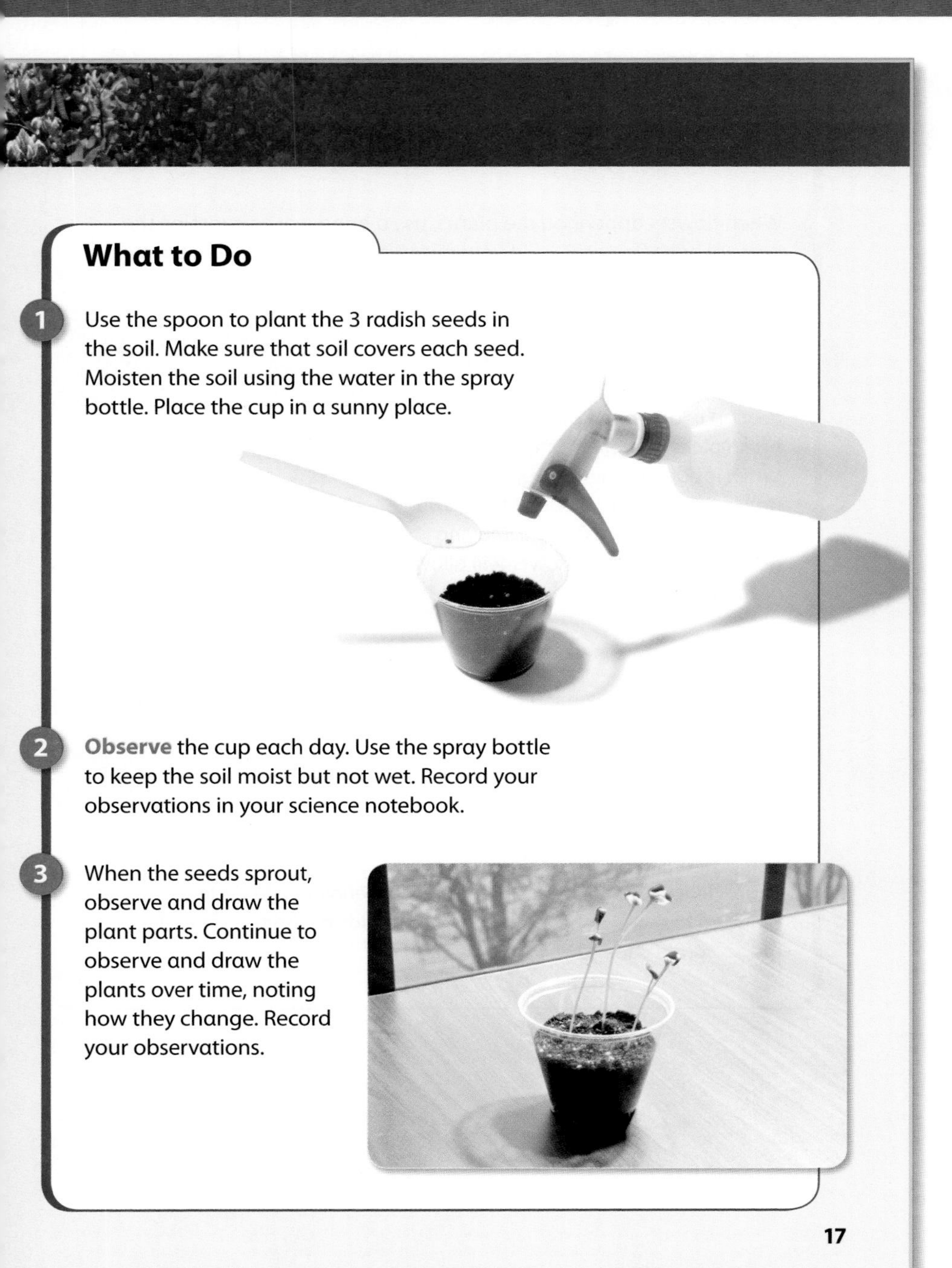

What to Do

1. Use the spoon to plant the 3 radish seeds in the soil. Make sure that soil covers each seed. Moisten the soil using the water in the spray bottle. Place the cup in a sunny place.

2. **Observe** the cup each day. Use the spray bottle to keep the soil moist but not wet. Record your observations in your science notebook.

3. When the seeds sprout, observe and draw the plant parts. Continue to observe and draw the plants over time, noting how they change. Record your observations.

17

Teaching Tips

- Use Rapid Radish™ seeds for this activity. The Rapid Radish™ plant used in this activity completes a full life cycle in about 5 weeks.
- Students' seedlings should sprout after 2–5 days. Most plants should have open flowers after 15 days. Growing time may have to be extended depending on growing conditions.

What to Expect

- Students will observe the life cycle of a radish plant. They will pollinate the flowers and analyze the life cycle.

Introduce, continued

Connect to the Big Idea

- Review the Big Idea Question, *How do plants grow and reproduce?* Explain to students that this activity will allow them to observe how a radish plant lives, grows, and reproduces.
- Have students open their Science Inquiry and Writing Books to page 16. Read the Question and invite students to share ideas about different stages in the life cycles of plants.

❷ Build Vocabulary

Science Process Words: observe, compare

Use this routine to introduce the words.

1. **Pronounce the Word** Say **observe**. Have students repeat it in syllables.
2. **Explain Its Meaning** Choral read the sentence. Ask students to define **observe**. (look closely, often with equipment) Discuss how a hand lens could help a scientist make observations.
3. **Encourage Elaboration** Ask: **What other equipment could you use to observe something?** (microscope, telescope, camera)

ELL Use Cognates

Spanish speakers may know the word **observar** in their home language. Use it to help them access the English word **observe**.

Repeat for the word **compare**. To encourage elaboration, ask: **How could you compare seeds?** (You could observe what is alike and different about the shape, size, and color.)

❸ Guide the Investigation

- Distribute materials. Read the activity steps on pages 17–18. Clarify steps if necessary.
- In step 1, make sure that students don't plant their seeds too deeply in the soil. Seeds should be about ⅛ inch from the surface and not too close together.
- Guide students to record their observations and draw and label the plant at each of the key stages of development: seed, sprout, mature leaves and flowers, seed production.
- As students observe the radish flowers, explain that flowers are the part of the plant where seeds are formed for reproduction.

Guide the Investigation, continued

- Students may need assistance when completing step 5. Further information and instructions are available in the booklet that accompanies the Rapid Radish™ seeds.
- As students pollinate the plants, explain that pollen allows seeds to form and plants to reproduce.
- In step 7, encourage students to combine the drawings they made at each stage of development to complete their life cycle drawing.

❹ Explain and Conclude

- Have each group share their observations and drawings with the class. Ask: **What is the advantage of using a drawing to record observations?** (Drawings give a visual representation of what was observed.) Point out that scientists record observations in various ways, including making labeled drawings.
- Ask: **How did the hand lens help you make detailed observations?** (The hand lens helped me see more details of the radish plant.)
- Ask: **Why is it important to keep accurate and detailed records of your investigation?** (to support conclusions and explanations; so others can compare results)
- Have students compare the life cycle of their plants with the life cycle of the purple coneflower. Discuss the similarities and differences.
- Explain that pollen grains are very small. They are usually measured in micrometers. Write *1/1,000,000 meter* on the board and tell students that a micrometer is equal to one-millionth of a meter. Have students visualize a meterstick divided into a million parts. Ask: **How do you think the pollen's size helps it function?** (It can easily be moved by wind or insects to other plants.)
- Ask: **Is this an experiment where variables are manipulated?** (no) Discuss with students how some scientific investigations are careful observations or modeling activities.

What to Do, continued

4. When flowers appear on the plants, use a hand lens to examine the parts of each flower. Record your observations and draw the flowers.

5. Use the brush to move the pollen between the flowers. Swirl the brush around in an open flower so that the brush picks up pollen. Then swirl the brush inside another flower so that the pollen sticks to the inside of the flower. Repeat for all open flowers. Moving pollen from flower to flower will allow seeds to form.

6. Continue to observe the plants and flowers until seeds form. Use the hand lens to observe the seeds.

7. Use your observations to draw the life cycle of the radish plant in your science notebook. Use the Purple Coneflower Life Cycle diagram to **compare** the life cycle of the radish with the life cycle of the purple coneflower.

18

Differentiated Instruction

ELL During the Investigation

BEGINNING

- In step 4, say: **The flower is the part of the plant that makes seeds.** Point to the flower of the plant as students echo your words.
- Repeat with similar statements about other parts of plants.

INTERMEDIATE

During steps 1–4, refer to the Purple Coneflower Life Cycle Learning Master. Say: **How are they alike and different?** (The radish plant leaves and flowers are smaller. Both have stems, flowers, and green leaves.)

ADVANCED

After step 7, invite students to present and describe their life cycle drawings.

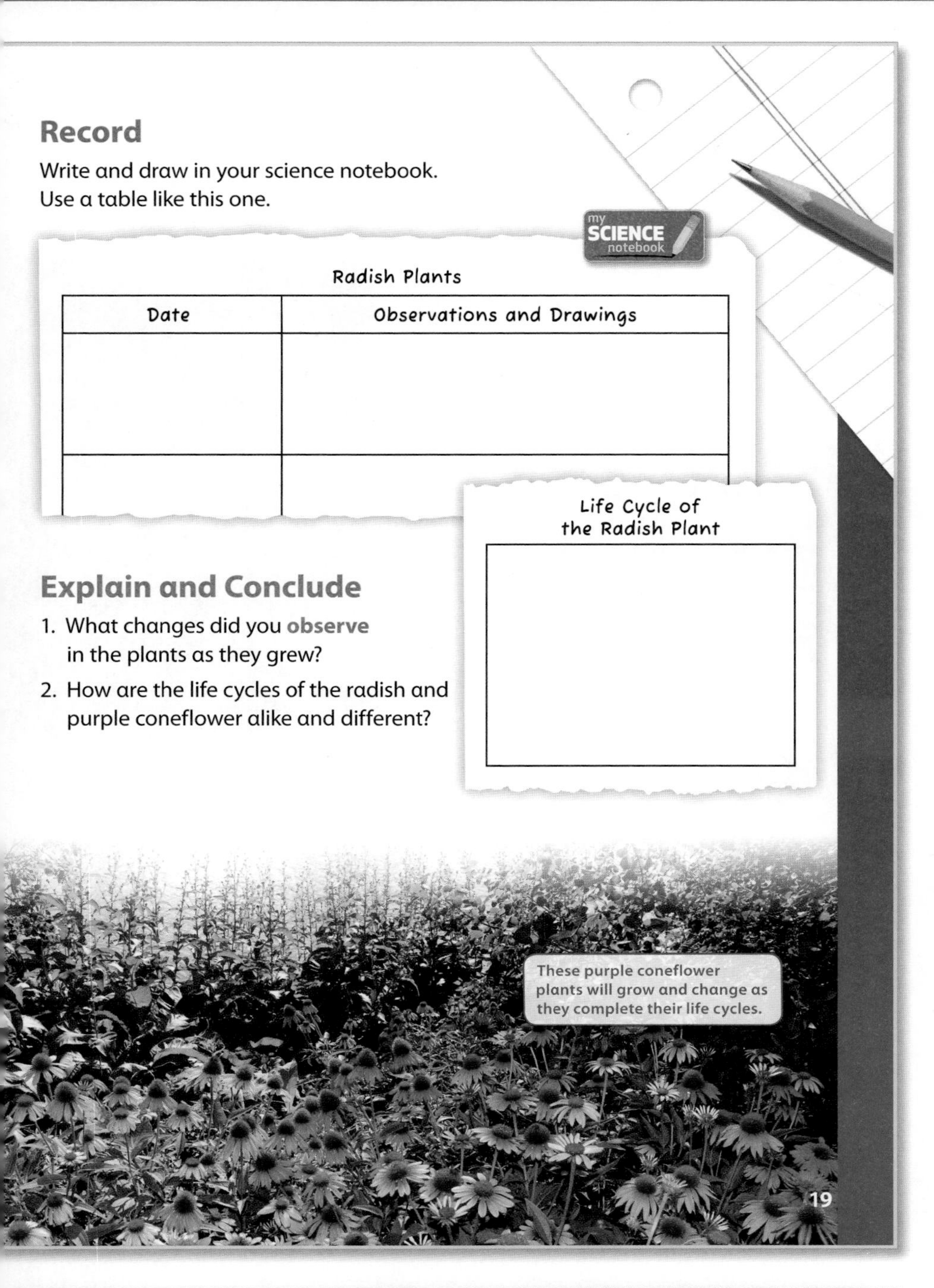

Record

Write and draw in your science notebook. Use a table like this one.

my SCIENCE notebook

Radish Plants

Date	Observations and Drawings

Life Cycle of the Radish Plant

Explain and Conclude

1. What changes did you **observe** in the plants as they grew?
2. How are the life cycles of the radish and purple coneflower alike and different?

These purple coneflower plants will grow and change as they complete their life cycles.

19

Inquiry Rubric at myNGconnect.com	Scale			
The student planted radish seeds and **observed** them as they grew.	4	3	2	1
The student collected and recorded **data** about the life cycle of a radish plant.	4	3	2	1
The student **compared** the life cycles of radish plants and purple coneflowers.	4	3	2	1
The student compared observations and explained any variations.	4	3	2	1
The student **shared conclusions** about the life cycle of a radish plant.	4	3	2	1
Overall Score	**4**	**3**	**2**	**1**

Explain and Conclude, continued

Answers

1. Answers will vary, but students should explain that the plants grew in size and developed leaves, flowers, and seeds.

2. Answers will vary, but students should highlight general similarities and differences between the life cycles of the radish and the purple coneflower.

5 Reflect and Assess

- To assess student work with the Inquiry Rubric shown below, see Assessment Handbook, page 182, or go online at myNGconnect.com
- Score each item. Decide on one overall score.
- Have students use the Inquiry Self-Reflection on Assessment Handbook, page 189, or at myNGconnect.com

6 Extend

- Have students compare the life cycle of a radish plant with a non-flowering plant such as a long leaf pine, fern, or moss.
- Repeat using bean and radish seeds. Students can compare the life cycle stages of each plant.

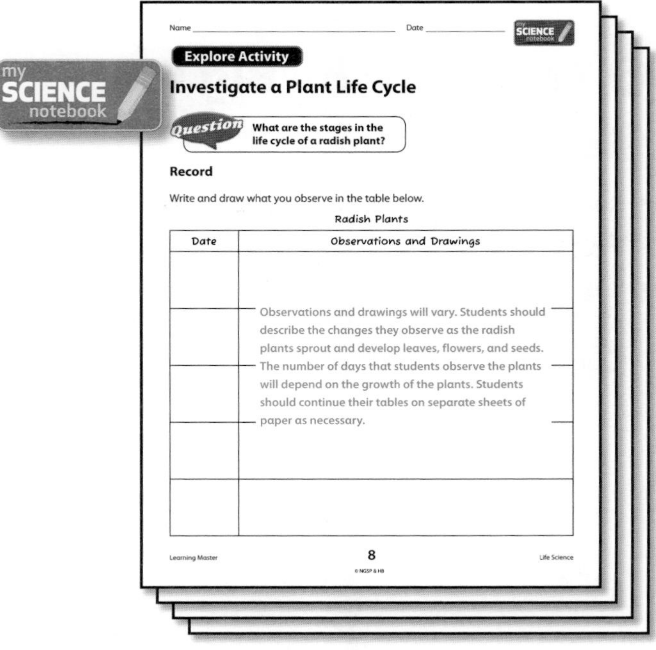

my SCIENCE notebook

Name ______ Date ______

Explore Activity

Investigate a Plant Life Cycle

Question: What are the stages in the life cycle of a radish plant?

Record

Write and draw what you observe in the table below.

Radish Plants

Date	Observations and Drawings
	Observations and drawings will vary. Students should describe the changes they observe as the radish plants sprout and develop leaves, flowers, and seeds. The number of days that students observe the plants will depend on the growth of the plants. Students should continue their tables on separate sheets of paper as necessary.

Learning Master 8 Life Science

Learning Masters 8–11 or at myNGconnect.com

LESSON 2 ▫ Big Idea Question and Vocabulary

Objectives

Students will be able to:

- Identify the major structures of a seed plant and relate them to their functions.

Science Academic Vocabulary

photosynthesis, conifer, pollination, fertilization, seed dispersal, inherited

PROGRAM RESOURCES

- Big Ideas Book: *Life Science*
- Big Ideas Book: *Life Science* eEdition at myNGconnect.com
- Vocabulary Games at myNGconnect.com
- Digital Library at myNGconnect.com
- Enrichment Activities at myNGconnect.com
- Read with Me at myNGconnect.com
- Learning Masters Book, page 12, or at myNGconnect.com

CHAPTER
1
HOW DO PLANTS GROW AND REPRODUCE?
Have you ever looked closely at a dandelion? Every yellow "flower" is made up of many smaller flowers. Each of these flowers forms a seed. A silky parachute is attached to each seed. When the wind blows, the parachutes carry the seeds away. How do these parachutes help the dandelion reproduce?
TECHTREK
myNGconnect.com
Each dandelion seed holds a new plant and food to help it start to grow.
6
7

❶ Introduce

Tap Prior Knowledge

- Have students share their experiences with gardening or observing plants grow. Write the word *garden* on the board. Then during the discussion add the descriptive words that students use, such as *seeds*, *plants*, *water*, *soil*, and *flowers*.

❷ Focus on the Big Idea

Big Idea Question

- Read the Big Idea Question aloud and have students echo it.
- Preview pages 10–13, 14–17, 18–23, 26–31, and 32–33, linking the headings with the Big Idea Question.

Differentiated Instruction

ELL Vocabulary Support

BEGINNING

Have students make flash cards for each of the vocabulary words. Partners may use their cards to quiz one another.

INTERMEDIATE

Provide Academic Language Frames for the vocabulary words.

- *A ____ makes seeds in cones.*
- *____ is the joining of an egg and a sperm cell.*
- *A living things passes through the stages of its ____.*

ADVANCED

Have students write new captions for each of the photos on pages 8–9. Remind students that the purpose of the captions is to explain the vocabulary words.

Focus on the Big Idea, continued

- With student input, post a chart that displays the chapter headings. Have students share what they expect to find in each section. Then read page 6.

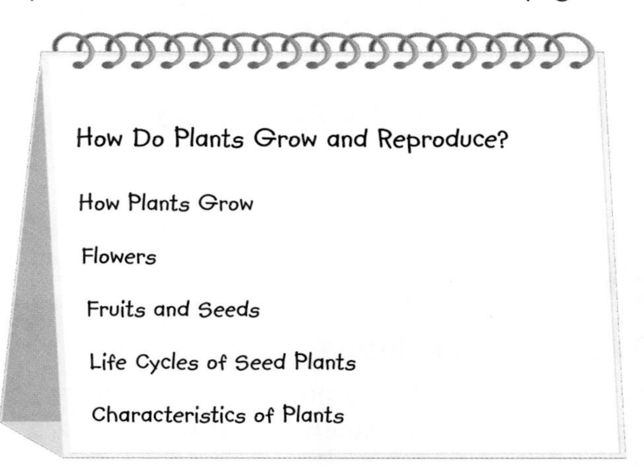

How Do Plants Grow and Reproduce?
How Plants Grow
Flowers
Fruits and Seeds
Life Cycles of Seed Plants
Characteristics of Plants

❸ Teach Vocabulary

Have students look at pages 8–9, and use this routine to teach each word. For example:

1. **Pronounce the Word** Say **photosynthesis** and have students repeat it.
2. **Explain Its Meaning** Say: **In photosynthesis, a gas from the air is combined with water to make food. Sunlight provides the energy.**
3. **Encourage Elaboration** Ask: **Plants hold leaves up in the air. How does this help leaves carry out photosynthesis?** (The leaves face the sun, and sunlight provides the energy for photosynthesis.)

Repeat for the words **conifer**, **pollination**, **fertilization**, **seed dispersal**, and **inherited** using the following Elaboration Prompts:

- **Along with making cones, how else are conifers different from other plants?** (Their leaves look like needles.)
- **What happens during pollination?** (Pollen moves from a stamen to a pistil, which are parts of a flower, or from a male cone to female cone.)
- **What happens during fertilization?** (An egg and a sperm cell join.)
- **Why is seed dispersal important to a plant?** (It helps plants spread to new places.)
- **What are some of the inherited characteristics of tulips?** (Answers include color, height, and shapes of petals and leaves.)

SCIENCE VOCABULARY

photosynthesis (FŌ-tō-SIN-thuh-sis)
In **photosynthesis**, plants use the energy of sunlight to make food. (p. 11)
Green leaves carry out photosynthesis.

conifer (CON-uh-fur)
A **conifer** is a seed plant that reproduces with cones. (p. 12)
A pine tree is a conifer.

pollination (POL-uh-NĀ-shun)
Pollination is the movement of pollen from a stamen to a pistil or from a male cone to a female cone. (p. 16)
When a butterfly carries pollen from flower to flower, pollination occurs.

8

my Science Vocabulary

conifer (CON-uh-fur)
fertilization (FUR-til-i-ZĀ-shun)
inherited (in-HAIR-it-ed)
photosynthesis (FŌ-tō-SIN-thuh-sis)
pollination (POL-uh-NĀ-shun)
seed dispersal (SĒD di-SPUR-suhl)

TECHTREK myNGconnect.com
Vocabulary Games

fertilization (FUR-til-i-ZĀ-shun)
In **fertilization**, an egg and a sperm cell join. (p. 17)
Fertilization takes place in the ovule of a flower.

seed dispersal (SĒD di-SPUR-suhl)
In **seed dispersal**, the seeds of a plant are carried to a new place. (p. 20)
Wind helps in the seed dispersal of dandelions.

inherited (in-HAIR-it-ed)
An **inherited** characteristic is passed down from parents to offspring. (p. 32)
The color of each tulip was inherited from its parents.

9

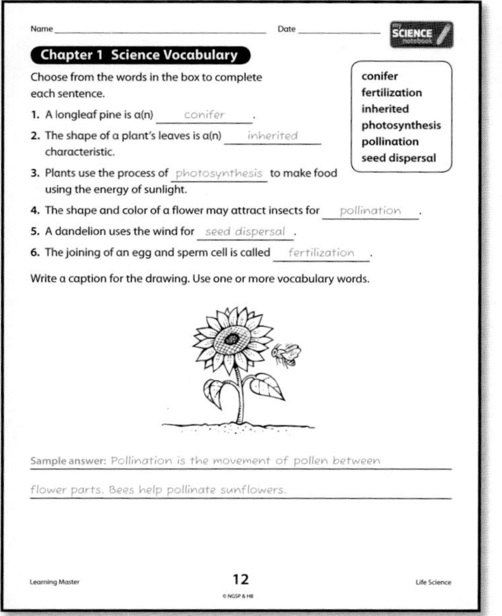

Name ____________ Date ____________

my SCIENCE notebook

Chapter 1 Science Vocabulary

Choose from the words in the box to complete each sentence.

conifer
fertilization
inherited
photosynthesis
pollination
seed dispersal

1. A longleaf pine is a(n) __conifer__.
2. The shape of a plant's leaves is a(n) __inherited__ characteristic.
3. Plants use the process of __photosynthesis__ to make food using the energy of sunlight.
4. The shape and color of a flower may attract insects for __pollination__.
5. A dandelion uses the wind for __seed dispersal__.
6. The joining of an egg and sperm cell is called __fertilization__.

Write a caption for the drawing. Use one or more vocabulary words.

Sample answer: Pollination is the movement of pollen between flower parts. Bees help pollinate sunflowers.

Learning Master 12 Life Science
© NGSP & HB

Learning Master 12 or at myNGconnect.com

LESSON 2 ▫ How Plants Grow

Objectives

Students will be able to:

- Identify the major structures of a seed plant and relate them to their functions, including photosynthesis.

Science Academic Vocabulary

photosynthesis

❶ Introduce

Tap Prior Knowledge

- Have students identify objects that have many parts that work together, such as bicycles, cars, and computers. For each object, discuss the different parts and their functions, and why all the parts are necessary for the object to work properly.

Set a Purpose and Read

- Read the headings. Tell students that they will learn how plants grow and how they are classified.
- Have students read pages 10–13.

❷ Teach

Academic Vocabulary: *photosynthesis*

- Write the word **photosynthesis** on the board, and then pronounce it. Draw a line between the two word parts. Explain that *photo* comes from a Greek word that means "light," and *synthesis* means "to make something."
- Say: **Photosynthesis is the process that plants use to make food using the energy of light.**

Identify Plant Parts and Their Functions

- Have students study the photo of the pepper plant on page 10 and read the caption and labels. Ask: **How do water and nutrients travel through the plant? Describe the path they follow.** (Water and nutrients get into the plant through the roots. Then they travel through stems and into leaves.)
- Ask: **What is the role of stems?** (Stems carry water and nutrients to the leaves, carry food from the leaves to other parts of the plant, and hold up leaves, flowers, and fruits.)

How Plants Grow

Have you ever looked closely at a plant? If so, you know it is made up of leaves, stems, and roots. It may also have flowers and fruits. Each of these parts carries out a different job. But they all work together to help the plant grow and reproduce.

Like all living things, plants need food for energy. But unlike animals, most plants are able to make their own food. Plants use food and other nutrients to live and grow larger.

The different parts of this pepper plant work together to help it grow and reproduce.

Leaves make food.
Peppers are fruits that hold seeds. They help the plant reproduce.
Stems support the plant and move water, nutrients, and food through the plant.
Roots take in water and nutrients from the soil.

10

Differentiated Instruction

ELL Language Support for Identifying Plant Parts and Their Functions

BEGINNING

Ask either/or questions to have students identify plant parts and their functions.
Do stems or roots take in water? Do leaves or flowers make food for the plant? Do flowers help plants grow larger or reproduce?

INTERMEDIATE

Provide Academic Language Frames to help students identify plant parts and their functions:

- *The roots of a plant take in water and ____.*
- *Plants make food in a process called ____.*
- *____ captures the energy of sunlight.*

ADVANCED

Help students use Academic Language Stems to write sentences that describe the role of each part of the plant. For example:

- *The role of leaves is . . .*
- *The role of stems is . . .*
- *The role of roots is . . .*

Plants make food by **photosynthesis**. Photosynthesis uses the energy of sunlight to change water and carbon dioxide into food. The water comes from the soil. The carbon dioxide comes from the air.

To carry out photosynthesis, leaves use a green substance called chlorophyll. Chlorophyll is able to capture the energy of sunlight. Leaves use this energy to combine water and carbon dioxide, making food. This process gives off oxygen. The oxygen goes into the air.

Leaves make food for a plant by photosynthesis.

NATIONAL GEOGRAPHIC Raise Your SciQ!

Gas Exchange Carbon dioxide and oxygen are both invisible gases that are part of Earth's atmosphere, and both are essential for nearly all living things. During photosynthesis, plants and related organisms take in carbon dioxide and release oxygen. The energy of food is released in a process called respiration, which both plants and animals perform. During respiration, the organism takes in oxygen and releases carbon dioxide—the reverse of the exchange that photosynthesis performs. Together, photosynthesis and respiration have been keeping the atmospheric levels of carbon dioxide and oxygen relatively stable. However, levels of carbon dioxide have been gradually rising due to the burning of fossil fuels and other activities.

Teach, continued

Describe Photosynthesis

- Have students study the equation that describes **photosynthesis.** Explain that each plus sign means "and" or "combined with" and that the arrow means "yields" or "makes." Then invite a volunteer to read the equation aloud to the class.
- Ask: **How do leaves get the sunlight, water, and carbon dioxide that they need to perform photosynthesis?** (Sunlight shines on leaves from above, water travels into leaves from roots and stems, and carbon dioxide enters leaves from the air.) Explain that the undersides of many leaves have pores, or tiny holes, through which gases enter and leave.
- Ask: **What do leaves do with the food and oxygen that they make during photosynthesis?** (The food travels to other plant parts, where it is used to help the plant live and grow. The oxygen is released into the air.) Discuss the importance of both products. Humans and other animals eat some of the food that plants make, and we breathe the oxygen that they release into the air. All animal life depends on the **photosynthesis** that plants and similar organisms perform.
- Ask: **Why is it important for leaves to be held in the air?** (They need sunlight to do their job, which is to make food.) Explain that flowers and fruits must also be held in the air. Many flowers rely on visits from bees, butterflies, birds, or other animals to do their jobs, which is to make seeds.

Draw an Illustration

Have students draw a picture of a plant leaf in their science notebook, such as the leaf pictured on page 11. Then have them add and label illustrations that show from where the leaf receives the sunlight, water, and carbon dioxide it needs for **photosynthesis.** Students may also write a caption for their illustration.

Objectives

Students will be able to:

- Classify plants according to those with seeds and those without seeds.

Science Academic Vocabulary

conifers

Teach, continued

Academic Vocabulary: *conifers*

- Point out the word **conifers.** Reread the sentence to define it. If possible, display pictures of different kinds of cones, or invite volunteers to bring examples of cones to class.

Compare and Contrast Groups of Plants

- Have students observe the different plants shown in the photos on pages 12 and 13. Ask: **What plants do you observe?** (Possible answers: wildflowers, trees, fern, grass, ivy) **What do all these plants have in common?** (Possible answers: All have green leaves; all have roots that hold them in place in the ground.)
- Ask: **How are the plants different?** (They have different sizes and shapes, and some make colorful flowers while others do not. The fern reproduces with spores, not seeds.) Discuss how scientists study the differences and similarities among plants in order to classify them.

Plants with Seeds Scientists classify plants according to the way they reproduce. Most plants reproduce with seeds.

The largest group of seed plants are the flowering plants. Roses and daisies are flowering plants. You may be surprised to learn that grasses have flowers, too. Trees that have broad leaves, such as oaks, also have flowers. The seeds of all flowering plants grow in fruits.

A second group of seed plants do not have flowers. These plants form their seeds in cones. Plants that have cones are called **conifers**. Pine trees are conifers. Conifers usually have leaves shaped like needles or scales.

12

Science Misconceptions

Plants with Cones Some students may believe that conifers are the only kinds of plants that make cones. Although conifers make up the majority of cone-producing plants, or gymnosperms, other plants that reproduce with cones include cycads, ginkgoes, and gnetophytes. The cones of each species are at least slightly different from one another.

Plants without Seeds Ferns and mosses do not form seeds. These plants reproduce with spores. Spores are much smaller than seeds. Spores do not carry food for the young plant.

You may have seen ferns growing on the forest floor. Like seed plants, ferns have roots, stems, and leaves.

Mosses are usually much smaller than ferns. They do not have roots or stems to carry water and nutrients to the rest of the plant. Instead, mosses grow low to the ground.

This tree fern grows in the damp woodlands of Tasmania.

Before You Move On

1. What two groups of plants reproduce with seeds?
2. How do leaves make food for the plant?
3. **Infer** Spruce trees have leaves that are shaped like needles. Do you think spruce trees reproduce with flowers or cones? Explain your reasoning.

13

Differentiated Instruction

ELL Language Support for Classifying Plants

BEGINNING

Have students sketch and label a plant shown on pages 12 or 13. Ask either/or questions to help them classify it: **Is your plant a flowering plant, a conifer, or a fern? Does it make seeds in flowers or cones or does it make spores?**

INTERMEDIATE

Provide Academic Language Frames to help students classify plants.

- *Scientists classify plants according to how they ____.*
- *____ plants make seeds in fruits.*
- *____ make seeds in cones.*

ADVANCED

Provide Academic Language Stems to help students classify plants.

- *Plants are classified according to . . .*
- *Three groups of plants are . . .*
- *Conifers reproduce by . . .*

Teach, continued

Classify Plants

- Remind students that scientists classify plants according to how they reproduce. Explain that to reproduce means to have offspring, or to make new plants. Ask: **What are three ways that plants can reproduce?** (with spores, with seeds made in flowers, with seeds made in cones) **Which of these ways is most common?** (seeds made in flowers) Discuss how the vast majority of familiar plants reproduce with flowers. These include garden plants, wildflowers, and grasses.
- Ask: **Which group of plants reproduces by making seeds in cones?** (conifers) Explain how cones serve a similar function to flowers. Both make seeds.
- Ask: **What are some plants that reproduce with spores?** (ferns and mosses) **How do spores and seeds compare?** (Spores are smaller than seeds and do not carry food for the young plant.)

3 Assess

» Before You Move On

1. **Name What two groups of plants reproduce with seeds?** (Flowering plants and conifers reproduce with seeds.)
2. **Explain How do leaves make food for the plant?** (Leaves carry out photosynthesis, a process that uses sunlight to combine carbon dioxide and water.)
3. **Infer Spruce trees have leaves that are shaped like needles. Do you think that the spruce tree reproduces with flowers or cones? Explain your reasoning.** (Since spruce trees have needles, they are probably conifers. Conifers do not have flowers—their seeds grow in cones.)

LESSON 3 ▫ Flowers

Objectives

Students will be able to:

- Describe the parts of a flower.

PROGRAM RESOURCES

- Learning Masters Book, pages 13–14, or at myNGconnect.com

❶ Introduce

Tap Prior Knowledge

- Ask students to describe flowers they are familiar with. Discuss how flowers may come in different sizes, shapes, and colors, and that many have attractive appearances or odors.

Preview and Read

- Read the heading. Then preview the pictures on pages 14–17. Tell students that they will learn about the purpose of a flower's parts and attractive appearance. Have students read pages 14–17.

❷ Teach

Explain the Purpose of Flowers

- Ask: **What is the purpose of flowers in a flowering plant?** (to make seeds and fruits) Remind students that plants reproduce by making seeds. Ask: **What parts help a flower make seeds and fruits?** (petals, stamens, and pistils) Have students identify these parts in the photo on page 14.

Flowers

Parts of a Flower Flowers such as the lily below are beautiful. But plants do not have flowers to look pretty. Plants have flowers to make seeds and fruits. To understand how, you need to learn the parts of a flower.

The first thing you notice about most flowers is their petals. The color and shape of the petals attract insects and other animals to the flower.

The bright color and sweet smell of this lily attract insects.

14

Extend Learning

MANAGING THE INVESTIGATION

Time

20 minutes

PROGRAM RESOURCES

- Learning Masters 13–14

Groups

Partners

MATERIALS

- various plants (optional)

Classify Flowering Plants

Preview	What To Do
 Question: **How can flowering plants be classified by observing their characteristics?** Students will observe the illustrations of flowering plants. Then, by following a series of guided questions, they will classify each plant into one of several families. Students will learn that differences among plant parts help classify plants.	1. Have students observe and compare the illustrations of plants in the Learning Master. 2. Work together as a class to classify one of the plants. Then have students work with partners to classify the remaining plants. 3. If appropriate plant specimens are available, have students use the descriptions of plant families to classify them.

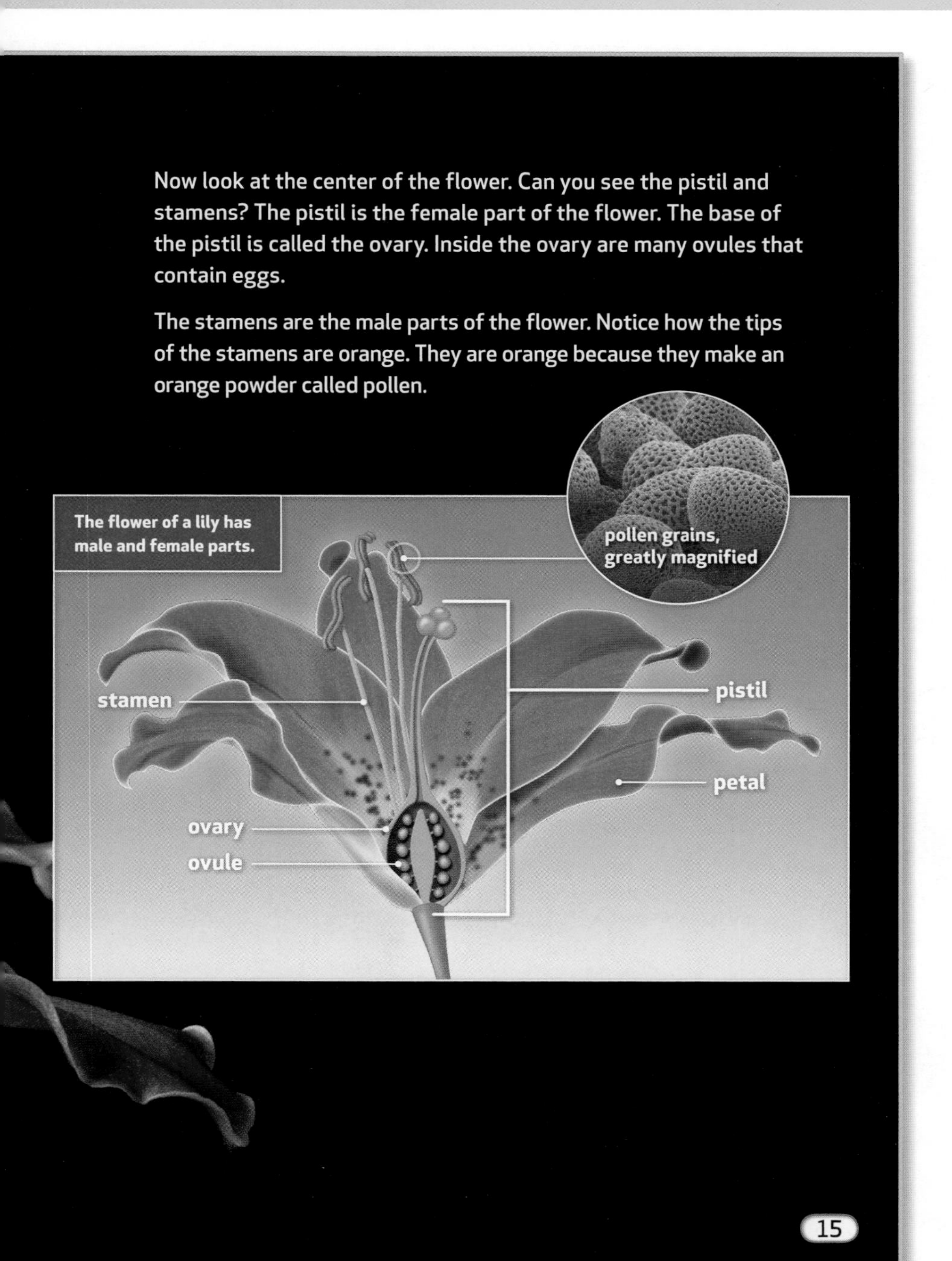

Now look at the center of the flower. Can you see the pistil and stamens? The pistil is the female part of the flower. The base of the pistil is called the ovary. Inside the ovary are many ovules that contain eggs.

The stamens are the male parts of the flower. Notice how the tips of the stamens are orange. They are orange because they make an orange powder called pollen.

The flower of a lily has male and female parts.

Teach, continued

Text Feature: Diagram and Labels

- Have students study the diagram on page 15 and identify each labeled part of the flower. Then using the diagram as a reference, have them identify the petals, stamens, and pistil in the photo of the lily on page 14.
- Ask: **Where does a flower make pollen?** (at the tips of the stamens) Have students identify the pollen in both the diagram and photo of the flower.
- Point to the inset photo of the pollen on page 15. Explain that the photo has been highly magnified. Ask: **What does pollen look like in this photo?** (tiny orange grains, with many pores on the surface.) Discuss how flowers typically make a huge number of pollen grains, each very tiny. Nearly all of the pollen will be wasted. But a few grains may be transferred to the female part of the flower and help start a new seed.

Making and Recording My Observations

Have students observe the diagram of the flower on page 15 and copy it in their science notebook. Students may then observe and copy the structures of flowers that they see in nature. Encourage students to label the flower parts that they can identify. Explain that not all flowers have the same set of parts shown in the diagram.

earning Masters 13–14, or
t myNGconnect.com

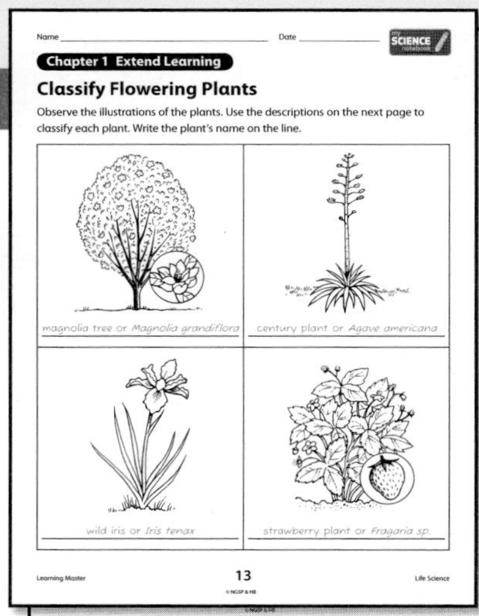

Chapter 1 Extend Learning

Classify Flowering Plants

Observe the illustrations of the plants. Use the descriptions on the next page to classify each plant. Write the plant's name on the line.

magnolia tree or *Magnolia grandiflora*	century plant or *Agave americana*
wild iris or *Iris tenax*	strawberry plant or *Fragaria sp.*

13

Explain Results

Students should explain that:

- Flowering plants can be classified according to observations of their roots, stems, leaves, and flowers.
- Many kinds of plants make flowers, and they differ in a wide variety of ways.

Explain that classification schemes help scientists identify new kinds of plants and understand their properties.

Objectives

Students will be able to:

- Describe the processes of pollination and fertilization in flowering plants.

Science Academic Vocabulary

pollination, fertilization

PROGRAM RESOURCES

- Enrichment Activities at myNGconnect.com

Teach, continued

Academic Vocabulary: *pollination, fertilization*

- Pronounce **pollination** and write it. Remind students that pollen is a powder that flowers make. Ask: **How is pollination related to pollen?** (Pollination is the transfer of pollen between flower parts.)
- Pronounce **fertilization** and write it. Ask: **What word part do you hear in the word fertilization?** (fertilize or fertile) Discuss how the words are related. Explain that one meaning of *fertilize* is "to make something able to produce much," and *fertile* means "able to make seeds."

Describe Pollination

- Ask: **What happens during pollination?** (Pollen moves from a stamen to a pistil.) **How does pollination happen?** (Animals, such as insects, birds, or bats, may carry pollen from the stamen of one flower to the pistil of another flower. Or the wind can pollinate flowers.)
- Have students observe the photo of the butterfly and flower on page 16 and then read the caption. Ask: **Why is the butterfly visiting the flower?** (to drink nectar, a sweet liquid that flowers make) Emphasize that nectar is the butterfly's reward for visiting the flower. Both the butterfly and flower benefit from the visit.
- Ask: **Why don't all flowers depend on animals for pollination?** (Some flowers are pollinated by the wind.) **What plants make flowers that are wind pollinated?** (most grasses and trees) Explain that wind-pollinated flowers are typically not very showy or attractive and that animals are not attracted to them.

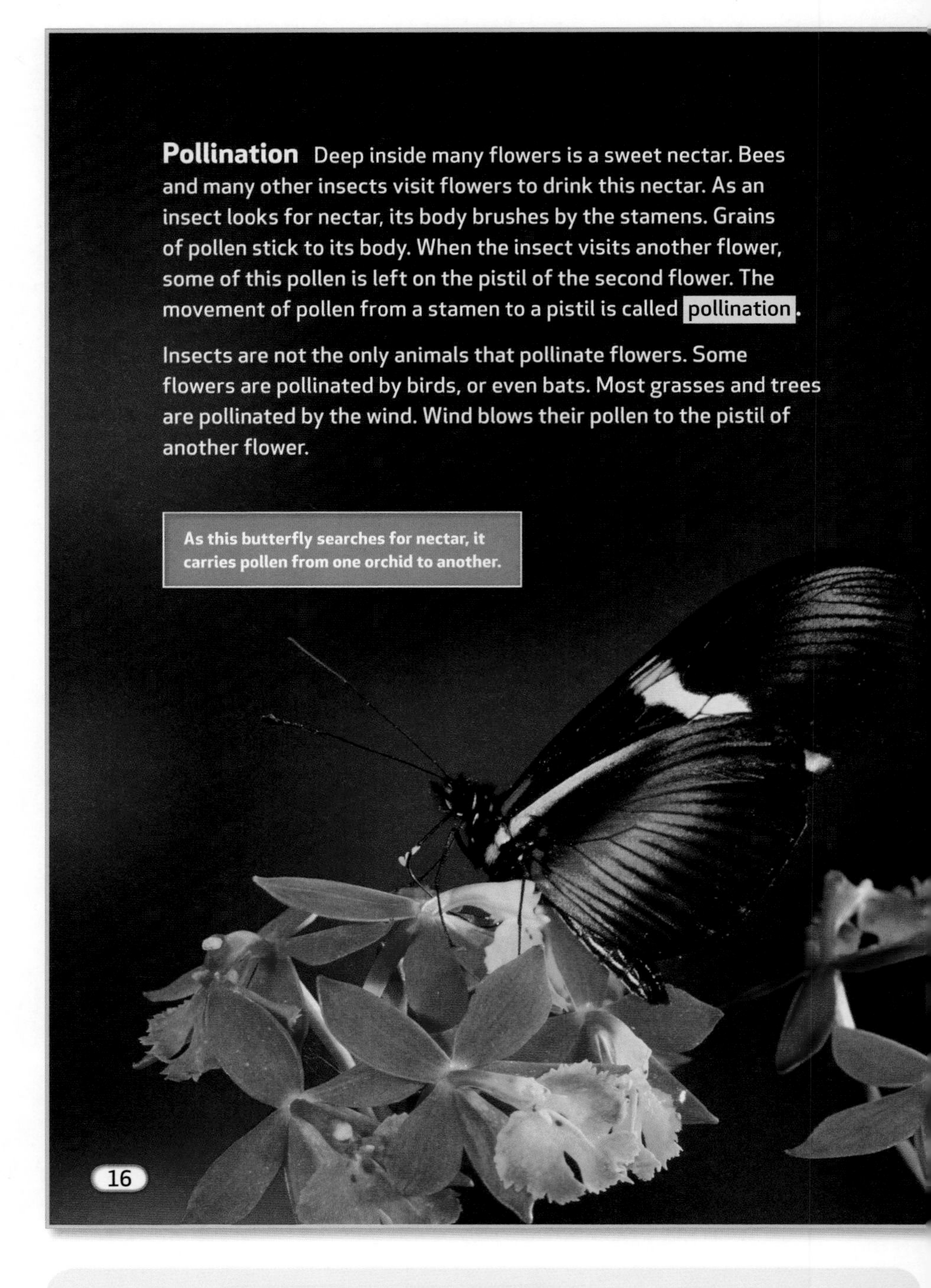

Pollination Deep inside many flowers is a sweet nectar. Bees and many other insects visit flowers to drink this nectar. As an insect looks for nectar, its body brushes by the stamens. Grains of pollen stick to its body. When the insect visits another flower, some of this pollen is left on the pistil of the second flower. The movement of pollen from a stamen to a pistil is called **pollination**.
Insects are not the only animals that pollinate flowers. Some flowers are pollinated by birds, or even bats. Most grasses and trees are pollinated by the wind. Wind blows their pollen to the pistil of another flower.
As this butterfly searches for nectar, it carries pollen from one orchid to another.
16

Science Misconceptions

Many Flowers Some students may think that all flowers are large and colorful, much like the flowers shown on pages 12, 14, and 16. In fact, many flowers are very plain, especially those that are pollinated by the wind. Some animal-pollinated flowers are colorful but very small, such as the flowers of clover. In addition, while some flowers have pleasant, sweet odors, some make foul odors that smell like rotting meat. These flowers attract beetles, flies, or other meat-feeding insects that act as pollinators.

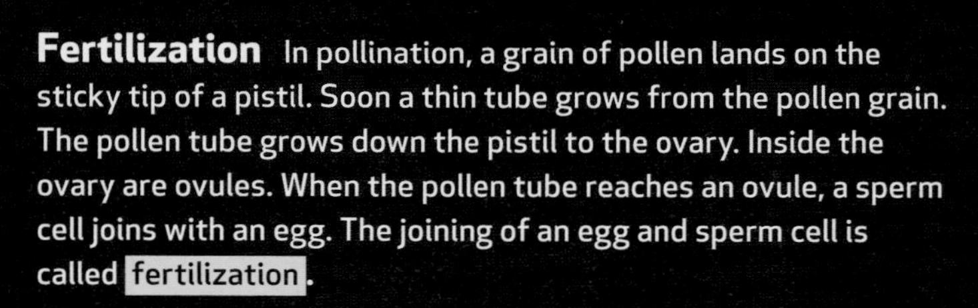

Fertilization In pollination, a grain of pollen lands on the sticky tip of a pistil. Soon a thin tube grows from the pollen grain. The pollen tube grows down the pistil to the ovary. Inside the ovary are ovules. When the pollen tube reaches an ovule, a sperm cell joins with an egg. The joining of an egg and sperm cell is called **fertilization**.

After fertilization, the flower begins to change. Seeds and fruits begin to form. The petals dry up and fall off.

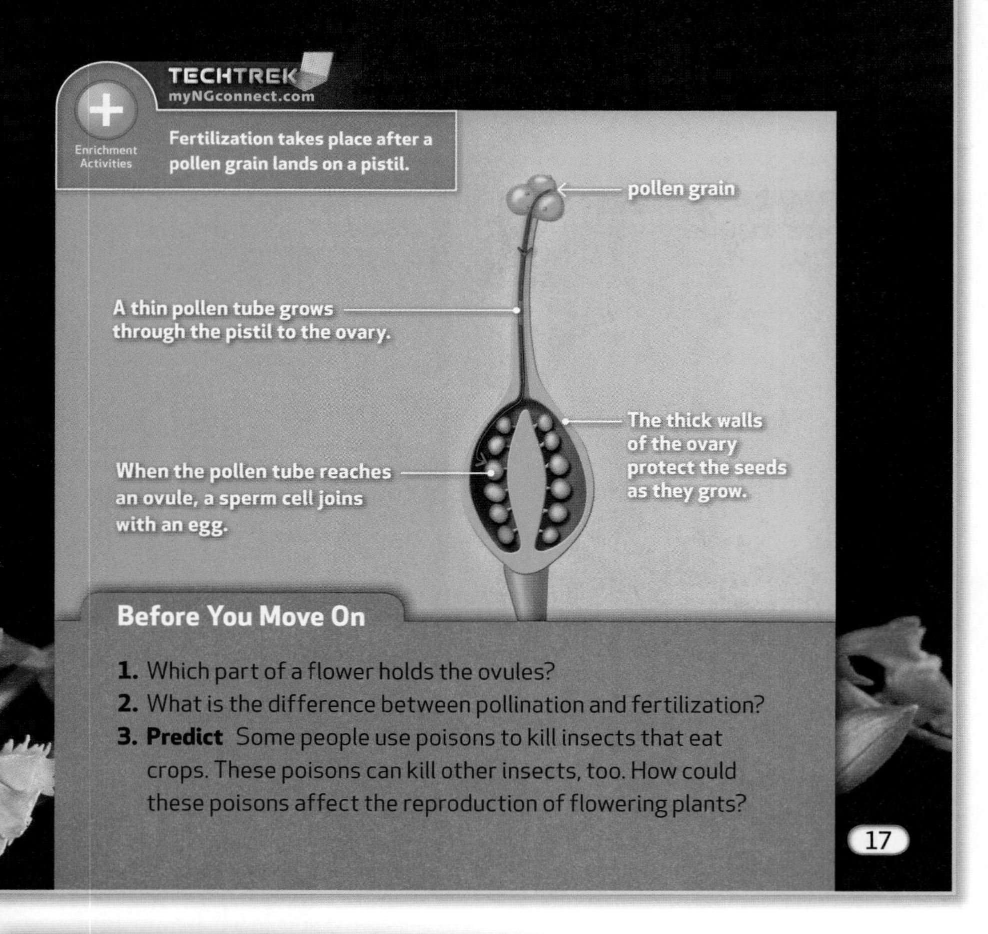

TECHTREK
myNGconnect.com
Enrichment Activities
Fertilization takes place after a pollen grain lands on a pistil.
pollen grain
A thin pollen tube grows through the pistil to the ovary.
The thick walls of the ovary protect the seeds as they grow.
When the pollen tube reaches an ovule, a sperm cell joins with an egg.

Before You Move On

1. Which part of a flower holds the ovules?
2. What is the difference between pollination and fertilization?
3. **Predict** Some people use poisons to kill insects that eat crops. These poisons can kill other insects, too. How could these poisons affect the reproduction of flowering plants?

17

Differentiated Instruction

ELL Language Support for Describing Pollination and Fertilization

BEGINNING

Help students explain pollination and fertilization by asking yes/no questions, for example:

Does pollen move from one part of a plant to another during pollination?

Do flower petals form after fertilization?

INTERMEDIATE

Provide Academic Language Frames to help students correctly use the terms *pollination* and *fertilization.*

- *A bird or insect may ____ a flower.*
- *A sperm cell joining with an egg cell is called ____.*

ADVANCED

Have students write new captions for the photos and illustrations on pages 14–17. Tell them that their captions should correctly describe the processes of pollination and fertilization.

Teach, continued

Enrichment Activities

myNGconnect.com

Have students use Enrichment Activities to find out more about flower fertilization.

Integrated Technology

Whiteboard Presentation Students can use the information to make a computer presentation about flower fertilization. You can help them show the presentation on a whiteboard.

Monitor and Fix Up

Ask students if anything they read was confusing. Students may confuse the processes of **pollination** and **fertilization.** Discuss possible fix-up strategies. For example, students may rewrite the definition of each word using their own words. They may also include a drawing to illustrate each word.

3 Assess

Before You Move On

1. **Identify Which part of a flower holds the ovules?** (The ovary holds the ovules.)
2. **Contrast What is the difference between pollination and fertilization?** (In pollination, pollen is transferred from the stamens to the pistil. In fertilization, sperm from the pollen unites with the egg.)
3. **Predict Some people use poisons to kill insects that eat crops. These poisons can kill other insects, too. How could these poisons affect the reproduction of flowering plants?** (If all the insects were killed, there would be no insects to pollinate the plants. The plants would not be able to make seeds.)

LESSON 4 ▫ Fruits and Seeds

Objectives

Students will be able to:

- Describe the main processes of seed and fruit formation in flowering plants.

PROGRAM RESOURCES

- Science Inquiry and Writing Book: *Life Science*
- Science Inquiry and Writing Book eEdition at myNGconnect.com

❶ Introduce

Tap Prior Knowledge

- Have students list and describe their favorite fruits as you write the names on the board. Lead them to recognize that most fruits have one or more seeds inside them. (Note that some fruits, such as bananas and seedless grapes, have been cultivated to produce either small, infertile seeds or no seeds.)

Set a Purpose and Read

- Read the heading. Tell students that they will read about how seeds and fruits form and how seeds are dispersed and germinate.
- Have students read pages 18–23.

❷ Teach

Compare and Contrast Seeds

- Have students study the illustrations, labels, and captions of the bean seed and corn seed. Remind students that the illustrations show what the seeds would look like when cut open.
- Ask: **How are these seeds alike?** (They each have an embryo, a seed coat, and stored food; they each develop from an ovule inside an ovary.) **How are these seeds different?** (A bean seed can be split into two parts and has two seed leaves, but a corn seed does not split into two parts and has only one seed leaf.)

Fruits and Seeds

Seeds After fertilization, seeds and fruits begin to grow. Seeds develop from ovules inside the ovaries.

Each seed holds everything that a new plant needs to grow. If you cut a seed open, you will find three main parts. These are the embryo, the seed coat, and stored food. The embryo will grow into a new plant. The seed coat protects the embryo and keeps it from drying out. The stored food helps the embryo to grow.

Look at the pictures of a corn seed and a bean seed. Both of these seeds have the three main parts. But the bean seed can split into two parts. The corn seed does not split into two parts. What other differences do you see?

A bean seed has two seed leaves. These thick leaves store food.

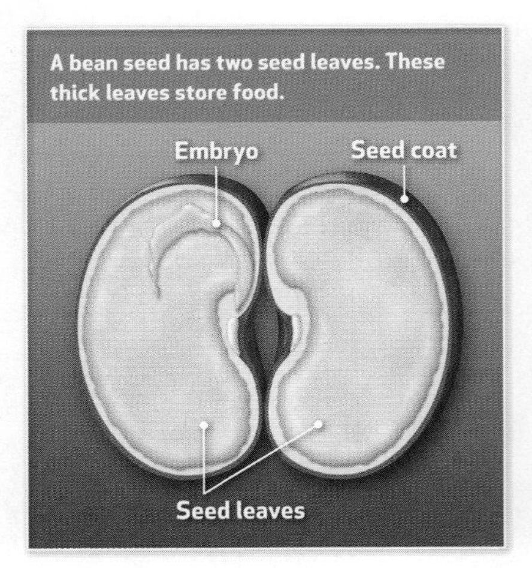

Corn seeds have only one seed leaf. Food is stored in another part of the seed.

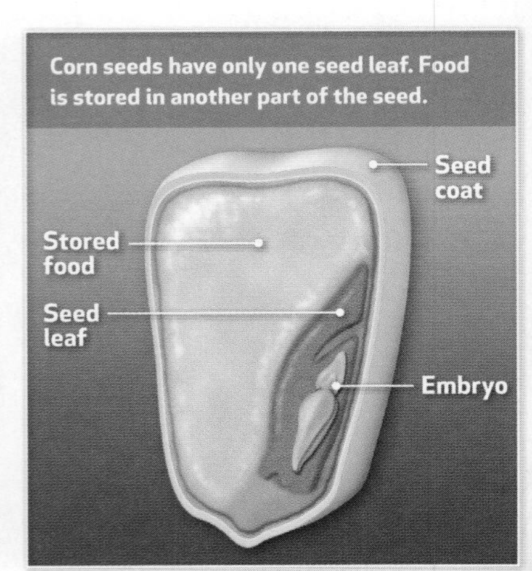

Raise Your SciQ!

Monocots and Dicots The term *monocot* is shortened from the word *monocotyledon*. Monocots are the group of flowering plants in which the embryo has only one seed leaf, a structure formally known as a cotyledon. Dicots are the group in which the embryos have two seed leaves. Examples of monocots include corn, lilies, grasses, orchids, irises, onions, and daffodils. Examples of dicots include beans, oaks, maples, roses, daisies, walnuts, and poppies. With some exceptions, members of each group also share other characteristics, such as certain patterns of veins in their leaves, organization of flower petals, and organization of vascular tissue in their stems.

Fruits Fruits grow from the walls of the ovaries that surround the seeds. Fruits protect and help to spread the seeds.

There are many different kinds of fruits. Some fruits, such as cherries, hold one seed. Other fruits, such as apples and pears, hold several seeds. Pumpkins have many seeds.

Apples and peaches are sweet and fleshy. Other fruits are dry. The shell of a walnut is a dry fruit. The nut that you eat is the seed. The pods that surround beans and peas are also fruits. The beans and peas are seeds.

Fruits come in many shapes and sizes. Even tomatoes are fruits, although you may call them vegetables!

19

Differentiated Instruction

Extra Support

Have students draw a picture of the parts of a bean seed and a corn seed and label each part. Have them write a brief caption for each seed part.

Challenge

Ask students to write a paragraph that explains how fruits and seeds form. They should include drawings to help illustrate the text.

Teach, continued

- Ask: **What is an embryo?** (the part of the seed that will grow into a new plant) **What is the function of a seed coat?** (It protects the embryo and keeps it from drying out.)
- Say: **Stored food is also known as a *seed leaf*. What is the function of stored food in a seed?** (It is the food supply that helps the embryo grow.)

Compare and Contrast Fruits

- Have students turn back to page 17 and review the structure of the pistil. Ask: **Where does a fruit form?** (A fruit forms from the wall of the ovary and thus surrounds seeds.) Ask them to find the ovaries at the bottom of the pistil.
- Have students observe the photos of fruits on page 19 and read the accompanying caption. Ask: **What is the function of a fruit?** (to protect the seeds and to help spread the seeds) Tell students that later in the chapter they will learn how seeds are spread.
- Ask: **How are the fruits the same?** (All contain seeds.) **How are the fruits different from one another?** (Some fruits are especially juicy, while others are dry; they contain different numbers of seeds.) Discuss how not all fruits are especially tasty or good for humans to eat. Different animals eat different kinds of fruit, while some fruits are not eaten but help spread seeds in other ways.

Sorting Seeds

Science Inquiry and Writing Book: *Life Science*, page 12

Materials aluminum foil, construction paper

Read and follow instructions together with students. Guide students to sort the seeds into groups based on their size, shape, and color. Students should record their observations and ideas in their science notebook.

What to Expect Students should sort like seeds into a group. Seeds that are the same kind may have a slightly different shape, color, or size. Other kinds of seeds may be a different color, size, or shape. Seeds are a part of reproduction in a plant's life cycle.

Objectives
Students will be able to:
- Describe how the seeds of flowering plants are dispersed.

Science Academic Vocabulary
seed dispersal

Teach, continued

Academic Vocabulary: *seed dispersal*
- Say the term **seed dispersal** and write it.
- Ask: **What does the word *disperse* mean?** (to scatter or spread) Discuss how **seed dispersal** means the scattering or spreading of seeds.
- Ask: **Why do you think seed dispersal is important to a plant?** (Plants cannot move from place to place on their own and neither can their seeds. They rely on animals or the wind to spread seeds to new places where they can grow.)

Seed Dispersal Seeds cannot move from place to place by themselves. For seeds to reach new places, something must carry them. In **seed dispersal**, the seeds of a plant are carried to a new place. Fruits help in seed dispersal.

Many seeds are spread by animals. When birds eat berries and other fleshy fruits, they also eat the hard seeds inside. Later the seeds drop to the ground. Squirrels gather nuts and bury them in the ground. If a squirrel does not find all the nuts, some may grow into new trees.

Some fruits have hooks that stick in an animal's fur. Later the seeds drop from the animal.

This cedar waxwing is helping to spread the seeds in the pyracantha berries.

Observing Fruits 

Materials pre-cut slices of strawberry, maple fruit, hand lens

Have students work in pairs. Distribute materials to each pair of students. Have students read and follow the instructions on page 21 of the Big Ideas Book. Students should record their observations and results in their science notebook.

Other seeds are carried by the wind. Dandelions have silky fruits that float on the wind. The seeds of some plants have tiny wings. These wings help the seeds spin through the air.

Some seeds are spread by water. Coconuts are the fruits of palm trees. Because coconuts float, ocean water can carry them far away. When coconuts wash onshore, they grow into new trees.

The fruits of coconuts are spread by water.

Dandelion seeds are spread by the wind.

Science in a Snap! Observing Fruits

Observe a strawberry that has been cut in half. Draw what you see. Label the seeds.

Use a hand lens to carefully observe a maple fruit. Carefully pull the thick part of the fruit open. Draw and label what you see inside.

Compare the fruits and seeds. Which seeds are dispersed by animals? Which are dispersed by wind? Explain your reasoning.

21

What to Expect Students should observe and record that both fruits have seeds inside. The fruits differ in that one is soft and fleshy, while the other is hard and not edible. Animals disperse the seeds of the fleshy fruit, while the wind disperses the maple fruit. The wing-like parts of the maple fruit help the wind carry it great distances.

Quick Questions Ask students the following questions:

- **How does it help a plant to make a large, sweet-tasting fruit?** (This type of fruit attracts animals that may carry the fruit and disperse the seeds.)
- **How do wing-like structures help a fruit do its job?** (Wings help the wind disperse the fruit and the seeds it carries.)

Teach, continued

Explain How Animals May Disperse Seeds

- Ask: **How does it help a plant for its fruits to travel to new places?** (Fruits contain seeds, and the seeds may begin growing in new places.)
- Say: **When seeds are far away from a plant, they might have more space in which to grow. If all of a plant's seeds stayed in one place, they would be crowded together and many seeds would not get enough of the sunlight or nutrients they need to grow.**
- Point to the photo on page 20 of the cedar waxwing with the berries. Ask: **How is this bird helping to disperse seeds?** (The berries hold pyracantha seeds. The bird may carry the berry far away, and when the berry is eaten the seeds may drop to the ground.) Explain that the cedar waxwing does not intend to disperse pyracantha seeds but may do so as a side effect of eating the berry.

Explain How Wind and Water May Disperse Seeds

- Ask: **Along with the actions of animals, how else are seeds dispersed?** (by wind and water) Point to the photos on page 21 and have students read the captions. Ask: **Why can coconuts be dispersed by water?** (They float in water.) Discuss how floating coconuts have allowed palm trees to spread to ocean islands all over the world.
- Ask: **Why can dandelion seeds be dispersed by the wind?** (They are small and lightweight and have feather-like parts that help the wind carry them.) Explain that some dandelion seeds might travel very little, especially on calm days. But in a strong wind, the seeds could travel a great distance.

Objectives

Students will be able to:

- Describe the germination and growth of flowering plants.

PROGRAM RESOURCES

- Digital Library at myNGconnect.com

Teach, continued

Describe Germination and Growth

- Ask: **What does *germinate* mean?** (to grow or sprout) **What must happen in order for a seed to germinate?** (Conditions must be right. Typically the seed must receive enough water and be at the right temperature.)
- Have students observe the diagram of the germinating bean seed on page 22. Ask: **What does the diagram show?** (the steps by which a bean seed germinates and becomes a young plant) Have students describe in their own words the sequence of events that the diagram shows.
- Ask: **When a seed germinates, where does the new plant get the energy it needs to grow?** (from food stored in the seed) Explain that with time and the proper conditions for growth, the young plant will grow leaves and be able to make its own food, just like mature plants do.

myNGconnect.com

Have students use the Digital Library to find images of the stages of a bean plant's life cycle.

Computer Presentation Students can use the information to make a computer presentation about the stages of a bean plant's life cycle. You can help them show the presentation on a whiteboard.

Germination and Growth Have you ever planted a garden? When the weather is warm enough, you put seeds in the soil. Then you water them. Seeds will not grow unless they have enough water. But when conditions are right, a seed germinates, or starts to grow.

First, the seed takes in water and begins to swell. Next, the seed coat breaks open. The root starts to grow down. Then the stem and young leaves grow upward.

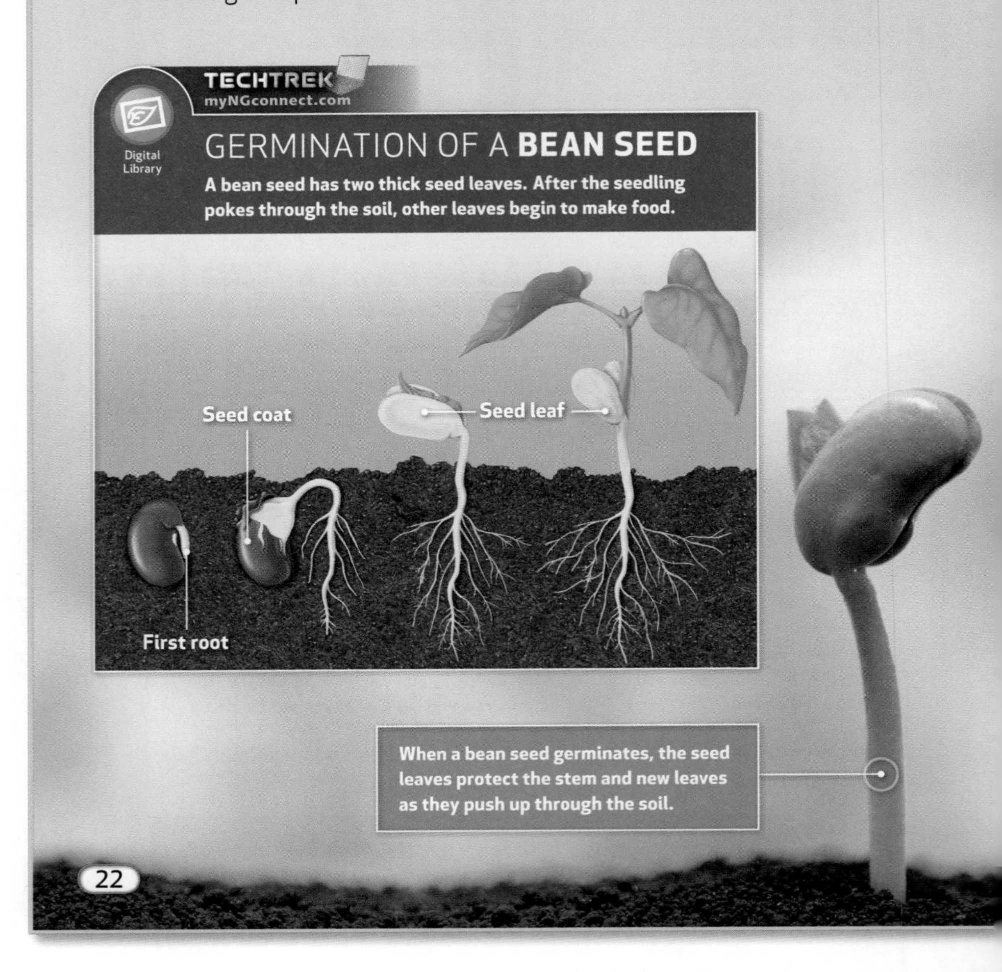

TECHTREK
myNGconnect.com
Digital Library
GERMINATION OF A BEAN SEED
A bean seed has two thick seed leaves. After the seedling pokes through the soil, other leaves begin to make food.
Seed coat
Seed leaf
First root
When a bean seed germinates, the seed leaves protect the stem and new leaves as they push up through the soil.
22

Science Misconceptions

Germination Requirements Some students may think that all seeds need warm temperatures to germinate. Many seeds require relatively warm temperatures, such as 21°C (70°F) or higher. However, the seeds of peas, lettuce, radish, broccoli, and wheat each germinate at temperatures near 10°C (50°F). Each type of seed also has unique moisture requirements in order to germinate.

A young plant is called a seedling. The seedling uses the food that was stored in the seed to start growing. When the new leaves form, they use sunlight to make food for the plant.

The seedling continues to increase in size. The stem grows taller and more leaves grow. In the soil, the roots grow deeper and wider. Over time, the seedling grows into a mature plant that is able to reproduce.

GERMINATION OF A **CORN SEED**

The seed leaf of a corn plant stays underground. A tough covering protects the new leaves as they grow through the soil.

Before You Move On

1. What part of a flower becomes a seed?
2. How does a fruit help a plant reproduce?
3. **Analyze** Why is seed dispersal important to a plant? What would happen if its seeds were not dispersed?

23

Teach, continued

Text Feature: Diagrams and Captions

- Have students study and compare the diagrams of the germinating bean seed and corn seed and then read the labels and captions.
- Say: **Both the corn seed and bean seed begin growing in a process of germination.** Ask: **How is germination similar in the two seeds?** (Possible answers: both germinate in the soil; both seedlings emerge from a seed coat; both send out roots that grow down and stems that grow upward.)
- Ask: **What are some differences between the germinating seeds?** (Possible answers: the seed coats are different colors and shapes; the roots grow in different patterns; the corn seed grows one first leaf while the bean seed grows two first leaves.)

3 Assess

» Before You Move On

1. **Identify What part of a flower becomes a seed?** (The ovule of a flowering plant becomes a seed.)
2. **Explain How does a fruit help a plant reproduce?** (Fruits protect the seeds and help disperse the seeds so that the seeds inside can grow into new plants.)
3. **Analyze Why is seed dispersal so important to a plant? What would happen if its seeds were not dispersed?** (Seed dispersal helps new plants spread to new places. If all the seeds fell under the parent plant, they would be very crowded and have no room to grow.)

Differentiated Instruction

ELL Language Support for Describing Germination

BEGINNING

Help students explain the germination process by asking either/or questions. For example:

Do seeds need water or light to germinate? Do bean seeds germinate in spring or winter? Does a germinating seed get food from the seed or the soil?

INTERMEDIATE

Help students describe germination using Academic Language Frames.

- *Seeds spread to new places in the process of ____.*
- *A seed begins growing when it ____.*
- *A young plant is called a ____.*

ADVANCED

Provide Academic Language Stems to help students explain seed dispersal and germination. For example:

Seeds may be dispersed by . . .

When a plant germinates it first . . . and then . . .

LESSON 5 ▫ We Need Honeybees!

Objectives

Students will be able to:

- Describe the importance of honeybees in the pollination of flowering plants.

❶ Introduce

Tap Prior Knowledge

- Ask students to describe their observations of bees. Students may discuss observing bees visiting flowers and flying around a beehive and their experiences with bee stings.

Preview and Read

- Read the heading. Then preview the pictures on pages 24–25. Tell students that they will read about the importance of honeybees to the pollination of crops.
- Have students read pages 24–25.

❷ Teach

Text Feature: Photos and Captions

- Have students observe the photo of the bee and flower on page 24. Ask: **Why do bees visit flowers?** (to gather nectar) Explain that nectar is a sweet liquid that many flowers make. Ask: **What do bees do with the nectar they gather?** (They use it to make honey.)
- Ask: **How do flowers benefit from the bees' visits?** (The bees spread pollen from flower to flower.) Discuss how the spread of pollen is a key step in the formation of many fruits, including those listed on page 24 of the Big Ideas Book.

NATIONAL GEOGRAPHIC
WE NEED HONEYBEES!
Have you ever seen honeybees buzzing around an orchard? Honeybees visit flowers to get nectar. What do the bees do with the nectar they gather? They take it back to their hives to make honey.
As honeybees move from flower to flower, they carry pollen. This pollinates the flowers. Honeybees help pollinate many crops, such as almonds, apples, and strawberries. Without honeybees, these plants could not make much fruit.
Bees visit flowers to get nectar and pollen.
24

NATIONAL GEOGRAPHIC Raise Your SciQ!

Pollinator Decline Scientists and farmers are concerned about the lowering populations of many important pollinators, especially European honeybees. Scientists have identified several possible causes for the decline of honeybees. Among the causes are pesticide overuse or misuse; pathogenic microbes, including viruses and fungi; and parasites, such as the mites presented in the Big Ideas Book. Although other insects species can and do pollinate flowers, few work as effectively as the honeybee.

Beekeepers raise honeybees to help pollinate crops and to harvest honey. But lately many bees have died. Beekeepers are having trouble raising enough bees. What is causing the problem?

One answer is the spread of mites. Mites are tiny animals that live on the bodies of bees. Mites make it hard for bees to fly. Mites can also kill bees. Scientists are trying to find ways to protect bees from mites. Our food depends on bees!

When almonds blossom, beekeepers bring hives of bees to help pollinate the trees.

These mites live on the bodies of bees. Too many mites can kill a bee.

An almond grove in blossom.

25

Differentiated Instruction

Extra Support

Have students write a new caption for the photos on pages 24–25. Encourage students to use the new words they learned in the chapter. Have students work in small groups and to read their captions to one another.

Challenge

Have students write a paragraph about the importance of honeybees. The paragraph should include a description of the pollination process using appropriate vocabulary terms.

Teach, continued

Describe Beekeeping

- Have students observe the photo of the orchard in bloom on page 25. Ask: **What is happening to these trees?** (Their flowers are blooming.) Explain that on many trees, flowers bloom in the spring before new leaves grow.
- Ask: **How do these trees illustrate the importance of honeybees?** (A large fruit crop depends on honeybees to pollinate the flowers.) Remind students that most flowers last only a short time. Pollination must occur within that time for seeds and fruits to grow.
- Have students observe the photo of the beekeeper on page 25 and read the caption. Ask: **Why do you think beekeepers bring whole hives into an orchard, not just a few bees?** (A large number of bees can pollinate many flowers quickly, and bees always return to their hives.) Discuss how visiting an orchard helps both the beekeeper, whose bees make honey, and the orchard owner, whose trees make fruit.
- Ask: **What is making the beekeepers' job more challenging than in the past?** (Beekeepers are having trouble raising enough bees, in part due the spread of mites.) Discuss how much of our food supply depends on honeybees for pollination.

❸ Assess

1. **Recall What kinds of crops are pollinated by honeybees?** (Answers include apples, almonds, strawberries, and oranges.)
2. **Explain Why are people concerned about dying bees?** (More bees have been dying now than in the past, due to the spread of mites and other causes. Fruit crops depend on healthy bees, so scientists, farmers, and beekeepers are concerned about the problem.)
3. **Analyze What new scientific discoveries might help farmers and beekeepers in the future? Discuss your ideas.** (Possible answers: scientists might discover ways to stop or kill the mites that are harming bees, ways to increase the bees' resistance to mites, or new pollinators that could replace bees.)

CHAPTER 1
LESSON 6 ▫ Life Cycles of Seed Plants

Objectives

Students will be able to:

- Compare and contrast the life cycles of flowering plants and conifers.

❶ Introduce

Tap Prior Knowledge

- Have students define the term *cycle* in their own words. Encourage them to include specific examples, such as a bicycle, programs to recycle paper or plastic, and events in nature such as the cycles of the seasons or day and night.

Set a Purpose and Read

- Remind students that they have just learned how flowering plants and conifers make seeds. Explain that they now will learn about the ways these plants live and grow from seeds to adults, which are stages in their life cycles.
- Have students read pages 26–27.

❷ Teach

Describe a Life Cycle

- Ask: **Why does a cycle have no beginning and no end?** (because it is like a circle, with one stage leading to another) Discuss how the life of an individual plant has a beginning and end, but making seeds allows the life cycle to continue.
- Point to the photo of the orange tree on pages 26–27 and have students read the caption. Ask: **How do fruits, such as these oranges, help a flowering plant complete its life cycle?** (Fruits contain seeds. The seeds may germinate and grow into new plants.)

Life Cycles of Seed Plants

Life Cycles of Flowering Plants Oak trees can live for hundreds of years. Some wildflowers live for only a few months. But all flowering plants go through similar stages of life.

Follow the diagram as you read about the life cycle of an orange tree. An orange tree begins life as a seed. Then the seed germinates and the seedling begins to grow. Over time, the seedling grows into a mature tree that can flower. Bees and other insects pollinate the orange blossoms. After fertilization, seeds and fruit form. When these seeds germinate, the cycle begins again.

Orange trees blossom in the spring. Their fruits are harvested in the fall.

26

Social Studies in Science

Citrus Fruits Oranges, grapefruits, lemons, and limes are called citrus fruits because of the citric acid they contain. These crops grow well in warm, wet climates. Have students research the economic importance of citrus crops in Florida, Texas, and California. Or have them research the important crops in your state or region.

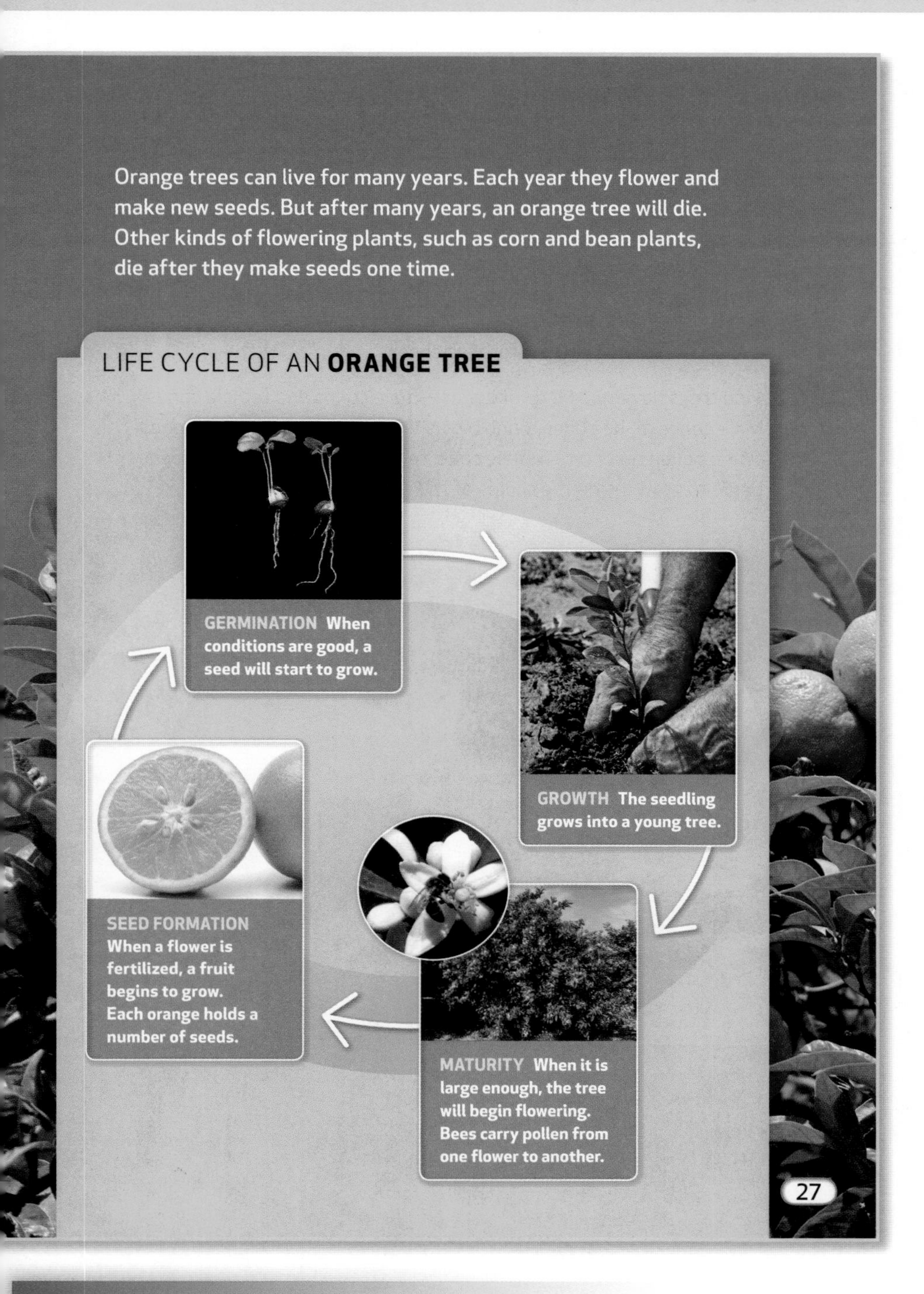

Orange trees can live for many years. Each year they flower and make new seeds. But after many years, an orange tree will die. Other kinds of flowering plants, such as corn and bean plants, die after they make seeds one time.
LIFE CYCLE OF AN ORANGE TREE
GERMINATION When conditions are good, a seed will start to grow.
GROWTH The seedling grows into a young tree.
MATURITY When it is large enough, the tree will begin flowering. Bees carry pollen from one flower to another.
SEED FORMATION When a flower is fertilized, a fruit begins to grow. Each orange holds a number of seeds.
27

Teach, continued

Text Feature: Diagram and Captions

- Have students study the diagram on page 27 and read the captions. Ask: **Why is the diagram of a life cycle shown as a circle?** (The stages in a life cycle repeat like a circle, with one stage following the other over and over again in the same order.) Have students name and identify the four stages: germination, growth, maturity, seed formation.
- Point to one of the seeds in the photo of the cut-open orange. Ask: **Do you think every orange seed completes the life cycle and becomes a mature tree? Why or why not?** (No. Most orange seeds never get the chance to germinate. Even those that do germinate may not grow into a mature tree.) Discuss how most plants, including orange trees, produce a huge number of seeds, and that very few seeds typically germinate.
- Say: **Individual orange trees will die after many years. Why will the life cycle continue?** (Seeds can grow into new orange trees, replacing those that have died.) Explain that death is a natural part of every life cycle. In some kinds of flowering plants, the plants die after making seeds the first year. Other flowering plants, such as oak trees, make seeds and may live for 200 years or more.
- Ask: **What might happen if plants reproduced but never died?** (There would be no room for new plants to grow.)

Differentiated Instruction

Extra Support

Have students fold a sheet of paper into four sections and then write the terms *maturity*, *germination*, *seed formation*, and *growth* in one section each. Have them illustrate each term with a drawing or tracing and then connect the pictures with arrows to show the life cycle of a plant.

Challenge

Have students write new captions for the stages of the life cycle of an orange tree. Tell them to use appropriate vocabulary to describe what happens at each stage of the life cycle.

Objectives

Students will be able to:

- Compare and contrast the life cycles of flowering plants and conifers.

Teach, continued

Describe Life Cycles of Conifers

- Have students read pages 28–29. As they read about each stage of the life cycle, have them point to each appropriate part of the diagram on page 29.
- Ask: **Why is a longleaf pine classified as a conifer?** (It makes seeds in cones, not flowers.) If necessary, review the definition of conifer from page 12.
- Ask: **How does the longleaf pine use two kinds of cones?** (The male cone makes pollen. The female cone receives pollen and carries the developing seeds.) Remind students that flowering plants also have male and female structures and that their male structures make pollen as well.
- Ask: **Do you think all of the fertilized seeds inside a cone grow into pine trees? Why or why not?** (No; some seeds may be eaten by animals; other seeds may land in water or on rocks and will not germinate.) Explain that only a small fraction of the seeds in a tree's cones typically will ever germinate, and an even smaller fraction will grow into maturity.

Life Cycles of Conifers The picture below shows different stages in the life cycle of a longleaf pine. How many different stages can you see in the picture?

Longleaf pines are conifers. Like flowering plants, conifers begin life as seeds. The seedlings of longleaf pines grow slowly. When the trees are mature, they reproduce by forming cones.

Pine trees have two kinds of cones—male cones and female cones. Male cones make pollen. Wind blows the pollen to the female cones. When pollen reaches a female cone, fertilization takes place. Seeds begin to grow inside the scales of the cone.

Longleaf pines grow in open woodlands called pine savannas.

28

NATIONAL GEOGRAPHIC Raise Your SciQ!

Conifers In flowering plants, seeds grow and develop inside a fruit. Conifers, however, are members of a group of plants called the gymnosperms, a term that means "naked seeds." The seeds of a conifer develop on the scales of cones, where they are relatively exposed.

Although many pine trees and other conifers grow in Florida and elsewhere in the southeast United States, conifers are the dominant plant of the forests of Canada and other far-northern places. Conifers have many adaptations that help them survive cold winters, including branches that bend but do not break when covered in snow.

The female cone protects the seeds as they grow. When the seeds are fully grown, the cone opens. Some seeds fall to the ground. Others are carried away by the wind. If a seed lands in a good place, it will germinate.

LIFE CYCLE OF A **LONGLEAF PINE**

GERMINATION **If a seed lands in a good place, it will start to grow.**

GROWTH **The seedling grows into a young tree.**

MATURITY **When it is large enough, the tree makes cones. Male cones produce pollen that is spread by the wind.**

SEED FORMATION **When pollen reaches the female cones, the ovules are fertilized. Seeds form inside the female cones.**

29

Teach, continued

Text Feature: Diagrams

- Have students study the diagram of the life cycle of the longleaf pine. Then have them compare it to the diagram of the flowering plant shown on page 27. Ask: **How is the life cycle of the longleaf pine like that of a flowering plant?** (The four stages have the same names, and they occur in the same order.)
- Ask: **How are the life cycles different?** (The longleaf pine makes seeds in cones, while the flowering plant makes seeds in fruits. Also, only the flowering plant makes flowers and depends on bees or other insects for pollination.) Remind students that conifers rely on the wind for pollination. Some flowering plants, including grasses, also are wind pollinated.
- Point out the caption for germination. Say: **Remember that germinating seeds need water and the proper temperature in order to grow.** Ask: **What else do you think makes a "good place" for a pine seed to start growing?** (Possible answer: a region of soil that is reasonably far away from other trees, yet not where a seedling would be trampled or uprooted by humans or animals.) Discuss how the diagram shows the complete life cycle for a longleaf pine. Say: **The tree's seeds may or may not germinate, and a growing tree may die or be killed at any stage of its life cycle.**

Differentiated Instruction

ELL Language Support for Describing Life Cycles of Conifers

BEGINNING

Help students describe life cycles of conifers using either/or questions. For example: **Do pine seeds form inside flowers or inside cones? Is pollen made by male cones or female cones? Do bees or the wind pollinate female cones?**

INTERMEDIATE

Help students use Academic Language Frames to describe life cycles. For example:

- *A seed begins to grow in an event called ____.*
- *A very young plant is called a ____.*
- *A pine tree begins to make cones when it reaches ____.*

ADVANCED

Provide sequence words to help students describe the life cycle of conifers. For example:

The life cycle of conifers includes different stages. First, . . . Second, . . . Then . . . Finally, . . .

Objectives

Students will be able to:

- Compare and contrast the life cycles of flowering plants and conifers.

Teach, continued

Text Feature: Photos and Captions

- Have students read pages 30–31 and compare the two plants pictured on page 30. Ask: **How do you know which tree is a flowering plant and which tree is a conifer?** (The orange tree is a flowering plant because it has fruits. The pine tree is a conifer because it has cones.)
- Ask: **What other differences do you observe between the two plants?** (The conifer has needle-like leaves, while the orange tree has flat leaves.)
- Point to one of the oranges (fruits of the orange tree). Ask: **How did this orange grow? What structure made it?** (The orange grew from the base of a flower, or orange blossom.) Explain that by the time oranges are large and mature, the petals and other flowering parts of the orange blossom have long since fallen away.
- Point to one of the cones on the conifer. Ask: **Was this cone made by a flower? How do you know?** (No. Conifers do not make flowers.) Remind students of the two ways that plants make seeds: either in cones or with flowers and fruits.

Comparing Life Cycles Think about the life cycles of the orange tree and the longleaf pine. How are their life cycles alike? How are they different?
The seeds of an orange tree are protected by the tough skin of its fruit.
The seeds of a longleaf pine are protected by the female cones.
30

Science Misconceptions

Many Fruits Some students may believe that the term *fruit* describes only large, fleshy fruits that are good to eat, such as the familiar fruits sold in supermarkets. Explain that in plant science, the term *fruit* describes a wide variety of structures that hold seeds, and they may or may not be edible by humans or other animals. Examples of fruits include the hard, wing-like structures that encase maple seeds and the sticky burrs of the cockleburr plant.

Both kinds of plants reproduce with seeds. The seeds of conifers grow in cones. Conifer seeds are not protected by fruits. The seeds of flowering plants form in flowers. Their seeds are protected by fruits.

Conifers are pollinated by the wind. Some flowering plants are also wind pollinated. But most flowering plants are pollinated by insects or other animals.

All conifers are trees or shrubs that live for many years and reproduce many times. Some kinds of flowering plants also live for many years. But many kinds of flowering plants have much shorter life cycles. Some live for only a season. After they make seeds, they die.

COMPARING **SEED PLANTS**

	REPRODUCTIVE STRUCTURES	POLLINATION	LOCATION OF SEEDS
FLOWERING PLANTS	Flowers	Most are pollinated by animals. Some are pollinated by wind.	In fruits
CONIFERS	Cones	All are pollinated by wind.	In cones

Before You Move On

1. What are the four main stages in the life cycle of a flowering plant?
2. Compare pollination in conifers with pollination in flowering plants.
3. **Apply** Which do you think would grow more quickly, a garden of flowering plants or a garden of conifers? Explain why.

31

Differentiated Instruction

Extra Support

Have students draw pictures that show how flowering plants and conifers reproduce. Their drawings should include a caption to explain the illustration. For example, a picture of a fruit might include the caption *Flowering plants make fruits with seeds inside.* Have students share their drawings and captions with the class.

Challenge

Have students write two paragraphs that compare the way flowering plants and conifers reproduce. Have students read their paragraphs to the class.

Teach, continued

Text Feature: Chart

- Say: **A chart is a way to organize information in a way that is clear and simple.** Ask: **What information is being compared in the chart?** (The chart compares reproductive structures, methods of pollination, and location of seeds among different kinds of plants.)
- Have students read the title of the chart. Ask: **What are the two kinds of plants that make seeds?** (flowering plants and conifers) Remind students that a few plants, such as mosses and ferns, reproduce without seeds. All seed plants make either flowers or cones.
- Point to the second column of the chart. Ask: **What is pollination?** (the transfer of pollen from male parts to female parts) Remind students that male parts make pollen in both flowering plants and conifers.

Summarize Information

Have students copy the chart on page 31 into their science notebook. Then have them review the chapter and add additional information to the chart to help them compare and contrast the two types of seed plants.

❸ Assess

» Before You Move On

1. List **What are the four main stages in the life cycle of a flowering plant?** (germination, growth of the seedling, maturity, and seed formation or reproduction)
2. Compare and Contrast **Compare pollination in conifers with pollination in flowering plants.** (In conifers, male cones make pollen, which blows onto the female cones. In flowering plants, the stamens make pollen, which may be carried to the pistil by animals or by wind.)
3. Apply **Which do you think would grow more quickly, a garden of flowering plants or a garden of conifers? Explain why.** (A garden of flowering plants would grow more quickly. All conifers are shrubs or trees that live for many seasons. Many kinds of flowering plants grow for only one season.)

LESSON 7 ▫ Characteristics of Plants

Objectives
Students will be able to:
- Identify individual differences in organisms of the same kind.
- Explain that although most characteristics of plants are inherited, some characteristics can be affected by factors in a plant's environment.

Science Academic Vocabulary
inherited

PROGRAM RESOURCES
- Learning Masters Book, page 15, or at myNGconnect.com

❶ Introduce

Tap Prior Knowledge
- Show the class a familiar seed, such as a sunflower seed. Have students describe the kind of plant that the seed will grow into.

Set a Purpose and Read
- Read the heading. Tell students that they will learn how plants get their characteristics such as the color of a flower or the size of a stem. Have students read pages 32–33.

❷ Teach

Academic Vocabulary: *inherited*
- Write the word **inherited** and have students pronounce it. Have a volunteer read the sentence on page 32 that defines this term.
- Invite students to discuss their **inherited** characteristics, which include their gender, eye and hair color, and general physical appearance.

Identify Inherited Characteristics
- Have students observe the photo of tulips on pages 32–33. Ask: **What are some inherited characteristics among these tulips?** (Possible answers: shapes and colors of the petals and leaves, the position of the flower at the end of the stem)
- Ask: **Where did these inherited characteristics come from?** (from each tulip's parent plants)

Characteristics of Plants

As you know, young plants are similar to their parents. The leaves of an orange tree are the same shape as those of its parents. The shape of an orange leaf is an **inherited** characteristic. Characteristics that are inherited have been passed from parents to their offspring.

Look at the tulips in the picture below. Why are the tulips different colors? The color of each flower has been inherited from different parent flowers.

The color of each of these tulips was inherited from the plant's parents.

32

Science Misconceptions

Parents and Offspring Some students may think that offspring must look exactly like one of their parents. Or they may think that the characteristics of offspring are a blend or average of the characteristics of both parents. While these ideas may be true in certain cases, the inheritance of characteristics is often very complex. A certain color for flowers, for example, may be hidden for many generations, then suddenly appear in an offspring. Interested students may research the work of Gregor Mendel, the scientist whose work pioneered the study of genetics in the late 1800s.

Most characteristics of plants are inherited. But things around a plant can affect the way it grows. Characteristics that have been changed by the environment cannot be passed to offspring.

The environment can change plants in many ways. If a plant does not get enough sunlight, its leaves may turn yellow or grow very large. Nutrients in the soil also affect how plants grow. The plants below were grown in different kinds of soil. How do these plants differ?

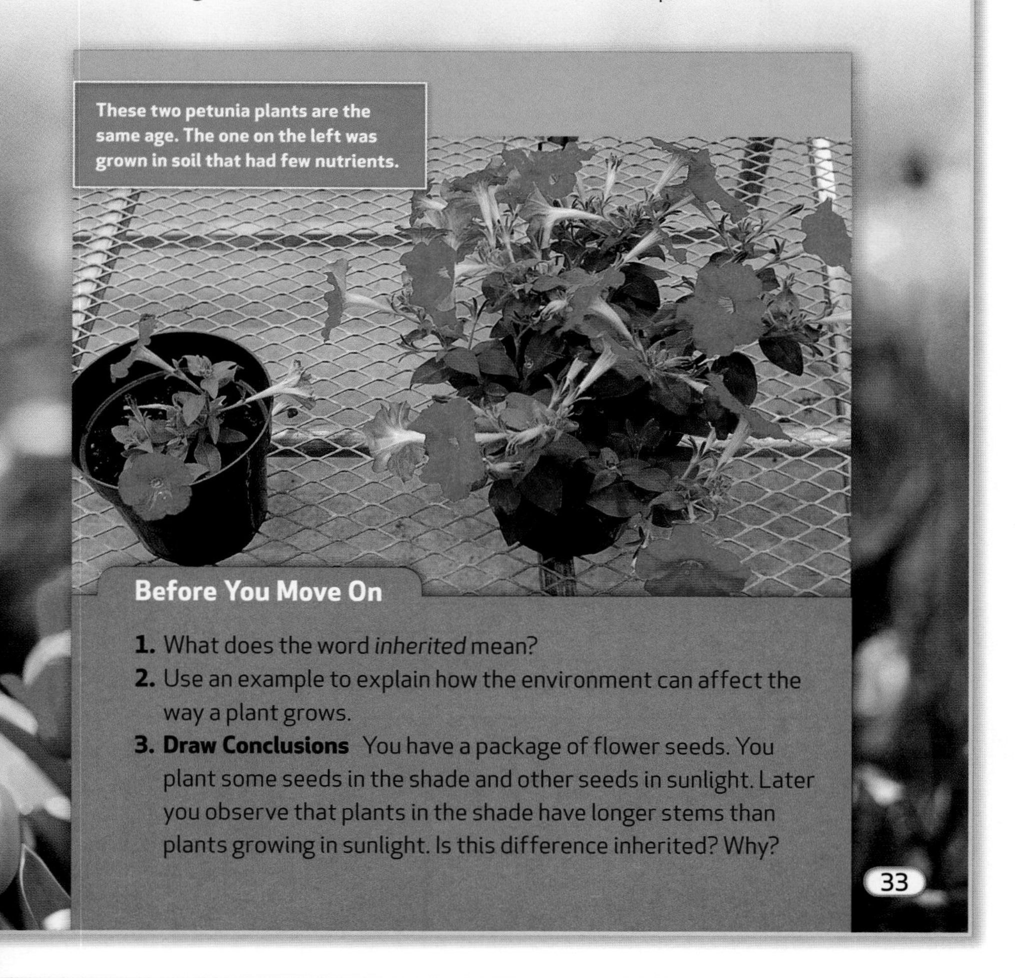

These two petunia plants are the same age. The one on the left was grown in soil that had few nutrients.

Before You Move On

1. What does the word *inherited* mean?
2. Use an example to explain how the environment can affect the way a plant grows.
3. **Draw Conclusions** You have a package of flower seeds. You plant some seeds in the shade and other seeds in sunlight. Later you observe that plants in the shade have longer stems than plants growing in sunlight. Is this difference inherited? Why?

Share and Compare

Flowering Plants and Conifers Give students the Learning Master. Have partners fill in the chart to compare flowering plants and conifers and then identify and label the parts of a flower. Check their work, and then encourage them to discuss the differences among plants and the ways that plants reproduce.

my SCIENCE notebook

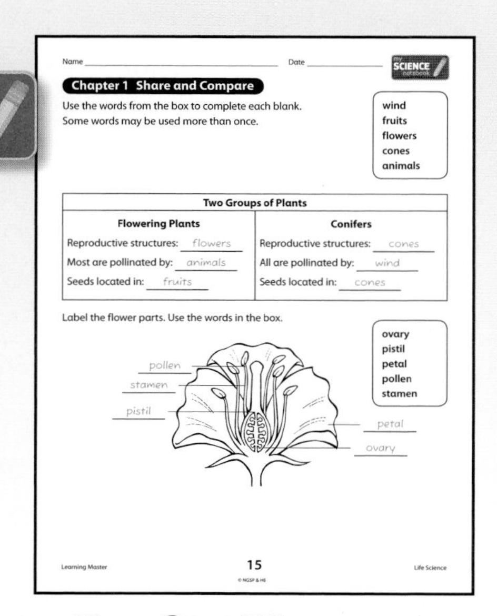

Name ____________ Date ____________

Chapter 1 Share and Compare

Use the words from the box to complete each blank. Some words may be used more than once.

wind
fruits
flowers
cones
animals

Two Groups of Plants	
Flowering Plants	**Conifers**
Reproductive structures: flowers	Reproductive structures: cones
Most are pollinated by: animals	All are pollinated by: wind
Seeds located in: fruits	Seeds located in: cones

Label the flower parts. Use the words in the box.

ovary
pistil
petal
pollen
stamen

Learning Master 15 Life Science

© NGSP & HB

Learning Master 15 at myNGconnect.com

Teach, continued

Text Feature: Photos and Captions

- Have students observe and compare the two photos of petunia plants on page 33. Ask: **What inherited characteristics do you observe in both petunia plants?** (Answers include leaf shape and the colors and shapes of the flowers.) Discuss how the **inherited** characteristics of both plants are nearly identical.
- Say: **As the caption tells you, both plants are the same age, and both are petunia plants.** Ask: **So why is one plant much larger and have more flowers than the other plant?** (The two plants were grown in different types of soil. The small plant received fewer nutrients than the larger plant.)
- Ask: **Infer what other factors in the environment might affect the way a plant looks and grows.** (Possible answers: amount of sunlight, amount of water, available space for growth, wind, and human actions.) Discuss how each of these factors is a common reason why plants of the same type can appear very different from one another.

3 Assess

» Before You Move On

1. **Define** **What does the word *inherited* mean?** (Something that is inherited is passed on from parent to offspring.)
2. **Explain** **Use an example to explain how the environment can affect the way a plant grows.** (Factors that affect plant growth include soil quality; amount of water or sunlight; and human actions, such as mowing a lawn or pruning a bush.)
3. **Draw Conclusions** **You have a package of flower seeds. You plant some seeds in the shade and other seeds in full sunlight. Later you observe that plants growing in the shade have longer stems than plants growing in sunlight. Is this difference between the plants inherited? Explain why.** (The differences are probably not inherited, since the seeds were planted at the same time and were planted from the same packet.)

Objectives

Students will be able to:

- Summarize information from charts and graphs to answer scientific questions.
- Compare and contrast the stages in the life cycles of plants.
- Interpret organized observations and measurements, recognizing simple patterns, sequences, and relationships.
- Explain that science focuses solely on the natural world.

PROGRAM RESOURCES

- Learning Masters Book, pages 16–17, or at myNGconnect.com

1 Introduce

Tap Prior Knowledge

- Ask: **What kinds of diagrams have you seen?** Encourage students to talk about diagrams they have seen, especially diagrams they have seen in science class. Discuss how students have used diagrams and what information they included.
- Ask: **How do you think diagrams are used in scientific investigations?** (to show information in an organized way)

2 Teach

Recording Observations

- Read page 20 aloud with students. Say: **The photographs and the table both give information about how plants grow.**
- Ask: **What information do you get from the photographs? How is that different from the information you get from the table?** (The photographs show what a plant looks like in three different stages as it sprouts from a seed. The table shows how the plant's height changed over time.)

Math in Science

Recording Observations

Making a diagram during an investigation is an important way that scientists keep accurate records of their observations. Scientists use different kinds of diagrams.

Table and Graph Diagrams Sometimes scientists record observations as tables with numbers. This kind of diagram is especially helpful when scientists make measurements.

Plant Growth

Day	Plant Height (cm)
1	0
3	1
5	3
7	5
9	8

20

Sometimes scientists graph the data from their tables.

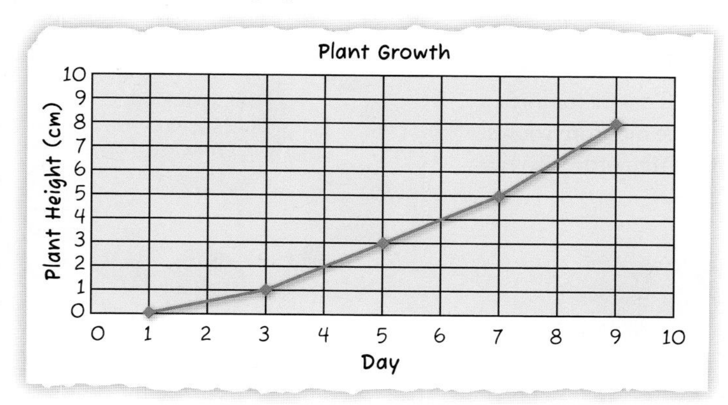

Scale Diagrams Often when scientists make a drawing of an object that is very large or very small, they compare each part of the object to the size of the real thing. Sometimes they make actual measurements. Then they draw the parts so that the relative sizes remain the same.

A scientist observed a leaf through a microscope. She saw tiny openings in the leaf. When the scientist drew what she saw, she made the parts larger but she drew them to scale.

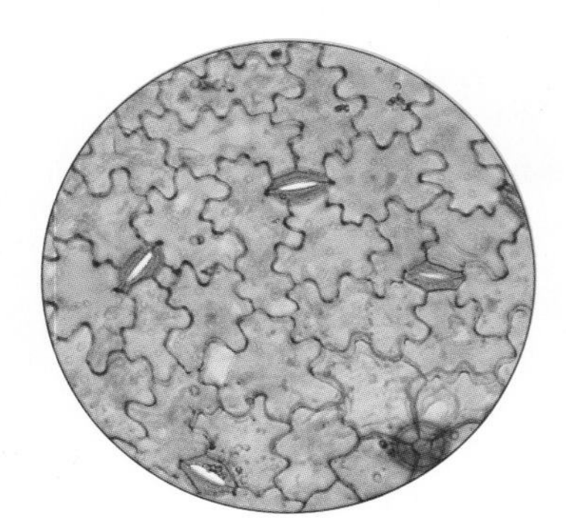

View of leaf through a microscope

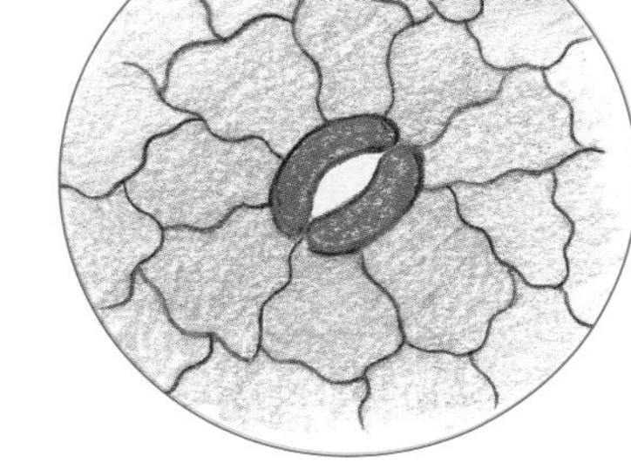

Scientist's drawing of what she saw

Raise Your SciQ!

NATIONAL GEOGRAPHIC

What Are Stomata? To prevent loss of moisture, many plants have waxy coatings on their leaves. These waxy coatings also block the transfer of gases. Tiny holes called stomata are present in leaves. These small openings allow plants to take in carbon dioxide and give off oxygen for photosynthesis and respiration.

Stomata help control the loss of moisture, but some water is still lost as gases pass in and out of the leaf. Cottonwood trees growing in very dry climates may lose about 100 gallons (about 379 liters) of water an hour through their stomata!

Teach, continued

- Read the first sentence on page 21 aloud with students and then have them study the graph. Ask: **What information is shown in the Plant Growth graph?** (how a plant's height changes over time) Point out that the data in the graph matches the data in the table.
- Help students interpret the line graph. Ask: **What information is on the x-axis?** (days) **What information is on the y-axis?** (plant height in cm)
- Check that students are able to read the line graph. Ask: **How tall was the plant on Day 3?** (1 cm tall) **On which day was the plant about 7 cm tall?** (day 8)
- Explain that a graph can show trends in data. Ask: **What trend does this graph show?** (that the plant grows over time) **What new information do you get from the graph that may not have been clear in the table?** (I can see that the plant did not grow the same amount each day because the line is not straight.)
- Read about scale diagrams aloud with students. Have students study the two pictures and point to the one that is a scale drawing. Ask: **How are the two pictures similar?** (Both show the surface of the leaf.) **How are they different?** (The scale drawing shows a smaller portion of the leaf in more detail.)
- Ask: **Why do you think scientists use scale diagrams?** (to accurately represent something on a larger or smaller scale)
- If students ask, explain that the tiny openings in the leaf are called stomata. These openings allow gases such as oxygen, carbon dioxide, and water vapor to pass in and out of the leaf.
- Ask: **What other things might scientists show in scale diagrams?** Accept all reasonable answers. Students may suggest using scale diagrams to show parts of the human body or very small animals.

Teach, continued

- Have students read page 22 and look at the photographs and the sequence diagram. Ask: **What is a life cycle?** (the series of stages that a plant, animal, or other living thing goes through during its life)
- Have students study the photograph of the tree. Explain to students that bald cypress trees grow in wetland habitats.
- Read aloud with students the stages in the life cycle of the bald cypress tree. Explain that the arrows show what stage comes next. To reinforce this concept, point to the *Tree matures* stage. Ask: **Which stage comes next?** (Pollen moves from male cone to female cone.)
- Ask: **What other kinds of information could be shown in a sequence diagram?** Students may suggest showing the water cycle or the rock cycle in a sequence diagram.
- Point to the photographs on pages 22–23. Ask: **What do these things have in common?** (They are plants; they are part of the natural world.) Help students understand that science is based solely on the natural world. Point out that science focuses on objects and events in nature that can be studied and observed.
- Have students work with a partner to answer the questions on page 22.

What Did You Find Out?

1. Scientists use diagrams to keep accurate records of their observations.

2. Tables and graphs can show numbers, such as measurements. Scale diagrams show the relative sizes of the parts of an object that is very large or very small. Sequence diagrams show the order in which something happens.

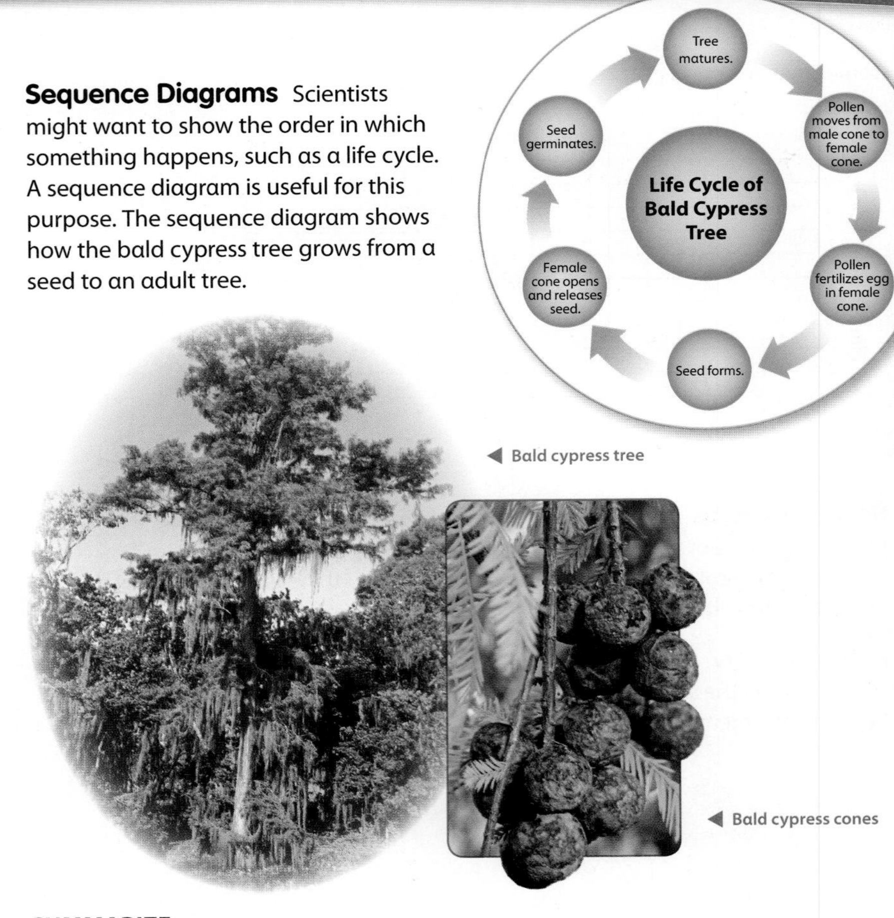
Sequence Diagrams Scientists might want to show the order in which something happens, such as a life cycle. A sequence diagram is useful for this purpose. The sequence diagram shows how the bald cypress tree grows from a seed to an adult tree.

Life Cycle of Bald Cypress Tree
Tree matures.
Pollen moves from male cone to female cone.
Pollen fertilizes egg in female cone.
Seed forms.
Female cone opens and releases seed.
Seed germinates.

◀ Bald cypress tree

◀ Bald cypress cones

SUMMARIZE
What Did You Find Out?

1. Why do scientists use diagrams?
2. What are three kinds of diagrams? What kind of information does each diagram show?

22

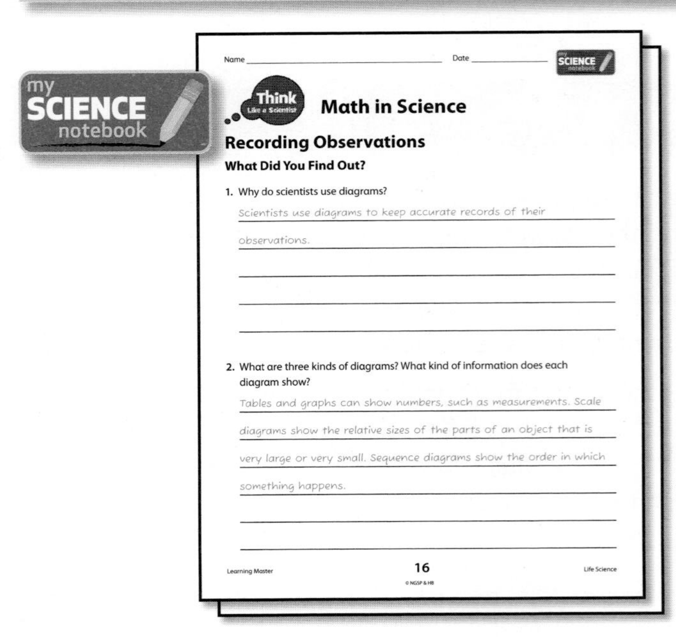
my SCIENCE notebook

Name ______ Date ______

Think Like a Scientist **Math in Science**

Recording Observations

What Did You Find Out?

1. Why do scientists use diagrams?

Scientists use diagrams to keep accurate records of their observations.

2. What are three kinds of diagrams? What kind of information does each diagram show?

Tables and graphs can show numbers, such as measurements. Scale diagrams show the relative sizes of the parts of an object that is very large or very small. Sequence diagrams show the order in which something happens.

Learning Master 16 Life Science

Learning Masters 16–17 or at myNGconnect.com

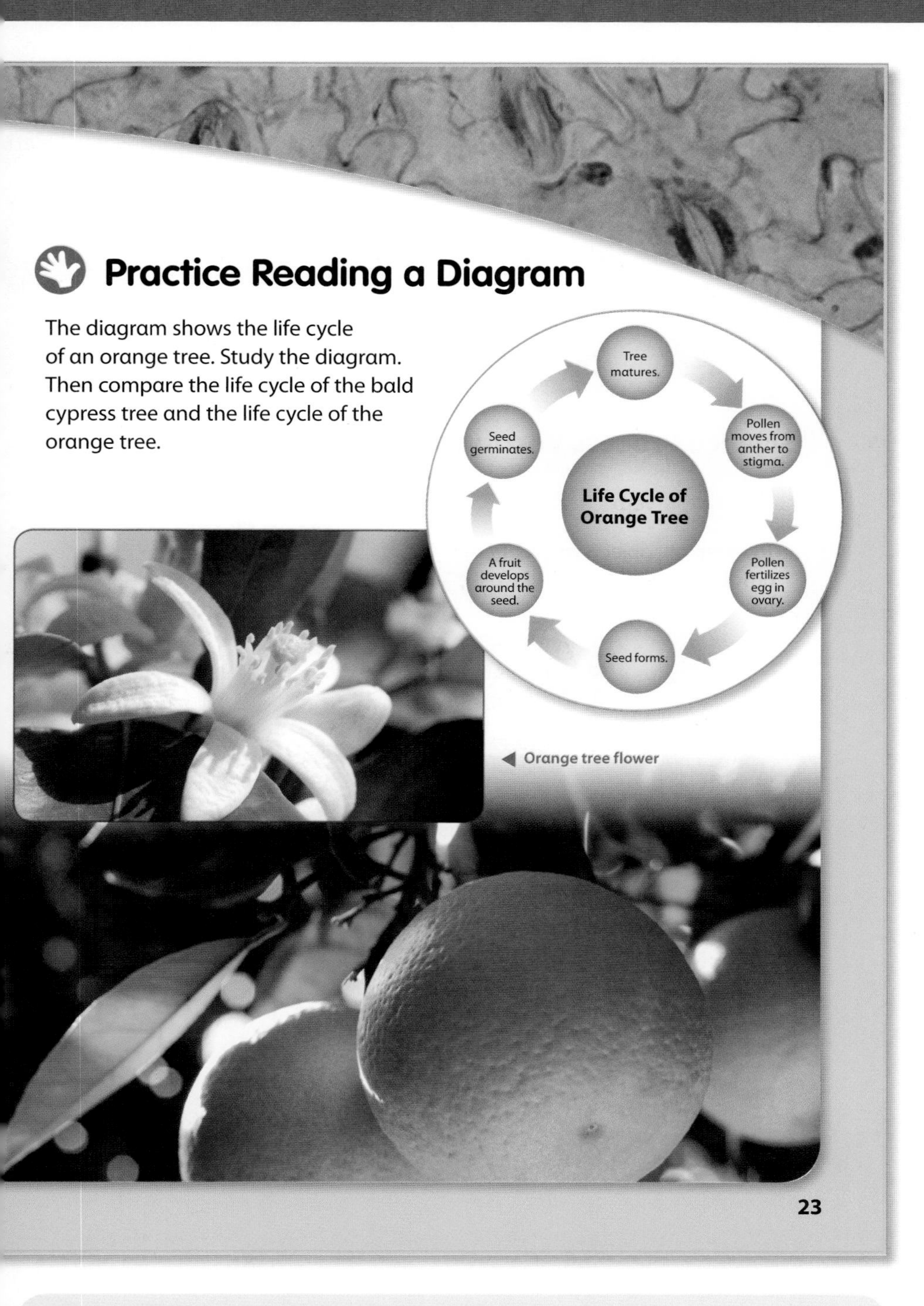
Practice Reading a Diagram
The diagram shows the life cycle of an orange tree. Study the diagram. Then compare the life cycle of the bald cypress tree and the life cycle of the orange tree.
Life Cycle of Orange Tree
Tree matures.
Pollen moves from anther to stigma.
Pollen fertilizes egg in ovary.
Seed forms.
A fruit develops around the seed.
Seed germinates.
Orange tree flower
23

NATIONAL GEOGRAPHIC Raise Your SciQ!

What's the Difference? Cypress trees are conifers, and orange trees are not. In the bald cypress, the male cone releases pollen. It fertilizes an egg in a female cone.

In an orange tree, pollen moves from the anther of a flower to the stigma at the top of the pistil. Then pollen fertilizes an egg in the ovary. In the bald cypress, a seed forms in the cone. In the orange, the seed forms in a fruit.

Teach, continued

Practice Reading a Diagram

- Read aloud with students the stages in the life cycle of an orange tree on page 23.
- Make sure students understand the words *anther, stigma,* and *ovary.* Explain that these are parts of flowers. The anther produces and contains pollen, the stigma is the structure at the top of the pistil that receives the pollen, and the ovary is the structure in which seeds are produced. The ovary eventually can develop into a fruit like the orange shown on page 23.
- Have students compare the life cycles of an orange tree and a bald cypress tree. Encourage them to write down how they are similar and how they are different. Students can work individually or in pairs.
- Ask: **What would happen next if the pollen from an orange flower moved to the stigma of another orange flower?** (A seed would form.)
- To enhance students' understanding of life cycles, have them make observations once a week over a few months of the changes in a plant outside. Have them record their measurements and observations in tables, graphs, scale diagrams, or sequence diagrams.
- After observations are complete, invite groups to share their diagrams and tell how the plant changed.
- Ask: **Why is it important to distinguish your observations from your inferences?** Discuss with students how observations and inferences are not the same things. An observation is something you find out about by using the senses. An inference is made by using observations and what you already know to make a decision or conclusion. Inferences can be correct or incorrect. Different inferences may be made based on the same observations.

3 Assess

- Use Learning Masters Book pages 16–17. Students should be able to identify how scientists keep records to describe observations, and compare and contrast the major stages in the life cycles of plants.

LESSON 9 ▫ Directed Inquiry

Objectives

Students will be able to:

- Investigate through Directed Inquiry (answer a question; make and compare observations; collect and record data and observations; generate explanations and conclusions based on evidence; share findings; ask questions based on observations to increase understanding).
- Identify processes of sexual reproduction in flowering plants, including pollination.
- Make observations using simple tools and use relevant safety procedures to measure and describe organisms by basic characteristics and their uses.

Science Process Vocabulary

compare, infer

PROGRAM RESOURCES

- Science Inquiry and Writing Book: *Life Science*
- Science Inquiry and Writing Book eEdition at myNGconnect.com
- Inquiry eHelp at myNGconnect.com
- Science Inquiry Kit: *Life Science*
- Learning Masters Book, pages 18–23, or at myNGconnect.com
- Inquiry Rubric: Assessment Handbook, page 182, or at myNGconnect.com
- Inquiry Self-Reflection: Assessment Handbook, page 190, or at myNGconnect.com

MATERIALS

Kit materials are listed in italics.

flower; Parts of a Flower Diagram Learning Master; scissors; *hand lens;* ruler; *microscope; slide;* white paper

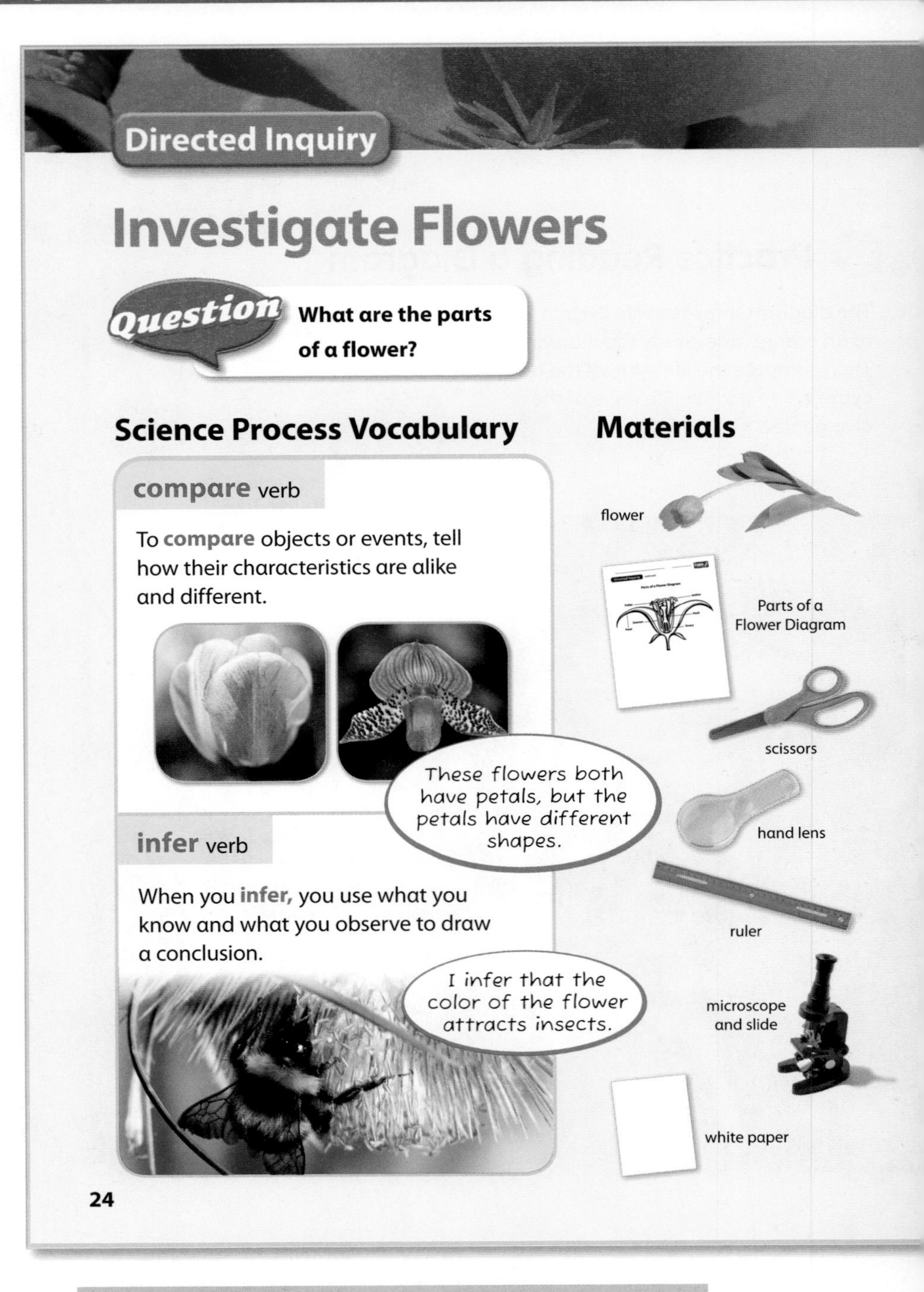

Directed Inquiry

Investigate Flowers

Question What are the parts of a flower?

Science Process Vocabulary

compare verb

To **compare** objects or events, tell how their characteristics are alike and different.

These flowers both have petals, but the petals have different shapes.

infer verb

When you **infer**, you use what you know and what you observe to draw a conclusion.

I infer that the color of the flower attracts insects.

Materials

flower

Parts of a Flower Diagram

scissors

hand lens

ruler

microscope and slide

white paper

24

MANAGING THE INVESTIGATION

Time

30 minutes

Groups

Small groups of 4

Advance Preparation

- Set up a microscope for each group.
- Copy the Parts of a Flower Learning Master.
- Familiarize students with the parts of the microscope. Have students practice focusing the microscope using a small piece of thin cloth or newspaper.

❶ Introduce

Tap Prior Knowledge

- Have students name flowers they are familiar with. Make a list on the board. Ask students to describe some of the flowers they have seen, including the color, petal shape, size, and smell.

What to Do

1. **Observe** the flower. Identify the following parts: petal, anther, stamen, ovary, and pistil. Use the Parts of a Flower Diagram to help you. Draw the whole flower and label each part in your science notebook. Make sure you draw the parts to scale.

2. Use the scissors to carefully cut a stamen from the flower. Observe the stamen with the hand lens. **Measure** its length with the ruler. Record your observations.

3. Remove another stamen from the flower and gently tap it on the microscope slide. The powdery material on the slide is pollen. Observe the pollen under the microscope. Record your observations and draw the pollen.

25

Teaching Tips

- Use plants with large flowers, such as daffodils, tulips, lilies, or gladioluses. Have different groups look at different flowers.
- In steps 2 and 5, remind students to be cautious with the scissors. Tell them to cut gently to avoid crushing the plant parts.

What to Expect

- Students will identify and draw in detail the parts of a flower, including the petal, anther, stamen, pistil, and pollen. Students will find individual differences in the shapes, sizes, and colors of the parts of different flowers.

Introduce, continued

Connect to the Big Idea

- Review the Big Idea Question, *How do plants grow and reproduce?* Explain to students that this inquiry will help them understand how pollination works in different flowering plants.
- Have students open their Science Inquiry and Writing Books to page 24. Read the Question and invite students to share what they know about the parts of flowers.

❷ Build Vocabulary

Science Process Words: compare, infer

Use this routine to introduce the words.

1. **Pronounce the Word** Say **compare**. Have students repeat it in syllables.
2. **Explain Its Meaning** Choral read the sentence. Ask students to define **compare**. (tell how things are similar and different)
3. **Encourage Elaboration** Ask: **How would you compare a sunflower and a tulip?** (You would tell how they are alike and different. For example, both plants have roots, stems, leaves, and flowers, but a sunflower is bigger than a tulip and has a different shape and color.)

ELL Language Production

Lay some flowers on a desk and ask students to **compare** them. Ask: **How are the flowers alike? How are they different?** Have students answer using the words *alike* and *different*.

Repeat for the word **infer**. To encourage elaboration, ask: **What information do you need to infer?** (observations and prior knowledge)

❸ Guide the Investigation

- Distribute materials. Read the inquiry steps on pages 25–26. Clarify steps if necessary.
- Review the parts of the flower with students before they begin the investigation. Have students use the Parts of a Flower Diagram Learning Master or the Big Ideas Book for reference.
- If students cannot participate in dissecting a flower for religious reasons, have them complete step 1 and then use clay or construction paper to make a model of the flower.

Guide the Investigation, continued

- Before step 3, demonstrate how to prepare a slide. Explain that students only need a very small amount of pollen on the slide.
- If students have difficulty using the microscope, encourage them to adjust the mirror to let in enough light. They can also flip down the yellow lens attachment for a more magnified view, or they can adjust the height of the slide.
- Ask: **How does the microscope help you see the pollen better?** (The microscope shows details that cannot be seen with the eye alone.)
- In step 5, guide students as they record their observations using a table similar to the one on page 27. Encourage students to write about each plant part, record their measurements in metric units, and make sketches.

❹ Explain and Conclude

- Have students compare and discuss with other groups the observations, drawings, and measurements they made.
- Discuss with students the parts of the flower. Ask: **Which part of a flowering plant is involved in reproduction?** (the flower) **What is the role of the pollen and how does it find its way to other flowers?** (Pollen is needed to produce seeds for reproduction. It moves to different plants by wind, water, insects, or other animals.)
- Tell students that in scientific investigations metric units are used for length measurements. Discuss English units for length such as inches, feet, or miles.
- Have students summarize their inferences and generalizations about plant parts. Encourage them to actively listen for other ideas and interpretations and use evidence to support their explanations.

SafetyFirst

Remind students to wash their hands. Students with allergies or respiratory problems should not handle pollen.

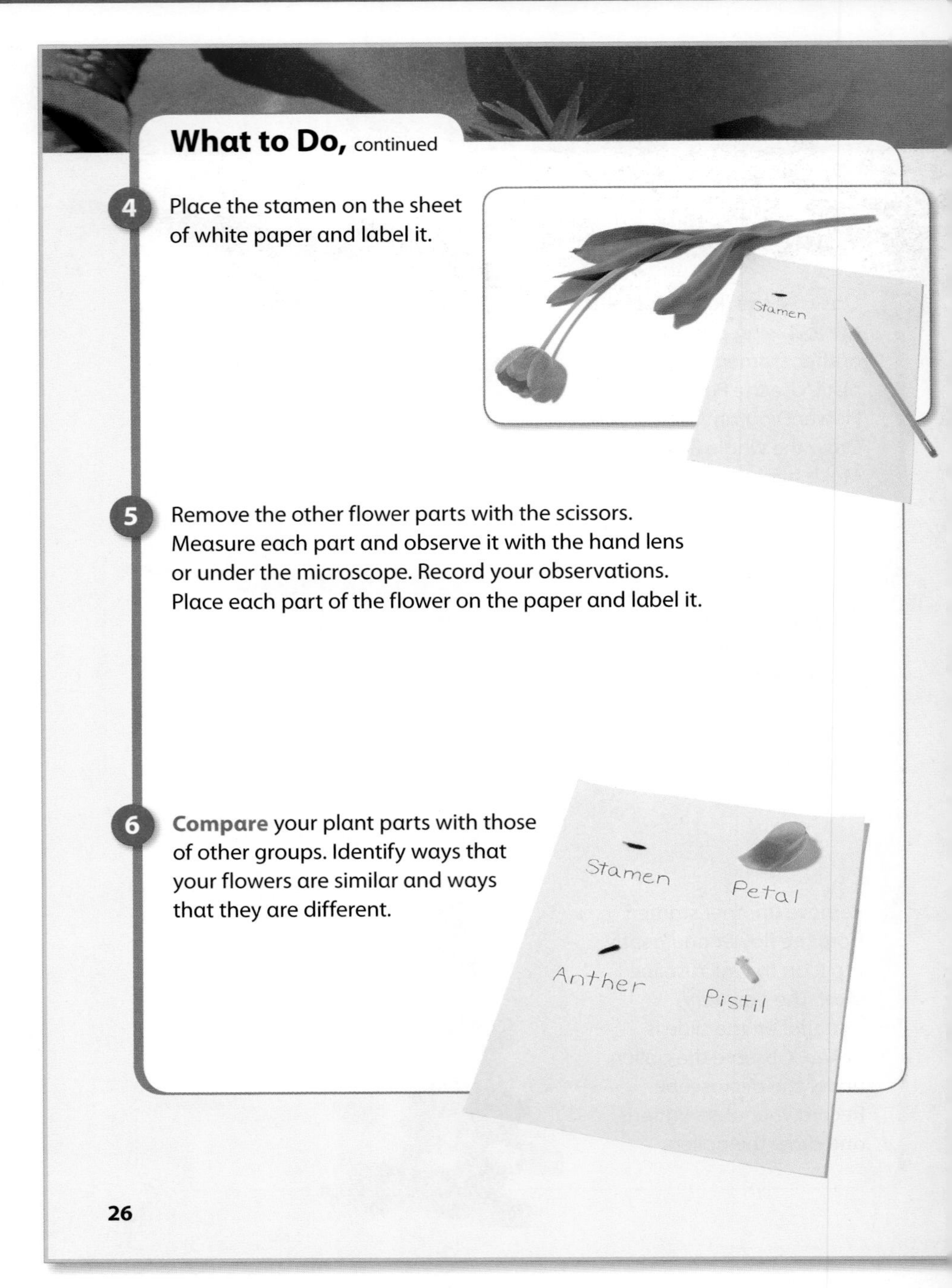

What to Do, continued
4 Place the stamen on the sheet of white paper and label it.
Stamen
5 Remove the other flower parts with the scissors. Measure each part and observe it with the hand lens or under the microscope. Record your observations. Place each part of the flower on the paper and label it.
6 Compare your plant parts with those of other groups. Identify ways that your flowers are similar and ways that they are different.
Stamen
Petal
Anther
Pistil
26

NATIONAL GEOGRAPHIC Raise Your SciQ!

Hay Fever Allergy to pollen is called hay fever. Generally, pollens that cause allergies are those of anemophilous plants, or plants in which the pollen is dispersed by air currents. In the United States, people often mistakenly blame the conspicuous goldenrod flower for allergies. Since this plant is entomophilous, with pollen dispersed by animals, its heavy, sticky pollen does not become independently airborne. Most late summer and fall pollen allergies are probably caused by ragweed, a widespread anemophilous plant. Anemophilous spring blooming plants such as oak, birch, hickory, pecan, and early summer grasses may also induce pollen allergies.

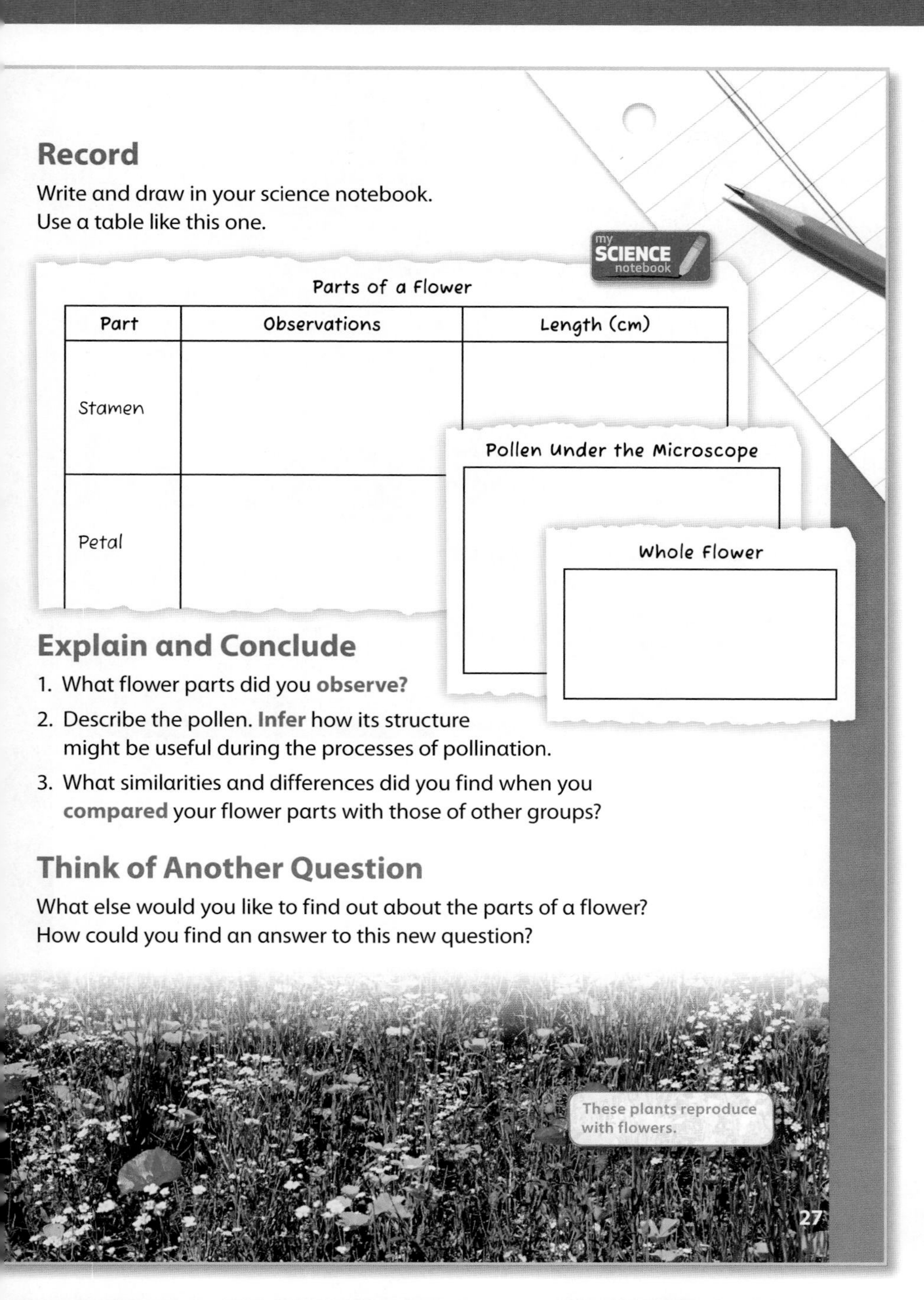

Record

Write and draw in your science notebook.
Use a table like this one.

my SCIENCE notebook

Parts of a Flower

Part	Observations	Length (cm)
Stamen		
Petal		

Pollen Under the Microscope

Whole Flower

Explain and Conclude

1. What flower parts did you **observe?**
2. Describe the pollen. **Infer** how its structure might be useful during the processes of pollination.
3. What similarities and differences did you find when you **compared** your flower parts with those of other groups?

Think of Another Question

What else would you like to find out about the parts of a flower? How could you find an answer to this new question?

These plants reproduce with flowers.

27

Inquiry Rubric at myNGconnect.com	Scale			
The student **observed** the parts of a flower using a hand lens and a microscope.	4	3	2	1
The student dissected a flower and identified each part.	4	3	2	1
The student **compared** the plant parts from different flowers.	4	3	2	1
The student collected **data** and **shared** results.	4	3	2	1
The student made an **inference** about how the structure of pollen aids in pollination and fertilization.	4	3	2	1
Overall Score	**4**	**3**	**2**	**1**

Explain and Conclude, continued

Answers

1. Possible answer: I observed the petal, anther, stamen, ovary, and pistil.

2. Possible answer: The pollen is small so it can easily move from the stamen to the pistil. The pollen is rough so it can stick to the tip of the pistil.

3. Answers will vary, but students should find that the general structure of each part is the same in various flowers of the same kind. Students might notice differences in size, color or shape of the parts.

5 Find Out More

Think of Another Question

- Students should use their observations to generate new questions from their investigations.

6 Reflect and Assess

- To assess student work with the Inquiry Rubric shown below, see Assessment Handbook, page 182, or go online at myNGconnect.com
- Have students use the Inquiry Self-Reflection on Assessment Handbook, page 190, or at myNGconnect.com

my SCIENCE notebook

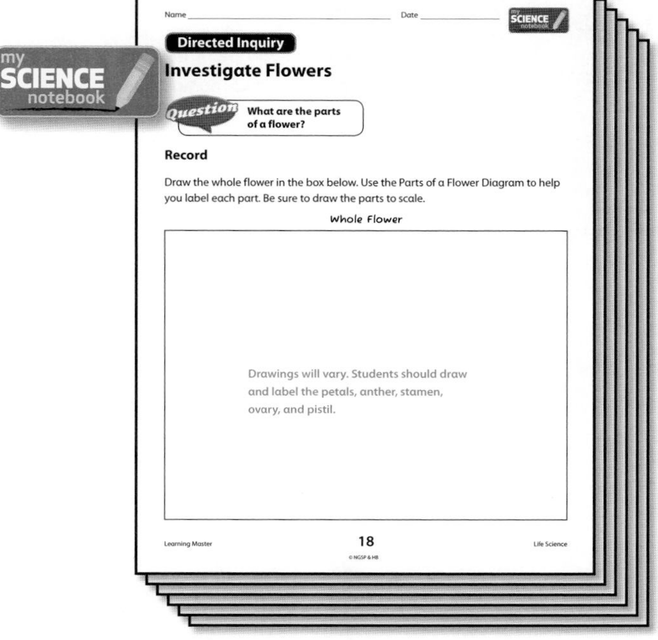

Name ______ Date ______

Directed Inquiry

Investigate Flowers

Question: What are the parts of a flower?

Record

Draw the whole flower in the box below. Use the Parts of a Flower Diagram to help you label each part. Be sure to draw the parts to scale.

Whole Flower

Drawings will vary. Students should draw and label the petals, anther, stamen, ovary, and pistil.

Learning Master 18 Life Science

Learning Masters 18–23 or at myNGconnect.com

PROGRAM RESOURCES

- Chapter 1 Test, Assessment Handbook, pages 4–6, or at myNGconnect.com
- NGSP ExamView CD-ROM

❶ Sum Up the Big Idea

- Display the chart from page T8–T9. Read the Big Idea Question. Then add new information to the chart.
- Ask: **What did you find out in each section? Is this what you expected to find?**

How Do Plants Grow and Reproduce?

How Plants Grow

Sandeep: Plants make their own food using photosynthesis. Plants can reproduce with or without seeds.

Flowers

Concetta: Flowers must be pollinated and fertilized before fruits and seeds can grow.

Fruits and Seeds

Ellen: Seeds have three main parts: an embryo, a seed coat, and stored food. Fruits protect the seeds inside them.

Life Cycles of Seed Plants

Ernesto: The life cycle of a seed plant involves germination, growth, maturity, and making seeds.

Characteristics of Plants

Amiko: A plant's characteristics can be inherited or affected by the environment.

Conclusion

The two main groups of plants are those that reproduce with seeds and those that do not have seeds. Flowering plants and conifers are seed plants. Flowering plants reproduce with seeds that grow in fruits. Conifers reproduce by making seeds in cones. Most of a plant's characteristics are inherited. But some characteristics are changed by its environment.

Big Idea The life cycle of a plant includes the germination of a seed, growth, maturity, reproduction, and finally death.

Seeds Germinate

Plants Grow and Reproduce

Plants Die

Vocabulary Review

Match the following terms with the correct definition.

A. conifer
B. photosynthesis
C. pollination
D. fertilization
E. inherited
F. seed dispersal

1. The joining of an egg and a sperm cell
2. The carrying of seeds to a new place
3. A seed plant that reproduces with cones
4. The movement of pollen from a stamen to a pistil
5. A characteristic that is passed from parents to offspring
6. The way plants use the energy of sunlight to make food

34

Review Academic Vocabulary

Academic Vocabulary Tell students that scientists use special words when describing their work to others. For example:

photosynthesis	conifer	pollination
fertilization	seed dispersal	inherited

Have students write the list in their science notebook and use each term in a sentence that describes how plants grow and reproduce. They should then share their sentences with a partner.

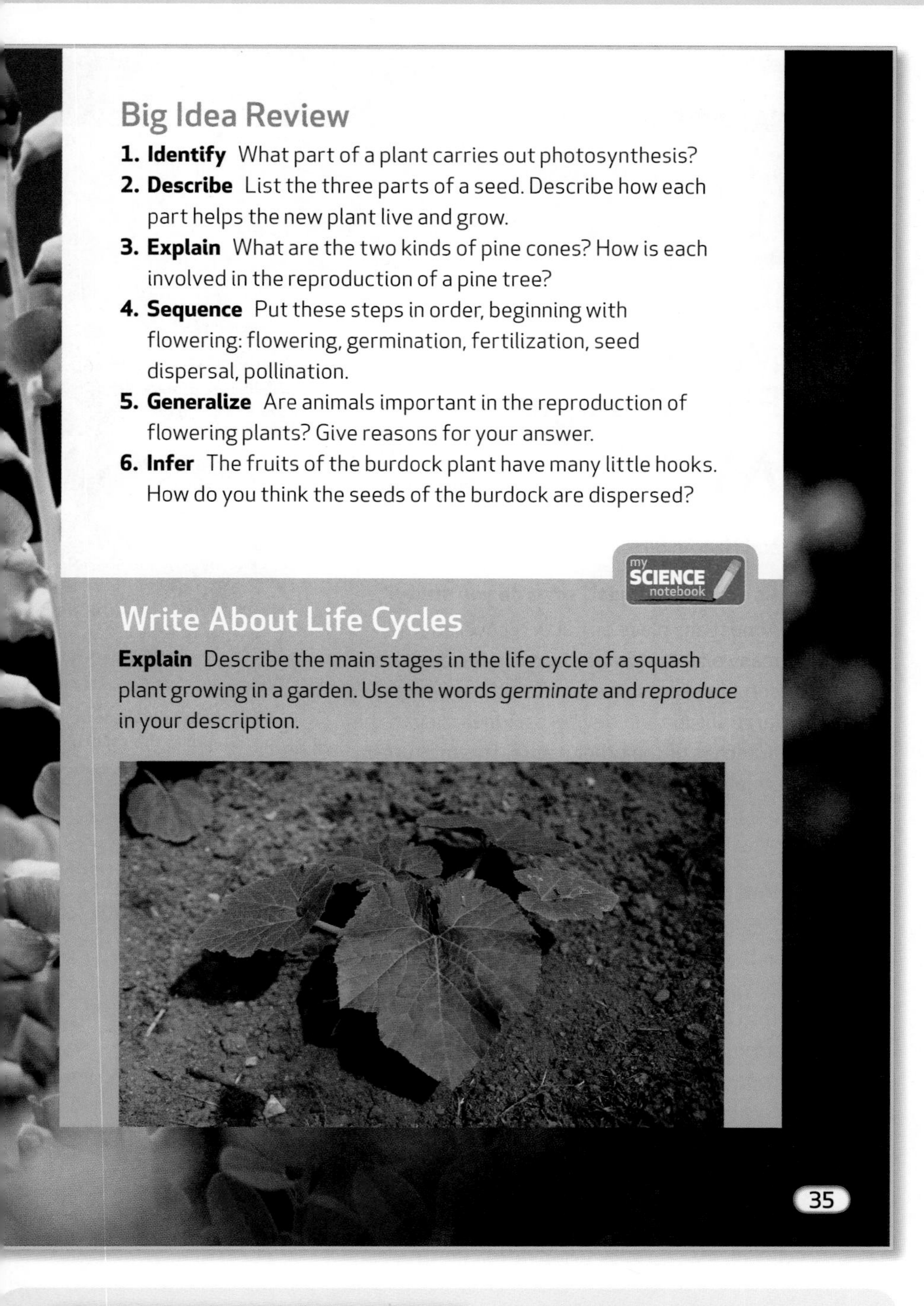
Big Idea Review

1. **Identify** What part of a plant carries out photosynthesis?
2. **Describe** List the three parts of a seed. Describe how each part helps the new plant live and grow.
3. **Explain** What are the two kinds of pine cones? How is each involved in the reproduction of a pine tree?
4. **Sequence** Put these steps in order, beginning with flowering: flowering, germination, fertilization, seed dispersal, pollination.
5. **Generalize** Are animals important in the reproduction of flowering plants? Give reasons for your answer.
6. **Infer** The fruits of the burdock plant have many little hooks. How do you think the seeds of the burdock are dispersed?

my SCIENCE notebook

Write About Life Cycles

Explain Describe the main stages in the life cycle of a squash plant growing in a garden. Use the words *germinate* and *reproduce* in your description.

35

❷ Discuss the Big Idea

Notetaking

Have students write in their science notebook to show what they know about the Big Idea. Have them:

1. Write about how flowering seed plants reproduce.
2. Write about the similarities and differences in fruits and seeds.
3. Write about the similarities and differences in the life cycles of flowering plants and conifers.

❸ Assess the Big Idea

Vocabulary Review

1. D **2.** F **3.** A **4.** C **5.** E **6.** B

Big Idea Review

1. The leaves of a plant carry out photosynthesis.
2. The three parts of a seed are the embryo, stored food, and seed coat. The embryo will grow into a new plant. The stored food helps the embryo grow until the tiny plant can make its own food. The seed coat protects the embryo.
3. Pine trees have male and female cones. The male cones make pollen. The female cones protect the seeds as they grow.
4. flowering, pollination, fertilization, seed dispersal, germination
5. Animals are important to the reproduction of flowering plants in two ways. First, they pollinate flowers by carrying pollen from one flower to another. Second, they help to disperse seeds.
6. The seeds are probably dispersed when the fruits of the burdock attach to the fur of an animal.

Write About Life Cycles

The life of a squash plant begins as a seed that germinates. The seed grows into a seedling, then a fully grown, or mature, plant. The mature plant then reproduces by making flowers. After the flower is fertilized, new seeds begin to grow inside fruits. The squash plant dies, but some of the new seeds will grow into new plants next season.

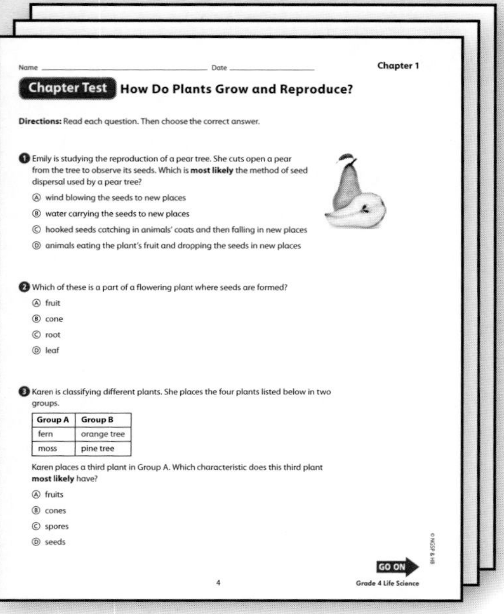
Name ________________ Date ________ Chapter 1

Chapter Test How Do Plants Grow and Reproduce?

Directions: Read each question. Then choose the correct answer.

1. Emily is studying the reproduction of a pear tree. She cuts open a pear from the tree to observe its seeds. Which is **most likely** the method of seed dispersal used by a pear tree?
   - Ⓐ wind blowing the seeds to new places
   - Ⓑ water carrying the seeds to new places
   - Ⓒ hooked seeds catching in animals' coats and then falling in new places
   - Ⓓ animals eating the plant's fruit and dropping the seeds in new places
2. Which of these is a part of a flowering plant where seeds are formed?
   - Ⓐ fruit
   - Ⓑ cone
   - Ⓒ root
   - Ⓓ leaf
3. Karen is classifying different plants. She places the four plants listed below in two groups.

Group A	Group B
fern	orange tree
moss	pine tree

Karen places a third plant in Group A. Which characteristic does this third plant **most likely** have?

- Ⓐ fruits
- Ⓑ cones
- Ⓒ spores
- Ⓓ seeds

© NGSP & HB

GO ON

4 Grade 4 Life Science

Assess Student Progress Chapter 1 Test

Have students complete their Chapter 1 Test to assess their progress in this chapter.

Chapter 1 Test, Assessment Handbook, pages 4–6, or at myNGconnect.com

or NGSP ExamView CD-ROM

Objectives

Students will be able to:

- Describe how plant biologists investigate the characteristics of plants.

PROGRAM RESOURCES

- Big Ideas Book: *Life Science*
- Big Ideas Book: *Life Science* eEdition at myNGconnect.com
- Digital Library at myNGconnect.com

❶ Introduce

Tap Prior Knowledge

- Ask students if they are familiar with the terms *biology* or *biologist*. Encourage volunteers to define either term in their own words, or to give examples of the kinds of subjects a biologist would study.

Preview and Read

- Read the heading. Have students preview the photos on pages 36–37. Ask them to think about what kind of work a plant biologist might do.
- Have students read pages 36–37.

❷ Teach

Analyze Word Parts

- Write *bio-*, *biology*, and *biologist* on the board and underline the suffixes. Say: ***-logy* means "the study of something." *Bio-* means "life." So *biology* is the study of life, and a *biologist* is a person who studies life. Mark Olson is a biologist who studies plant life.**

Discuss Plant Diversity

- Ask: **What is the main question that Mark Olson studies?** (Where does all the diversity of plants come from?) **What is diversity?** (It is all the different kinds of life forms that exist on Earth.)
- Explain: **Mark Olson studies all kinds of plants. When he observes a plant with a unique or unusual characteristic, he wants to know the cause.**

NATIONAL GEOGRAPHIC

CHAPTER 1 LIFE SCIENCE EXPERT: PLANT BIOLOGIST

Mark Olson: Plant Biologist

Mark Olson is a plant biologist who teaches at Mexico's National University. He studies how plants develop different shapes and sizes in different parts of the world.

As a plant biologist, what do you study?

Almost any place on Earth is covered with many different living things. In the dry tropical shrubland outside my office, there are very large shrubs and tiny herbs. Vines cling to the branches of the other plants. This mixture in shapes, sizes, and ways of living is what makes our Earth so interesting. The question I study is: Where did all this diversity come from?

How do you study plants in the field?

I spend a little more than half of my time working in the field. We may spend the day measuring tiny cacti on a desert plain, collecting plants on the cliffs above a tropical beach, or climbing a mountain to find a rare plant. We might use ropes to climb rain forest trees. Sometimes we fly above the rain forest in a glider to learn how the trees collect light from the sun.

Mark Olson is wearing a helmet to protect him as he flies in the glider.

Mark studies the shape of plants such as this cactus.

36

NATIONAL GEOGRAPHIC Raise Your SciQ!

Plants in the Desert A desert is an ecosystem that receives very little annual rainfall—less than 25 centimeters (10 inches) per year. Deserts also may experience large temperature extremes between day and night. The temperatures may be very hot all day but near freezing at night. Cacti, succulents, and other desert plants have adaptations that help them survive desert conditions. Cacti have thick, waterproof stems that help them retain water. Their leaves are modified into spines that protect them from animals. Their roots often spread widely and deeply into the sandy soil, allowing them to take in the water they need after the rare rainfall.

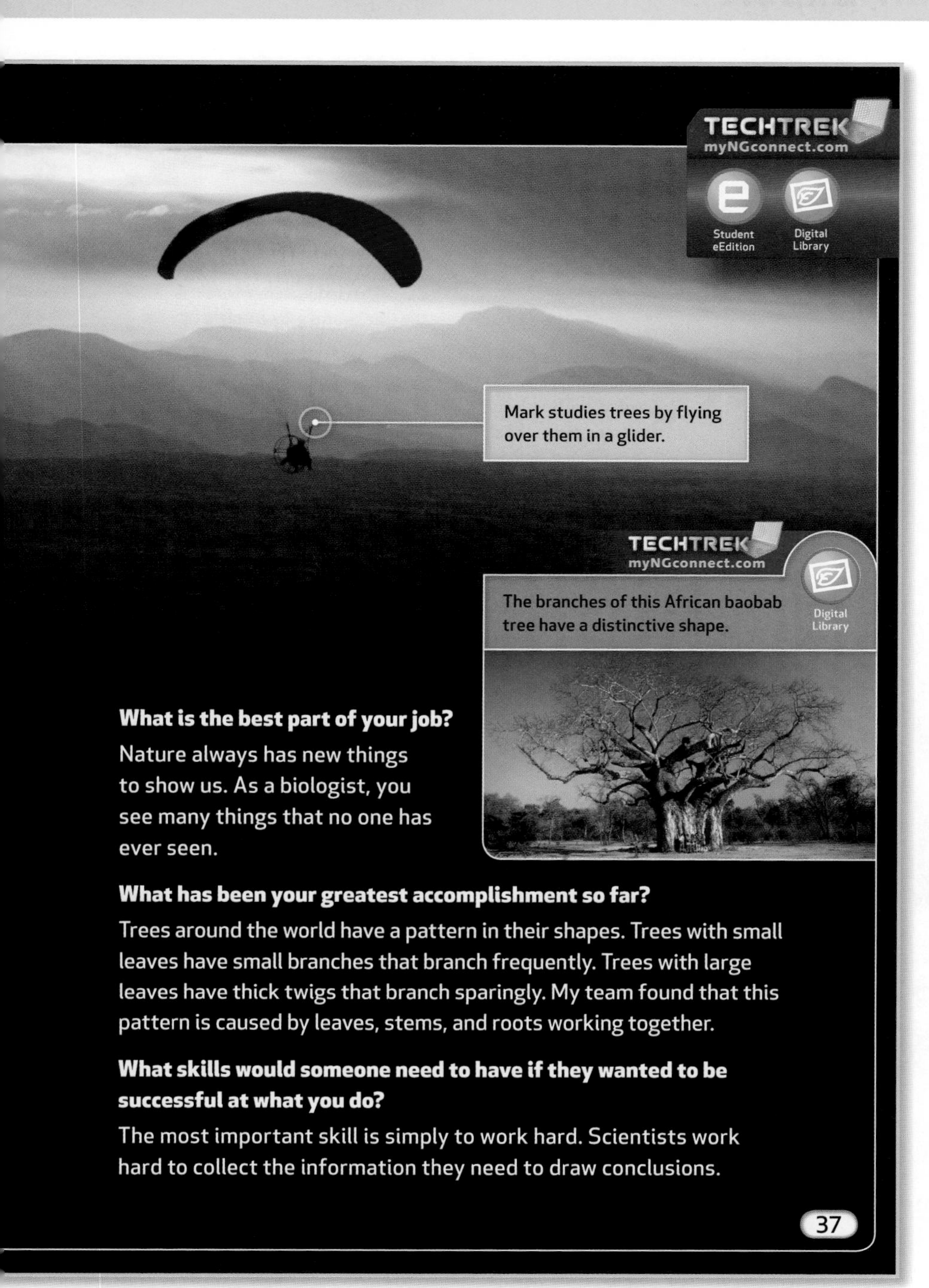

Mark studies trees by flying over them in a glider.

The branches of this African baobab tree have a distinctive shape.

What is the best part of your job?
Nature always has new things to show us. As a biologist, you see many things that no one has ever seen.

What has been your greatest accomplishment so far?
Trees around the world have a pattern in their shapes. Trees with small leaves have small branches that branch frequently. Trees with large leaves have thick twigs that branch sparingly. My team found that this pattern is caused by leaves, stems, and roots working together.

What skills would someone need to have if they wanted to be successful at what you do?
The most important skill is simply to work hard. Scientists work hard to collect the information they need to draw conclusions.

37

Teach, continued

Describe the Work of a Plant Biologist

- Ask: **Why does Mark Olson travel to different places around the world to study plants?** (By traveling widely, he can compare different plants and find out why plants are so diverse.)
- Ask: **Does Mark Olson work outdoors or indoors?** (both) Explain: **After Mark makes observations in the field, he needs to organize his data and analyze the results.**
- Have students observe the photo of Mark and the glider on page 37. Ask: **Why is Mark flying in a glider?** (to observe trees from above)

Find Out More

- Ask: **Is this a career you would find interesting? Why or why not?** (Accept all answers, positive or negative.) Encourage interested students to research what other questions plant biologists study.

myNGconnect.com

Have students use the **Digital Library** to find images of various types of trees and other plants.

Digital Booklet Students can use the information to make a digital booklet with captions that describes observations or inferences about various types of trees and other plants.

Differentiated Instruction

ELL Language Support for Describing Plant Biologists

BEGINNING

Help students use words to describe a plant biologist. Give them choices, such as:

Do plant biologists study plants or grow them for food? Do plant biologists study plants all over the world or just outside their offices?

INTERMEDIATE

Help students write new captions for the photos on pages 36–37. Encourage students to include vocabulary words from the chapter when appropriate.

ADVANCED

Help students describe plant biologists by using Academic Language Stems, such as:

- *Mark Olson is a plant biologist. He works . . .*
- *I learned . . .*

3 Assess

1. **Recall What is a plant biologist?** (a person who studies plant life)
2. **Explain What skills of science inquiry do plant biologists use when they study plants in nature?** (Possible answers: observing, measuring, collecting samples, synthesizing)
3. **Draw Conclusions Why might being a plant biologist be a rewarding career?** (Possible answer: plants are important living things around the world, and plant biologists help people conserve and protect plants and to use plants wisely.)

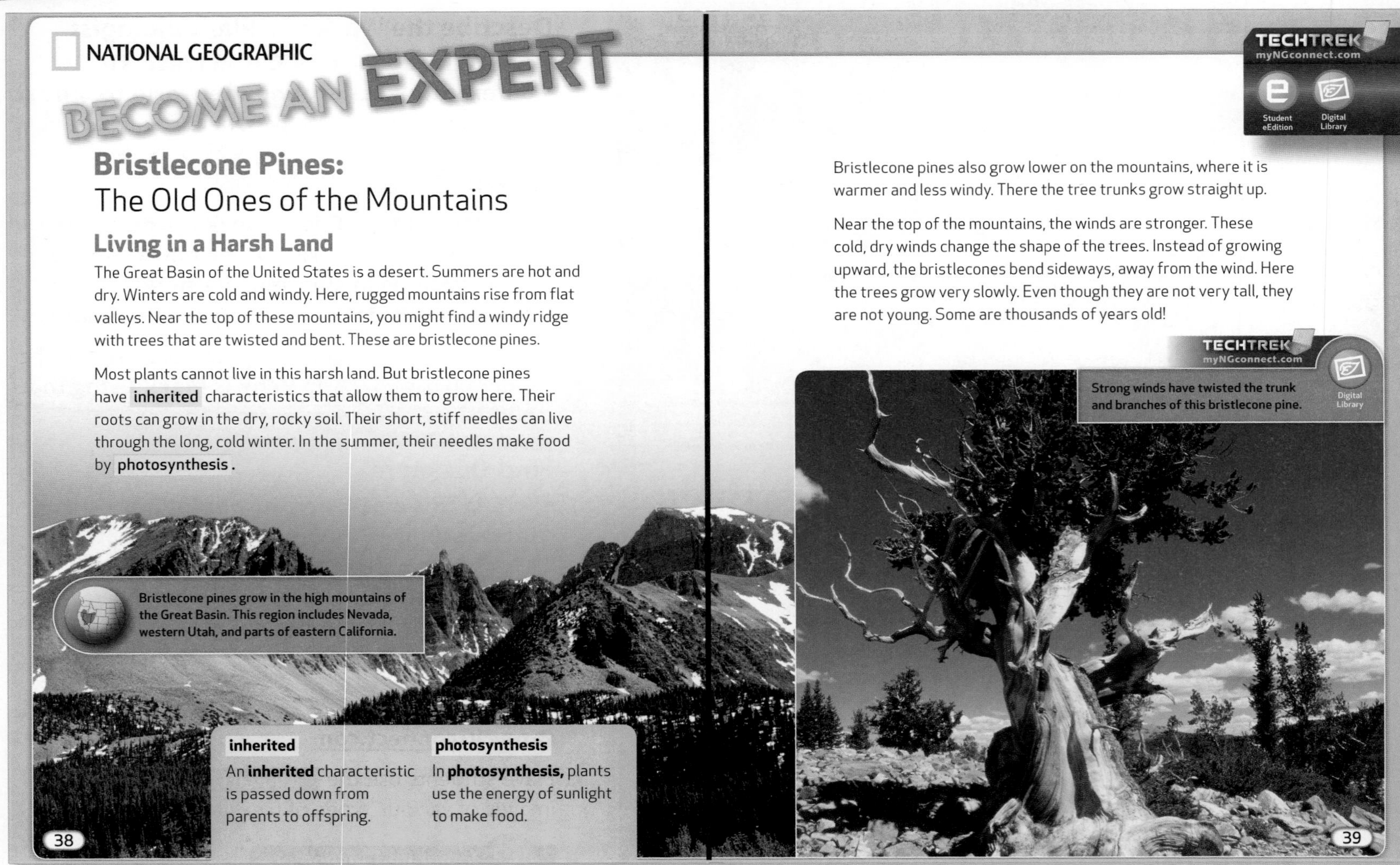
NATIONAL GEOGRAPHIC

BECOME AN EXPERT

Bristlecone Pines:
The Old Ones of the Mountains

Living in a Harsh Land

The Great Basin of the United States is a desert. Summers are hot and dry. Winters are cold and windy. Here, rugged mountains rise from flat valleys. Near the top of these mountains, you might find a windy ridge with trees that are twisted and bent. These are bristlecone pines.

Most plants cannot live in this harsh land. But bristlecone pines have **inherited** characteristics that allow them to grow here. Their roots can grow in the dry, rocky soil. Their short, stiff needles can live through the long, cold winter. In the summer, their needles make food by **photosynthesis**.

Bristlecone pines grow in the high mountains of the Great Basin. This region includes Nevada, western Utah, and parts of eastern California.

inherited An **inherited** characteristic is passed down from parents to offspring.

photosynthesis In **photosynthesis**, plants use the energy of sunlight to make food.

38

TECHTREK myNGconnect.com Student eEdition Digital Library

Bristlecone pines also grow lower on the mountains, where it is warmer and less windy. There the tree trunks grow straight up.

Near the top of the mountains, the winds are stronger. These cold, dry winds change the shape of the trees. Instead of growing upward, the bristlecones bend sideways, away from the wind. Here the trees grow very slowly. Even though they are not very tall, they are not young. Some are thousands of years old!

TECHTREK myNGconnect.com Digital Library

Strong winds have twisted the trunk and branches of this bristlecone pine.

39

PROGRAM RESOURCES

- Big Ideas Book: *Life Science*
- Big Ideas Book: *Life Science* eEdition at myNGconnect.com
- Digital Library at myNGconnect.com

Access Science Content

Describe Inherited Characteristics

- Have students read pages 38–39, study the photographs, and read the captions. Point out the word **inherited** on page 38 and have a volunteer read the definition at the bottom of the page. Ask: **What inherited characteristics could make one kind of tree different from another?** (Possible answers: the shape and thickness of the trunk, shape and size of leaves, network of roots) Ask: **What inherited characteristics make the bristlecone pine unique?** (short, stiff needles; roots that can grow in dry, rocky soil) Discuss how these characteristics help bristlecone pines survive in the Great Basin.
- Ask: **Why is photosynthesis important to bristlecone pines?** (Their needles use photosynthesis to make food for the tree.) Review the definition of **photosynthesis** at the bottom of page 38.

Digital Library

myNGconnect.com

Have students use the Digital Library to find images of bristlecone pines and other plants.

Integrated Technology

Whiteboard Presentation Students can use the information to make a computer presentation about bristlecone pines and other plants. You can help them show the presentation on a whiteboard.

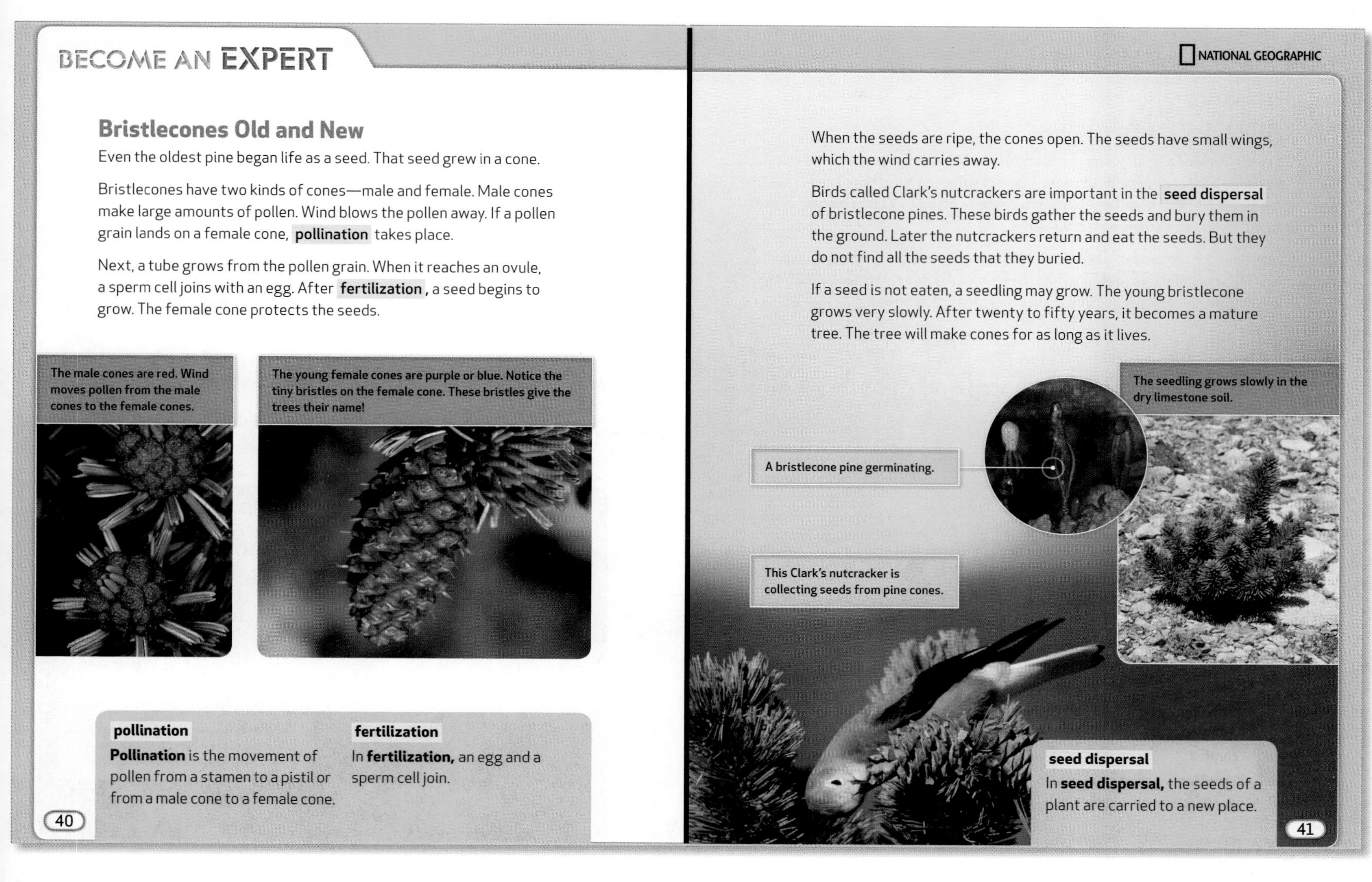
BECOME AN EXPERT

NATIONAL GEOGRAPHIC

Bristlecones Old and New

Even the oldest pine began life as a seed. That seed grew in a cone.

Bristlecones have two kinds of cones—male and female. Male cones make large amounts of pollen. Wind blows the pollen away. If a pollen grain lands on a female cone, **pollination** takes place.

Next, a tube grows from the pollen grain. When it reaches an ovule, a sperm cell joins with an egg. After **fertilization**, a seed begins to grow. The female cone protects the seeds.

The male cones are red. Wind moves pollen from the male cones to the female cones.

The young female cones are purple or blue. Notice the tiny bristles on the female cone. These bristles give the trees their name!

pollination
Pollination is the movement of pollen from a stamen to a pistil or from a male cone to a female cone.

fertilization
In **fertilization,** an egg and a sperm cell join.

40

When the seeds are ripe, the cones open. The seeds have small wings, which the wind carries away.

Birds called Clark's nutcrackers are important in the **seed dispersal** of bristlecone pines. These birds gather the seeds and bury them in the ground. Later the nutcrackers return and eat the seeds. But they do not find all the seeds that they buried.

If a seed is not eaten, a seedling may grow. The young bristlecone grows very slowly. After twenty to fifty years, it becomes a mature tree. The tree will make cones for as long as it lives.

The seedling grows slowly in the dry limestone soil.

A bristlecone pine germinating.

This Clark's nutcracker is collecting seeds from pine cones.

seed dispersal
In **seed dispersal,** the seeds of a plant are carried to a new place.

41

Access Science Content

Describe Pollination, Fertilization, and Seed Dispersal

- Have students read pages 40–41, observe the photos, and read the captions. Ask: **How are the female cones different from the male cones?** (Female cones are larger, they have a different color, and they have bristles.) **What do the male and female cones do?** (Male cones make pollen. Female cones receive pollen and carry the developing seeds.) Discuss how **pollination** and **fertilization** both take place on the female cones.
- Ask: **How does the Clark's nutcracker help bristlecone pines?** (It buries some of its seeds, allowing them to germinate and become seedlings.) Review the definition of **seed dispersal** and its importance to plants.

Make Inferences

Point to the photo of the Clark's nutcracker on page 41 and read the caption. Ask: **Why is this bird collecting pine seeds?** (to store them and then eat them) Explain that the bird does not truly intend to help disperse pine tree seeds. Instead, **seed dispersal** is a side effect of its feeding behavior.

Assess

1. **Recall What role does the wind play in the life cycle of the bristlecone pine?** (The wind blows the pollen from the male cone to the female cone. It also helps in seed dispersal by blowing the ripened seed out of the cone and onto the ground.)
2. **Draw Conclusions How does it help a bristlecone pine to produce many cones every year?** (Only a small number of its seeds may ever get the chance to germinate, and only a few seedlings may grow to maturity. By making many cones, a bristlecone pine increases the chances that its life cycle will continue after it dies.)

BECOME AN EXPERT

NATIONAL GEOGRAPHIC

How Long Do They Live?

The life cycle of a bristlecone starts with a seed. But once it is mature, a tree may live for thousands of years. Some bristlecone pines have lived for nearly 5,000 years. These ancient trees have only a tiny bit of living wood. Finally, even the oldest bristlecones die.

How does the life cycle of a bristlecone pine compare with that of other trees? Look at the chart below to find the answer. Are you surprised to see that these **conifers** live longer than trees that have flowers?

WHICH TREES LIVE THE LONGEST?

The chart compares the life spans of some American trees.

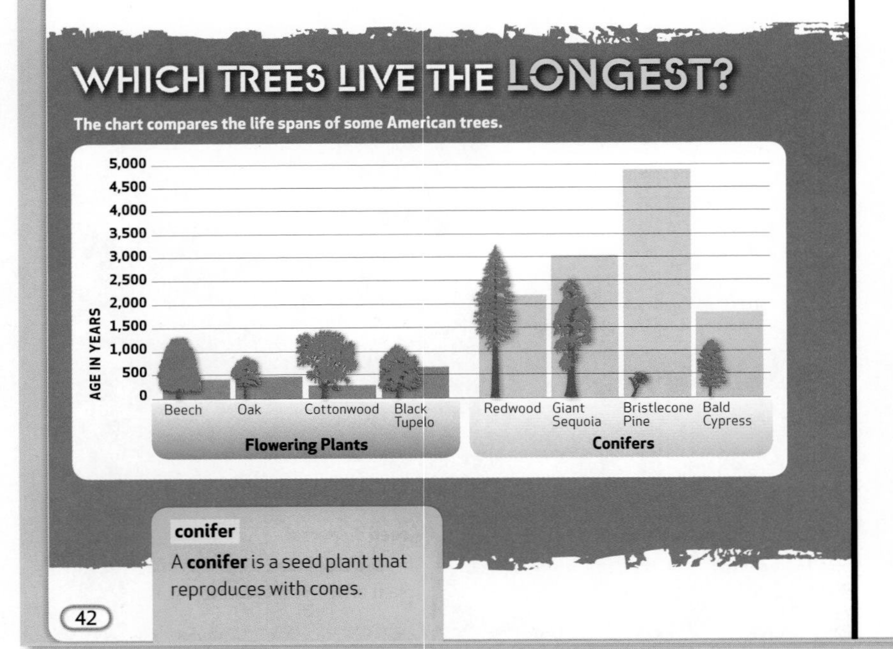

conifer
A **conifer** is a seed plant that reproduces with cones.

How do scientists learn how old a tree is? They look at its growth rings. Each year, the trunk of a tree grows a tiny bit wider. This growth forms a ring. Scientists count these rings to find how many years the tree has lived. If a tree is cut down, it is easy to see the rings. But scientists can also take a tiny core from a tree trunk and count the rings.

Scientists counted the rings on one bristlecone pine that had lived for 4,950 years. That tree was already 300 years old when the Pyramids of Egypt were being built!

This scientist is using a drill to take a core from the trunk of a bristlecone.

The growth rings of this bristlecone tell how many years it lived. Because the tree grew slowly, its rings are close together.

42 43

Access Science Content

Compare Life Cycles of Trees

- Have students read pages 42–43 and review the definition of **conifer** as they read. Then have them study the bar graph on page 42. Ask: **How does the life of a conifer compare with the life of a flowering tree?** (Conifers tend to live to a much older age.) Define the life span of a plant as the length of time between a seed's germination and the plant's death. Explain that the chart shows the average or possible life spans. Not every individual bristlecone pine tree may live for thousands of years.

Differentiated Instruction

ELL Language Support for Describing Bristlecone Pines

BEGINNING	INTERMEDIATE	ADVANCED
Ask either/or questions about bristlecone pines, such as: **Are bristlecone pines conifers or flowering plants? Are its seeds made before or after fertilization?**	Provide Academic Language Frames, such as: • *Bristlecone pines are classified as ____.* • *Seeds begin to form after ____.*	Provide Academic Language Stems, such as: • *Conifers are plants that . . .* • *Seeds begin to form after . . .* • *Animals help plants by . . .*

Assess

1. **Recall How is the life of bristlecone pine trees different from those of redwoods, sequoias, and other conifers?** (The life span of the bristlecone pine is nearly 5,000 years, which is much longer than other trees.)
2. **Predict Is a bristlecone pine in nature more likely to die as a seedling or as a mature tree? Explain your prediction.** (Possible answer: The tree may be more likely to die as a seedling, when it is small and more vulnerable to harm from animals or other factors in the environment. A mature tree is large and stable enough to survive most environmental events and changes.)

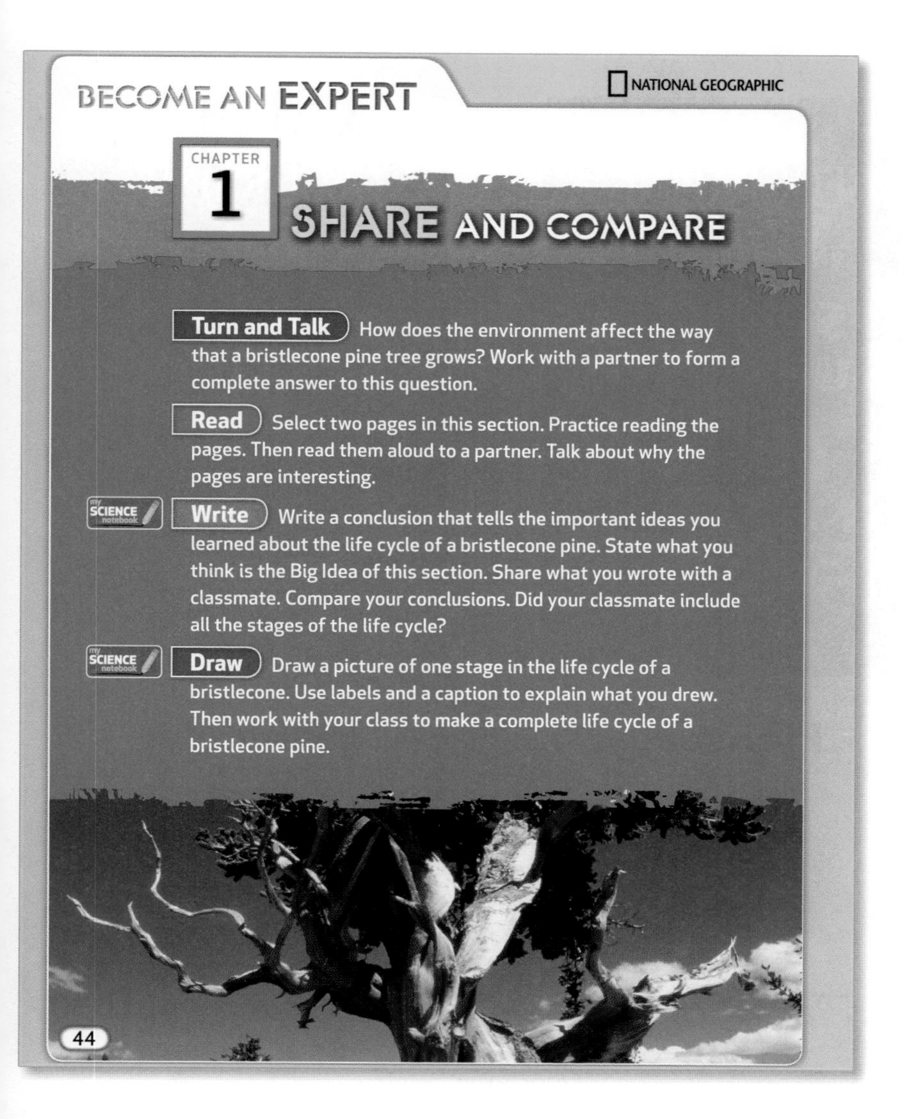
BECOME AN EXPERT

NATIONAL GEOGRAPHIC

CHAPTER 1

SHARE AND COMPARE

Turn and Talk How does the environment affect the way that a bristlecone pine tree grows? Work with a partner to form a complete answer to this question.

Read Select two pages in this section. Practice reading the pages. Then read them aloud to a partner. Talk about why the pages are interesting.

my SCIENCE notebook **Write** Write a conclusion that tells the important ideas you learned about the life cycle of a bristlecone pine. State what you think is the Big Idea of this section. Share what you wrote with a classmate. Compare your conclusions. Did your classmate include all the stages of the life cycle?

my SCIENCE notebook **Draw** Draw a picture of one stage in the life cycle of a bristlecone. Use labels and a caption to explain what you drew. Then work with your class to make a complete life cycle of a bristlecone pine.

44

Sum Up

Tell students that to sum up a text helps them pull together the ideas of the text. Remind students that they summed up the Become an Expert lesson in the Write section of Share and Compare. Have students take turns reading the conclusions they wrote to a partner. Invite them to compare and contrast their conclusions.

Share and Compare

Turn and Talk

Ask students to turn to partners and talk about what they learned about bristlecone pines and the life cycles of trees. Prompt students by asking:

1. **Recall Name some inherited characteristics of bristlecone pine trees.** (short, stiff needles; roots that can grow in dry, rocky soil)
2. **Sequence What are the stages in the life cycle of a bristlecone pine?** (pollination in cones, fertilization, seed growth, seed dispersal, germination, seedling, adult tree)
3. **Summarize How does the environment affect the way that a bristlecone pine tree grows?** (Strong winds can cause the trunk and branches to grow in twisted shapes; animals, weather, and changes in the soil (such as a landslide) can damage or kill trees.)

Read

Before students begin reading aloud, have them review the definitions and pronunciations of unfamiliar words. Encourage students to be honest as they discuss their opinions.

Have students write about the life cycle of a bristlecone pine tree. Ask them to summarize what they learned and to state the Big Idea of the section. Have students compare what they wrote with a partner. Ask if partners stated the same Big Idea.

Encourage students to include labels and captions with their drawings. Ask students to share their drawing with a classmate. They may describe which stage they drew and where it fits into the overall life cycle. Have partners or small groups work together to draw a complete life cycle diagram of a bristlecone pine.

Read Informational Text

Explore on Your Own books provide additional opportunities for your students to:

- deepen their science content knowledge even more as they focus on one Big Idea.
- independently apply multiple reading comprehension strategies as they read.

After applying the strategies throughout the chapter, students will independently read Explore on Your Own books. Facsimiles of the Explore on Your Own pages are shown on pages T44a–T44h.

Pioneer

Life Science
PATHFINDER EDITION
NATIONAL GEOGRAPHIC
Explore
ON YOUR OWN
SERIOUS SURVIVORS
By Susan Halko

Pathfinder

It's Time to Explore on Your Own!

Good readers use multiple strategies as they read on their own. Use the four key reading comprehension strategies below:

1 PREVIEW AND PREDICT
- Look over the text.
- Form ideas about how the text is organized and what it says.
- Confirm ideas about how the text is organized and what it says.

2 MONITOR AND FIX UP
- Think about whether the text is making sense and how it relates to what you know.
- Identify comprehension problems and clear up the problems.

3 MAKE INFERENCES
- Use what you know to figure out what is not said or shown directly.

4 SUM UP
- Pull together the text's big ideas.

Remember that you can choose different strategies at different times to help you understand what you are reading.

PIONEER EDITION

By Susan Halko

CONTENTS

A deer grazes in the woods. Suddenly, it hears a noise. It sees a coyote.

The deer runs. It jumps over logs and streams. The coyote runs after it. Then the coyote gets tired. It gives up. The deer survives!

Most animals can run or hide to escape an attack. But plants are stuck in the ground. How can they survive attacks from their enemies?

Plants have amazing ways to survive. Some discourage animals from attacking. Others fight back. Some plants even call for help!

Let's find out about these serious survivors!

Self-protection. This plant can't run from danger, but it still has ways to protect itself.

Beware of Thorns

Some plants warn plant eaters not to attack. They use mechanical defenses. These are parts of the plant that help protect the plant. Some **mechanical defenses** include spines, thorns, and prickles.

The holly plant has spines on its leaves. The spines can make it hard for animals to eat the leaves.

The stem of some raspberry plants is covered with thorns. Sharp thorns can pierce an animal's skin.

Holly Plant

Playing Tricks

Some plants trick their enemies. The passionflower's enemy is the caterpillar. Butterflies lay eggs on the leaves. When the butterfly eggs hatch into caterpillars, the caterpillars eat the leaves.

Fake eggs on a passionflower leaf

Real butterfly eggs on a leaf

Living Rocks. Can you spot the pebble plants in these rocks?

The plant has yellow bumps that **mimic**, or copy, butterfly eggs. A butterfly will see the bumps and look for another spot to land.

The pebble plant uses mimicry, too. It looks just like the other desert rocks. Desert animals crawl right over the plant. They don't know that they just gave up a nice snack.

Tiny Bodyguards

Some plants team up with safe insects. These insects then protect the plant, like tiny bodyguards.

Ants live inside the thorns of acacia trees. The ants are fierce. They protect the tree by stinging plant eaters.

Plant Protectors. These ants protect the tree against anything that comes near it.

Fighting Back

Stop and think about the coyote. What if it had attacked a skunk? The skunk would have fought back with its spray. P.U.!

Some plants use **chemical defenses** to fight back, just like animals do. But they don't spray them like a skunk.

Instead, plants hide **toxins** in their leaves or stems. Toxins are chemicals that are harmful to animals.

Persimmon trees have a chemical that can make their fruit taste bitter. Foxgloves have toxins that are deadly to some animals.

Pool Cover. The "lid" on a pitcher plant prevents rainwater from overflowing its pool.

Foxglove

Bug-Eating Plants

The pitcher plant has an easy way to get rid of insects—it eats them!

The pitcher plant grows in swampy areas. The soil does not have enough nutrients. So the pitcher plant gets nutrients from insects.

The pitcher plant has a long tube. The tube has a pool of water and enzymes at the bottom. An insect slips off the rim and drowns in the pool. The plant has a good meal.

Calling for Help

Some plants call for help when they are attacked.

When caterpillars chew on a cotton plant, the plant gives off smelly chemicals. The odor attracts wasps. The wasps lay eggs on the caterpillar. In a few days, the eggs hatch and eat the caterpillar.

6

LIFE CYCLE

Trickster or fighter, all plants grow and change. Look at the pictures below to review the **life cycle** of a plant.

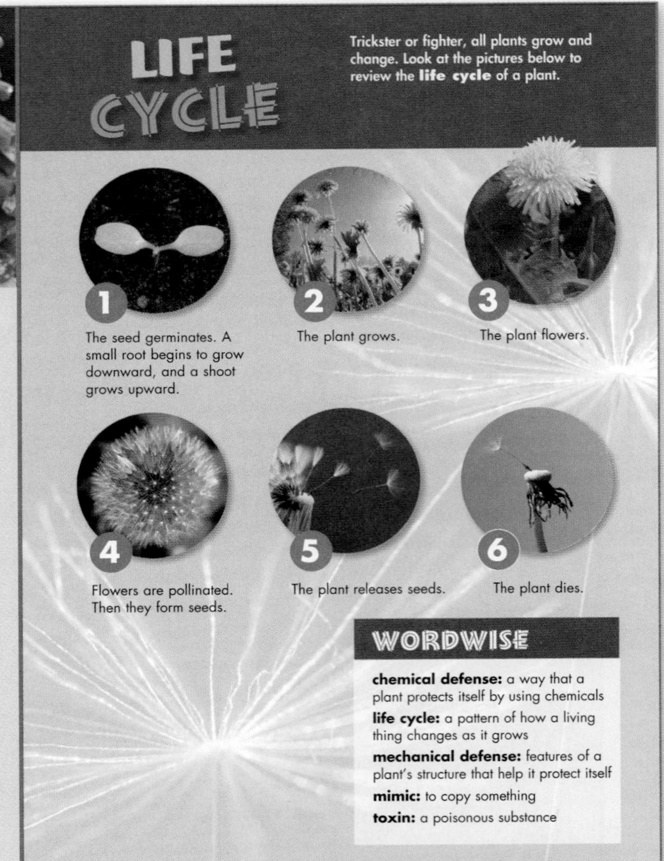

1 The seed germinates. A small root begins to grow downward, and a shoot grows upward.

2 The plant grows.

3 The plant flowers.

4 Flowers are pollinated. Then they form seeds.

5 The plant releases seeds.

6 The plant dies.

WORDWISE

chemical defense: a way that a plant protects itself by using chemicals

life cycle: a pattern of how a living thing changes as it grows

mechanical defense: features of a plant's structure that help it protect itself

mimic: to copy something

toxin: a poisonous substance

7

SNEAKY PLANTS

To stay alive, plants must do more than survive enemy attacks. They need to reproduce!

But plants are stuck. How do they get their pollen to other plants?

They get sneaky. Some plants trick insects to collect their pollen and take it with them.

The mirror orchid is one of these sneaky plants. It uses mimicry to attract insects that pollinate plants.

The mirror orchid looks just like a female wasp. A male wasp thinks it has found a mate!

So the male wasp lands on the flower. Then it flies away and carries the orchid's pollen to another flower.

Mirror Image. This wasp's wings look similar to the orchid's petals.

Mirror Orchid

8

9

Good Smells

A male bee can't resist the smell of a bucket orchid. The bee slips and falls into the orchid's bucket. It gets trapped in a pool of liquid at the bottom.

To escape, the bee has to squeeze through a tight tunnel. As it crawls out, a pollen sac gets stuck on the bee's back.

Sometimes the bee is tricked again by another bucket orchid of the same species. Again, the bee falls into the bucket and has to climb out.

But this time, the flower grabs the pollen as the bee climbs out. The pollen sac now fits into the pistil, or female part of the flower. The orchid is pollinated!

The Great Escape. A male bee climbs out of the bucket orchid with a pollen sac.

Surprising Smell. This skunk cabbage might look pretty, but it smells bad.

Offensive Odor. Some orchids have a rotten odor that attracts flies.

Bad Smells

Skunk cabbage smells bad—like dead animals. It uses its strong odor to attract pollinators.

Hungry flies and beetles can't resist the smell. They fly from one flower to another, spreading pollen as they go.

Staying Alive

Plants might not be able to run from an enemy or go looking for a mate, but they have amazing ways of surviving.

They are one of nature's greatest success stories.

10

11

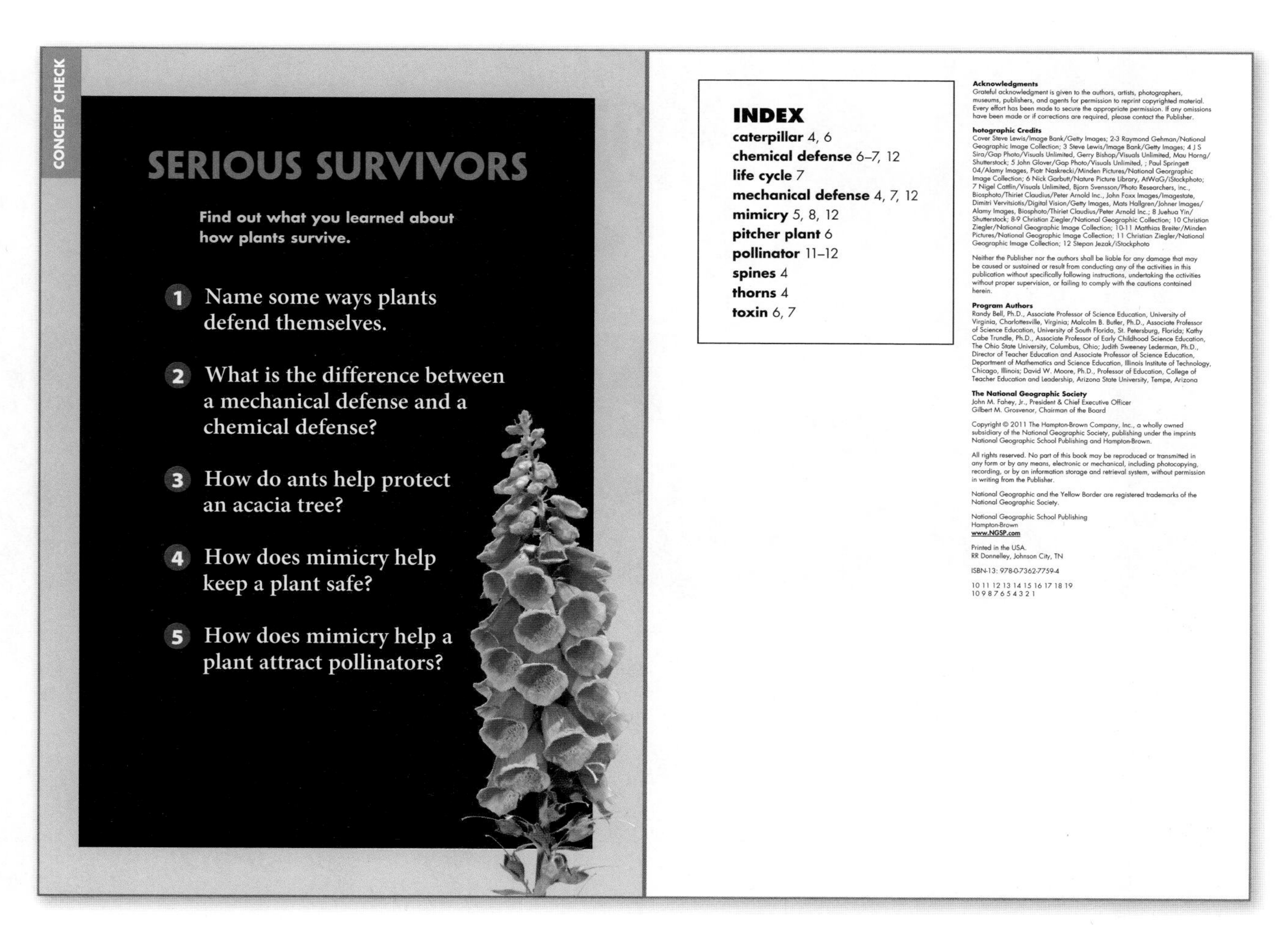

CONCEPT CHECK

SERIOUS SURVIVORS

Find out what you learned about how plants survive.

1. Name some ways plants defend themselves.
2. What is the difference between a mechanical defense and a chemical defense?
3. How do ants help protect an acacia tree?
4. How does mimicry help keep a plant safe?
5. How does mimicry help a plant attract pollinators?

INDEX

Acknowledgments
Grateful acknowledgment is given to the authors, artists, photographers, museums, publishers, and agents for permission to reprint copyrighted material. Every effort has been made to secure the appropriate permission. If any omissions have been made or if corrections are required, please contact the Publisher.

hotographic Credits
Cover Steve Lewis/Image Bank/Getty Images; 2-3 Raymond Gehman/National Geographic Image Collection; 3 Steve Lewis/Image Bank/Getty Images; 4 J S Sira/Gap Photo/Visuals Unlimited, Gerry Bishop/Visuals Unlimited, Mau Horng/Shutterstock; 5 John Glover/Gap Photo/Visuals Unlimited, ; Paul Springett 04/Alamy Images, Piotr Naskrecki/Minden Pictures/National Georgraphic Image Collection; 6 Nick Garbutt/Nature Picture Library, AtWaG/iStockphoto; 7 Nigel Cattlin/Visuals Unlimited, Bjorn Svensson/Photo Researchers, Inc., Biosphoto/Thiriet Claudius/Peter Arnold Inc., John Foxx Images/Imagestate, Dimitri Vervitsiotis/Digital Vision/Getty Images, Mats Hallgren/Johner Images/Alamy Images, Biosphoto/Thiriet Claudius/Peter Arnold Inc.; 8 Juehua Yin/Shutterstock; 8-9 Christian Ziegler/National Geographic Collection; 10 Christian Ziegler/National Geographic Image Collection; 10-11 Matthias Breiter/Minden Pictures/National Geographic Image Collection; 11 Christian Ziegler/National Geographic Image Collection; 12 Stepan Jezak/iStockphoto

Program Authors
Randy Bell, Ph.D., Associate Professor of Science Education, University of Virginia, Charlottesville, Virginia; Malcolm B. Butler, Ph.D., Associate Professor of Science Education, University of South Florida, St. Petersburg, Florida; Kathy Cabe Trundle, Ph.D., Associate Professor of Early Childhood Science Education, The Ohio State University, Columbus, Ohio; Judith Sweeney Lederman, Ph.D., Director of Teacher Education and Associate Professor of Science Education, Department of Mathematics and Science Education, Illinois Institute of Technology, Chicago, Illinois; David W. Moore, Ph.D., Professor of Education, College of Teacher Education and Leadership, Arizona State University, Tempe, Arizona

National Geographic School Publishing
Hampton-Brown
www.NGSP.com

Printed in the USA.
RR Donnelley, Johnson City, TN

ISBN-13: 978-0-7362-7759-4

10 11 12 13 14 15 16 17 18 19
10 9 8 7 6 5 4 3 2 1

Life Science
PATHFINDER EDITION

It's Time to Explore on Your Own!

Good readers use multiple strategies as they read on their own. Use the four key reading comprehension strategies below:

1 PREVIEW AND PREDICT

- Look over the text.
- Form ideas about how the text is organized and what it says.
- Confirm ideas about how the text is organized and what it says.

2 MONITOR AND FIX UP

- Think about whether the text is making sense and how it relates to what you know.
- Identify comprehension problems and clear up the problems.

3 MAKE INFERENCES

- Use what you know to figure out what is not said or shown directly.

4 SUM UP

- Pull together the text's big ideas.

Remember that you can choose different strategies at different times to help you understand what you are reading.

SERIOUS SURVIVORS

PATHFINDER EDITION

By Susan Halko

CONTENTS

SERIOUS SURVIVORS

By Susan Halko

A deer peacefully grazes in the woods. Suddenly, its white tail flips straight up. It's a warning sign to other deer. A hungry coyote is nearby. The deer bolts. It runs through the woods, jumping over logs and streams. The coyote runs after it. The deer darts left, and then right, trying to lose the coyote. After a while, the coyote gets tired and gives up. The deer escapes. It survives!

Most animals can run or hide to escape an enemy. But plants are rooted in place. How can they survive attacks from plant-eating animals and insects?

You might think that plants are helpless because they are stuck in one place. Think again. Plants have some amazing ways to survive.

Some plants discourage animals from attacking in the first place. They might have sharp thorns or spines that turn away enemies. Or they might trick insects and animals into not attacking them.

Other plants fight back, using built-in defenses. Some plants even call for help! Let's find out about these serious survivors.

Self-protection. This plant can't run from danger, but it still has ways to protect itself.

2 3

BEWARE OF THORNS

Some plants convince hungry plant eaters not to attack. They are equipped with **mechanical defenses**, or parts such as spines, thorns, or prickles. These razor-sharp features can pierce an animal's skin.

The leaves of a holly plant have spines around the edges that can make it hard for animals to eat.

The sharp thorns on the stem of some raspberry plants send a clear message: "Beware. You will get hurt if you try to attack."

Holly Plant

PLAYING TRICKS

Some plants trick their enemies into not eating them. The passionflower's worst enemy is the caterpillar. Caterpillars love to eat it up! When a butterfly looks for a spot to lay its eggs, it often lands on a passionflower. When the butterfly eggs hatch into caterpillars, the caterpillars eat the leaves.

The passionflower tricks the butterfly as protection against these future attackers. It has small, yellow bumps that **mimic**, or copy, the look of butterfly eggs. A butterfly sees the bumps and looks for another place to land.

Fake eggs on a passionflower leaf

Real butterfly eggs on a leaf

4

Living Rocks. Pebble plants hide easily among rocks. Can you spot the plants?

When a living thing looks like something else to fool enemies, it's called mimicry. Another example of mimicry is found in the desert of South Africa. The pebble plant mimics the ground around it. It looks just like—you guessed it—a pebble. Desert animals crawl right over the plant, never knowing that they just passed up a satisfying snack.

TINY BODYGUARDS

Sometimes plants team up with insects to fend off plant eaters. For example, fierce ants live inside the hollow thorns of acacia trees.

The ants, in turn, act as tiny bodyguards. They protect the tree by stinging intruders that would munch on the tree. The tree also makes food for the ants.

Plant Protectors. These ants protect the tree against anything that comes near it, including people!

5

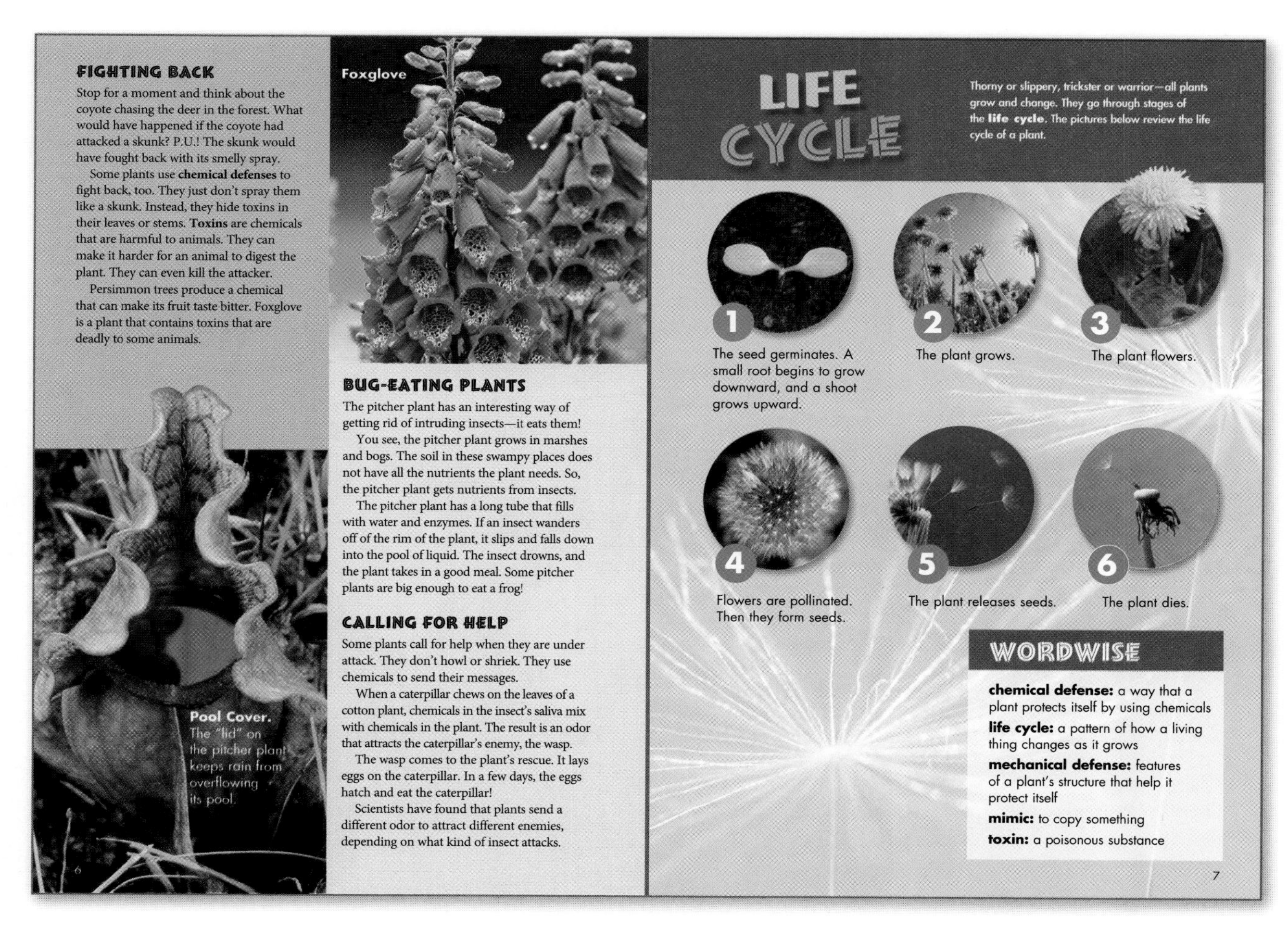

FIGHTING BACK

Stop for a moment and think about the coyote chasing the deer in the forest. What would have happened if the coyote had attacked a skunk? P.U.! The skunk would have fought back with its smelly spray.

Some plants use **chemical defenses** to fight back, too. They just don't spray them like a skunk. Instead, they hide toxins in their leaves or stems. **Toxins** are chemicals that are harmful to animals. They can make it harder for an animal to digest the plant. They can even kill the attacker.

Persimmon trees produce a chemical that can make its fruit taste bitter. Foxglove is a plant that contains toxins that are deadly to some animals.

Foxglove

Pool Cover. The "lid" on the pitcher plant keeps rain from overflowing its pool.

BUG-EATING PLANTS

The pitcher plant has an interesting way of getting rid of intruding insects—it eats them!

You see, the pitcher plant grows in marshes and bogs. The soil in these swampy places does not have all the nutrients the plant needs. So, the pitcher plant gets nutrients from insects.

The pitcher plant has a long tube that fills with water and enzymes. If an insect wanders off of the rim of the plant, it slips and falls down into the pool of liquid. The insect drowns, and the plant takes in a good meal. Some pitcher plants are big enough to eat a frog!

CALLING FOR HELP

Some plants call for help when they are under attack. They don't howl or shriek. They use chemicals to send their messages.

When a caterpillar chews on the leaves of a cotton plant, chemicals in the insect's saliva mix with chemicals in the plant. The result is an odor that attracts the caterpillar's enemy, the wasp.

The wasp comes to the plant's rescue. It lays eggs on the caterpillar. In a few days, the eggs hatch and eat the caterpillar!

Scientists have found that plants send a different odor to attract different enemies, depending on what kind of insect attacks.

6

LIFE CYCLE

Thorny or slippery, trickster or warrior—all plants grow and change. They go through stages of the **life cycle**. The pictures below review the life cycle of a plant.

1 The seed germinates. A small root begins to grow downward, and a shoot grows upward.

2 The plant grows.

3 The plant flowers.

4 Flowers are pollinated. Then they form seeds.

5 The plant releases seeds.

6 The plant dies.

WORDWISE

chemical defense: a way that a plant protects itself by using chemicals

life cycle: a pattern of how a living thing changes as it grows

mechanical defense: features of a plant's structure that help it protect itself

mimic: to copy something

toxin: a poisonous substance

7

SNEAKY PLANTS

To keep the cycle of life going, plants must do more than survive enemy attacks. They need to reproduce!

Plants can't go looking for a mate like animals can. So how do they get their pollen to other plants?

They get sneaky. Some plants trick insects' sense of sight, smell, and touch. They lure insects to collect their pollen and take it with them.

The mirror orchid is one of these sneaky plants. It uses mimicry to attract pollinators. Pollinators are animals that spread pollen from one plant to another.

The blue part on the flower looks just like the sky reflecting on a female wasp's wings. When a male wasp sees it, he thinks he has found a mate. So he lands on the flower. Then he flies away and carries that orchid's pollen to another orchid flower.

Mirror Image. This wasp's wings look similar to the orchid's petals.

Mirror Orchid

8

9

DELICIOUS SMELLS

Imagine walking past a bakery and catching a whiff of fresh bread. Pretty hard to resist, right? That's how the bucket orchid smells to male bees.

A male bee can't resist the smell of a bucket orchid. The bee lands on the orchid and slips on its smooth surface. He falls into the orchid's deep bucket, where he becomes stuck in a pool of a liquid at the bottom.

To escape, the bee has to squeeze through a tight tunnel. As he crawls out, a pollen sac gets stuck on the bee's back.

Then the bee is tricked again by another bucket orchid of the same species, and the same thing happens. The bee falls into the bucket and has to climb out.

But this time the bee is carrying the pollen sac from another orchid. As the bee climbs out, the pollen sac gets trapped in the female part of the flower. The bee pollinates the orchid, and he doesn't even know it!

The Great Escape. A bee climbs out of the bucket orchid with a pollen sac on his back.

Surprising Smell. This skunk cabbage might look pretty, but it has a bad smell.

Offensive Odor. Some orchids have a rotten odor that attracts flies.

DISGUSTING SMELLS

Skunk cabbage smells bad—like dead animals. It doesn't use the strong odor to defend itself, though. It uses its stench to attract pollinators.

Its strong odor smells like good food to hungry flies and beetles. They can't resist. They swarm over from one flower to another, spreading pollen as they go.

STAYING ALIVE

Plants might not be able to run from an enemy or go looking for a mate, but they have amazing ways of surviving. They are one of nature's greatest success stories.

10

11

CONCEPT CHECK

SERIOUS SURVIVORS

Find out what you learned about how plants survive.

1. What is an example of a plant's mechanical defense?
2. What is an example of a plant's chemical defense?
3. How do ants help protect an acacia tree?
4. How does mimicry help a plant discourage animals from attacking?
5. How does mimicry help a plant attract pollinators?

INDEX

Acknowledgments
Grateful acknowledgment is given to the authors, artists, photographers, museums, publishers, and agents for permission to reprint copyrighted material. Every effort has been made to secure the appropriate permission. If any omissions have been made or if corrections are required, please contact the Publisher.

Photographic Credits
Cover Steve Lewis/Image Bank/Getty Images; 2-3 Raymond Gehman/National Geographic Image Collection; 3 Steve Lewis/Image Bank/Getty Images; 4 J S Sira/Gap Photo/Visuals Unlimited, Gerry Bishop/Visuals Unlimited, Mau Horng/Shutterstock; 5 John Glover/Gap Photo/Visuals Unlimited, ; Paul Springett 04/Alamy Images, Piotr Naskrecki/Minden Pictures/National Georgraphic Image Collection; 6 Nick Garbutt/Nature Picture Library, AtWaG/iStockphoto; 7 Nigel Cattlin/Visuals Unlimited, Bjorn Svensson/Photo Researchers, Inc., Biosphoto/Thiriet Claudius/Peter Arnold Inc., John Foxx Images/Imagestate, Dimitri Vervitsiotis/Digital Vision/Getty Images, Mats Hallgren/Johner Images/Alamy Images, Biosphoto/Thiriet Claudius/Peter Arnold Inc.; 8 Juehua Yin/Shutterstock; 8-9 Christian Ziegler/National Geographic Collection; 10 Christian Ziegler/National Geographic Image Collection; 10-11 Matthias Breiter/Minden Pictures/National Geographic Image Collection; 11 Christian Ziegler/National Geographic Image Collection; 12 Stepan Jezak/iStockphoto

Program Authors
Randy Bell, Ph.D., Associate Professor of Science Education, University of Virginia, Charlottesville, Virginia; Malcolm B. Butler, Ph.D., Associate Professor of Science Education, University of South Florida, St. Petersburg, Florida; Kathy Cabe Trundle, Ph.D., Associate Professor of Early Childhood Science Education, The Ohio State University, Columbus, Ohio; Judith Sweeney Lederman, Ph.D., Director of Teacher Education and Associate Professor of Science Education, Department of Mathematics and Science Education, Illinois Institute of Technology, Chicago, Illinois; David W. Moore, Ph.D., Professor of Education, College of Teacher Education and Leadership, Arizona State University, Tempe, Arizona

National Geographic School Publishing
Hampton-Brown
www.NGSP.com

Printed in the USA.
RR Donnelley, Johnson City, TN

ISBN-13: 978-0-7362-7762-4

10 11 12 13 14 15 16 17 18 19
10 9 8 7 6 5 4 3 2 1

Notes

How Do Animals Grow and Change?

LIFE SCIENCE

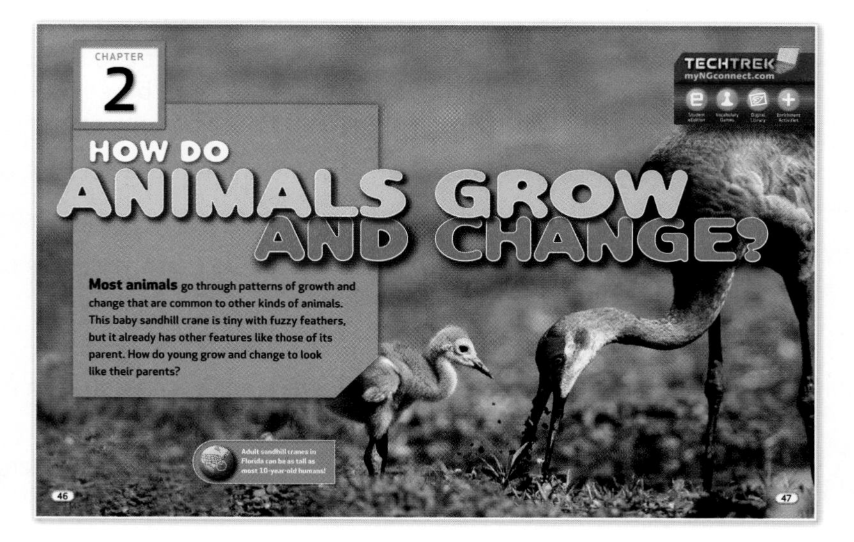

Chapter 2 Contents

After reading Chapter 2, you will be able to:

- Describe animal life cycles and list characteristics that are common to all living things. **ANIMAL LIFE CYCLES**
- Compare and contrast complete and incomplete metamorphosis. **METAMORPHOSIS**
- Generalize the kinds of traits that are inherited and identify examples. **INHERITED TRAITS**
- Generalize the kinds of traits that are acquired and identify examples. **ACQUIRED TRAITS**
- Science in a Snap! Generalize the kinds of traits that are inherited and identify examples. **INHERITED TRAITS**

45

Preview and Predict

Tell students that they can preview a text by looking over it to get an idea of what the text is about and how it is organized. Tell students that predicting is forming ideas about what they will read. Students should confirm their ideas with others.

Have students preview and make predictions about the chapter. Remind them to use section heads, vocabulary words, pictures, and their background knowledge to make predictions.

CHAPTER 2 ▫ How Do Animals Grow and Change?

LESSON	PACING	OBJECTIVES
1 **Directed Inquiry** *Investigate Earthworm Behavior* pages T45e–T45h **Science Inquiry and Writing Book** pages 28–31	**30** minutes	Investigate through Directed Inquiry (answer a question; make and compare observations; collect and record data and observations; generate explanations and conclusions based on evidence; share findings; ask questions based on observations to increase understanding). Observe that some animal behaviors are probably inherited characteristics. Keep careful records that distinguish observations from inferences. Identify and evaluate the reasonableness of explanations and measurements.
2 **Big Idea Question and Vocabulary** pages T46–T47—T48–T49 **Animal Life Cycles** pages T50–T53	**35** minutes	Describe the different life cycles of animals and identify the different stages of these life cycles. Recognize that living things grow and change, and need water, nutrients, and air to survive. Compare and contrast the similarities and differences among offspring of different animal life cycles.
3 **Metamorphosis** pages T54–T59	**30** minutes	Describe the different life cycles of animals and identify the different stages of these life cycles. Compare and contrast the similarities and differences among offspring of different animal life cycles. Describe the process of molting during the life cycles of incomplete metamorphosis.
4 **Inherited Traits** pages T60–T63	**25** minutes	Observe that traits of animals are inherited from their parents and that individuals vary within every species. Describe animal behaviors that result from heredity.
5 **Acquired Traits** pages T64–T67	**25** minutes	Recognize that acquired traits in animals can result from the environment. Describe animal behaviors that result from learning.
6 NATIONAL GEOGRAPHIC **On the Job with Service Dogs** pages T68–T69	**20** minutes	Describe animal behaviors that result from learning.
7 **Guided Inquiry** *Investigate Insect Life Cycles* pages T69a–T69d **Science Inquiry and Writing Book** pages 32–35	**20** minutes	Investigate through Guided Inquiry (answer a question; make and compare observations; collect and record data and observations; generate explanations and conclusions based on evidence; share findings; ask questions based on observations to increase understanding; adjust explanations based on findings and new ideas). Compare and contrast the major stages in the life cycles of insects. Explain why keeping accurate records of observations and investigations is important.

VOCABULARY	RESOURCES	ASSESSMENT
predict **conclude**	Science Inquiry and Writing Book: *Life Science* Science Inquiry Kit: *Life Science* Directed Inquiry: Learning Masters 24–26	Inquiry Rubric: Assessment Handbook, page 183 Inquiry Self-Reflection: Assessment Handbook, page 191 Reflect and Assess, page T45h
	Vocabulary: Learning Master 27 *Life Science* Big Ideas Book	Assess, page T53
metamorphosis **larva** **pupa** **nymph**		Assess, page T59
heredity		Assess, page T63
	Science Inquiry and Writing Book: *Life Science*	Assess, page T67
	Share and Compare: Learning Master 28	Assess, page T69
observe **compare**	Science Inquiry and Writing Book: *Life Science* Science Inquiry Kit: *Life Science* Guided Inquiry: Learning Masters 29–34	Inquiry Rubric: Assessment Handbook, page 183 Inquiry Self-Reflection: Assessment Handbook, page 192 Reflect and Assess, page T69d

TECHNOLOGY RESOURCES

STUDENT RESOURCES

myNGconnect.com

- Student eEdition
- Big Ideas Book
- Science Inquiry and Writing Book
- Explore on Your Own Books
- Read with Me
- Vocabulary Games
- Digital Library
- Enrichment Activities

National Geographic Kids

National Geographic Explorer!

TEACHER RESOURCES

myNGconnect.com

- Teacher eEdition
- Teacher's Edition
- Science Inquiry and Writing Book
- Explore on Your Own Books
- Online Lesson Planner
- National Geographic Unit Launch Videos
- Assessment Handbook
- Presentation Tool
- Digital Library

NGSP ExamView CD-ROM

IF TIME IS SHORT...

FAST FORWARD.

CHAPTER 2 ▫ How Do Animals Grow and Change?

LESSON	PACING	OBJECTIVES
8 Conclusion and Review pages T70–T71	**15** minutes	
9 NATIONAL GEOGRAPHIC LIFE SCIENCE EXPERT *Lepidopterist* pages T72–T73 NATIONAL GEOGRAPHIC BECOME AN EXPERT *Nature's Transformers* pages T74–T75—T84	**35** minutes	Research biographical information about various scientists and inventors from different gender and ethnic backgrounds, and describe how their work contributed to science and technology. Describe the different life cycles of animals and identify the different stages of these life cycles.

FAST FORWARD
ACCELERATED PACING GUIDE

DAY 1 30 minutes

Directed Inquiry

Investigate Earthworm Behavior, page T45e

DAY 2 35 minutes

NATIONAL GEOGRAPHIC LIFE SCIENCE EXPERT *Lepidopterist,* page T72

NATIONAL GEOGRAPHIC BECOME AN EXPERT *Nature's Transformers,* page T74–T75

VOCABULARY	RESOURCES	ASSESSMENT
		Assess the Big Idea, page T71 Chapter 2 Test: Assessment Handbook, pages 8–10 NGSP ExamView CD-ROM
		Assess, pages T73, T78–T79, T82–T83

TECHNOLOGY RESOURCES

STUDENT RESOURCES

myNGconnect.com

- Student eEdition
- Big Ideas Book
- Science Inquiry and Writing Book
- Explore on Your Own Books
- Read with Me
- Vocabulary Games
- Digital Library
- Enrichment Activities

National Geographic Kids

National Geographic Explorer!

TEACHER RESOURCES

myNGconnect.com

- Teacher eEdition
- Teacher's Edition
- Science Inquiry and Writing Book
- Explore on Your Own Books
- Online Lesson Planner
- National Geographic Unit Launch Videos
- Assessment Handbook
- Presentation Tool
- Digital Library

NGSP ExamView CD-ROM

DAY 3 **20** minutes

Guided Inquiry

Investigate Insect Life Cycles, page T69a

Objectives

Students will be able to:

- Investigate through Directed Inquiry (answer a question; make and compare observations; collect and record data and observations; generate explanations and conclusions based on evidence; share findings; ask questions based on observations to increase understanding).
- Observe that some animal behaviors are probably inherited characteristics.
- Keep careful records that distinguish observations from inferences.
- Identify and evaluate the reasonableness of explanations and measurements.

Science Process Vocabulary

predict, conclude

PROGRAM RESOURCES

- Science Inquiry and Writing Book: *Life Science*
- Science Inquiry and Writing Book eEdition at myNGconnect.com
- Inquiry eHelp at myNGconnect.com
- Science Inquiry Kit: *Life Science*
- Learning Masters Book, pages 24–26, or at myNGconnect.com
- Inquiry Rubric: Assessment Handbook, page 183, or at myNGconnect.com
- Inquiry Self-Reflection: Assessment Handbook, page 191, or at myNGconnect.com

MATERIALS

Kit materials are listed in italics.

plastic pan; 2 paper towels; *spray bottle (16 oz);* water; *2 sheets of black construction paper;* masking tape; *plastic spoon; 3 earthworms (live materials coupon); flashlight, stopwatch*

For teacher use: *medium-sized plastic container; loamy soil; cheesecloth; rubber band;* food scraps

❶ Introduce

Tap Prior Knowledge

- Ask: **Have you seen earthworms? When and where?** Students may recall seeing earthworms after a rainy day, or while digging in a garden.

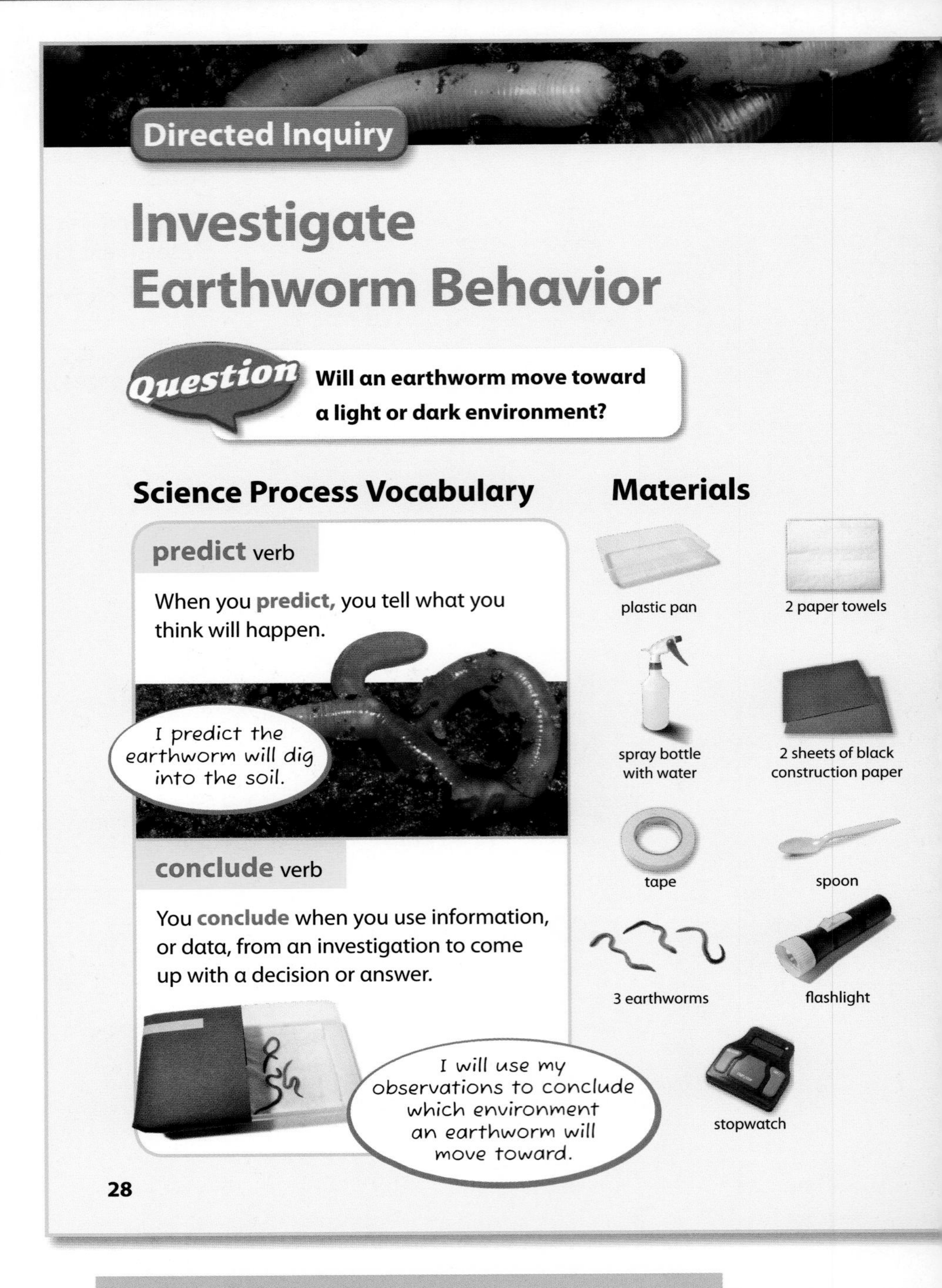

Directed Inquiry

Investigate Earthworm Behavior

Question Will an earthworm move toward a light or dark environment?

Science Process Vocabulary

predict verb

When you **predict,** you tell what you think will happen.

conclude verb

You **conclude** when you use information, or data, from an investigation to come up with a decision or answer.

Materials

28

MANAGING THE INVESTIGATION

Time

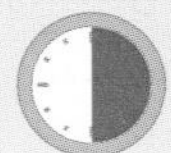

30 minutes

Groups

Small groups of 4

Advance Preparation

- Use the live material coupon to send away for the earthworms. Allow 2–3 weeks for delivery.
- Upon arrival, transfer earthworms to a plastic container with damp, loamy soil. Wrap the sides of the container in black paper. Cover with cheesecloth and a large rubber band.
- Add fresh loamy soil to the worms' habitat every week. Add kitchen scraps, such as bread, pasta, fruits, vegetables, and eggshells. Do not add meat or dairy products.

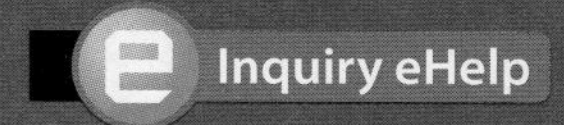

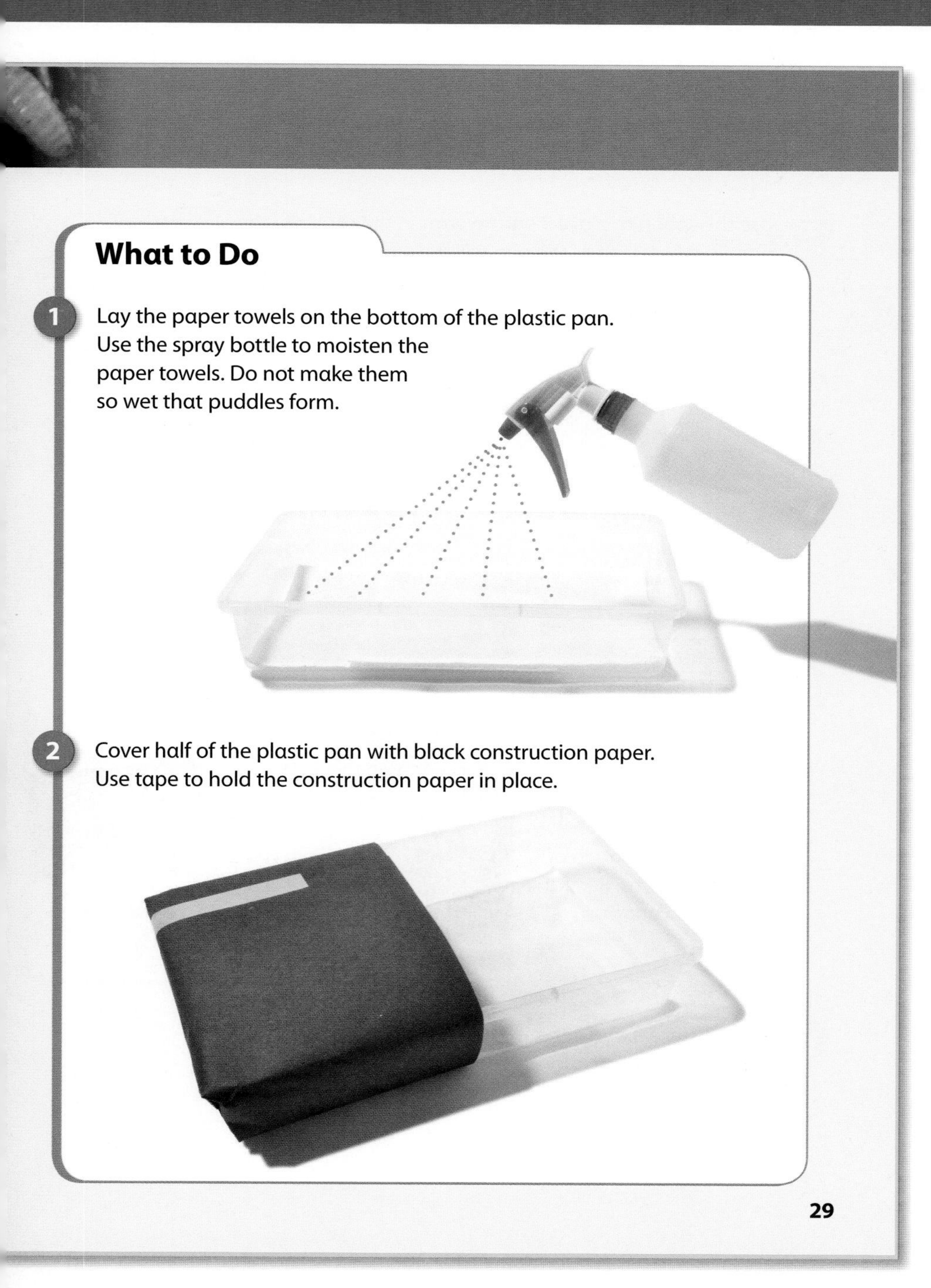

What to Do
1
Lay the paper towels on the bottom of the plastic pan. Use the spray bottle to moisten the paper towels. Do not make them so wet that puddles form.
2
Cover half of the plastic pan with black construction paper. Use tape to hold the construction paper in place.
29

Introduce, continued

Connect to the Big Idea

- Review the Big Idea Question, *How do animals grow and change?* Explain that this inquiry will help students understand how earthworms respond to their environment.
- Have students open their Science Inquiry and Writing Books to page 28. Read the Question and discuss how some animals live in dark places and others may live in places where there is more light.

❷ Build Vocabulary

Science Process Words: predict, conclude

Use this routine to introduce the words.

1. **Pronounce the Word** Say **predict**. Have students repeat it in syllables.
2. **Explain Its Meaning** Choral read the sentence. Ask students to define **predict**. (make a statement about what will happen)
3. **Encourage Elaboration** Ask: **What do you predict will happen if you place an earthworm on soil?** (It will burrow in the soil.)

ELL Language Production

Write: *I predict* ____. Ask: **Do you think the earthworms will move slowly or quickly?** (I **predict** the earthworms will move slowly.)

Repeat for the word **conclude**. To encourage elaboration, ask: **What can you conclude about how earthworms move?** (I can conclude that the earthworms move slowly.)

❸ Guide the Investigation

- Distribute materials. Read the inquiry steps on pages 29–30. Clarify steps if necessary. Discuss the safe practices, such as washing hands, that are necessary during this inquiry.
- Students may have difficulty wrapping the pan in step 2. Tell them to wrap the pan like they would wrap a gift. On the dark half of the pan, they should wrap the sides as well as the top.
- If the earthworms are not moving in step 4, have students hold the flashlight closer to the earthworms. Students should be careful not to touch the earthworms with the flashlight.

Teaching Tips

- Remind students to handle the earthworms gently.
- Make sure the damp paper towels fit inside the pan, with no edges sticking up. The earthworms may try to crawl under the paper towels.
- After the investigation, earthworms should not be released into the environment. Earthworms may be kept in the classroom, used as fishing bait, or added to a compost pile. Save these earthworms for use in the Open Inquiry Investigation Model.

What to Expect

- Students should observe the earthworms move away from the light.

Guide the Investigation, continued

- When students repeat the trial, instruct them to use the spoon to gently move the earthworms back to the center of the pan. If one group of students has especially inactive earthworms, they may choose to observe another group's habitat.
- If students observe that a worm is located in both light and dark areas or that it has not moved, they can record its location under the general *Observations* area in the table.

Remind students to wash their hands.

❹ Explain and Conclude

- Have students compare their observations. They should discuss how their data varied between trials or in different groups of earthworms and actively seek interpretations and ideas of others.
- Ask: **Was your prediction supported by your data and results?** Have students share their answers and evaluate each other's explanations. Encourage students to analyze whether the explanations and measurements are reasonable.
- Discuss with students why it is important to keep accurate records of their observations and results. Have them explain the difference between their predictions, observations, and any conclusions they may draw from their results.
- Ask students to summarize their conclusions about how earthworms react to light both aloud and in writing. Have them identify and use evidence to defend their conclusions.
- Remind students that scientists do not rely on claims or conclusions unless they are supported by observations that can be confirmed. Discuss with students why their answer would be considered a fact or an opinion.

Answers

1. Answers will vary based on student predictions. Students may have predicted that the earthworms would move to the darker area.

What to Do, continued

3. Use the spoon to place 3 earthworms in the center of the pan. To which part of the pan will the earthworms move? Write your **prediction** in your science notebook.

4. Turn off the classroom lights. Shine the flashlight directly over the uncovered side of the pan. Use the stopwatch to time 5 minutes. Then **observe** the location of the earthworms. Record your observations.

5. Repeat steps 3 and 4 three more times. Record your observations.

30

NATIONAL GEOGRAPHIC Raise Your SciQ!

Segmented Worms Earthworms belong to the Phylum *Annelida*, which also includes leeches and marine worms. Annelids are commonly referred to as segmented worms. This is because the body is made of individual segments, each containing elements of the circulatory, nervous, and excretory tracts. A segmented body allows the worm to contract its muscles to move, segment by segment.

Record

Write in your science notebook.
Use a table like this one.

my SCIENCE notebook

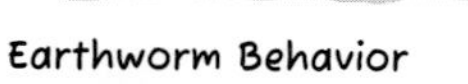

Earthworm Behavior

Trial	Prediction	Observations	
		Number of Worms in the Dark After 5 Minutes	Number of Worms in the Light After 5 Minutes
1			
2			

Explain and Conclude

1. Do your results support your **predictions?** Explain.
2. **Compare** the behavior of all 3 earthworms. What do you think caused any differences?
3. Use the results of your **investigation** to draw a **conclusion** about earthworm behavior. Do you think the earthworms inherited this behavior or was this behavior affected by the environment? Explain.

Think of Another Question

What else would you like to find out about whether earthworms will move toward light or dark environments? How could you find an answer to this new question?

Earthworms are affected by changes in their environment.

31

Inquiry Rubric at myNGconnect.com	Scale			
The student made a **prediction** about earthworm behavior in light and dark environments.	4	3	2	1
The student made **observations** about how the earthworms reacted to different environments.	4	3	2	1
The student repeated the trial several times and recorded any differences in observations.	4	3	2	1
The student **compared** data and observations from multiple trials.	4	3	2	1
The student made a **conclusion** about earthworm behavior.	4	3	2	1
Overall Score	4	3	2	1

Explain and Conclude, continued

2. Possible answer: Two of the three earthworms moved quickly to the dark side of the pan. The third earthworm moved much more slowly. I think this was because the third earthworm wasn't able to move as fast as the others.

3. Possible answer: I conclude that earthworms prefer dark environments to light environments. I think this behavior is inherited because most of the earthworms acted the same way.

5 Find Out More

Think of Another Question

Students should use their observations to generate new questions. Students may ask: *What would happen if half of the pan were cold or warm?* Discuss what students could do to find answers.

6 Reflect and Assess

- To assess student work with the Inquiry Rubric shown below, see Assessment Handbook, page 183, or go online at myNGconnect.com
- Have students use the Inquiry Self-Reflection on Assessment Handbook, page 191, or at myNGconnect.com

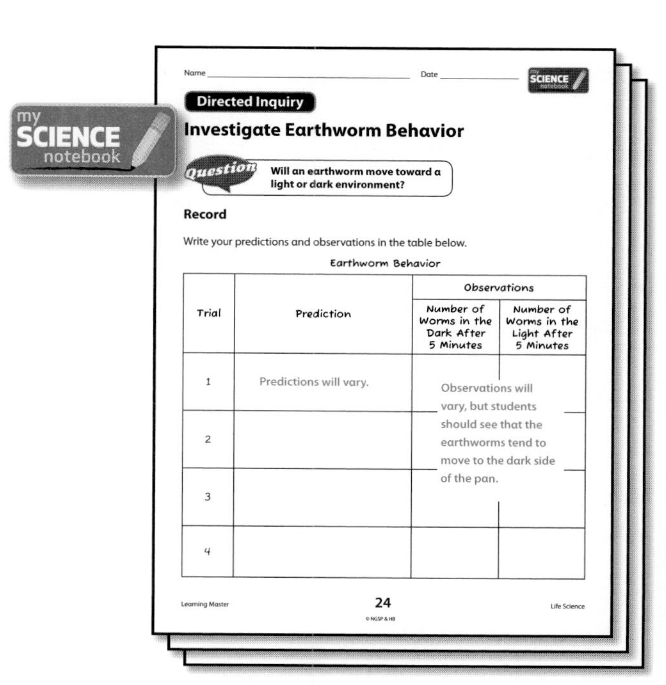

my SCIENCE notebook

Name ______ Date ______

Directed Inquiry

Investigate Earthworm Behavior

Question: Will an earthworm move toward a light or dark environment?

Record

Write your predictions and observations in the table below.

Earthworm Behavior

Trial	Prediction	Observations	
		Number of Worms in the Dark After 5 Minutes	Number of Worms in the Light After 5 Minutes
1	Predictions will vary.	Observations will vary, but students should see that the earthworms tend to move to the dark side of the pan.	
2			
3			
4			

Learning Master 24 Life Science

Learning Masters 24–26 or at myNGconnect.com

Objectives

Students will be able to:

- Describe the different life cycles of animals and identify the different stages of these life cycles.

Science Academic Vocabulary

metamorphosis, larva, pupa, nymph, heredity

PROGRAM RESOURCES

- Big Ideas Book: *Life Science*
- Big Ideas Book: *Life Science* eEdition at myNGconnect.com
- Vocabulary Games at myNGconnect.com
- Digital Library at myNGconnect.com
- Enrichment Activities at myNGconnect.com
- Read with Me at myNGconnect.com
- Learning Masters Book, page 27, or at myNGconnect.com

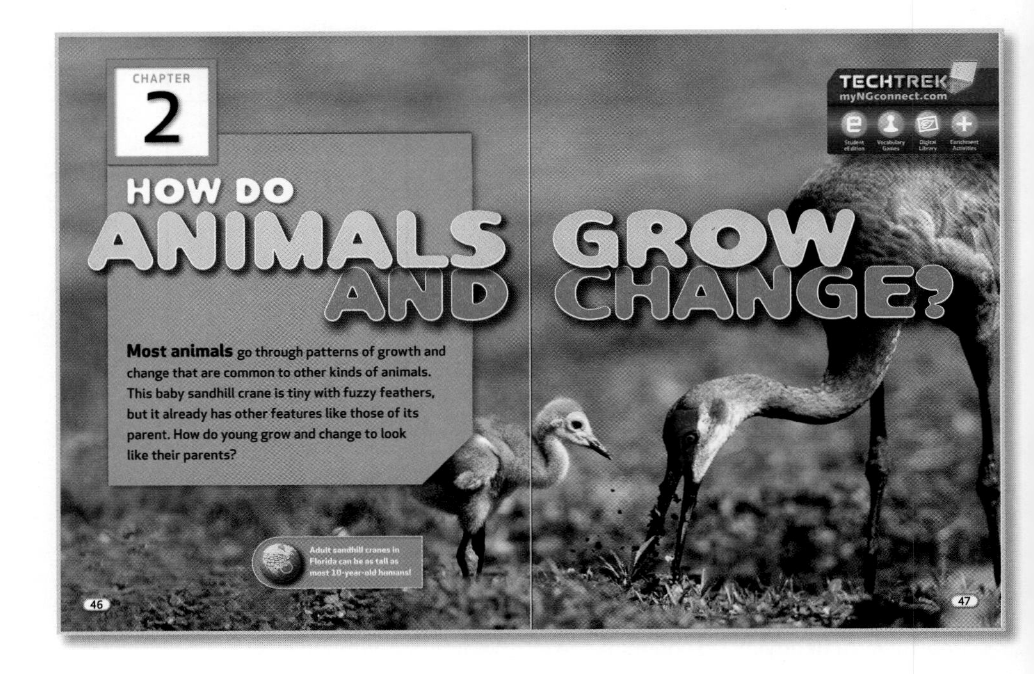

CHAPTER 2

HOW DO ANIMALS GROW AND CHANGE?

Most animals go through patterns of growth and change that are common to other kinds of animals. This baby sandhill crane is tiny with fuzzy feathers, but it already has other features like those of its parent. How do young grow and change to look like their parents?

❶ Introduce

Tap Prior Knowledge

- Work with students to name three animals that have many differences from one another, such as a dog, a fish, and an insect. Discuss how the animals are alike and different.

❷ Focus on the Big Idea

Big Idea Question

- Read the Big Idea Question aloud and have students echo it.
- Preview pages 50–53, 54–59, 60–63, and 64–67, linking the headings with the Big Idea Question.

Differentiated Instruction

ELL Vocabulary Support

BEGINNING

Have partners practice pronouncing the vocabulary words, reading the definitions in the text, and spelling them correctly.

INTERMEDIATE

Help students learn the vocabulary words by providing Academic Language Frames:

- *Major changes are a part of ____.*
- *A caterpillar is a kind of ____.*
- *The passing of traits to offspring is called ____.*

ADVANCED

Have students write sentences that use each of the vocabulary words. Students may refer to lesson pages to help them learn the words and write their sentences.

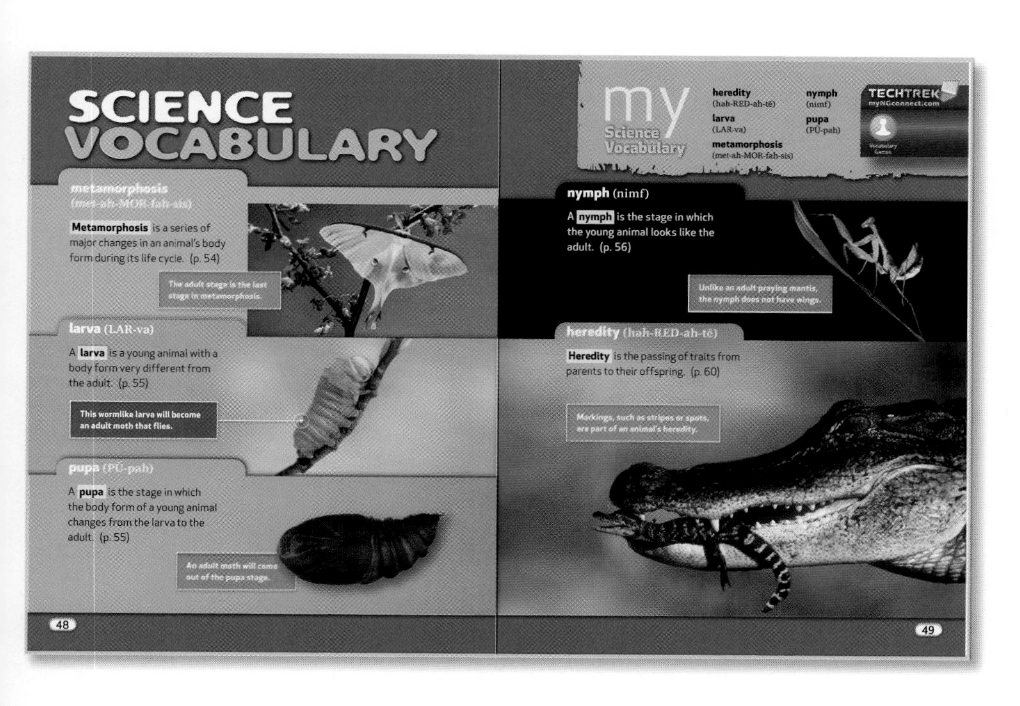
SCIENCE VOCABULARY

metamorphosis (met-ah-MOR-fah-sis)

Metamorphosis is a series of major changes in an animal's body form during its life cycle. (p. 54)

The adult stage is the last stage in metamorphosis.

larva (LAR-va)

A **larva** is a young animal with a body form very different from the adult. (p. 55)

This wormlike larva will become an adult moth that flies.

pupa (PŪ-pah)

A **pupa** is the stage in which the body form of a young animal changes from the larva to the adult. (p. 55)

An adult moth will come out of the pupa stage.

48

my Science Vocabulary

heredity (hah-RED-ah-tē)
larva (LAR-va)
metamorphosis (met-ah-MOR-fah-sis)
nymph (nimf)
pupa (PŪ-pah)

TECHTREK myNGconnect.com

nymph (nimf)

A **nymph** is the stage in which the young animal looks like the adult. (p. 56)

Unlike an adult praying mantis, the nymph does not have wings.

heredity (hah-RED-ah-tē)

Heredity is the passing of traits from parents to their offspring. (p. 60)

Markings, such as stripes or spots, are part of an animal's heredity.

49

Focus on the Big Idea, continued

- With student input, post a chart that displays the chapter headings. Have students share what they expect to find in each section. Then read page 46 aloud.

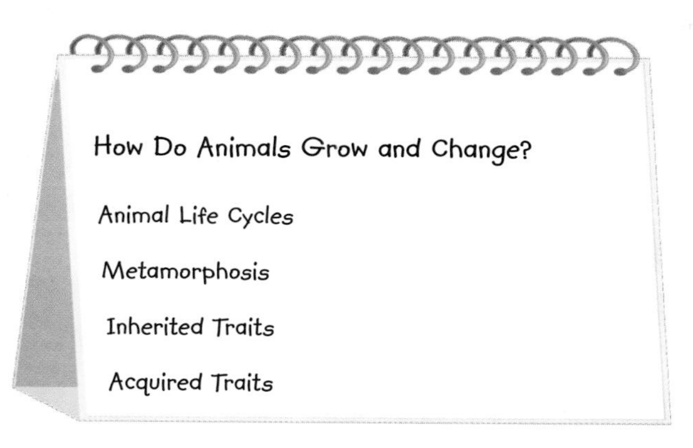

How Do Animals Grow and Change?
Animal Life Cycles
Metamorphosis
Inherited Traits
Acquired Traits

❸ Teach Vocabulary

Have students look at pages 48–49, and use this routine to teach each word. For example:

1. **Pronounce the Word** Say **metamorphosis** and have students repeat it.
2. **Explain Its Meaning** Read the word, definition, and sample sentence, and use the photos on page 48 to explain the word's meaning. Say: **In metamorphosis, an animal's body form changes during its life.**
3. **Encourage Elaboration** Point to the **larva** and ask: **What will the larva become when it is an adult?** (an adult moth)

Repeat for the words **larva**, **pupa**, **nymph**, and **heredity** using the following Elaboration Prompts:

- **What does the moth larva look like?** (a fat worm with many legs)
- **What happens during the pupa stage of metamorphosis?** (The body form of a young animal changes from the larva to the adult.)
- **What is the nymph stage?** (a stage in which the young animal looks like the adult)
- **What is heredity?** (the passing of traits from parents to their offspring)

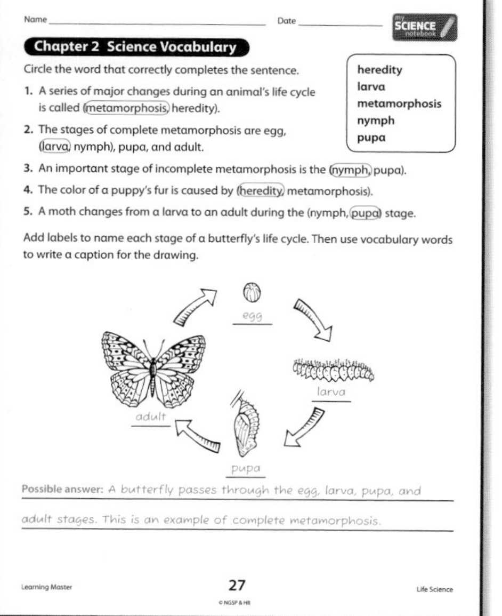
Name ______________ Date ______________

Chapter 2 Science Vocabulary

Circle the word that correctly completes the sentence.

heredity
larva
metamorphosis
nymph
pupa

1. A series of major changes during an animal's life cycle is called (metamorphosis, heredity).
2. The stages of complete metamorphosis are egg, (larva, nymph), pupa, and adult.
3. An important stage of incomplete metamorphosis is the (nymph, pupa).
4. The color of a puppy's fur is caused by (heredity, metamorphosis).
5. A moth changes from a larva to an adult during the (nymph, pupa) stage.

Add labels to name each stage of a butterfly's life cycle. Then use vocabulary words to write a caption for the drawing.

egg
larva
pupa
adult

Possible answer: A butterfly passes through the egg, larva, pupa, and adult stages. This is an example of complete metamorphosis.

Learning Master 27 Life Science

© NGSP & HB

Learning Master 27 or at myNGconnect.com

Objectives

Students will be able to:

- Describe the different life cycles of animals and identify the different stages of these life cycles.
- Recognize that living things grow and change, and need water, nutrients, and air to survive.

PROGRAM RESOURCES

- Digital Library at myNGconnect.com

❶ Introduce

Tap Prior Knowledge

- Explain that most of the milk sold in supermarkets comes from cows. Ask: **Why do cows make milk?** (to feed their young, or calves) Discuss how cows, like many other animals, care for their young until the young can care for themselves.

Set a Purpose and Read

- Read the heading. Then point out that the cattle in the photo show different stages in the life cycle of cattle, meaning that some are adults and some are young cows, or calves. Tell students that they will read about different life cycles of animals.
- Have students read pages 50–53.

❷ Teach

Compare Calves and Adult Cattle

- Have students study the photo of the adult cow and calf. Have them compare the calf with the adult.
- Ask: **How are calves like adult cattle?** (Both have similar shapes and body parts.) **How are they different?** (Calves are smaller, and their bodies are less developed.) Discuss how newborn calves are especially fragile and need the help of an adult to survive.

Animal Life Cycles
Observe the cows in the photograph. Find the baby cow. Like all living things, cows reproduce more of their own kind. An animal's life cycle tells how the animal is born, grows, changes into an adult, and reproduces. In a cow's life cycle, the young are born alive and drink milk until they can eat other foods. They grow into adults and have young, repeating the life cycle.
TECHTREK
myNGconnect.com
Digital Library
Name some ways the calf looks like the parent.
50

Science Misconceptions

Cows and Bulls Students may confuse the words *cattle*, *cows*, and *bulls*. The word *cattle* is the name of the species. Among cattle, cows are females, and bulls are males. The words *cow* and *bull* also describe the two genders in elephants, oxen, whales, and moose. Often, however, the word *cow* is used to refer to either gender of cattle.

Cows also share more characteristics with other living things. They take in nutrients, or food, and give off wastes. They take in oxygen from the air. Other living things take in oxygen from water, or take in other gases. Cows also get larger as they grow, just like other living things. A cow can live around 20 years before it dies.

Cows share certain characteristics with other living things.

Differentiated Instruction

Extra Support

Have students draw a picture to illustrate the life cycle of cattle. Then they may add pictures to illustrate other types of cycles, such as a bicycle, recycling, and the cycle of the seasons. Have them discuss with partners what all the cycles have in common.

Challenge

Have students write a paragraph that defines the term *life cycle* in their own words and that describes the life cycle of cattle. Students may illustrate their paragraphs and read them aloud to the class.

Teach, continued

Describe the Life Cycle of Cattle

- Have students observe the photos on pages 50–51 and read the captions. Ask: **How do calves change during their life cycle?** (They grow larger and stronger until they reach adulthood.)
- Ask: **When a cow reaches adulthood, how can the life cycle continue?** (An adult cow may have a calf.) Remind students that a life cycle involves young growing into adults, who then have young of their own, and the process repeats over and over again.
- Ask: **What other animals pass through stages of a life cycle?** (All animals pass through stages of a life cycle.)
- Discuss how all animals live only for a certain period of time. Their lives begin, they grow and develop, and eventually they die. The length of time from an animal's birth to its death is called its life span. Life spans vary among different kinds of animals.

Recognize the Characteristics of Living Things

- Ask: **How can you tell that cows are living things?** (Possible answer: They take in nutrients and oxygen, give off wastes, and grow larger as they age.) Discuss how these characteristics apply to other living things in the photo, such as the grasses and trees, as well as other plants and animals that students might name.

myNGconnect.com

Have students use the Digital Library to find images of parents and their offspring.

Whiteboard Presentation Students can use the information to make a computer presentation about parents and their offspring. You can help them show the presentation on a whiteboard.

Objectives

Students will be able to:

- Describe the different life cycles of animals and identify the different stages of these life cycles.
- Compare and contrast the similarities and differences among offspring of different animal life cycles.
- Recognize that living things grow and change, and need water, nutrients, and air to survive.

Teach, continued

Describe the Life Cycle of Dolphins

- Ask students to observe the photo of the dolphins on page 52 and read the caption. Ask: **How does the mother dolphin compare to the dolphin calf?** (The dolphin calf looks like its mother, but it is smaller.)
- Have students turn back to page 50 and review the photo of cattle and the cattle's life cycle. Ask: **How are the life cycles of cattle and dolphins similar?** (Both cattle and dolphins give birth to live young; they produce milk for their young; the young look like smaller versions of the adults.)
- Ask: **How are the life cycles of cattle and dolphins different?** (Cattle give birth to their young on dry land. Dolphins give birth to live young underwater.)

The dolphin calf was born alive, but underwater! Soon after birth, the mother dolphin quickly pushes the calf to the water's surface for air. After three to six years, the calf will leave its mother to live on its own. As an adult, it will continue the life cycle of dolphins by having offspring.
The Atlantic spotted dolphin calf has the same body shape, fins, and pointed nose as its mother. What is different?
calf
adult
52

Raise Your SciQ!

Dolphins Although dolphin calves are born underwater, they must quickly get to the surface to breathe. Like all mammals, dolphins have lungs and must breathe air, not water. Dolphins breathe through a blowhole, or modified nostril, located at the top of their heads. Whales also breathe through blowholes.

Dolphin calves develop inside the mother for about 12 months. Mothers nurse their young for up to 18 months and care for them for three to six years. Dolphins often live for about 30 years.

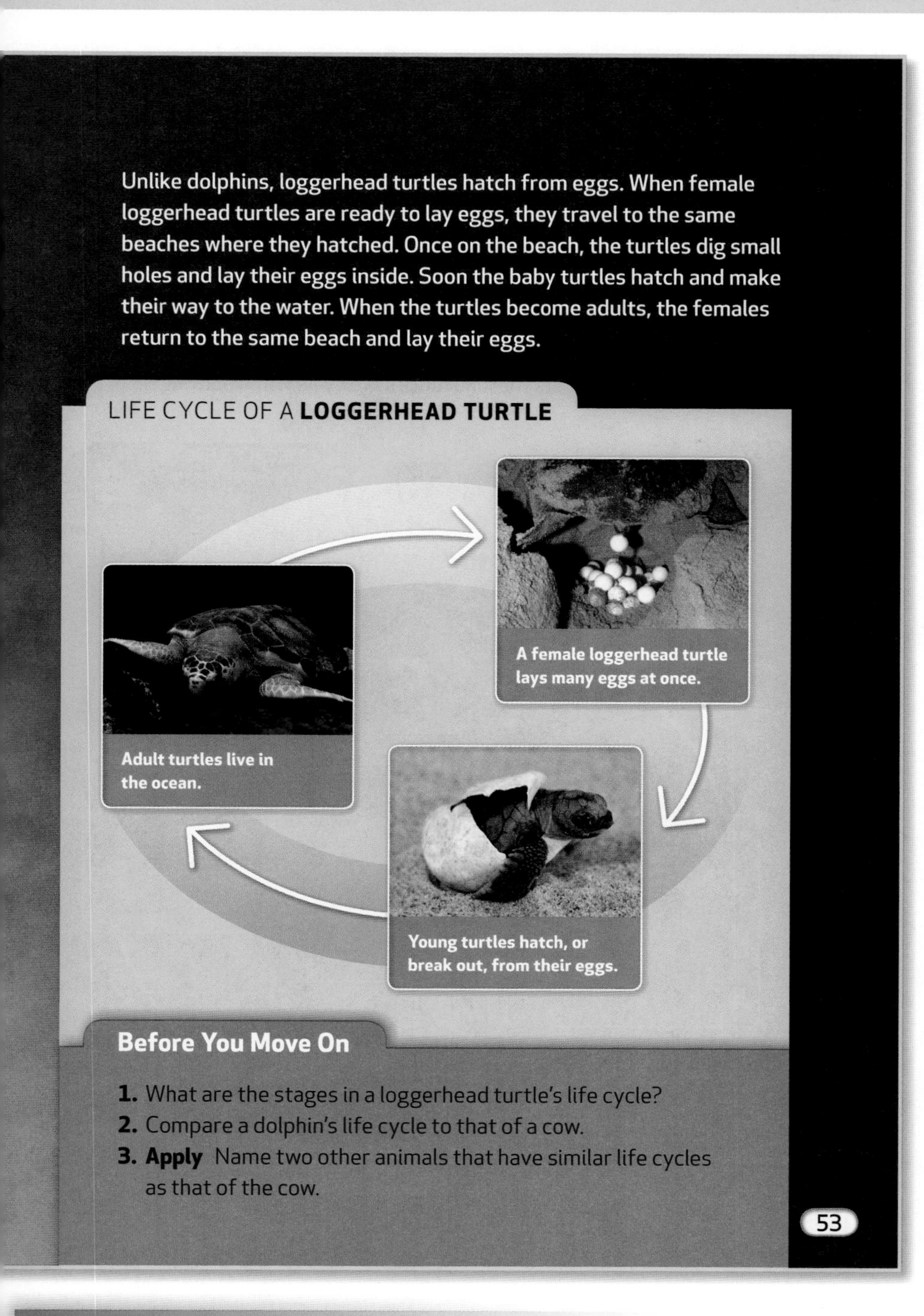
Unlike dolphins, loggerhead turtles hatch from eggs. When female loggerhead turtles are ready to lay eggs, they travel to the same beaches where they hatched. Once on the beach, the turtles dig small holes and lay their eggs inside. Soon the baby turtles hatch and make their way to the water. When the turtles become adults, the females return to the same beach and lay their eggs.

LIFE CYCLE OF A **LOGGERHEAD TURTLE**

A female loggerhead turtle lays many eggs at once.

Young turtles hatch, or break out, from their eggs.

Adult turtles live in the ocean.

Before You Move On

1. What are the stages in a loggerhead turtle's life cycle?
2. Compare a dolphin's life cycle to that of a cow.
3. **Apply** Name two other animals that have similar life cycles as that of the cow.

53

Teach, continued

Text Feature: Diagrams

- Ask students to observe the photos and read the captions that describe the life cycle of the loggerhead turtle. Ask: **Why do you think it is helpful for a female turtle to bury eggs in the sand?** (The sand keeps the eggs warm and relatively safe.) Explain that mother turtles leave their eggs behind. They do not guard or sit upon the eggs as many birds do.
- Ask: **How are the life cycles of loggerhead turtles, dolphins, and cattle similar?** (All of the life cycles produce offspring that look like small versions of the parents.) **How are the life cycles different?** (Cattle and dolphins give birth to live young. Loggerhead turtles lay eggs that hatch.)
- Discuss the stages of the turtle's life cycle. Ask: **What does a pregnant female turtle do when she arrives on a beach?** (She digs a small hole and lays eggs in the hole.) **Then what happens?** (After some time, the baby turtles hatch.) **What do the baby turtles do after they hatch?** (They crawl to the water and swim away.) Discuss how this life cycle continues over and over again as turtles hatch, grow into adults, and return to the beaches to lay eggs.

3 Assess

» Before You Move On

1. Sequence **What are the stages in a loggerhead turtle's life cycle?** (The mother lays eggs; the eggs hatch; the baby turtles go back to the ocean; they become adults and return to the beach to lay more eggs.)
2. Compare **Compare a dolphin's life cycle to that of a cow.** (They both have young that are born alive. The young feed on milk, grow bigger, and eventually have young of their own.)
3. Apply **Name two other animals that have similar life cycles as that of the cow.** (Students will likely name dolphins and humans. Accept any other mammal, such as dogs or cats.)

Differentiated Instruction

ELL Language Support for Describing the Life Cycles of Dolphins and Loggerhead Turtles

BEGINNING

Have students copy each stage of the loggerhead turtle's life cycle on separate cards. Students then share their three cards with a partner who puts them in order and describes the order.

INTERMEDIATE

Have students prepare a chart to compare and contrast the life cycles of dolphins and loggerhead turtles. Students may illustrate important facts in their charts, such as how dolphins give birth to live young while turtles lay eggs.

ADVANCED

Provide sequence words to help students describe the life cycle of loggerhead turtles. For example:

The life cycle of loggerhead turtles includes different stages. First, . . . Second, . . . Then . . .

LESSON 3 ▫ Metamorphosis

Objectives

Students will be able to:

- Describe the different life cycles of animals and identify the different stages of these life cycles.
- Compare and contrast the similarities and differences among offspring of different animal life cycles.

Science Academic Vocabulary

metamorphosis, larva, pupa

PROGRAM RESOURCES

- Enrichment Activities at myNGconnect.com

❶ Introduce

Tap Prior Knowledge

- Ask: **Have you ever seen caterpillars and butterflies? What do they look like?** Lead students to recognize that caterpillars and butterflies look very different from one another.

Set a Purpose and Read

- Read the headings. Tell students that they will read about the life cycles of insects such as luna moths, praying mantises, and cicadas.
- Have students read pages 54–59.

❷ Teach

Academic Vocabulary: *metamorphosis, larva, pupa*

- Pronounce **metamorphosis** and write it. Have a student read the definition. Say: **Complete metamorphosis is a series of changes that a moth and butterfly go through during their life cycle.**
- Pronounce **larva** and **pupa,** and then write them. Point to the photos that illustrate these words on page 54. Have a student read the definitions.

Metamorphosis

Some kinds of animals, such as amphibians and insects, go through **metamorphosis** during their life cycles. Metamorphosis is a series of major changes in an animal's body form as it moves through its life cycle.

Complete Metamorphosis Insects with four life cycle stages undergo complete metamorphosis. Find the egg stage of the luna moth in the diagram. Trace the luna moth's life cycle with your finger.

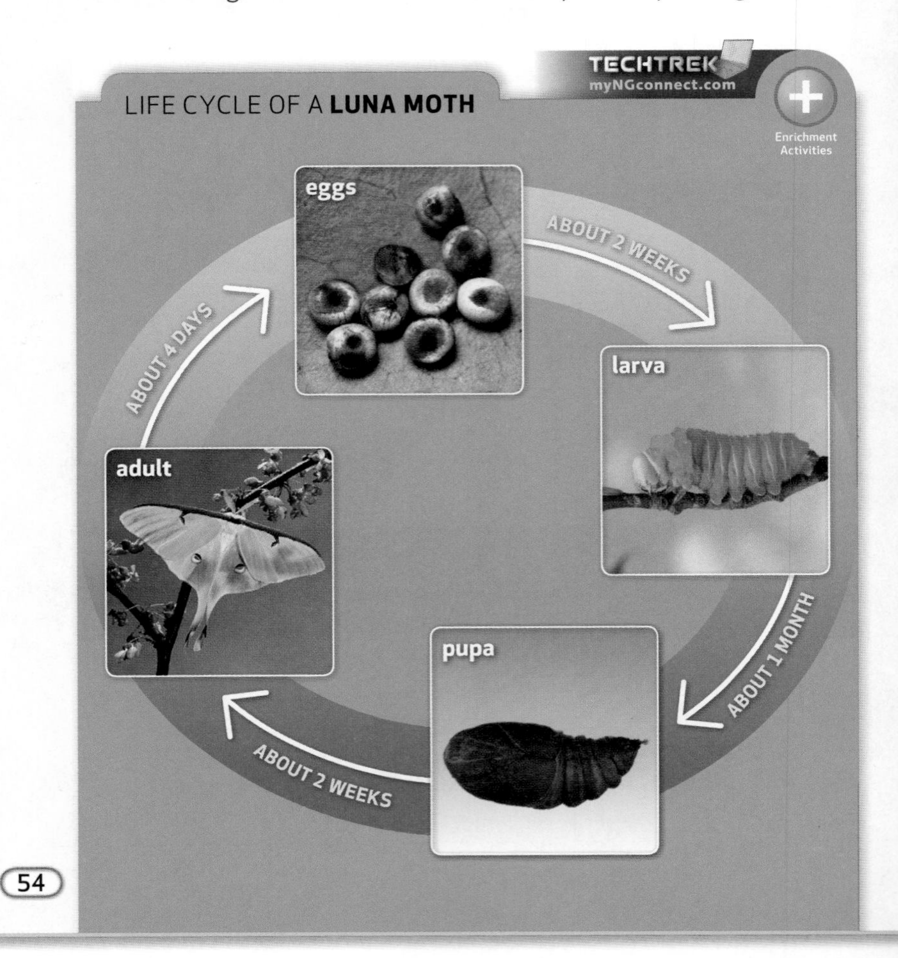

Think Like a Scientist: Math in Science

Measure Luna moths are unusually large moths. A common size of their body is a length of 15 centimeters (6 inches) and a wingspan of 10 centimeters (4 inches). Have students use a ruler to draw a rectangle that is 10 centimeters wide and 15 centimeters long. Then have them draw a luna moth inside the box. Have students compare the size of the luna moth to other familiar objects or animals.

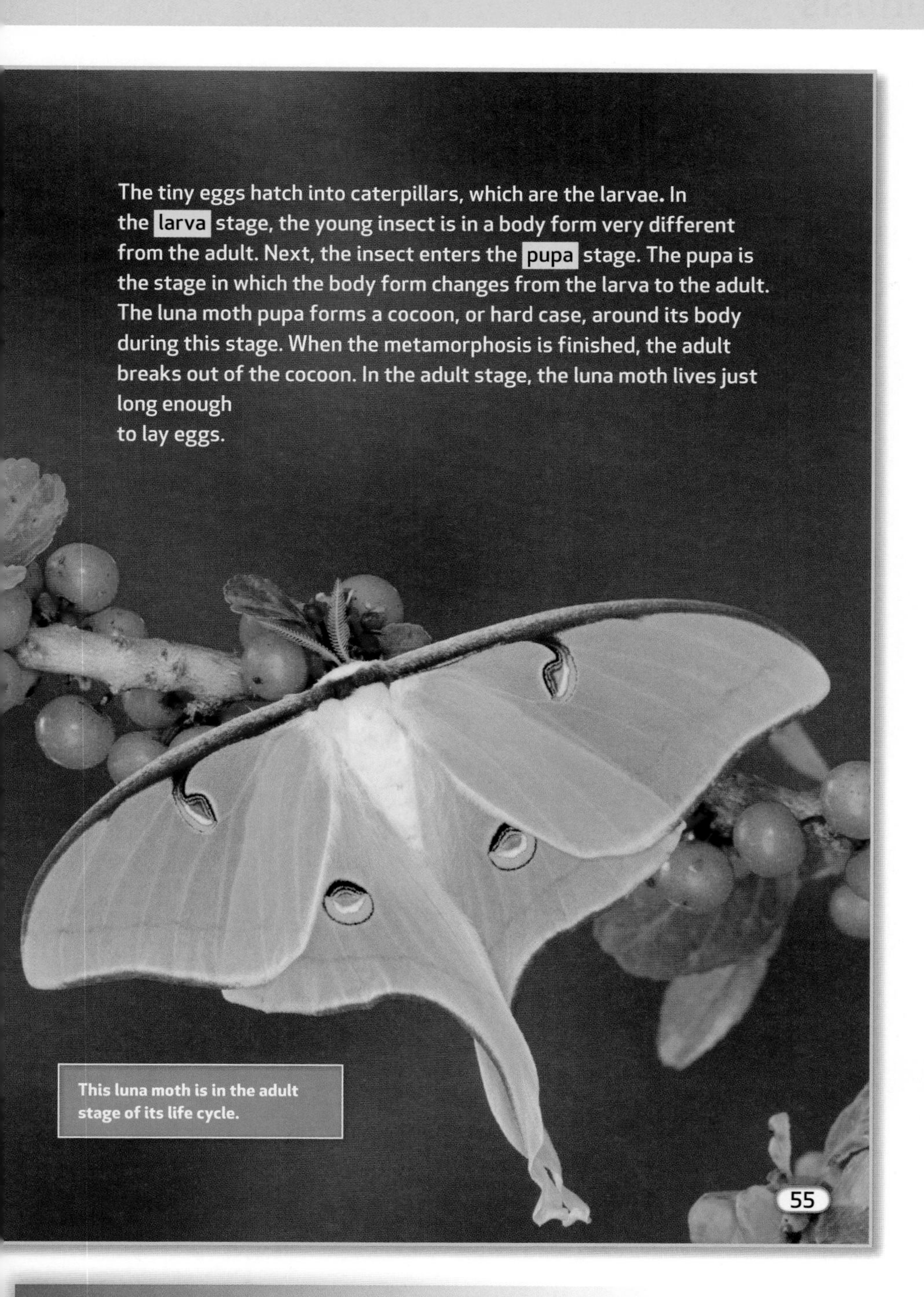

The tiny eggs hatch into caterpillars, which are the larvae. In the larva stage, the young insect is in a body form very different from the adult. Next, the insect enters the pupa stage. The pupa is the stage in which the body form changes from the larva to the adult. The luna moth pupa forms a cocoon, or hard case, around its body during this stage. When the metamorphosis is finished, the adult breaks out of the cocoon. In the adult stage, the luna moth lives just long enough
to lay eggs.
55

This luna moth is in the adult stage of its life cycle.

Teach, continued

- Ask: **What does the luna moth larva look like?** (Possible answer: a fat worm) Ask: **What does the luna moth pupa look like?** (Possible answer: a pod in a hard shell) Discuss how the **larva** turns into a **pupa** as part of the luna moth life cycle.

Describe Complete Metamorphosis

- Have students observe the diagram of the life cycle of the luna moth on page 54 and read the labels. Point out the four arrows that connect the photos and discuss how the labels show the passage of time.
- Ask: **Would you describe the changes of the luna moth as small changes, moderate changes, or huge changes?** (huge changes) **Why?** (Each stage of the moth's life cycle is very different from the other stages.)
- Ask: **Which of the stages lasts the longest time?** (the larva stage, which lasts about 1 month) Explain that the larva grows larger as it eats food, which is necessary for it to change into an adult.
- Ask: **Which of the stages lasts the shortest time?** (the adult stage, which lasts only about 4 days) Explain that after the adults mate and the females lay eggs, their role in the life cycle is complete.

myNGconnect.com

Have students use Enrichment Activities to find out more about **metamorphosis**.

Computer Presentation Students can use the information to make a computer presentation that explains **metamorphosis**.

Differentiated Instruction

ELL Language Support for Describing Complete Metamorphosis

BEGINNING

Provide Academic Language Frames for students to use to describe complete metamorphosis.

- *An egg hatches into a ____.*
- *The next stage is the ____.*
- *Finally, the luna moth becomes an ____.*

INTERMEDIATE

Have students copy the illustration of the luna moth life cycle and then add labels or write captions to describe the changes that the luna moth undergoes.

ADVANCED

Provide sequence words to help students describe the life cycle of luna moths. For example:

The life cycle of luna moths includes different stages. First, . . . Second, . . . Then . . . Finally,

Objectives

Students will be able to:

- Describe the different life cycles of animals and identify the different stages of these life cycles.
- Compare and contrast the similarities and differences among offspring of different animal life cycles.

Science Academic Vocabulary

nymph

PROGRAM RESOURCES

- Digital Library at myNGconnect.com

Teach, continued

Academic Vocabulary: *nymph*

- Pronounce **nymph** and write it. Have a student read the definition. Emphasize that the **nymph** stage is a part of incomplete metamorphosis, not complete metamorphosis.

Describe Incomplete Metamorphosis

- Have students observe the diagram of the life cycle of the praying mantis on page 57 and read the labels. Point out the three arrows that connect the photos, and discuss how the labels show the passage of time.
- Ask: **How does the nymph compare to the adult?** (The nymph is similar in shape but smaller.) Discuss how, in about 6 months, the nymph will grow larger in many stages to become an adult.

Incomplete Metamorphosis Some animals, such as praying mantises, go through metamorphosis in a three-stage life cycle. This is called incomplete metamorphosis. The three stages are egg, nymph, and adult. A nymph is the stage of incomplete metamorphosis in which the young animal looks like the adult but may not have some features, such as wings. Praying mantis eggs soon hatch into the nymph stage. Over time, the nymph grows larger as it becomes an adult.
An adult praying mantis can lay hundreds of eggs.
56

Raise Your SciQ!

NATIONAL GEOGRAPHIC

Life Cycle of the Praying Mantis Praying mantises are carnivores that catch and eat other insects, including their own kind. Nymphs may eat their siblings, and females often eat their mates. In the fall, praying mantises lay their eggs in a kind of foam that hardens into a protective case. The adults die soon after laying eggs. In the spring, the eggs hatch, and the nymphs begin hunting other insects.

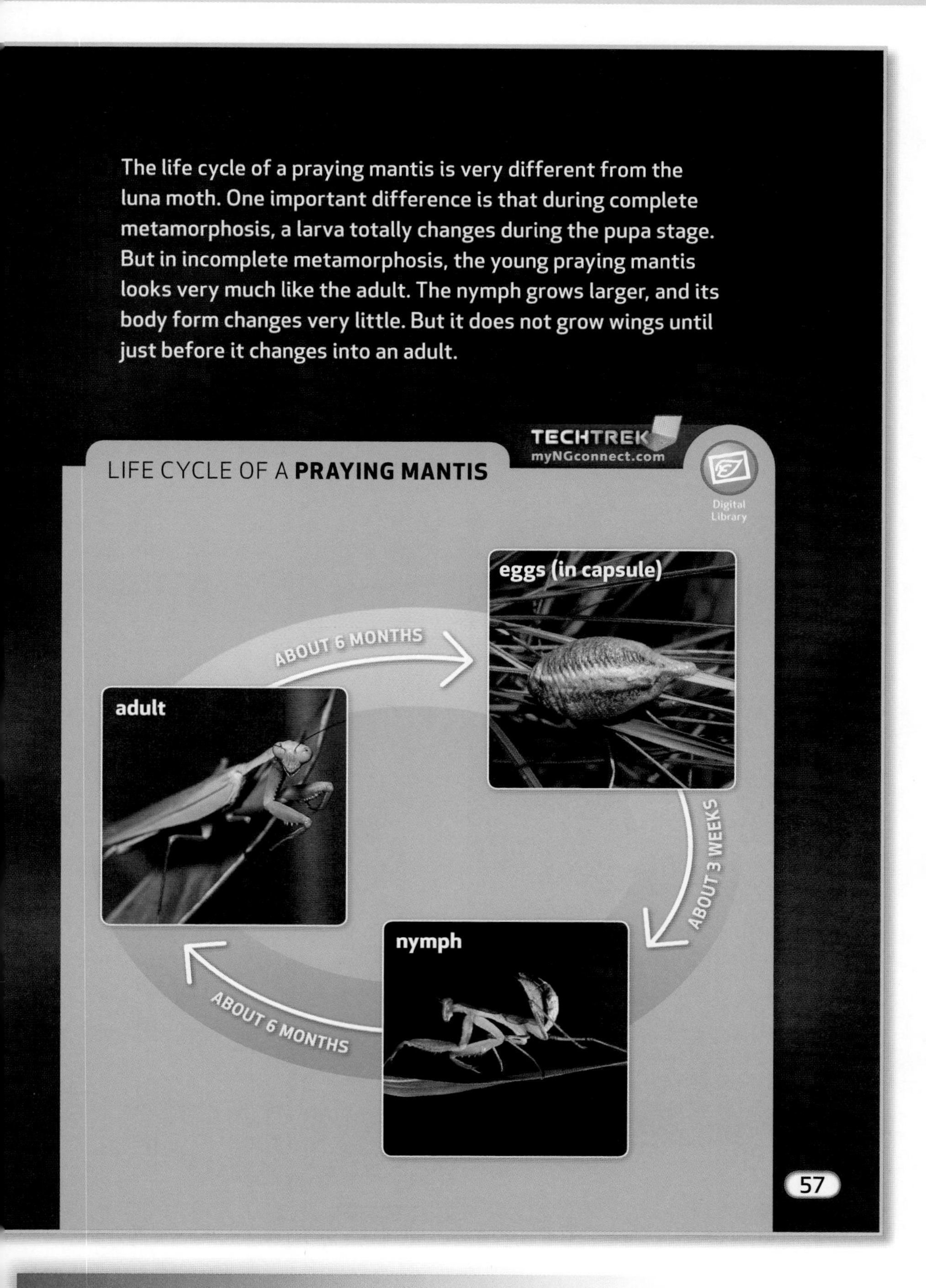

The life cycle of a praying mantis is very different from the luna moth. One important difference is that during complete metamorphosis, a larva totally changes during the pupa stage. But in incomplete metamorphosis, the young praying mantis looks very much like the adult. The nymph grows larger, and its body form changes very little. But it does not grow wings until just before it changes into an adult.
TECHTREK
myNGconnect.com
LIFE CYCLE OF A PRAYING MANTIS
Digital Library
eggs (in capsule)
ABOUT 6 MONTHS
adult
ABOUT 3 WEEKS
nymph
ABOUT 6 MONTHS
57

Teach, continued

Compare and Contrast Complete and Incomplete Metamorphosis

- Have students compare and contrast the diagrams on pages 54 and 57. Ask: **How are complete and incomplete metamorphosis alike?** (Each begins with an egg and ends with an adult, and it involves many changes to an animal's body.)
- Ask: **How are they different?** (They differ in the numbers and kinds of stages—larva and pupa in complete metamorphosis, nymph in incomplete metamorphosis. Also, the differences among the stages of complete metamorphosis are much greater.)

Notetaking 

Have students draw and label the life cycles of complete and incomplete metamorphosis in their science notebooks. Encourage students to refer to these drawings as they progress through their studies of life science.

myNGconnect.com

Have students use the **Digital Library** to find images of praying mantises.

Digital Booklet Students can use the information to make a digital booklet with captions that describes observations or inferences about praying mantises.

Differentiated Instruction

ELL Language Support for Describing Incomplete Metamorphosis

BEGINNING

Provide Academic Language Frames for students to use to describe incomplete metamorphosis.

- *An egg hatches into a ____.*
- *The next stage is the ____.*
- *Finally, the adult lays ____.*

INTERMEDIATE

Have students copy the illustration of the praying mantis cycle and then add labels or write captions to describe the changes that the praying mantis undergoes.

ADVANCED

Help students create a two-column chart that compares complete and incomplete metamorphosis. Then have students use their charts as aids to compare complete and incomplete metamorphosis.

Objectives

Students will be able to:

- Describe the process of molting during the life cycles of incomplete metamorphosis.

Teach, continued

Explain the Purpose of Molting

- Ask a volunteer to take off one shoe and hold it up to the class. Ask: **Why do you need to keep buying larger shoes as you get older?** (My feet keep growing larger.)
- Ask: **Have you ever tried to put on a shoe that was too small? How did it feel?** (Possible answer: It hurt my foot.) Discuss how an insect's hard outer covering is like a shoe around a foot. When the insect's body grows, its hard outer covering must be shed and replaced.

Molting As a nymph grows, its outer covering hardens. It becomes almost like a shell. As the nymph grows bigger, that hard outer covering becomes too small. The nymph molts, or sheds that covering. The nymph grows bigger and its covering hardens again. When this covering becomes too small, the nymph molts again. Many insects molt several times during the nymph stage.

Dogday harvestfly cicada nymphs live underground. They crawl into trees when they molt for the last time.

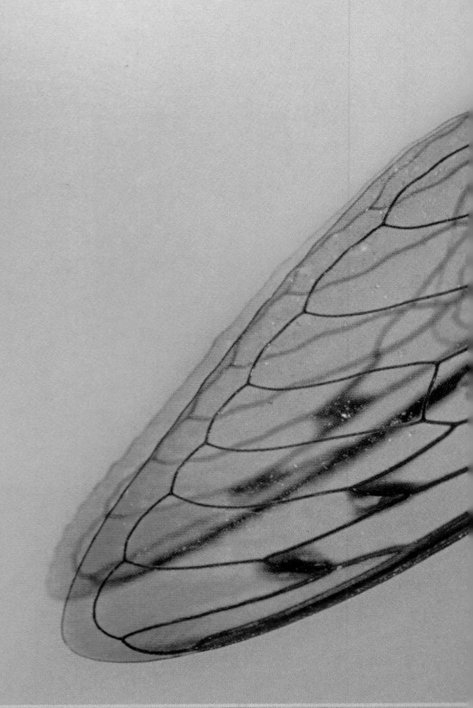

58

Raise Your SciQ!

Cicadas Cicadas live throughout the United States. Cicada nymphs live underground in burrows. They eat the sap from the roots of grasses, trees, and other kinds of plants. Cicadas molt several times underground. Then, before its last molt, a cicada climbs out of the ground onto a tree trunk or other woody stem. It anchors itself and emerges from its exoskeleton, or outer covering. The adult cicadas mate and reproduce above ground. An adult cicada usually lives only a few weeks.

Finally, the insect goes through its last molt. If you see an insect with wings shedding its outer shell, it is most likely the adult. So if you see a tiny fly with wings, for example, that kind of fly is just tiny. It will not grow into a bigger fly.

The adult female dogday harvestfly cicada cuts open a twig and lays eggs inside it. The eggs hatch in about six weeks.

Before You Move On

1. How does molting allow an animal to grow bigger?
2. How is complete metamorphosis different from incomplete metamorphosis?
3. **Generalize** How is a nymph different from a pupa?

59

Teach, continued

Text Feature: Photos and Captions

- Have students observe the photo of the molting cicada nymph on page 58 and read its caption. Point to the brown outer shell and ask: **What is this brown shell?** (the old outer covering of the nymph, which it just shed)
- Ask: **Why was it necessary for the cicada nymph to shed its outer covering?** (The nymph's body was growing larger, and the outer covering no longer fit it.) Explain that the cicada in the photo has just molted and that it has not yet grown a new outer covering.

Monitor and Fix Up

Ask students if anything they read was confusing. Students may be confused by the word *molt*. Discuss possible fix-up strategies. For example, students may rewrite the definition in their own words and study the photos on pages 58–59. If possible, collect the outer covering of a molted cicada to show students. Encourage students to bring one to class to show fellow students.

Differentiated Instruction

ELL Language Support for Explaining Molting

BEGINNING

Have students answer either/or questions about molting:

What molts, a pupa or a nymph?

When an insect molts, does it become larger or smaller?

INTERMEDIATE

Help students write a new caption for the photo of the molting cicada on page 58. Tell students to use the caption to define the word *molt* in their own words.

ADVANCED

Show students how to repeat questions in their answers. Ask: **Why is molting important for some animals?** (Molting is important for some animals because . . .)

3 Assess

Before You Move On

1. **Recall How does molting allow an animal to grow bigger?** (An animal's hard outer shell prevents it from becoming larger, so a growing animal with such a shell must molt, sometimes more than once.)
2. **Explain How is complete metamorphosis different from incomplete metamorphosis?** (Possible answer: In complete metamorphosis, the animal goes through four stages—egg, larva, pupa, and adult—during its life cycle. In incomplete metamorphosis, the animal goes through three stages—egg, nymph, and adult.)
3. **Generalize How is a nymph different from a pupa?** (By molting many times, a nymph can grow into an adult that looks much like it. By passing through a pupa stage, an insect's body can change very much between the younger and adult forms.)

LESSON 4 ▫ Inherited Traits

Objectives

Students will be able to:

- Observe that traits of animals are inherited from their parents and that individuals vary within every species.

Science Academic Vocabulary

heredity

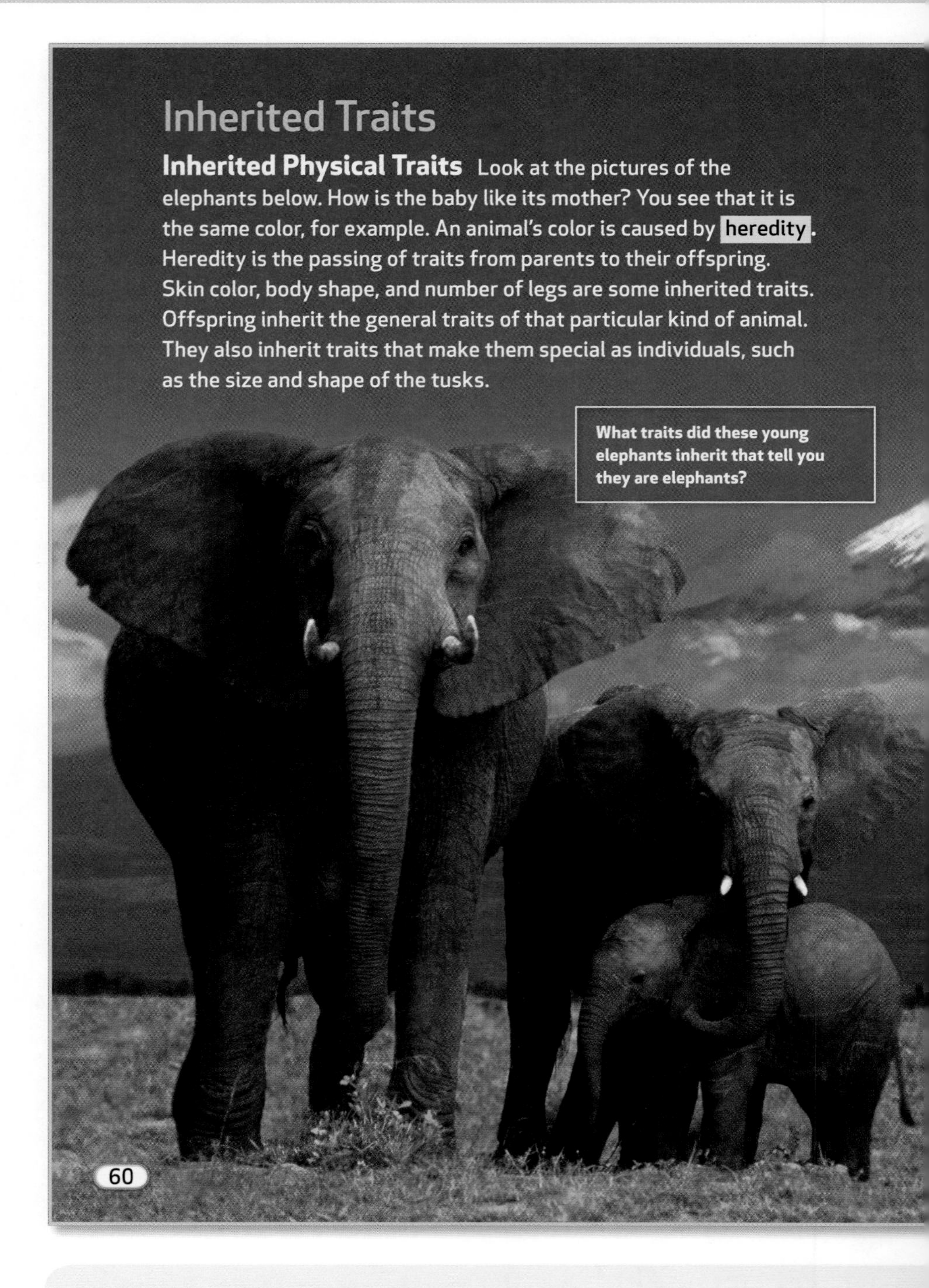

Inherited Traits
Inherited Physical Traits Look at the pictures of the elephants below. How is the baby like its mother? You see that it is the same color, for example. An animal's color is caused by heredity. Heredity is the passing of traits from parents to their offspring. Skin color, body shape, and number of legs are some inherited traits. Offspring inherit the general traits of that particular kind of animal. They also inherit traits that make them special as individuals, such as the size and shape of the tusks.
What traits did these young elephants inherit that tell you they are elephants?
60

❶ Introduce

Tap Prior Knowledge

- Have students describe young animals that they have observed, such as pet kittens or puppies, or young animals in the wild. Discuss how each animal will grow up to resemble its parents in many ways.

Set a Purpose and Read

- Read the heading. Then point out the elephants in the photo on page 60. Tell students that they will read about traits that animals receive from their parents.
- Have students read pages 60–63.

❷ Teach

Academic Vocabulary: *heredity*

- Pronounce **heredity** and write it. Have a student read the definition.
- Read aloud the caption to the photo on page 60 and invite students to answer it. (The young elephants have the body color and shape of their parents, as well as traits such as a trunk and wide ears.) Discuss how young animals look like their parents because of **heredity**.

Colorful Characteristics 

Read and follow the instructions together with students. Encourage students to observe each of the chicks and to compare their traits with the traits of the adult hen. Students should record their observations and results in their science notebook.

What to Expect The young inherited all of their physical traits from their parents, including their body shape, feathers, beaks, eyes, and legs.

Many kinds of animals inherit traits from two parents. The offspring will be similar to both parents. But it will not be exactly like either parent. In American alligators, the size of the head, button-shape scales on the chest, and the roundness of the snout are some differences you can see. These traits will be different from one alligator to the next. An individual's special traits are caused by how the parents' traits work together.

The baby alligator has stripes that fade as it grows. As an adult it will look much like its parents and other American alligators.

Science in a Snap! Colorful Characteristics

Find the adult in this photograph. Closely observe its traits. Then closely observe the traits of the young.

Which of the traits you observed in the young do you think are inherited from the parents?

61

Teach, continued

Describe Inherited Traits

- Have students observe the photo of the baby alligator and its parent on page 61 and then read the caption. Ask: **How can you tell from the photo that the young animal is a baby alligator?** (It looks like a small version of the adult alligator that is holding it.)
- Ask: **How is the baby alligator different from the adult?** (Differences include its small size and the stripes on its body.) Explain that the baby alligator inherited physical traits, such as body shape and striped markings, even if the parents lack these traits in their adult bodies.

Match Parents and Offspring

Provide students with separate photos of animal parents and their offspring. Have students match the parents to their offspring, and then record their answers and reasons in their science notebook.

Quick Questions Ask students the following questions.

- **How are the chicks' traits different from the traits of the adult hen?** (The chicks are smaller, their feathers are downy and of different colors, and they lack combs.) Explain that although chicks have different physical traits from their parents, they still inherited these traits.
- **How will the chicks change as they grow older?** (They will grow larger, grow new feathers and a comb, and begin to resemble their parents more.) Discuss how the way a young animal grows into an adult is also determined by heredity.

Objectives

Students will be able to:

- Describe animal behaviors that result from heredity.

Teach, continued

Recognize Inherited Behaviors

- Say: **An animal's behaviors are its actions. Give some examples of animal behaviors.** (Possible answers: Birds fly and build nests; lions hunt other animals for food; worms tunnel through the soil; spiders spin webs to catch insects.) Discuss how many animal behaviors are inherited. Eagles, for example, hatch with the knowledge of how to build a nest and where they should build it.
- Have students observe the photo of the spinybacked orbweaver and its web and then read the caption. Say: **Spinybacked orbweavers spin only one kind of web. Why are they able to spin one kind of web but not other kinds?** (Web spinning is an inherited behavior in spiders. Spiders do not learn to spin new kinds of webs.) Discuss how web spinning is an example of an *instinct*, a complex behavior that is inherited.

Inherited Behaviors Many animal behaviors, or actions, are also inherited. For example, a bird inherits the ability to fly. More complex inherited behaviors are instincts. Instincts help animals find or catch food, protect themselves, and reproduce. A spider inherits its ability to make a certain kind of web. The kind of web a spider weaves is related to the kind of food it eats.

Spinybacked orbweavers are common in Florida. They eat flies, moths, and beetles that get caught in the web.

62

Differentiated Instruction

Extra Support

Have students write new captions for the photos on pages 62–63. Encourage students to use the captions to describe inherited behaviors or instincts and how they help animals survive.

Challenge

Have students research the inherited behaviors of an animal of their choice. Interesting examples to choose include the migration of many birds and monarch butterflies, the seasonal feeding patterns and hibernation of chipmunks, and the tunneling of the star mole. Have students report their findings to the class.

Animals follow their instincts all through their lives. For example, young salmon hatch in streams. Then they follow their instincts to swim to the ocean. Some salmon swim more than 1,600 km (1,000 miles) in this journey to the ocean! Years later, as adults, they follow their instincts and return to the streams where they hatched. They swim upstream past dams, fishermen, and other hazards. Once they reach the stream where they hatched, the females lay their eggs.
These pink and chum salmon left Hartney Creek in Alaska when they were about the size of your finger. By instinct they have returned to lay eggs.
Before You Move On
1. What is heredity? List some inherited traits.
2. How do instincts help an animal survive? Give an example.
3. Draw Conclusions A bird knows how to build a nest without ever learning how. Explain how this is possible.
63

Science Misconceptions

Inherited Behaviors Some students may think that very intelligent animals do not follow instincts or other inherited behaviors. In fact, scientists have identified many instincts in a wide variety of animals, including humans. Human babies, for example, have instincts to nurse when they are hungry and to cry when they are in pain or when their needs are not met. However, humans have an enormous capacity to learn throughout their lives. Learning guides human behavior perhaps more so than it guides any other animal.

Teach, continued

Explain the Inherited Behaviors of Salmon

- Have students observe the photo of salmon on pages 62–63 and then read the caption. Ask: **What instincts do salmon follow during their lives?** (After hatching, they swim from streams to the ocean. As adults, they return to the streams where they hatched.) Explain that scientists believe that adult salmon can sense chemicals in the water that indicate which rivers and streams are the correct ones to return to.
- Say: **Salmon find food in the ocean, while streams are safe places for their eggs to hatch.** Discuss how salmon would not likely travel between their stream and ocean environments without instincts to guide them.

❸ Assess

» Before You Move On

1. List **What is heredity? List some inherited traits.** (Heredity is the passing of traits from parents to their offspring. Examples of inherited traits include skin color, body shape, number of legs, eye color, and hair color.)
2. Explain **How do instincts help an animal survive? Give an example.** (Instincts often help animals meet their basic needs, such as finding food, protecting themselves, and reproducing. Building webs is an instinct that helps spiders catch other animals for food. Because web-building is an instinct, young spiders can survive without any help or instruction from other spiders.)
3. Draw Conclusions **A bird knows how to build a nest without ever learning how. Explain how this is possible.** (The bird knows how to build a nest because of instinct. The instinct to build a nest is an inherited behavior.)

Objectives

Students will be able to:

- Recognize that acquired traits in animals can result from the environment.

PROGRAM RESOURCES

- Digital Library at myNGconnect.com

❶ Introduce

Tap Prior Knowledge

- Have students discuss minor injuries they have observed in friends or family, such as a sprained or broken limb or finger, or a scrape or bruise. Explain that animals can also become injured. Unlike humans, animals do not always receive expert help for healing their injuries.

Set a Purpose and Read

- Read the heading. Tell students that they will read about acquired traits, which are traits an animal gains during its life.
- Have students read pages 64–67.

❷ Teach

Text Feature: Photo and Caption

- Have students observe the photo of the topi on page 64 and read the caption. Point out the missing horn. Ask: **How might the topi have lost a horn?** (Possible answers: in a fight with another animal; in an accident with a fence or a gate) Explain that the topi inherited two horns and that its lost horn is an acquired trait.

Acquired Traits
Acquired Physical Traits Not all traits are inherited. Some are acquired, or gained. These traits can be acquired because of behaviors or factors in the environment. The topi below might have acquired the trait of one horn during a fight over its territory.
Having two horns is an inherited trait in topis. This topi has the acquired trait of one horn.
64

NATIONAL GEOGRAPHIC Raise Your SciQ!

Acquired Traits Animals often gain acquired traits because of human actions. For example, many manatees have scars on their backs from collisions with motor boats. Animals also gain acquired traits as part of the mating process. Males of deer, elk, and moose often injure their horns as they fight with one another over a female. Roosters use their spurs to slash at other roosters, which can cause gashes and other injuries. Roosters also may pluck feathers off female hens as part of the mating process.

Animals can also acquire physical traits from their diet. A diet is all the foods an animal eats. Diet affects an animal's body size, weight, and health. In some animals, diet can even change body color! Flamingos, for example, are born with white feathers. Flamingos have diets of algae, insect larvae, and shrimp. Substances in these foods cause a flamingo's feathers to turn pink. The more of these foods they eat, the pinker they get!

The colors of a flamingo's feathers can change from white to deep pink.

65

Teach, continued

Describe Acquired Traits from Diet

- Have students observe the photo of the flamingos on page 65 and read the caption. Point out the different colors of the flamingos. Ask: **Why do flamingos change color from white to pink?** (Foods in their diet cause their feathers to change colors.)
- Ask: **How else can diet affect an animal's body?** (With the proper diet, an animal can grow to its full size and develop a healthy body. Without enough food or the right kind of food, an animal may not grow and develop properly.) Discuss how the flamingos in the photo have grown from tiny chicks to large birds. Their food gave them the energy and nutrients needed to do this.

Recognize Similarities Between Self and Parents

Have students think about the similarities between themselves and their parents, and then list several similarities in their science notebook. Have students classify each similarity as an inherited trait or acquired trait.

myNGconnect.com

Have students use the Digital Library to find images that illustrate inherited or acquired traits.

Whiteboard Presentation Students can use the information to make a computer presentation about inherited and acquired traits. You can help them show the presentation on a whiteboard.

Science Misconceptions

Animal Nutrition When students study health and the human body, they learn that their bodies need a wide variety of nutrients and that they need them in the proper amounts. However, students may not recognize that many animals also rely on a specific set of nutrients to stay healthy and that they need these nutrients in the proper amounts. In zoos, nutritionists typically try to replicate the animals' diets in the wild. They also monitor the animals' development and health and alter their diets as necessary.

Objectives

Students will be able to:

- Describe animal behaviors that result from learning.

PROGRAM RESOURCES

- Science Inquiry and Writing Book: *Life Science*
- Science Inquiry and Writing Book eEdition at myNGconnect.com

Teach, continued

Identify Learned Traits

- Have students observe the sequence of photos on pages 66–67 and then read the captions. Explain that a capuchin is a kind of monkey. Ask: **What is the capuchin trying to do?** (break open a palm nut with a rock) **Why do you think it is trying to do this?** (It wants to eat the nut.)
- Ask: **Why is the capuchin's behavior an example of a learned trait, not an instinct?** (The capuchin was not born with the knowledge that dropping a rock can break a nut. Instead it learned this fact from its own experience.) Discuss how other capuchins might not necessarily learn this behavior, especially if they were raised in an environment where food was very easy to obtain.
- Ask: **What behaviors can pet dogs learn?** (Possible answers: Dogs can learn simple commands, such as *sit*, *stay*, *fetch*, and *roll over*. They also learn to recognize certain people as friends and others as strangers, and they may behave in certain ways when signals alert them that food or outdoor exercise is coming. They also learn to be housebroken.) Remind students that learned behaviors include behaviors that humans teach animals, as well as behaviors they learn by themselves.
- Say: **When you teach a dog to fetch or roll over, the dog gains an acquired trait.**

Learned Traits Some acquired traits result from learning. Many kinds of animals learn behaviors. Learning can be simply getting used to something happening. A squirrel in your backyard learns that you are not a predator. So it will allow you to stand closer to it than a squirrel in the wild would. The capuchin monkey below shows more complicated learning.

The capuchin finds a good place to put the palm nut.

The capuchin balances on its tail and gets ready to slam the rock down on the palm nut.

66

Social Studies in Science

Use a Map Show students a globe or map of the world and ask them to find Brazil in South America. Here capuchin monkeys live in tropical rainforests. Discuss how a variety of primates, including apes and monkeys, live in places around the world. The monkeys of Central and South America are quite different from those that live in Africa.

Young capuchin monkeys learn that they can crack open a nut using a heavy rock. Scientists have observed that the monkeys hold the rock in a certain way so they can hit the nut harder.

Adult birds such as cardinals and warblers learn the songs they sing to attract mates. Your pet dog or cat can learn where and when you will feed it. What other behaviors do you think a pet cat or dog can learn?

Notice that the capuchin is still balancing itself with its tail as it slams the rock down. Did the palm nut break open? What do you think the capuchin will do next?

Before You Move On

1. What is an acquired trait? Give an example.
2. How do animals acquire traits?
3. **Draw Conclusions** Do you think it would be easy to train a capuchin to use a hammer to hit a nail? Explain why or why not.

67

Differentiated Instruction

ELL Language Support for Identifying Traits

BEGINNING

Help students classify traits by asking yes/no questions, for example:

Are brown eyes a learned trait? Is a scar on your hand an acquired trait? Is changing classes when the bell rings a learned trait?

INTERMEDIATE

Provide Academic Language Frames to help students classify traits as inherited, instinctive, or learned.

- *The brown fur of a monkey is an ____ trait.*
- *Salmon swimming to and from the ocean is an ____ trait.*
- *Eating food with a fork is a ____ trait.*

ADVANCED

Have students write a paragraph about an animal's traits. The paragraph should identify traits as inherited, instinctive, or learned.

Teach, continued

Compare Inherited and Acquired Traits

- Have students observe the large photo of the capuchin on page 66. Ask: **What are some inherited traits of the capuchin?** (Possible answers: four limbs; dark brown fur on head and tail; light brown fur on body; a tail)
- Ask: **What is the difference between the inherited traits and the learned trait?** (The monkey was born with the inherited traits. The monkey acquired the learned trait after it was born.) Explain that all capuchins share certain physical traits that they inherited, but may or may not gain certain acquired traits.

Animal Observations

Science Inquiry and Writing Book: *Life Science*, page 13

Read and follow the instructions together with students. Encourage students to list a wide range of animal behaviors. Students should record their observations and ideas in their science notebook.

What to Expect Examples of inborn or instinctive behaviors may involve movement, choice of habitat, feeding styles and food preferences, grooming, and raising young. Examples of learned behaviors include a pet's tricks, excitement at feeding time, and recognition of specific friends or enemies.

3 Assess

» Before You Move On

1. **Recall What is an acquired trait? Give an example.** (An acquired trait is gained from learning or from the environment. Scars or scratches on an animal's body or learned behaviors are kinds of acquired traits.)
2. **Explain How do animals acquire traits?** (An acquired trait is the result of factors in the environment, behavior, or learning.)
3. **Draw Conclusions Do you think it would be easy to train a capuchin to use a hammer to hit a nail? Explain why or why not.** (Possible answer: It might be easy to train a capuchin to use a hammer since this action is similar to hitting a nut with a rock.)

LESSON 6 ▫ On the Job with Service Dogs

Objectives

Students will be able to:

- Describe animal behaviors that result from learning.

PROGRAM RESOURCES

- Learning Masters Book, page 28, or at myNGconnect.com
- Digital Library at myNGconnect.com

❶ Introduce

Tap Prior Knowledge

- Have students describe unusual behaviors that they have observed in dogs, such as catching a ball or flying disc, barking loudly at certain people, or leading their owner across the street. Discuss how dogs can learn a variety of behaviors from their interactions with people.

Preview and Read

- Read the heading. Then preview the pictures on pages 68–69. Tell students that they will read about how service dogs are used and trained.
- Have students read pages 68–69.

❷ Teach

myNGconnect.com

Have students use the Digital Library to find photos that illustrate dogs and their behaviors.

Whiteboard Presentation Students can use the information to make a computer presentation about the behaviors of dogs. You can help them show the presentation on a whiteboard.

NATIONAL GEOGRAPHIC
ON THE JOB WITH SERVICE DOGS
Did you know that many dogs are trained to help people? They guide people who cannot see, and they help find missing people. To do these jobs, dogs use both instincts and learned behaviors.
TECHTREK
myNGconnect.com
Digital Library
Border collies have a natural instinct for herding, which makes them easy to train to help people.
68

NATIONAL GEOGRAPHIC Raise Your SciQ!

Border Collies The border collie is a breed of dog that originated along the border between England and Scotland. Here these dogs have been herding sheep for over 300 years. Border collies often stare sheep in the eyes in order to frighten them. This is possibly an inherited trait that has been passed down from the wolf, the ancestor of all dogs.

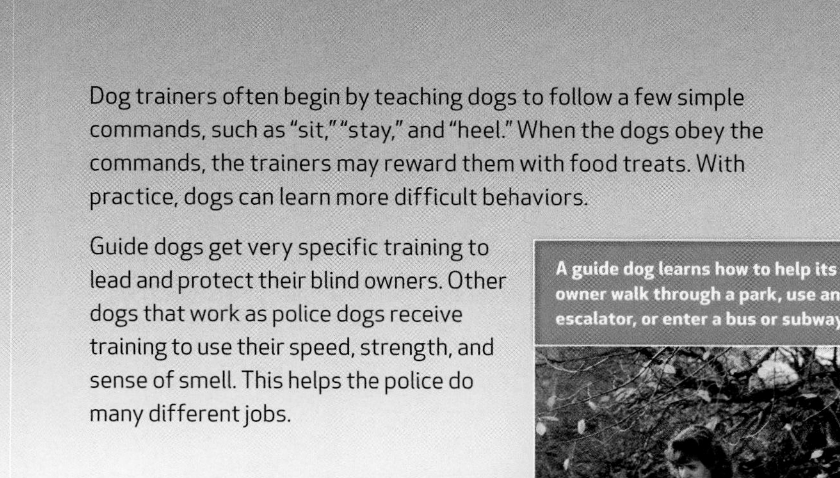

Dog trainers often begin by teaching dogs to follow a few simple commands, such as "sit," "stay," and "heel." When the dogs obey the commands, the trainers may reward them with food treats. With practice, dogs can learn more difficult behaviors.

Guide dogs get very specific training to lead and protect their blind owners. Other dogs that work as police dogs receive training to use their speed, strength, and sense of smell. This helps the police do many different jobs.

A guide dog learns how to help its owner walk through a park, use an escalator, or enter a bus or subway.

This golden retriever is being trained as a search-and-rescue dog. To rescue a person, the dog may need to walk across high, narrow paths.

69

Share and Compare

Classifying Traits Give students the Learning Master. Have them complete the chart by writing examples of inherited traits and acquired traits. Check their work, and then have partners share and compare the examples they cited.

my SCIENCE notebook

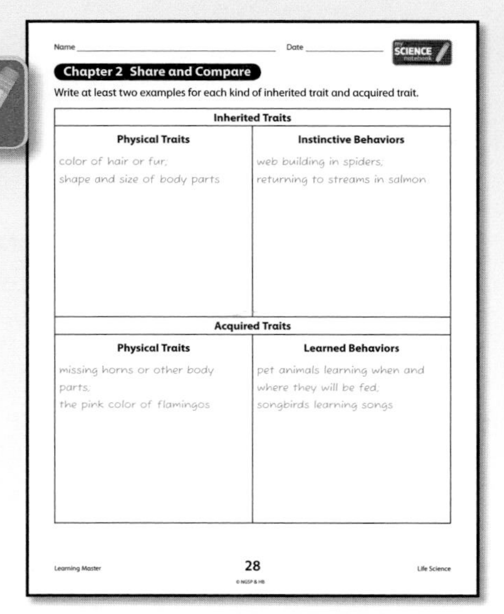

Name ______ Date ______

Chapter 2 Share and Compare

Write at least two examples for each kind of inherited trait and acquired trait.

Inherited Traits	
Physical Traits	**Instinctive Behaviors**
color of hair or fur; shape and size of body parts	web building in spiders; returning to streams in salmon
Acquired Traits	
Physical Traits	**Learned Behaviors**
missing horns or other body parts; the pink color of flamingos	pet animals learning when and where they will be fed; songbirds learning songs

Learning Master 28 Life Science

© NGSP & HB

Learning Master 28, or at myNGconnect.com

Teach, continued

Recognizing Behaviors Shaped by Behavior and Learning

- Have students observe the photo of the border collie and sheep on page 68 and then read the caption. Ask: **What is the border collie doing in this photo?** (chasing the sheep) Explain that the border collie is herding the sheep, which means it is trying to keep the sheep together.
- Ask: **How do sheep ranchers make the collie's behavior useful?** (They teach the dogs to follow commands for when and how to herd sheep.) Discuss how the collie's natural instincts for herding make the training possible. A trained border collie on a sheep ranch shows a combination of instinctive and learned behaviors.
- Have students observe the photo of the guide dog and its owner on page 69. Ask: **Why do guide dogs need an unusual amount of training?** (Guide dogs must both serve and protect their owners. This requires following many complex commands and responding to many events in the environment, such as a busy crosswalk.)
- Ask: **Why do you think dogs make better guides than other animals, like cats or pigs?** (Dogs have instincts that allow them to form bonds with humans, and they have the ability to learn many new behaviors. Other animals do not have this particular set of instincts and abilities.)

❸ Assess

1. **Describe How can a dog's behaviors help people?** (Possible answers: border collies can herd sheep and cattle; guide dogs help blind people; search-and-rescue dogs help emergency workers; guard dogs protect property.)
2. **Cause and Effect How does rewarding a dog help it learn new behaviors?** (When a dog is rewarded for behaving in a certain way, it learns that this behavior is desirable, and it will continue to behave that way.)
3. **Infer Why can't a dog be trained to do any task that a human can do?** (Dogs have a limited intelligence and ability to learn. Many of the new behaviors they learn are extensions of their natural instinctive behaviors, not entirely new behaviors.)

Objectives

Students will be able to:

- Investigate through Guided Inquiry (answer a question; make and compare observations; collect and record data and observations; generate explanations and conclusions based on evidence; share findings; ask questions based on observations to increase understanding; adjust explanations based on findings and new ideas).
- Compare and contrast the major stages in the life cycles of insects.
- Explain why keeping accurate records of observations and investigations is important.

Science Process Vocabulary

observe, compare

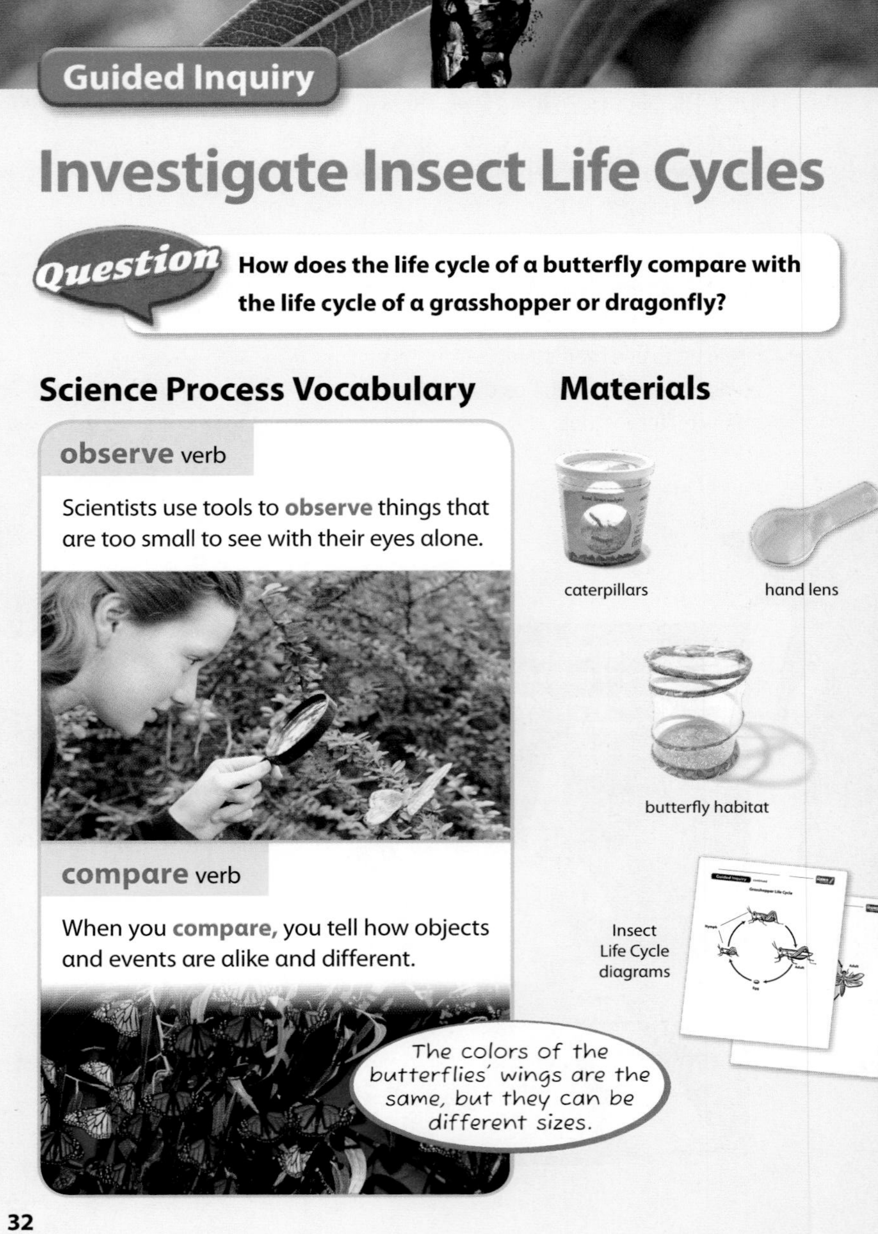

Guided Inquiry
Investigate Insect Life Cycles
Question
How does the life cycle of a butterfly compare with the life cycle of a grasshopper or dragonfly?
Science Process Vocabulary
observe verb
Scientists use tools to observe things that are too small to see with their eyes alone.
compare verb
When you compare, you tell how objects and events are alike and different.
The colors of the butterflies' wings are the same, but they can be different sizes.
Materials
caterpillars
hand lens
butterfly habitat
Insect Life Cycle diagrams
32

PROGRAM RESOURCES

- Science Inquiry and Writing Book: *Life Science*
- Science Inquiry and Writing Book eEdition at myNGconnect.com
- Inquiry eHelp at myNGconnect.com
- Science Inquiry Kit: *Life Science*
- Learning Masters Book, pages 29–34, or at myNGconnect.com
- Inquiry Rubric: Assessment Handbook, page 183, or at myNGconnect.com
- Inquiry Self-Reflection: Assessment Handbook, page 192, or at myNGconnect.com

MATERIALS

Kit materials are listed in italics.

caterpillars (live materials coupon); hand lens; butterfly habitat; Insect Life Cycle Diagram Learning Masters

For teacher use: *safety pin;* (optional): facial tissue or carnation; sugar (3 tsp); water (1 cup); orange

❶ Introduce

Tap Prior Knowledge

- Students have probably seen caterpillars crawling around in a flower or vegetable garden. Discuss how caterpillars will eventually turn into butterflies or moths. Ask: **What do you know about the life cycle of a butterfly?** Write student answers on the board.

MANAGING THE INVESTIGATION

Time

20 minutes for setup, then 10 minutes a day over several weeks for observation

Groups

Whole class

Advance Preparation

- Copy the Insect Life Cycle Diagram Learning Masters for the class.
- Send in the live materials coupon found inside the butterfly habitat package. Allow 2–3 weeks for delivery.
- Read and follow the instructions about caterpillar and butterfly care that come with the habitat.
- Plan in advance what you will do with the adult butterflies.

What to Do

1. **Observe** the photo of the butterfly egg. Draw the egg and record your observations in your science notebook.

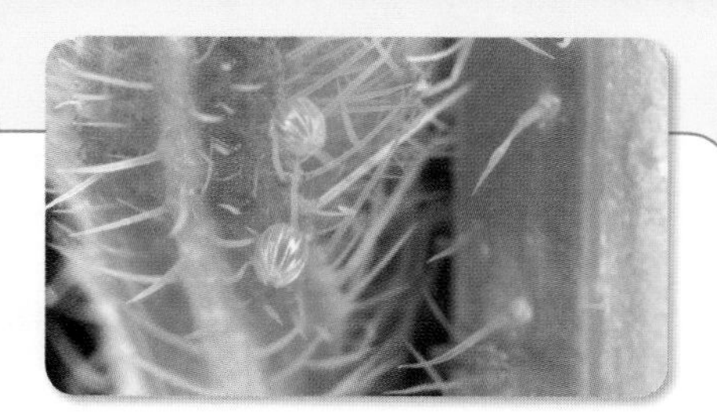

Butterflies reproduce by laying eggs. The eggs in this photograph appear larger than actual eggs, which are about the size of the head of a pin.

2. A caterpillar is the larva stage of a butterfly life cycle. Use the hand lens to observe the larvae in the container. Record your observations and draw the larvae.

3. Observe the larvae each day. Record any changes you observe. When pupae form, observe a pupa with a hand lens. Record your observations and draw the pupa.

33

What to Expect

- Students will observe caterpillars, or larvae, as they turn into pupae and then adult butterflies. There is some variation in how long each stage of the life cycle will take. Usually, it will take about 7–10 days for a larva to reach the pupa stage, and 7–10 days for the pupa to become an adult butterfly.

Introduce, continued

Connect to the Big Idea

- Review the Big Idea Question, *How do animals grow and change?* Explain to students that this inquiry will allow them to observe the complete metamorphosis of a butterfly.
- Have students open their Science Inquiry and Writing Books to page 32. Read the Question and invite students to share ideas about insect life cycles that they have observed.

❷ Build Vocabulary

Science Process Words: observe, compare

Use this routine to introduce the words.

1. **Pronounce the Word** Say **observe**. Have students repeat it in syllables.
2. **Explain Its Meaning** Choral read the sentence. Ask students for another word or phrase that means the same as **observe**. (use your senses to learn about objects or events)
3. **Encourage Elaboration** Ask: **How can you observe with your hands?** (I can use my hands to observe texture, temperature, and shape.)

ELL Vocabulary Support

Remind students that there are many ways to **observe**. Tell students to close their eyes. Ask: **What do you observe with your other senses?** (I feel the air with my skin. I feel vibrations with my body. I smell odors with my nose. I hear sounds with my ears.)

Repeat for the word **compare**. To encourage elaboration, ask students to **compare** a larva and an adult butterfly.

❸ Guide the Investigation

- Distribute materials. Have students match each pictured item to the real item.
- Read the inquiry steps on pages 33–34 together with students. Move from group to group and clarify steps if necessary.
- If students touch the larvae, pupae, or adult butterflies, they should wash their hands.
- Discuss with students that the plural form of larva is larvae and the plural form of pupa is pupae. Review the pronunciation with students for both.

Guide the Investigation, continued

- In step 1, guide students to understand that the egg is the first stage of the butterfly life cycle. Remind them to include this stage in their diagrams.
- In step 5, have students compare and contrast the adult butterflies and the larvae. Ask: **How is their behavior similar and different?** Guide students to discuss eating habits, movement, and activity level.
- In step 6, students should make detailed drawings of each stage. Make sure students accurately label the stages.

❹ Explain and Conclude

- Ask students to use evidence when communicating scientific ideas. To encourage this, ask them questions such as **What did you observe at the larva stage? the pupa stage? the adult stage? How were the stages alike and different?**
- Have students distinguish actual observations from ideas and speculations when they record their observations. Ask: **Why is it important to observe and record detailed observations every day during the life cycle of the butterfly?** (the life cycle continues and changes each day) Remind students that scientists use evidence to support their explanations.
- Discuss the difference between the complete metamorphosis of a butterfly and the incomplete metamorphosis of a grasshopper or a dragonfly. Guide students to understand that a dragonfly and grasshopper keep a similar shape throughout each stage of life. The butterfly changes completely.
- Encourage students to share and evaluate each other's comparisons and ask questions. Ask students to identify similarities and differences as they compare results and ideas. They should adjust their conclusions if necessary.

Answers

1. Possible answer: During the pupa stage, the larva grows wings and its body changes into an adult butterfly.

What to Do, continued

4. Your teacher will pin the paper circle with the pupae to the butterfly habitat. Observe the pupae each day. Record your observations.

5. When the pupae become adults, observe and draw the butterflies. Record your observations.

6. Use your observations and drawings to make a diagram of the life cycle of the butterfly. Label the stages **Egg, Larva, Pupa,** and **Adult.**
7. Choose one of the Insect Life Cycle diagrams. **Compare** the life cycle in the diagram you chose with the butterfly life cycle you drew.
8. **Share** your observations of the diagrams with other groups. Did you observe the same similarities and differences between the insect life cycles?

34

MANAGING THE INVESTIGATION

Teaching Tips

- Keep the butterflies away from classroom heaters and direct sunlight.
- After a chrysalis forms, carefully transfer it to the butterfly habitat. Remove the paper from the plastic cup, remove excess silk, and use a safety pin to secure the paper to the inside of the habitat.
- If you plan to keep the adult butterflies, place an orange wedge or a facial tissue or carnation soaked in sugar water in the butterfly habitat.
- Adult butterflies live for 2–4 weeks. To protect native wildlife, do not release the butterflies outdoors. Maintain the butterflies in the classroom or place them in a sealed container, freeze, then dispose.

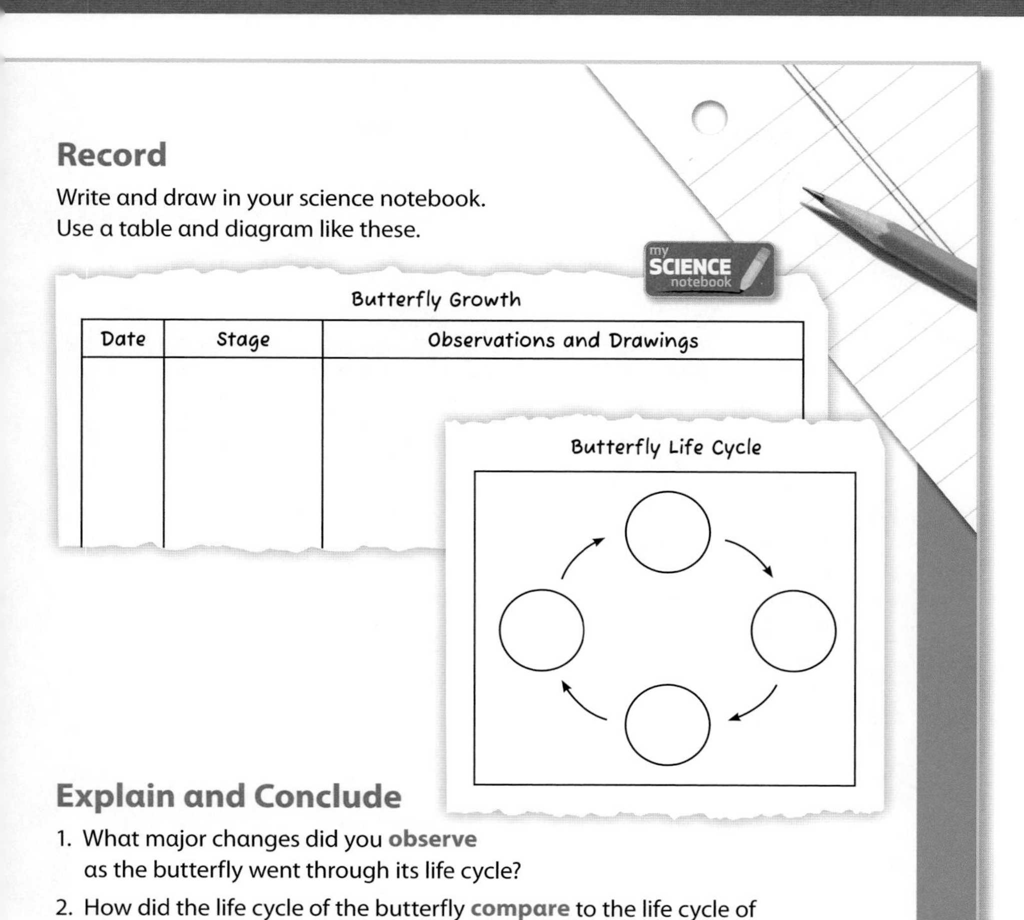
Record

Write and draw in your science notebook. Use a table and diagram like these.

Butterfly Growth

Date	Stage	Observations and Drawings

Explain and Conclude

1. What major changes did you **observe** as the butterfly went through its life cycle?
2. How did the life cycle of the butterfly **compare** to the life cycle of the grasshopper or dragonfly? Which insect undergoes complete metamorphosis? Which undergoes incomplete metamorphosis?

Think of Another Question

What else would you like to find out about the life cycles of butterflies, grasshoppers, and dragonflies?

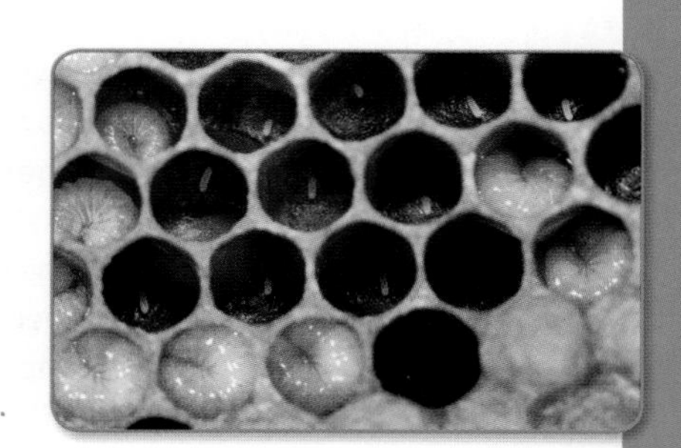

These honeybee larvae will change into pupae before becoming adult honeybees.

35

Explain and Conclude, continued

2. A butterfly larva that hatches from an egg looks completely different from its adult form. The grasshopper and dragonfly nymphs that hatch from the eggs look similar to the adult. The butterfly undergoes complete metamorphosis. The grasshopper and the dragonfly undergo incomplete metamorphosis.

5 Find Out More

Think of Another Question

Students should generate other questions that could be studied using readily available materials. Students may ask: *How can I study complete metamorphosis in other insects?* Record student questions. Discuss how to find answers.

6 Reflect and Assess

- To assess student work with the Inquiry Rubric shown below, see Assessment Handbook, page 183, or go online at myNGconnect.com
- Have students use the Inquiry Self-Reflection on Assessment Handbook, page 192, or at myNGconnect.com

Inquiry Rubric at myNGconnect.com	Scale			
The student recorded **observations** of butterfly growth and behavior.	4	3	2	1
The student made a diagram of a butterfly life cycle.	4	3	2	1
The student **compared** the life cycle of a butterfly with that of a grasshopper or a dragonfly.	4	3	2	1
The student compared observations and diagrams with other students.	4	3	2	1
The student **shared** results and **conclusions** with other students.	4	3	2	1
Overall Score	4	3	2	1

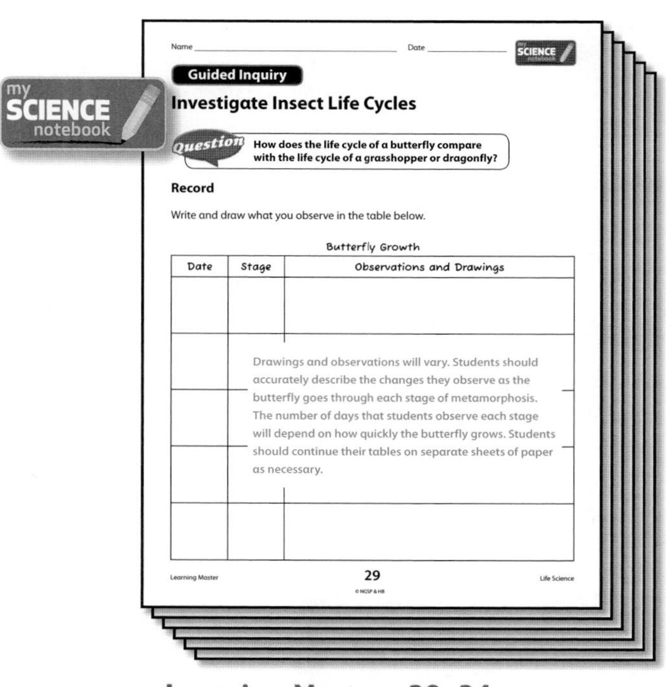
my SCIENCE notebook

Name ____ Date ____

Guided Inquiry

Investigate Insect Life Cycles

Question How does the life cycle of a butterfly compare with the life cycle of a grasshopper or dragonfly?

Record

Write and draw what you observe in the table below.

Butterfly Growth

Date	Stage	Observations and Drawings
		Drawings and observations will vary. Students should accurately describe the changes they observe as the butterfly goes through each stage of metamorphosis. The number of days that students observe each stage will depend on how quickly the butterfly grows. Students should continue their tables on separate sheets of paper as necessary.

Learning Master 29 Life Science

Learning Masters 29–34 or at myNGconnect.com

LESSON 8 ▫ Conclusion and Review

PROGRAM RESOURCES

- Chapter 2 Test, Assessment Handbook, pages 8–10, or at myNGconnect.com
- NGSP ExamView CD-ROM

1 Sum Up the Big Idea

- Display the chart from page T48–T49. Read the Big Idea Question. Then add new information to the chart.
- Ask: **What did you find out in each section? Is this what you expected to find?**

How Do Animals Grow and Change?

Animal Life Cycles

Mario: Animals pass through stages of growth and development during their life cycle.

Metamorphosis

Eve: Some animals undergo complete metamorphosis, while others undergo incomplete metamorphosis.

Inherited Traits

Jamil: Many animal traits are inherited, or passed from the parents to the offspring.

Acquired Traits

Michiko: Some traits are acquired or gained during an animal's life, such as a missing horn or a scar.

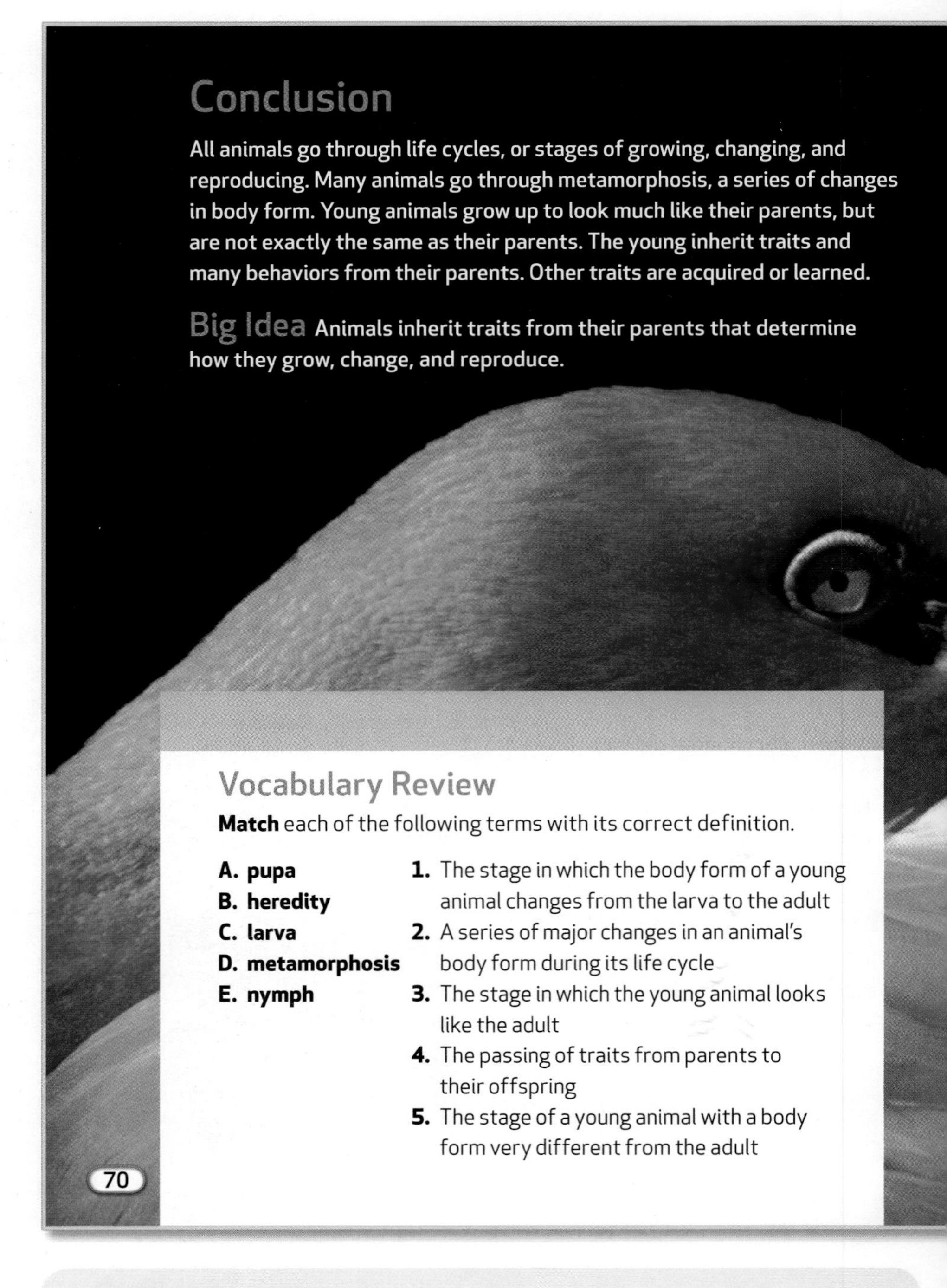

Conclusion

All animals go through life cycles, or stages of growing, changing, and reproducing. Many animals go through metamorphosis, a series of changes in body form. Young animals grow up to look much like their parents, but are not exactly the same as their parents. The young inherit traits and many behaviors from their parents. Other traits are acquired or learned.

Big Idea Animals inherit traits from their parents that determine how they grow, change, and reproduce.

Vocabulary Review

Match each of the following terms with its correct definition.

A. pupa
B. heredity
C. larva
D. metamorphosis
E. nymph

1. The stage in which the body form of a young animal changes from the larva to the adult
2. A series of major changes in an animal's body form during its life cycle
3. The stage in which the young animal looks like the adult
4. The passing of traits from parents to their offspring
5. The stage of a young animal with a body form very different from the adult

70

Review Academic Vocabulary

Academic Vocabulary Tell students that scientists use special words when describing their work to others. For example:

metamorphosis **larva** **pupa**
nymph **heredity**

Have students write the list in their science notebook and use each word in a sentence that tells about how animals grow and change. They should then share their sentences with a partner.

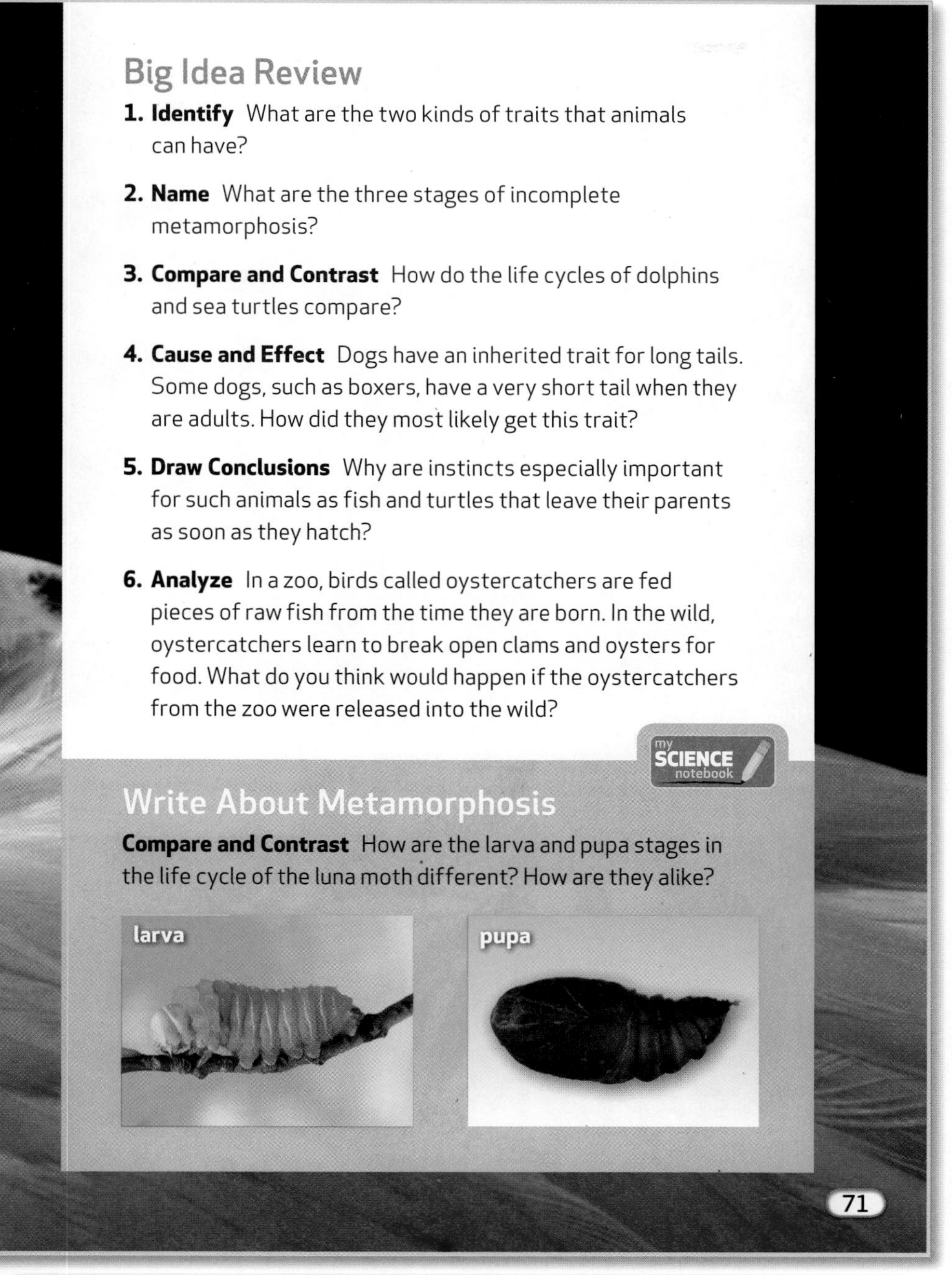
Big Idea Review

1. **Identify** What are the two kinds of traits that animals can have?
2. **Name** What are the three stages of incomplete metamorphosis?
3. **Compare and Contrast** How do the life cycles of dolphins and sea turtles compare?
4. **Cause and Effect** Dogs have an inherited trait for long tails. Some dogs, such as boxers, have a very short tail when they are adults. How did they most likely get this trait?
5. **Draw Conclusions** Why are instincts especially important for such animals as fish and turtles that leave their parents as soon as they hatch?
6. **Analyze** In a zoo, birds called oystercatchers are fed pieces of raw fish from the time they are born. In the wild, oystercatchers learn to break open clams and oysters for food. What do you think would happen if the oystercatchers from the zoo were released into the wild?

Write About Metamorphosis

Compare and Contrast How are the larva and pupa stages in the life cycle of the luna moth different? How are they alike?

71

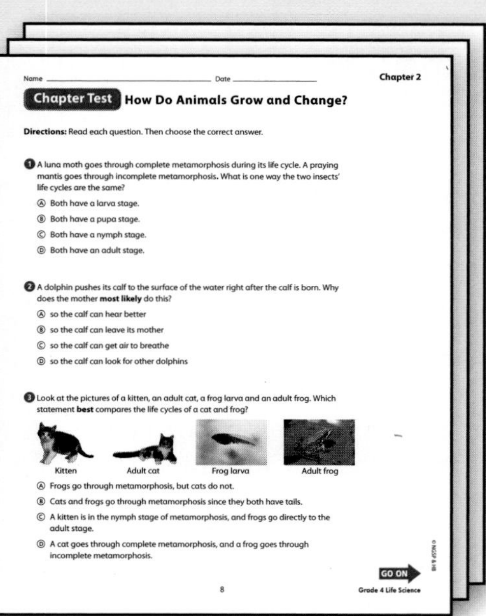
Name ____ Date ____ Chapter 2

Chapter Test How Do Animals Grow and Change?

Directions: Read each question. Then choose the correct answer.

1. A luna moth goes through complete metamorphosis during its life cycle. A praying mantis goes through incomplete metamorphosis. What is one way the two insects' life cycles are the same?
   - Ⓐ Both have a larva stage.
   - Ⓑ Both have a pupa stage.
   - Ⓒ Both have a nymph stage.
   - Ⓓ Both have an adult stage.

2. A dolphin pushes its calf to the surface of the water right after the calf is born. Why does the mother **most likely** do this?
   - Ⓐ so the calf can hear better
   - Ⓑ so the calf can leave its mother
   - Ⓒ so the calf can get air to breathe
   - Ⓓ so the calf can look for other dolphins

3. Look at the pictures of a kitten, an adult cat, a frog larva and an adult frog. Which statement **best** compares the life cycles of a cat and frog?

   Kitten Adult cat Frog larva Adult frog

   - Ⓐ Frogs go through metamorphosis, but cats do not.
   - Ⓑ Cats and frogs go through metamorphosis since they both have tails.
   - Ⓒ A kitten is in the nymph stage of metamorphosis, and frogs go directly to the adult stage.
   - Ⓓ A cat goes through complete metamorphosis, and a frog goes through incomplete metamorphosis.

GO ON

8 Grade 4 Life Science

© NGSP & HB

Assess Student Progress Chapter 2 Test

Have students complete their Chapter 2 Test to assess their progress in this chapter.

Chapter 2 Test, Assessment Handbook, pages 8–10, or at myNGconnect.com

or NGSP ExamView CD-ROM

❷ Discuss the Big Idea

Notetaking

Have students write in their science notebook to show what they know about the Big Idea. Have them:

1. Write about how animals pass through different stages as they complete their life cycles.
2. Explain the difference between inherited traits and behaviors and acquired traits and behaviors.
3. Write about how the traits of animals are used to help humans.

❸ Assess the Big Idea

Vocabulary Review

1. A **2.** D **3.** E **4.** B **5.** C

Big Idea Review

1. inherited and acquired
2. The three stages are egg, nymph, and adult.
3. In both life cycles, the young grow up in the ocean to look very much like their parents. They are different in that turtles lay eggs on land, and the parents do not care for their offspring. Dolphins are born live in the water, and the mother feeds and raises her young.
4. The very short tail is most likely an acquired trait.
5. Animals that are not cared for by their parents have little or no opportunity to learn behaviors that would help them survive. Instincts, or inborn behaviors, are very important.
6. They would probably die since they would not have learned this behavior.

Write About Metamorphosis

The larva and pupa might look about the same size. They also look like they have different parts or segments. But the caterpillar can move around while the pupa does not. The larva is a stage of the animal that looks very different from the adult. The pupa is the stage in which the larva changes form into the adult.

Objectives

Students will be able to:

- Research biographical information about various scientists and inventors from different gender and ethnic backgrounds, and describe how their work contributed to science and technology.

PROGRAM RESOURCES

- Big Ideas Book: *Life Science*
- Big Ideas Book: *Life Science* eEdition at myNGconnect.com
- Digital Library at myNGconnect.com

❶ Introduce

Tap Prior Knowledge

- Have students observe the photo of the adult monarch butterfly on page 72. Then ask them to compare the monarch to other butterflies and moths that they have observed. Lead students to recognize that there is a wide variety of different kinds of butterflies and moths.

Preview and Read

- Read the heading. Have students preview the photos on pages 72–73. Ask them to think about the butterflies they have seen.
- Have students read pages 72–73.

❷ Teach

Digital Library

myNGconnect.com

Have students use the Digital Library to find images of monarch butterflies.

Integrated Technology

Digital Booklet Students can use the information to make a digital booklet with captions that describes observations or inferences about monarch butterflies.

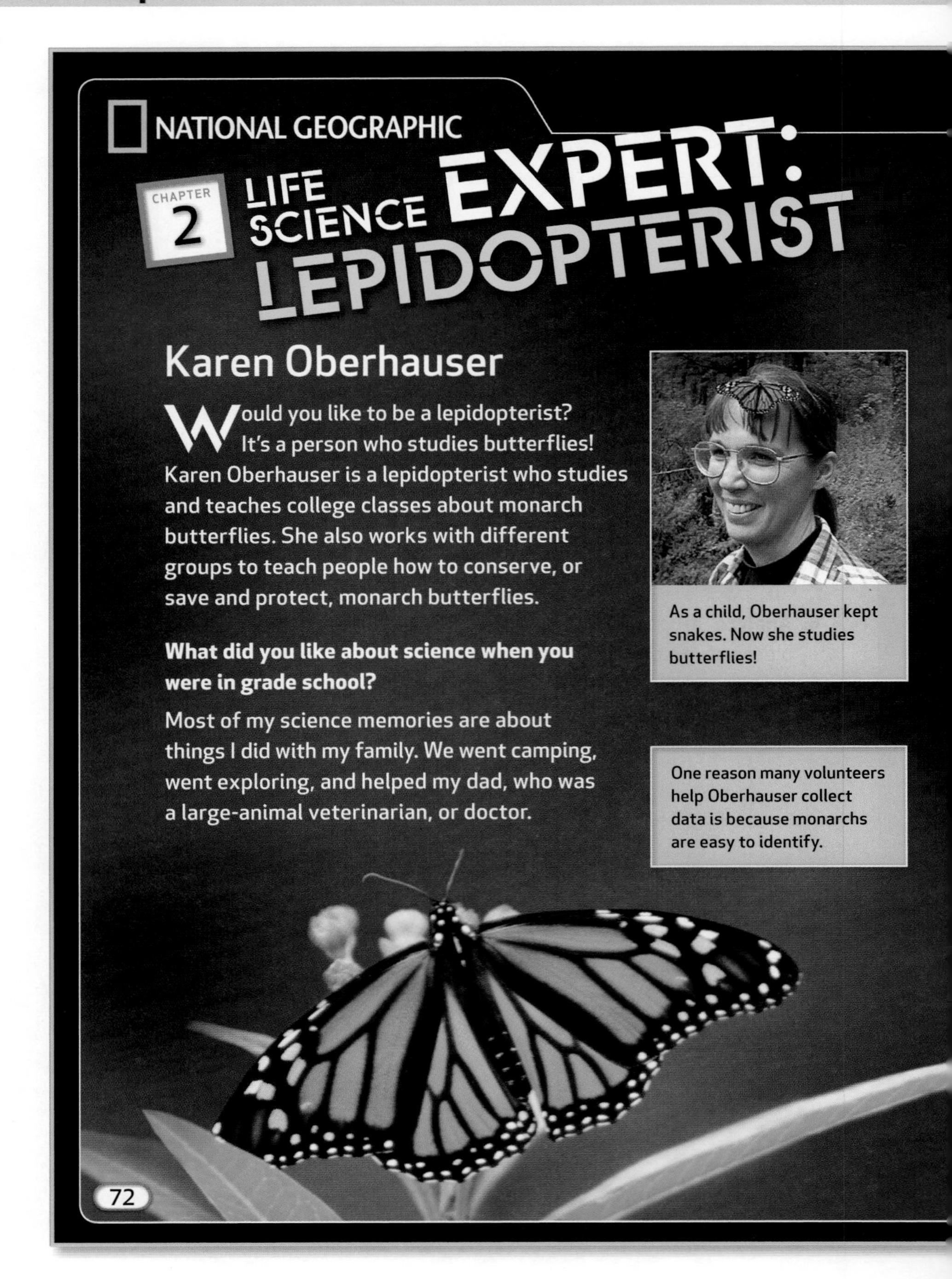

NATIONAL GEOGRAPHIC
CHAPTER 2
LIFE SCIENCE EXPERT: LEPIDOPTERIST
Karen Oberhauser
Would you like to be a lepidopterist? It's a person who studies butterflies! Karen Oberhauser is a lepidopterist who studies and teaches college classes about monarch butterflies. She also works with different groups to teach people how to conserve, or save and protect, monarch butterflies.
What did you like about science when you were in grade school?
Most of my science memories are about things I did with my family. We went camping, went exploring, and helped my dad, who was a large-animal veterinarian, or doctor.
As a child, Oberhauser kept snakes. Now she studies butterflies!
One reason many volunteers help Oberhauser collect data is because monarchs are easy to identify.
72

NATIONAL GEOGRAPHIC Raise Your SciQ!

Monarch Butterflies In summer, monarch butterflies live in open fields and meadows. Although adults feed on nectar from flowers, the larvae feed on milkweed. Milkweed contains chemicals that make the larvae taste horrible to predators. This helps monarchs avoid being eaten.

Monarchs migrate between their summer homes in the north and their winter homes in Mexico and other places in the south. But individual monarchs do not complete a round-trip migration journey. Instead, their offspring complete the trip. Migration and the migratory path among monarchs is an example of an instinctive behavior.

What would you say has been the most interesting part of your job?
I loved having a job that I could share with my daughters as they were growing up. When you study something such as monarch butterflies, it's fun for kids to be involved.

Have there been any big surprises in your job?
I've studied many features of monarch butterflies, from how they reproduce to how they migrate. It's been fun to become an expert on this butterfly and how it fits into its environment. I've also loved being a part of teaching people how important it is to preserve the natural world for the future.

What has been your greatest accomplishment in your job?
I'm proud that I have worked hard to plan programs that conserve habitats of monarch butterflies and other animals. I have worked to get people all over the United States to take part in collecting data about monarch butterflies. My goal has been to conserve as much of the natural world as possible. I feel that one way I have done that is by helping people see the wonder of the natural environments in this world.

Monarch wings are made of thousands of tiny overlapping scales, like shingles on a roof.

73

Differentiated Instruction

ELL Language Support for Exploring a Career in the Life Sciences

BEGINNING

Help students describe the work of Karen Oberhauser by asking questions, such as:

What do lepidopterists study? Where are monarch butterflies found?

INTERMEDIATE

Help students describe lepidopterists using Academic Language Frames.

- *Lepidopterists study ____.*
- *To ____ monarchs is to save and protect them.*
- *Monarchs travel from one place to another when they ____.*

ADVANCED

Have students write new captions for the photos on pages 72–73. Tell them that the captions should describe monarchs, their stages of growth and development, or the work of lepidopterists.

❷ Teach

Analyze Word Parts

- Write *lepido-*, *ptera*, and *lepidoptera* on the board. Under *lepido-*, write "means scale, like fish scale." Under *ptera*, write "means wing."
- *Say:* **So, *lepidoptera* means "scale wing." Insects with four scaly wings, such as butterflies, are classified as Lepidoptera. A person that studies these insects is called a lepidopterist.**

Explore a Career That Studies Life Cycles

- Ask: **What does Karen Oberhauser consider the most interesting part of her job?** (sharing her job with her daughters)
- Ask: **What surprised her about her job?** (She was surprised at how fun it was to study monarch butterflies and to teach other people about them.)
- Ask: **What does she consider her greatest accomplishment?** (conserving habitats of butterflies and getting other people to take part in the process)

Find Out More

- Ask: **Is this a career you would find interesting? Why or why not?** (Accept all answers, positive or negative.)
- Encourage interested students to research to find more information about science careers that involve animals.

❸ Assess

1. **Recall What does a lepidopterist do?** (studies butterflies)
2. **Explain Why does Karen Oberhauser share her work with her daughters?** (Both she and her daughters enjoy butterflies, and she wants to pass on her knowledge to young people so that they too can help protect the natural world.)
3. **Draw Conclusions Why is Karen Oberhauser interested in all living things, not just the monarch butterflies that she studies?** (She's been interested in living things since she was a young girl, and her goal today is to conserve as much of the natural world as possible. Butterflies are only one part of this world.)

NATIONAL GEOGRAPHIC

BECOME AN EXPERT

TECHTREK
myNGconnect.com
Student eEdition
Digital Library

Nature's Transformers

Many kinds of animals are transformers! They change during different stages in their lives. As young, they can look completely different than they do as adults. This metamorphosis occurs not just in insects such as beetles and dragonflies, but other kinds of animals, too. Frogs, sea stars, and lobsters are some other kinds of animals that go through metamorphosis.

Squat lobsters live in the ocean in many places around the world. Look at how the squat lobster changes as it grows from a young lobster to an adult. Find a feature that both forms have, such as claws. Then tell how that feature is different in the two forms.

Adult squat lobsters live under stones and rocks on the shallow areas of the ocean floor near coastlines.

Identify some of the features of this young squat lobster.

metamorphosis
Metamorphosis is a series of major changes in an animal's body form during its life cycle.

74

75

PROGRAM RESOURCES

- Big Ideas Book: *Life Science*
- Big Ideas Book: *Life Science* eEdition at myNGconnect.com
- Digital Library at myNGconnect.com

Access Science Content

Identify Animals That Undergo Metamorphosis

- Have students observe the photos on pages 74–75. Point out the two life cycle stages of the squat lobster. Ask: **How do the two stages of the squat lobster compare?** (The young squat lobster has a pinkish color and translucent legs and claws. The adult has a brown body and thicker legs and claws.) Discuss how the squat lobster undergoes **metamorphosis.** Have students read the definition at the bottom of page 74.
- Have students read pages 74–75. Discuss how the title of the feature—Nature's Transformers—refers to **metamorphosis.**

Notetaking

Have students study the photos on pages 74–75. Have them make drawings showing the two stages of **metamorphosis** in the life cycle of the squat lobster.

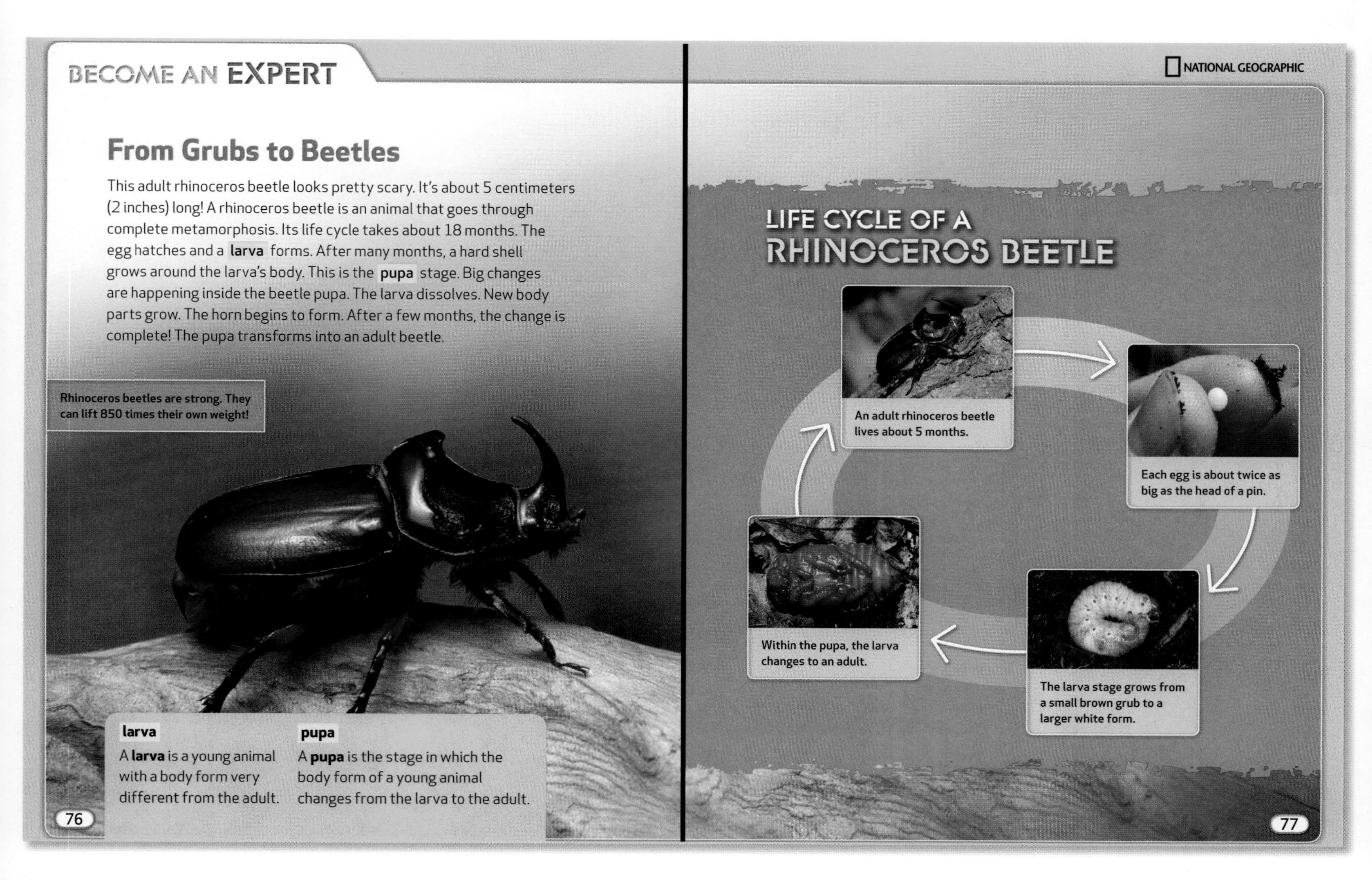
BECOME AN EXPERT

From Grubs to Beetles

This adult rhinoceros beetle looks pretty scary. It's about 5 centimeters (2 inches) long! A rhinoceros beetle is an animal that goes through complete metamorphosis. Its life cycle takes about 18 months. The egg hatches and a **larva** forms. After many months, a hard shell grows around the larva's body. This is the **pupa** stage. Big changes are happening inside the beetle pupa. The larva dissolves. New body parts grow. The horn begins to form. After a few months, the change is complete! The pupa transforms into an adult beetle.

Rhinoceros beetles are strong. They can lift 850 times their own weight!

larva
A **larva** is a young animal with a body form very different from the adult.

pupa
A **pupa** is the stage in which the body form of a young animal changes from the larva to the adult.

76

NATIONAL GEOGRAPHIC

LIFE CYCLE OF A RHINOCEROS BEETLE

An adult rhinoceros beetle lives about 5 months.

Each egg is about twice as big as the head of a pin.

The larva stage grows from a small brown grub to a larger white form.

Within the pupa, the larva changes to an adult.

77

Access Science Content

Sequence the Life Cycle of Rhinoceros Beetles

- Have students read pages 76–77, study the photographs, and read the captions. Point to the diagram and ask: **What are the stages of complete metamorphosis for the rhinoceros beetle?** (egg, larva, pupa, and adult) Have students read the definition of **larva** and **pupa** on the bottom of page 76. Ask: **Why is the life cycle of the rhinoceros beetle an example of metamorphosis?** (The beetle undergoes major changes as it passes through its life cycle.)

Differentiated Instruction

ELL Language Support for Metamorphosis

BEGINNING

Have students make flash cards for the egg, larva, pupa, and adult stages of the rhinoceros beetle. Ask partners to use the flash cards to quiz one another.

INTERMEDIATE

Provide Academic Language Frames, such as:

- *An egg hatches into a* ____.
- *The adult body forms during the* ____ *stage.*

ADVANCED

Have students write new captions for the four photos on page 77. Encourage students to use appropriate vocabulary words.

BECOME AN EXPERT
NATIONAL GEOGRAPHIC
From Nymphs to Dragonflies
Some insects, such as this dragonfly, go through incomplete metamorphosis. In this life cycle, there is no larva or pupa stage. A young dragonfly, or nymph, looks much like an adult, but it doesn't have wings. A nymph has a hard outer covering. As the nymph grows, it molts, or sheds, its covering. Some nymphs can live up to five years before they change into adults!
Metamorphosis is nearly complete. Soon a winged four-spotted chaser dragonfly will break out of its hard outer covering.
Notice how much longer the adult dragonfly is than the hard covering it just broke out of!
nymph
A nymph is the stage in which the young animal looks like the adult.
78
79

Access Science Content

Describe Metamorphosis in Dragonflies

- Have students read pages 78–79, observe the photos, and read the captions. Ask: **How do the life cycles of dragonflies and rhinoceros beetles compare?** (Both undergo metamorphosis, but of different types. The dragonfly undergoes incomplete metamorphosis, meaning it passes through a nymph stage but not the larva and pupa stages.) Discuss how the dragonfly **nymph** appears very similar to the adult, only smaller. Ask a volunteer to read the definition of **nymph** presented at the bottom of page 78.

Make Inferences

Ask students to closely observe the photo on page 79. Ask: **How could this large dragonfly fit into the hard brown shell?** (The body is soft, and the wings can fold up. These traits make it possible for the dragonfly to fit into such a tight space.)

Notetaking

Have students write down new words they encounter in their science notebook. For each word, ask them to write a descriptive sentence.

Assess

1. **Recall What kind of metamorphosis does a dragonfly undergo?** (incomplete metamorphosis)
2. **Compare and Contrast How is incomplete metamorphosis like complete metamorphosis? How is it different?** (Both involve major changes to an animal's body. Incomplete metamorphosis involves three stages, and the nymph stage resembles the adult. Complete metamorphosis involves four stages, each very different from the other.)

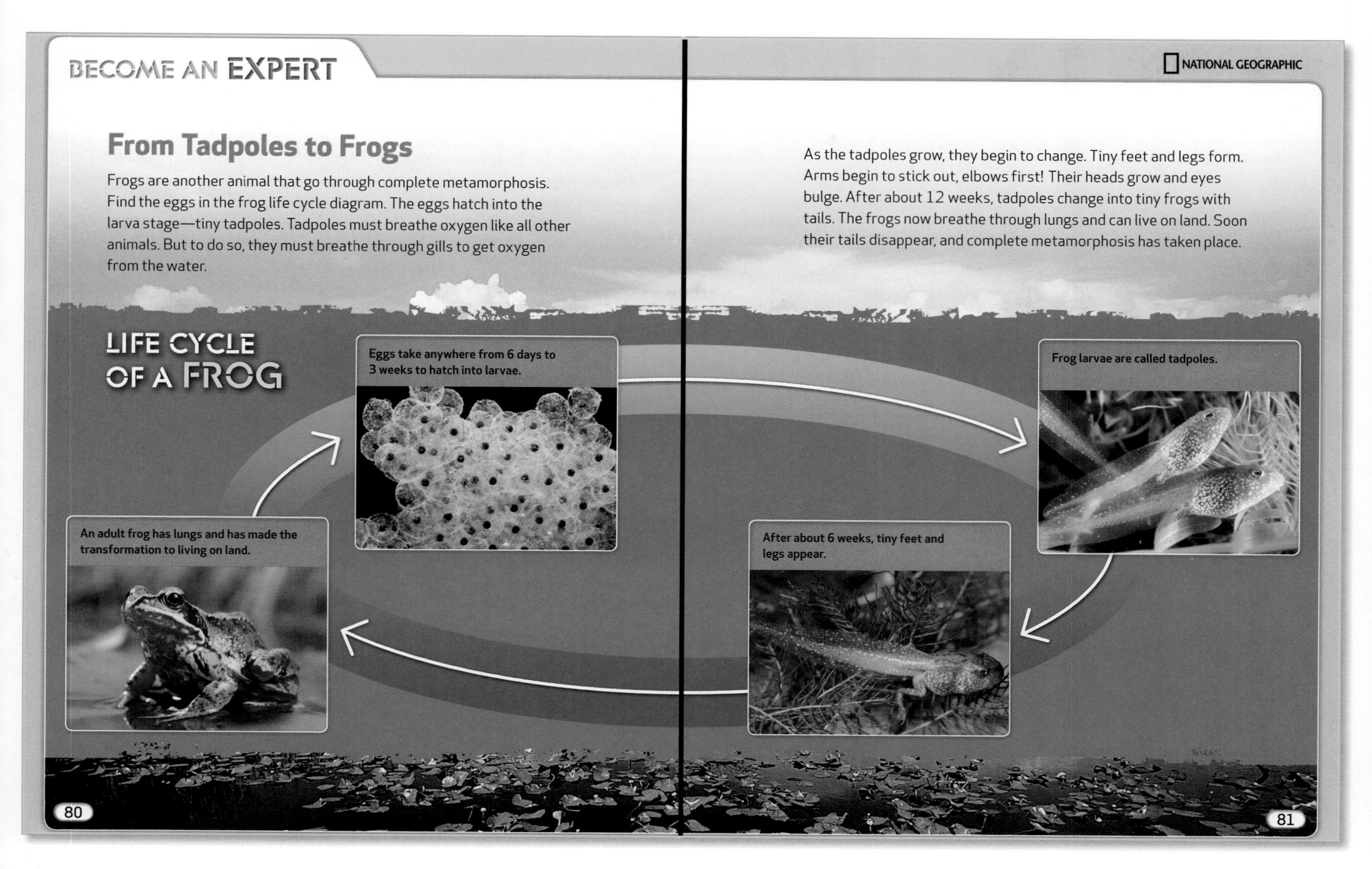

BECOME AN EXPERT
NATIONAL GEOGRAPHIC
From Tadpoles to Frogs
Frogs are another animal that go through complete metamorphosis. Find the eggs in the frog life cycle diagram. The eggs hatch into the larva stage—tiny tadpoles. Tadpoles must breathe oxygen like all other animals. But to do so, they must breathe through gills to get oxygen from the water.
As the tadpoles grow, they begin to change. Tiny feet and legs form. Arms begin to stick out, elbows first! Their heads grow and eyes bulge. After about 12 weeks, tadpoles change into tiny frogs with tails. The frogs now breathe through lungs and can live on land. Soon their tails disappear, and complete metamorphosis has taken place.
LIFE CYCLE OF A FROG
Eggs take anywhere from 6 days to 3 weeks to hatch into larvae.
Frog larvae are called tadpoles.
After about 6 weeks, tiny feet and legs appear.
An adult frog has lungs and has made the transformation to living on land.
80
81

Access Science Content

Sequence the Life Cycle of a Frog

- Have students read pages 80–81 and observe the diagram that spans the pages. Then have students describe each stage as they trace the life cycle with their fingers. Ask: **How are frog larvae different from adult frogs?** (Larvae breathe through gills, and they have tails without legs. Adults breathe through lungs and have legs.) Remind students of the definition of metamorphosis from page 74 and discuss how the frog life cycle meets this definition.

Monitor and Fix-Up

As the photos on pages 80–81 show, frog larvae live in water while adult frogs can breathe air and live on land. However, students may incorrectly believe that adult frogs avoid the water. Explain that many adult frogs spend much of their time in water, too. Female frogs must also return to the water to lay eggs.

LESSON 9 ▫ Become an Expert

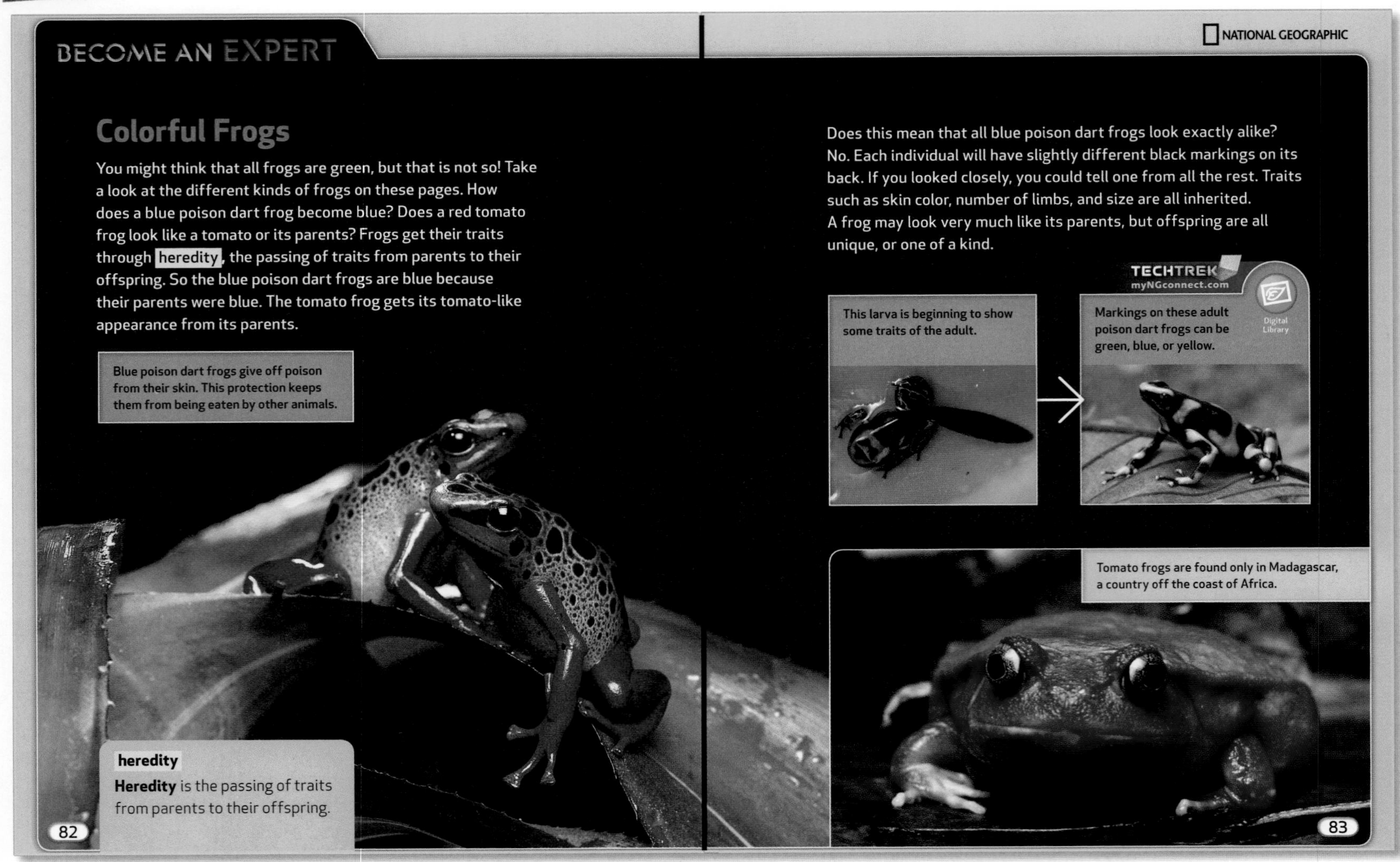

BECOME AN EXPERT

NATIONAL GEOGRAPHIC

Colorful Frogs

You might think that all frogs are green, but that is not so! Take a look at the different kinds of frogs on these pages. How does a blue poison dart frog become blue? Does a red tomato frog look like a tomato or its parents? Frogs get their traits through **heredity**, the passing of traits from parents to their offspring. So the blue poison dart frogs are blue because their parents were blue. The tomato frog gets its tomato-like appearance from its parents.

Blue poison dart frogs give off poison from their skin. This protection keeps them from being eaten by other animals.

heredity
Heredity is the passing of traits from parents to their offspring.

82

Does this mean that all blue poison dart frogs look exactly alike? No. Each individual will have slightly different black markings on its back. If you looked closely, you could tell one from all the rest. Traits such as skin color, number of limbs, and size are all inherited. A frog may look very much like its parents, but offspring are all unique, or one of a kind.

TECHTREK myNGconnect.com Digital Library

This larva is beginning to show some traits of the adult.

Markings on these adult poison dart frogs can be green, blue, or yellow.

Tomato frogs are found only in Madagascar, a country off the coast of Africa.

83

PROGRAM RESOURCES

- Digital Library at myNGconnect.com

Access Science Content

Explain Heredity in Frogs

- Have students read pages 82–83, study the photos, and read the captions. Point to the blue poison dart frogs and ask: **What traits did these frogs inherit from their parents?** (Possible answers: blue color and markings, body shape, number of eyes and limbs) Discuss how these traits make the blue poison dart frogs different from the other frogs shown in this section. Have a volunteer read the word **heredity** and its definition from the bottom of page 82.
- Point to the adult poison dart frog shown on page 83. Ask: **How can two adult poison dart frogs be different?** (Their markings may be different.) Discuss how in poison dart frogs, like other animal species, two individuals may have inherited many similar traits, but not all of their traits are the same.

myNGconnect.com

Have students use the Digital Library to find images of dart frogs and other amphibians.

Computer Presentation Students can use the information to make a computer presentation about amphibians.

Assess

1. **Sequence What are the stages in the life cycle of a frog?** (egg, larva or tadpole, tiny frogs with tails, and adult frogs)
2. **Infer At which stage of its life is a frog most like a small fish? Explain.** (during the larva or tadpole stage, when a frog has a thin body like a small fish and it breathes with gills)

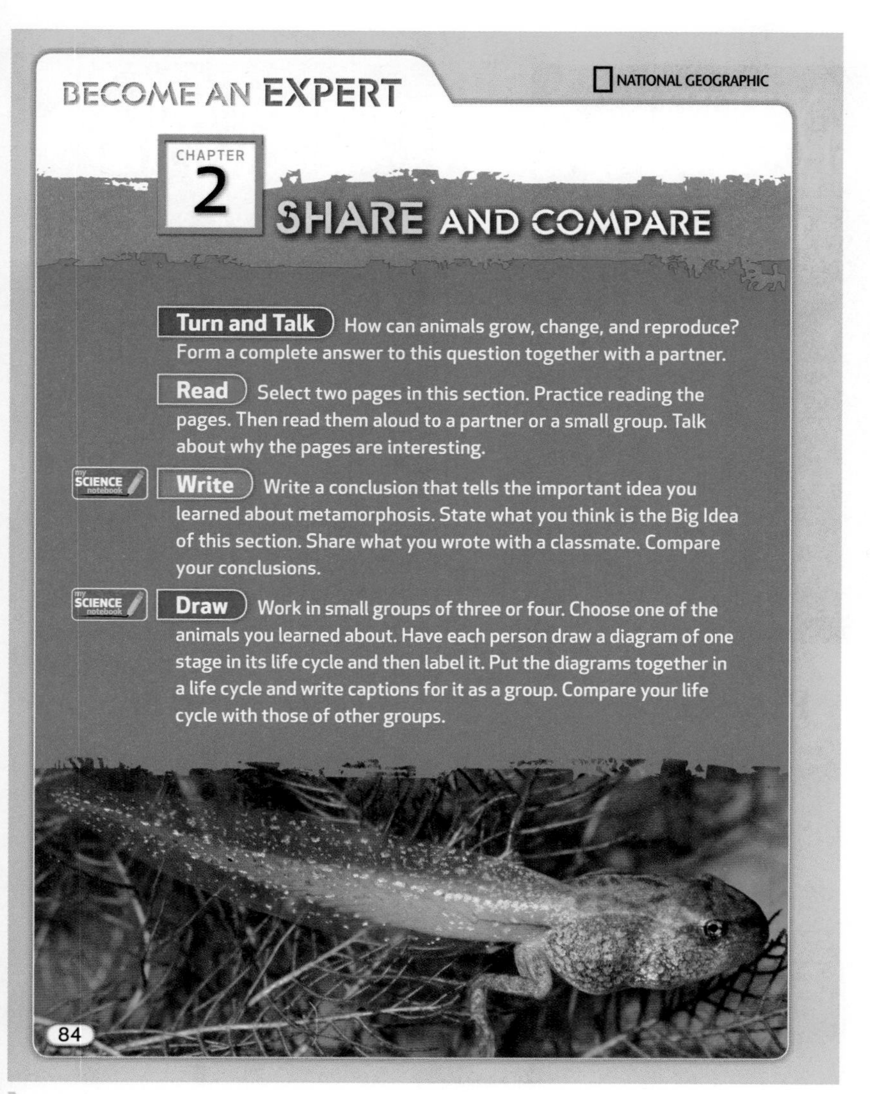

BECOME AN EXPERT
NATIONAL GEOGRAPHIC
CHAPTER
2
SHARE AND COMPARE
Turn and Talk How can animals grow, change, and reproduce? Form a complete answer to this question together with a partner.
Read Select two pages in this section. Practice reading the pages. Then read them aloud to a partner or a small group. Talk about why the pages are interesting.
Write Write a conclusion that tells the important idea you learned about metamorphosis. State what you think is the Big Idea of this section. Share what you wrote with a classmate. Compare your conclusions.
Draw Work in small groups of three or four. Choose one of the animals you learned about. Have each person draw a diagram of one stage in its life cycle and then label it. Put the diagrams together in a life cycle and write captions for it as a group. Compare your life cycle with those of other groups.
84

Sum Up

Tell students that to sum up a text helps them pull together the ideas of the text. Remind students that they summed up the Become an Expert lesson in the Write section of Share and Compare. Have students take turns reading the conclusions they wrote to a partner. Invite them to compare and contrast their conclusions.

Share and Compare

Turn and Talk

Ask students to turn to partners and talk about what they learned about the ways animals grow, change, and reproduce. Prompt students by asking:

1. **List What are the stages of complete and incomplete metamorphosis?** (complete: egg, larva, pupa, and adult; incomplete: egg, nymph, and adult)
2. **Contrast How are the nymph and larva stages different?** (A nymph resembles an adult, while a larva does not.)
3. **Summarize How can animals grow, change, and reproduce?** (Many animals undergo metamorphosis, a series of major changes in body form during their life cycle. Complete metamorphosis involves four stages that are very different from one another, while incomplete metamorphosis involves three stages in which the nymph resembles the adult.)

Read

Encourage students to familiarize themselves with any unfamiliar vocabulary words and then to speak these words slowly and carefully when they read them aloud. Also encourage students to speak honestly about why they liked or disliked the selections they read.

Students' conclusions should show an understanding of metamorphosis and the different types of changes it brings in an animal's life cycle.

Encourage group members to collaborate and cooperate so that their drawings work together to describe all stages of an animal's life cycle. Group members should also listen to each other's ideas before deciding on captions and labels to include.

Read Informational Text

Explore on Your Own books provide additional opportunities for your students to:

- deepen their science content knowledge even more as they focus on one Big Idea.
- independently apply multiple reading comprehension strategies as they read.

After applying the strategies throughout the chapter, students will independently read Explore on Your Own books. Facsimiles of the Explore on Your Own pages are shown on pages T84a–T84h.

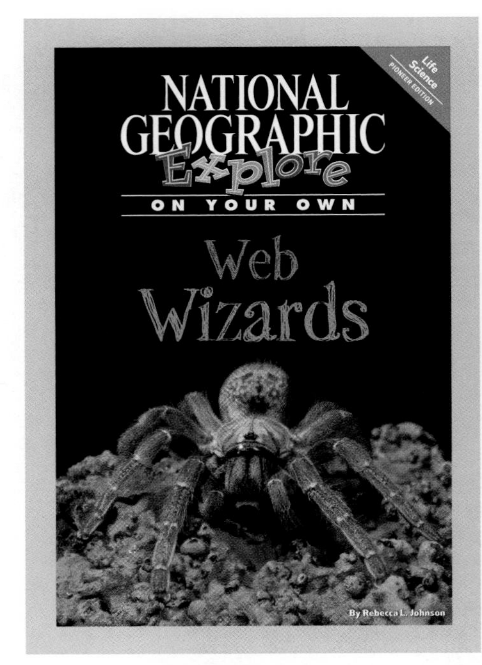

Pioneer

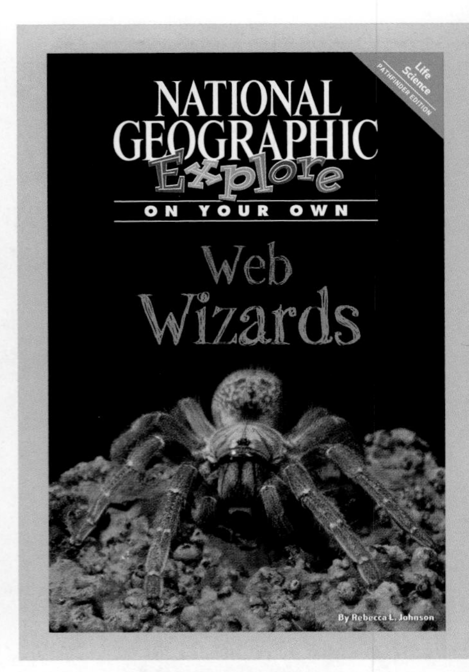

Pathfinder

Life Science
PIONEER EDITION

It's Time to Explore on Your Own!

Good readers use multiple strategies as they read on their own. Use the four key reading comprehension strategies below:

1 PREVIEW AND PREDICT
- Look over the text.
- Form ideas about how the text is organized and what it says.
- Confirm ideas about how the text is organized and what it says.

2 MONITOR AND FIX UP
- Think about whether the text is making sense and how it relates to what you know.
- Identify comprehension problems and clear up the problems.

3 MAKE INFERENCES
- Use what you know to figure out what is not said or shown directly.

4 SUM UP
- Pull together the text's big ideas.

Remember that you can choose different strategies at different times to help you understand what you are reading.

Web Wizards

PIONEER EDITION

By Rebecca L. Johnson

CONTENTS

Keeping Watch. A female fishing spider guards a sac filled with many tiny spider eggs.

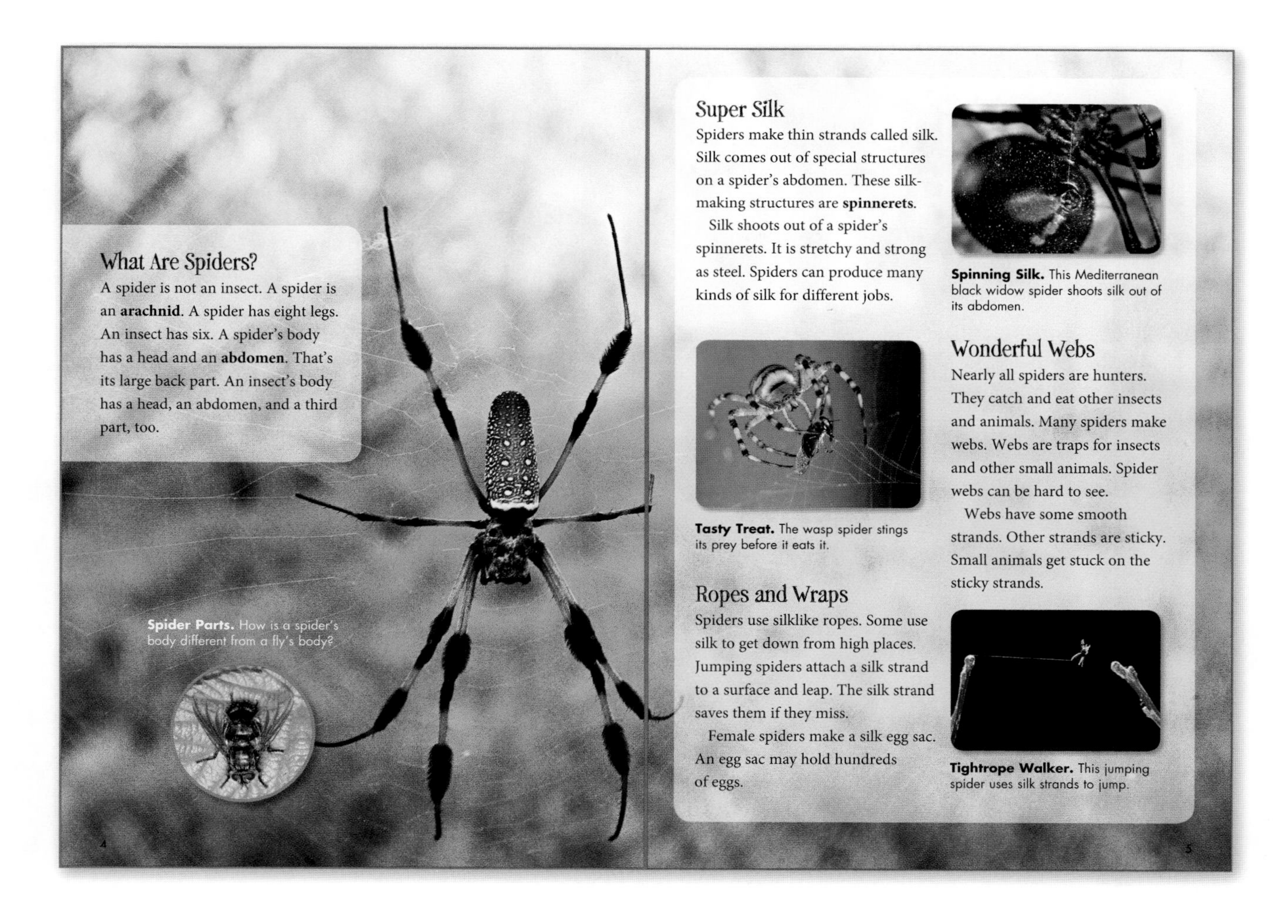

What Are Spiders?
A spider is not an insect. A spider is an arachnid. A spider has eight legs. An insect has six. A spider's body has a head and an abdomen. That's its large back part. An insect's body has a head, an abdomen, and a third part, too.
Spider Parts. How is a spider's body different from a fly's body?
Super Silk
Spiders make thin strands called silk. Silk comes out of special structures on a spider's abdomen. These silk-making structures are spinnerets.
Silk shoots out of a spider's spinnerets. It is stretchy and strong as steel. Spiders can produce many kinds of silk for different jobs.
Spinning Silk. This Mediterranean black widow spider shoots silk out of its abdomen.
Tasty Treat. The wasp spider stings its prey before it eats it.
Ropes and Wraps
Spiders use silklike ropes. Some use silk to get down from high places. Jumping spiders attach a silk strand to a surface and leap. The silk strand saves them if they miss.
Female spiders make a silk egg sac. An egg sac may hold hundreds of eggs.
Wonderful Webs
Nearly all spiders are hunters. They catch and eat other insects and animals. Many spiders make webs. Webs are traps for insects and other small animals. Spider webs can be hard to see.
Webs have some smooth strands. Other strands are sticky. Small animals get stuck on the sticky strands.
Tightrope Walker. This jumping spider uses silk strands to jump.
4
5

NATIONAL GEOGRAPHIC
SERIOUS SURVIVORS

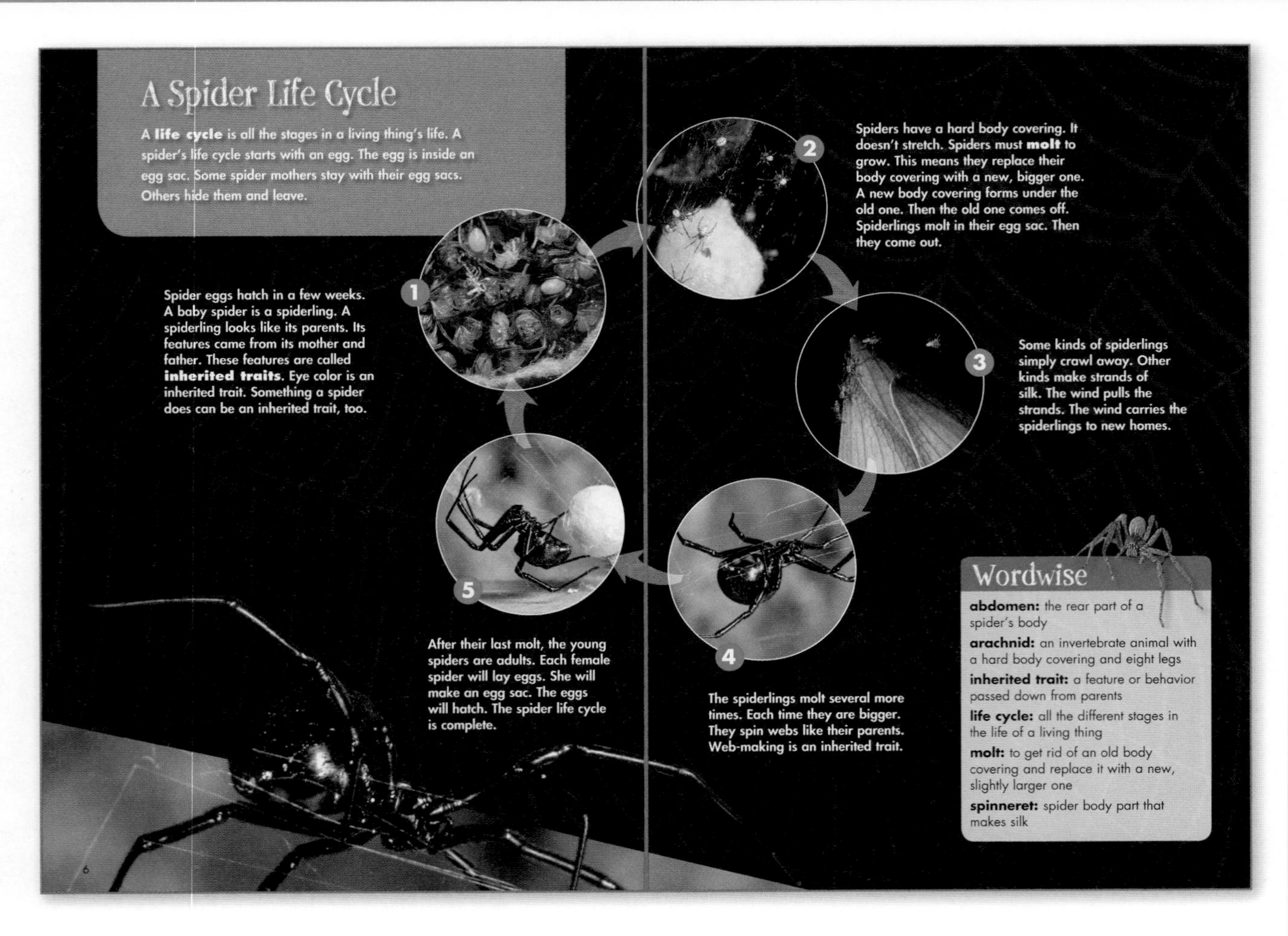
A Spider Life Cycle
A **life cycle** is all the stages in a living thing's life. A spider's life cycle starts with an egg. The egg is inside an egg sac. Some spider mothers stay with their egg sacs. Others hide them and leave.
1
Spider eggs hatch in a few weeks. A baby spider is a spiderling. A spiderling looks like its parents. Its features came from its mother and father. These features are called **inherited traits**. Eye color is an inherited trait. Something a spider does can be an inherited trait, too.
2
Spiders have a hard body covering. It doesn't stretch. Spiders must **molt** to grow. This means they replace their body covering with a new, bigger one. A new body covering forms under the old one. Then the old one comes off. Spiderlings molt in their egg sac. Then they come out.
3
Some kinds of spiderlings simply crawl away. Other kinds make strands of silk. The wind pulls the strands. The wind carries the spiderlings to new homes.
4
The spiderlings molt several more times. Each time they are bigger. They spin webs like their parents. Web-making is an inherited trait.
5
After their last molt, the young spiders are adults. Each female spider will lay eggs. She will make an egg sac. The eggs will hatch. The spider life cycle is complete.
Wordwise
abdomen: the rear part of a spider's body
arachnid: an invertebrate animal with a hard body covering and eight legs
inherited trait: a feature or behavior passed down from parents
life cycle: all the different stages in the life of a living thing
molt: to get rid of an old body covering and replace it with a new, slightly larger one
spinneret: spider body part that makes silk
6

A Spinning Sampler
Different spiders make different webs. They move silk strands around with their legs. They work quickly. They are careful not to touch the sticky strands!
Sheet Web
Sheet webs are flat sheets of silk strands. The spider hangs below the sheet. Insects land on top of it. The spider pulls the insects down through the sheet.
Orb Web
An orb web looks like a wheel. Some strands come out from the center. It has sticky strands that go around and around.
Tangle Web
Tangle webs look messy. The strands go every which way. Look for tangle webs in high corners.
Funnel Web
Funnel webs are wide on one end. The other end is narrow. The spider hides in the narrow part. Insects get caught in the wide part. Then the spider grabs them.
The Net
Net-casting spiders have long legs. They spin webs that they hold between their legs. Insects fly by. The spider tosses its net over them.
9

Veggie Spider!

Scientists thought that all spiders ate other animals. Then they discovered a small jumping spider in Central America. The spider eats plants.

The spider eats the tiny sweet tips of leaves. The leaves grow on a certain kind of plant. But ants guard it. They attack anything that tries to steal the leaf tips.

The spider watches the ants. Then it runs toward a leaf tip. The spider jumps away if ants move in. Or it leaps off the leaf and hangs on a silk strand. Usually, the spider gets away with a sweet leaf tip.

Standing Guard. This acacia ant is guarding the leaf tips on an acacia plant.

10

11

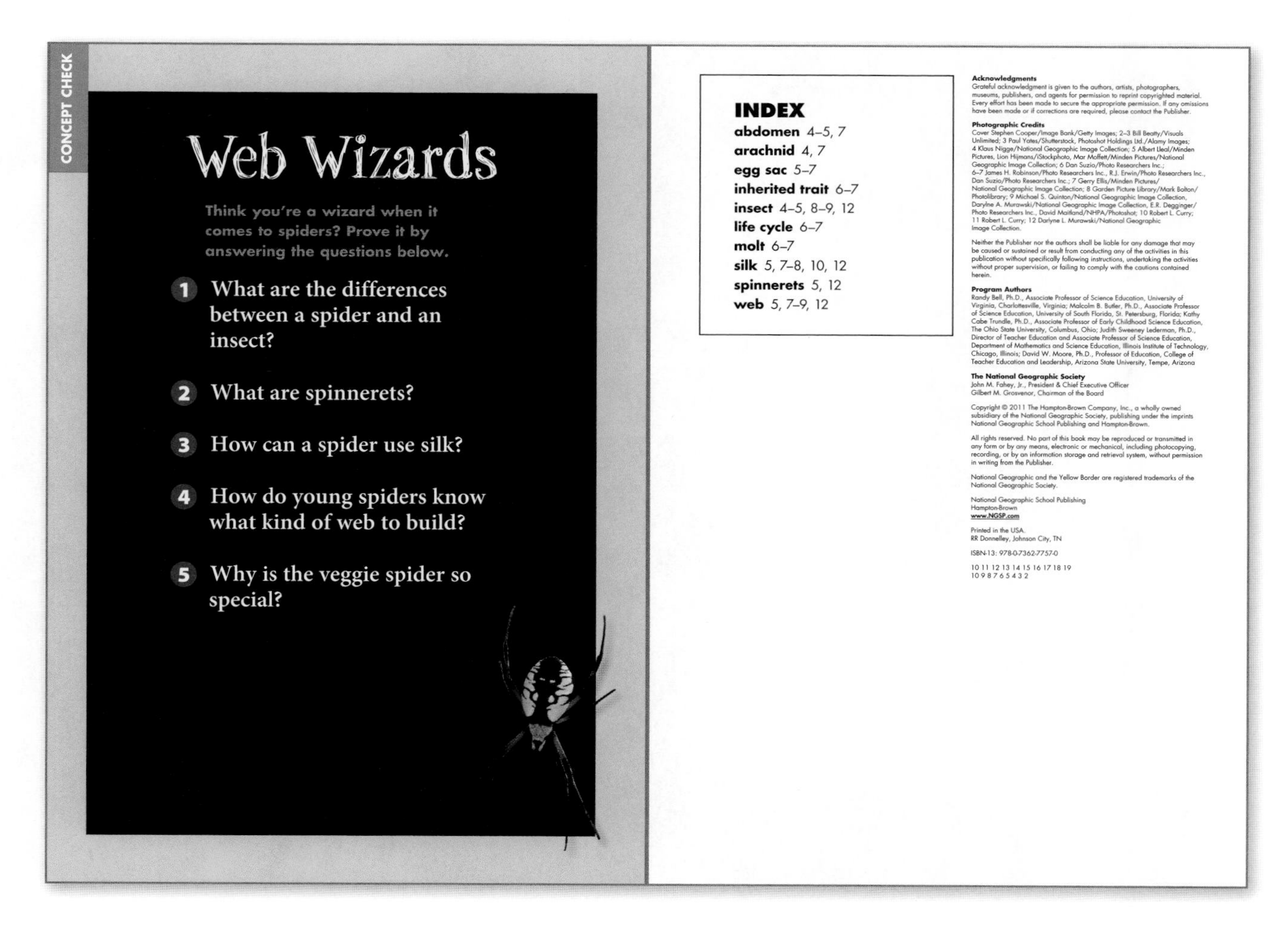

CONCEPT CHECK

Web Wizards

Think you're a wizard when it comes to spiders? Prove it by answering the questions below.

1. What are the differences between a spider and an insect?
2. What are spinnerets?
3. How can a spider use silk?
4. How do young spiders know what kind of web to build?
5. Why is the veggie spider so special?

INDEX

Acknowledgments
Grateful acknowledgment is given to the authors, artists, photographers, museums, publishers, and agents for permission to reprint copyrighted material. Every effort has been made to secure the appropriate permission. If any omissions have been made or if corrections are required, please contact the Publisher.

Photographic Credits
Cover Stephen Cooper/Image Bank/Getty Images; 2–3 Bill Beatty/Visuals Unlimited; 3 Paul Yates/Shutterstock, Photoshot Holdings Ltd./Alamy Images; 4 Klaus Nigge/National Geographic Image Collection; 5 Albert Lleal/Minden Pictures, Lion Hijmans/iStockphoto, Mar Moffett/Minden Pictures/National Geographic Image Collection; 6 Dan Suzio/Photo Researchers Inc.; 6–7 James H. Robinson/Photo Researchers Inc., R.J. Erwin/Photo Researchers Inc., Dan Suzio/Photo Researchers Inc.; 7 Gerry Ellis/Minden Pictures/ National Geographic Image Collection; 8 Garden Picture Library/Mark Bolton/ Photolibrary; 9 Michael S. Quinton/National Geographic Image Collection, Darylne A. Murawski/National Geographic Image Collection, E.R. Degginger/ Photo Researchers Inc., David Maitland/NHPA/Photoshot; 10 Robert L. Curry; 11 Robert L. Curry; 12 Darlyne L. Murawski/National Geographic Image Collection.

Program Authors
Randy Bell, Ph.D., Associate Professor of Science Education, University of Virginia, Charlottesville, Virginia; Malcolm B. Butler, Ph.D., Associate Professor of Science Education, University of South Florida, St. Petersburg, Florida; Kathy Cabe Trundle, Ph.D., Associate Professor of Early Childhood Science Education, The Ohio State University, Columbus, Ohio; Judith Sweeney Lederman, Ph.D., Director of Teacher Education and Associate Professor of Science Education, Department of Mathematics and Science Education, Illinois Institute of Technology, Chicago, Illinois; David W. Moore, Ph.D., Professor of Education, College of Teacher Education and Leadership, Arizona State University, Tempe, Arizona

National Geographic School Publishing
Hampton-Brown
www.NGSP.com

Printed in the USA.
RR Donnelley, Johnson City, TN

ISBN-13: 978-0-7362-7757-0

10 11 12 13 14 15 16 17 18 19
10 9 8 7 6 5 4 3 2

Life Science
PATHFINDER EDITION

It's Time to Explore on Your Own!

Good readers use multiple strategies as they read on their own. Use the four key reading comprehension strategies below:

1 PREVIEW AND PREDICT
- Look over the text.
- Form ideas about how the text is organized and what it says.
- Confirm ideas about how the text is organized and what it says.

2 MONITOR AND FIX UP
- Think about whether the text is making sense and how it relates to what you know.
- Identify comprehension problems and clear up the problems.

3 MAKE INFERENCES
- Use what you know to figure out what is not said or shown directly.

4 SUM UP
- Pull together the text's big ideas.

Remember that you can choose different strategies at different times to help you understand what you are reading.

NATIONAL GEOGRAPHIC
School Publishing

Web Wizards

PATHFINDER EDITION

By Rebecca L. Johnson

CONTENTS

Web Wizards
Spiders are everywhere. You can find them in dusty attics and damp basements. You can spot spiders in forests and fields. Earth is home to about 40,000 different kinds. They range from tiny jumping spiders to bird-eating spiders as big as your hand. Let's find out more about these amazing creatures.
By Rebecca L. Johnson
2
3

Keeping Watch. A female fishing spider guards a round, white sac. Inside the sac are many tiny spider eggs.

What Are Spiders?

Many people think spiders are insects. But a spider is an **arachnid**. How can you tell the difference? A spider has eight legs, while an insect has six. A spider's body has a head and an **abdomen**. That's its large back part. An insect's body has a head and an abdomen, too. But it also has a third part in between, called a thorax.

Spider Parts. You can see the spider's eight legs and two body parts. How is a spider's body different from a fly's body?

4

Super Silk

Spiders can make threadlike strands called silk. Silk comes out of special structures on a spider's abdomen. Each structure looks like a nozzle at the end of a hose. These silk-making structures are spinnerets.

Silk comes shooting out of a spider's **spinnerets**. It is both stretchy and strong. A strand of spider silk is about five times stronger than a steel wire with the same thickness. Spiders can produce as many as seven kinds of silk. They use different kinds for different jobs.

Spinning Silk. This Mediterranean black widow spider shoots silk out of its abdomen to spin a web.

Tasty Treat. The wasp spider stings its prey before eating it. The wasp spider's venom is poisonous to insects, but not people.

Wonderful Webs

Nearly all spiders are hunters. They need to catch and eat insects or other small animals for food. Many spiders use silk to make webs. Most webs are traps for insects and other small animals that spiders eat. Spiders are good at making webs that are hard to see. You know that's true if you've ever bumped into one!

Webs have some silk strands that are strong and smooth. But other strands are as sticky as glue. When insects or small animals fly into webs, they get stuck on these sticky strands.

Ropes and Wraps

Have you ever seen a spider hanging by a silk strand? Spiders also use silklike ropes. They use these silk ropes to get down from high places, almost like mountain climbers. Jumping spiders attach a silk strand to a surface before they leap. If they miss their target, this safety rope keeps them from crashing.

Female spiders use a special type of silk to wrap up their eggs. They make an egg sac in which their eggs will be safe. An egg sac can contain dozens, or even hundreds, of eggs.

Tightrope Walker. This jumping spider jumps from one plant to another using silk strands.

5

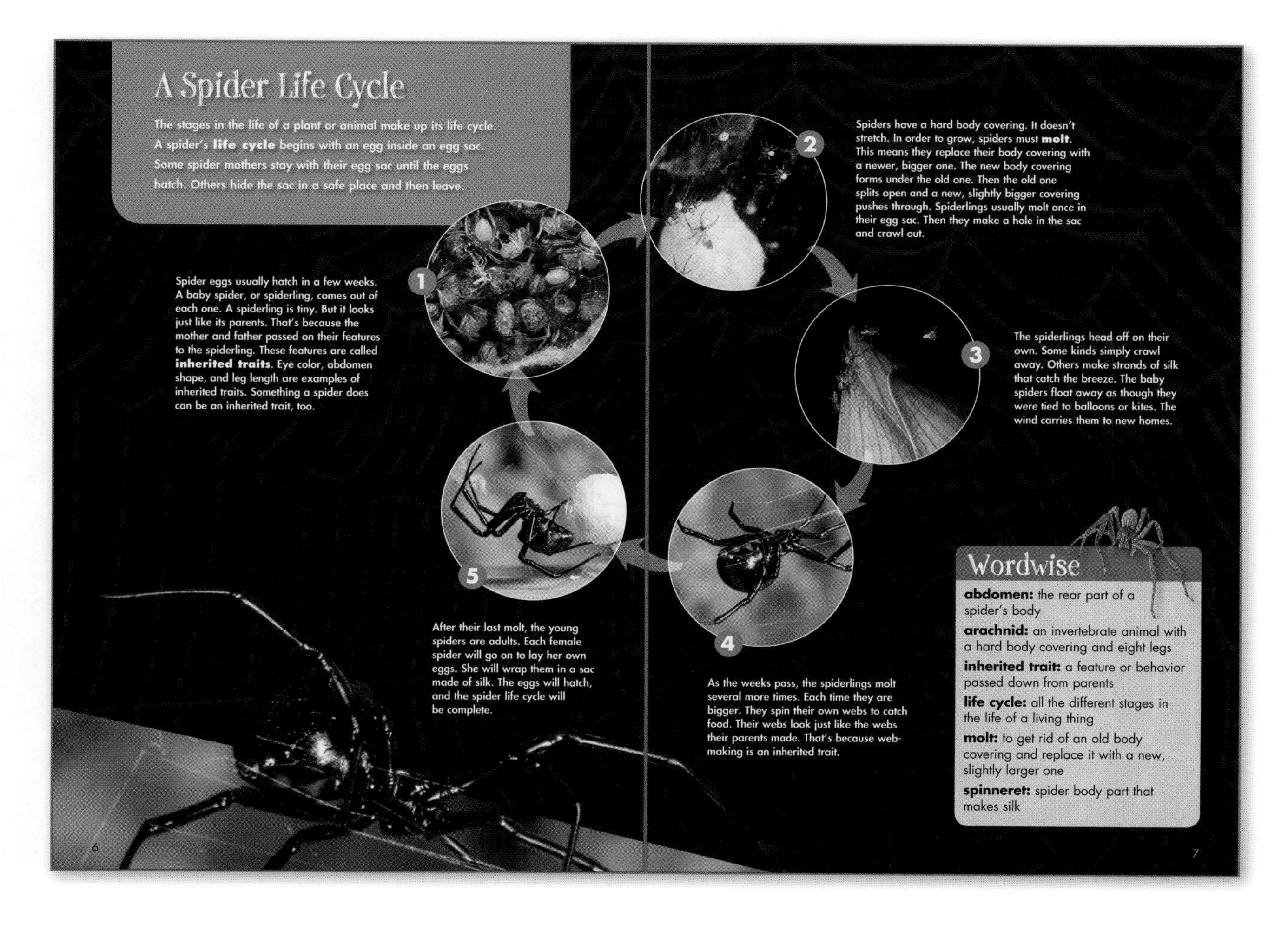

A Spider Life Cycle

The stages in the life of a plant or animal make up its life cycle. A spider's **life cycle** begins with an egg inside an egg sac. Some spider mothers stay with their egg sac until the eggs hatch. Others hide the sac in a safe place and then leave.

1 Spider eggs usually hatch in a few weeks. A baby spider, or spiderling, comes out of each one. A spiderling is tiny. But it looks just like its parents. That's because the mother and father passed on their features to the spiderling. These features are called **inherited traits**. Eye color, abdomen shape, and leg length are examples of inherited traits. Something a spider does can be an inherited trait, too.

2 Spiders have a hard body covering. It doesn't stretch. In order to grow, spiders must **molt**. This means they replace their body covering with a newer, bigger one. The new body covering forms under the old one. Then the old one splits open and a new, slightly bigger covering pushes through. Spiderlings usually molt once in their egg sac. Then they make a hole in the sac and crawl out.

3 The spiderlings head off on their own. Some kinds simply crawl away. Others make strands of silk that catch the breeze. The baby spiders float away as though they were tied to balloons or kites. The wind carries them to new homes.

4 As the weeks pass, the spiderlings molt several more times. Each time they are bigger. They spin their own webs to catch food. Their webs look just like the webs their parents made. That's because web-making is an inherited trait.

5 After their last molt, the young spiders are adults. Each female spider will go on to lay her own eggs. She will wrap them in a sac made of silk. The eggs will hatch, and the spider life cycle will be complete.

Wordwise

abdomen: the rear part of a spider's body

arachnid: an invertebrate animal with a hard body covering and eight legs

inherited trait: a feature or behavior passed down from parents

life cycle: all the different stages in the life of a living thing

molt: to get rid of an old body covering and replace it with a new, slightly larger one

spinneret: spider body part that makes silk

6

7

A Spinning Sampler

Different kinds of spiders make different kinds of webs. They twist and loop and weave silk strands with the tips of their legs. Web-making spiders work quickly. They are careful not to get caught on their own sticky strands!

Sheet Web

Sheet webs are flat sheets of silk strands. A spider that builds this kind of web hangs upside down beneath it. When an insect lands on top of the sheet, the spider moves fast. It grabs the insect and pulls it down through the sheet.

Orb Web

An orb web is very organized. It looks like a wheel with spokes. It also has sticky strands that circle around and around from the center to the edge. The spider often sits right in the middle of its web.

Tangle Web

Tangle webs look very messy. The silk strands go every which way. A good place to find tangle webs is in corners near the ceiling.

Funnel Web

Not surprisingly, funnel webs are shaped like a funnel. They are wide at one end and narrow at the other. The spider hides in the narrow part. Insects land on the wide part and get caught on sticky strands. The spider dashes out and grabs them.

The Net

Net-casting spiders have long legs and very good eyesight. They spin webs that they hold between their legs. The spider waits for insects to fly by. Then the spider tosses its net and catches them.

8 9

Veggie Spider!

For a long time, people thought all spiders ate other animals for food. Recently, scientists working in Central America got a surprise. They discovered a small jumping spider that eats plants. It's the first veggie spider known to science.

The spider eats the tiny sweet tips of leaves that grow on a certain kind of plant. Ants eat the sweet leaf tips, too. In fact, the ants guard the leaves. They will attack anything that tries to steal their favorite food.

The spider watches the ants from a safe spot. When the ants aren't paying attention, it runs toward a leaf tip. If the ants move in, the spider jumps out of reach. Or it may jump off the leaf and hang by a strand of silk. There, it is safely out of reach. But the veggie spider is a speedy spider. It is much faster than the ants. It gets away with a sweet leaf tip nearly every time.

Standing Guard. This acacia ant is guarding the leaf tips on an acacia plant. Will the veggie spider be able to outsmart the ant to get the leaf?

10 11

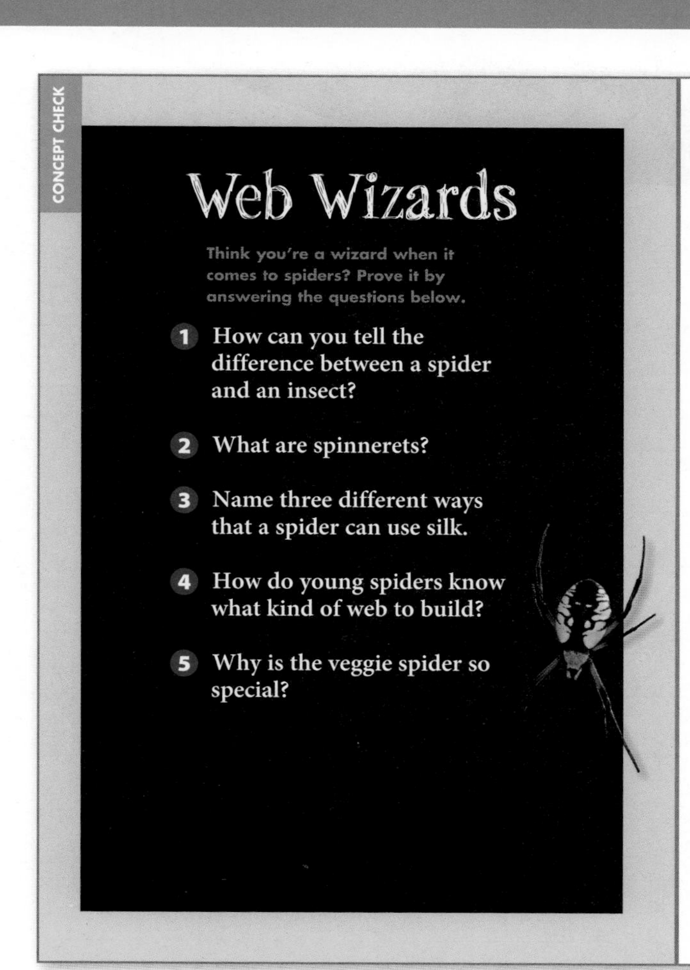
CONCEPT CHECK

Web Wizards

Think you're a wizard when it comes to spiders? Prove it by answering the questions below.

1. How can you tell the difference between a spider and an insect?
2. What are spinnerets?
3. Name three different ways that a spider can use silk.
4. How do young spiders know what kind of web to build?
5. Why is the veggie spider so special?

INDEX

Acknowledgments
Grateful acknowledgment is given to the authors, artists, photographers, museums, publishers, and agents for permission to reprint copyrighted material. Every effort has been made to secure the appropriate permission. If any omissions have been made or if corrections are required, please contact the Publisher.

Photographic Credits
Cover Stephen Cooper/Image Bank/Getty Images; 2–3 Bill Beatty/Visuals Unlimited; 3 Paul Yates/Shutterstock, Photoshot Holdings Ltd./Alamy Images; 4 Klaus Nigge/National Geographic Image Collection; 5 Albert Lleal/Minden Pictures, Lion Hijmans/iStockphoto, Mar Moffett/Minden Pictures/National Geographic Image Collection; 6 Dan Suzio/Photo Researchers Inc.; 6–7 James H. Robinson/Photo Researchers Inc., R.J. Erwin/Photo Researchers Inc., Dan Suzio/Photo Researchers Inc.; 7 Gerry Ellis/Minden Pictures/National Geographic Image Collection; 8 Garden Picture Library/Mark Bolton/Photolibrary; 9 Michael S. Quinton/National Geographic Image Collection, Darylne A. Murawski/National Geographic Image Collection, E.R. Degginger/Photo Researchers Inc., David Maitland/NHPA/Photoshot; 10 Robert L. Curry; 11 Robert L. Curry; 12 Darlyne L. Murawski/National Geographic Image Collection.

Program Authors
Randy Bell, Ph.D., Associate Professor of Science Education, University of Virginia, Charlottesville, Virginia; Malcolm B. Butler, Ph.D., Associate Professor of Science Education, University of South Florida, St. Petersburg, Florida; Kathy Cabe Trundle, Ph.D., Associate Professor of Early Childhood Science Education, The Ohio State University, Columbus, Ohio; Judith Sweeney Lederman, Ph.D., Director of Teacher Education and Associate Professor of Science Education, Department of Mathematics and Science Education, Illinois Institute of Technology, Chicago, Illinois; David W. Moore, Ph.D., Professor of Education, College of Teacher Education and Leadership, Arizona State University, Tempe, Arizona

National Geographic School Publishing
Hampton-Brown
www.NGSP.com

Printed in the USA.
RR Donnelley, Johnson City, TN

ISBN-13: 978-0-7362-7760-0

10 11 12 13 14 15 16 17 18 19
10 9 8 7 6 5 4 3 2

Notes

LIFE SCIENCE

CHAPTER
3
HOW DO LIVING THINGS
DEPEND ON THEIR
ENVIRONMENT?
TECHTREK
myNGconnect.com
The African savanna is rich in life. The cape buffalo, cattle egret, and grasses shown here are part of this richness. These living things depend on each other. Cape buffalo graze on the grasses. Cattle, egrets, and other birds find food by living closely with cape buffalo and other grazers. All of these living things depend on the sun. How?
86
87

After reading Chapter 3, you will be able to:

- Recognize and explain how animals and plants get the energy they need for life. **WHAT LIVING THINGS NEED FROM THEIR ENVIRONMENT**
- Classify living things as producers or different kinds of consumers. **COMMUNITIES OF LIVING THINGS**
- Distinguish among herbivores, carnivores, and omnivores, and give examples. **WHAT LIVING THINGS NEED FROM THEIR ENVIRONMENT**
- Describe how energy passes from one living thing to another in a community. **FOOD CHAINS AND FOOD WEBS IN COMMUNITIES**
- Identify the role that each organism in a food web plays. **FOOD CHAINS AND FOOD WEBS IN COMMUNITIES**
- Explain how living things depend on and compete with each other. **INTERACTIONS AMONG LIVING THINGS**
- Describe forest, prairie, tundra, pond, and ocean environments and the types of plants and animals they support. **DIFFERENT ENVIRONMENTS**
- Science in a Snap! Describe how energy passes from one living thing to another in a community. **FOOD CHAINS AND FOOD WEBS IN COMMUNITIES**

85

Chapter 3 Contents

Preview and Predict

Tell students that they can preview a text by looking over it to get an idea of what the text is about and how it is organized. Tell students that predicting is forming ideas about what they will read. Students should confirm their ideas with others.

Have students preview and make predictions about the chapter. Remind them to use section heads, vocabulary words, pictures, and their background knowledge to make predictions.

CHAPTER 3 ▫ How Do Living Things Depend On Their Environment?

LESSON	PACING	OBJECTIVES
1 **Directed Inquiry** *Investigate Owl Pellets* pages T85e–T85h **Science Inquiry and Writing Book** pages 36–39	**40** minutes	Investigate through Directed Inquiry (answer a question; make and compare observations; collect and record data and observations; generate explanations and conclusions based on evidence; share findings; ask questions based on observations to increase understanding). Explain that when animals eat plants or other animals, the energy stored in the food source is passed to them. Recognize that conducting science activities requires an awareness of potential hazards and the need for safe practices.
2 **Big Idea Question and Vocabulary** pages T86–T87—T88–T89 **What Living Things Need from Their Environment** pages T90–T91	**30** minutes	Identify needs that living things get from their environment. Identify the needs of living things. Recognize the way plants and animals get the energy they need for life.
3 **Communities of Living Things** pages T92–T97	**30** minutes	Recognize communities as groups of living things that depend on each other for food. Recognize that green plants are producers because they make food using sunlight, water, and air. Identify consumers as living things that must eat food to get energy. Identify examples of consumers as either herbivores or carnivores. Identify examples of consumers that are omnivores. Describe the role of decomposers in a community.
4 **Food Chains and Food Webs in Communities** pages T98–T101	**25** minutes	Describe how energy passes from one living thing to another in a community. Distinguish predators from prey. Recognize that several interacting food chains form a food web.
5 NATIONAL GEOGRAPHIC **Wolves Choose Fishing Over Hunting** pages T102–T103	**20** minutes	Describe how energy passes from one living thing to another in a community. Explain why wolves choose to hunt salmon over deer in some places during the fall.
6 **Interactions Among Living Things** pages T104–T111	**35** minutes	Describe ways that organisms interact with one another. Classify living things as predators or prey. Explain how parasites get their food. Recognize examples of parasites and their hosts. Describe examples of how plants and animals interact with one another. Explain how individuals within a species may compete with each other. Identify the kinds of resources for which living things compete.

VOCABULARY	RESOURCES	ASSESSMENT
count **data**	Science Inquiry and Writing Book: *Life Science* Science Inquiry Kit: *Life Science* Directed Inquiry: Learning Masters 35–38	Inquiry Rubric: Assessment Handbook, page 184 Inquiry Self-Reflection: Assessment Handbook, page 193 Reflect and Assess, page T85h
	Vocabulary: Learning Master 39 *Life Science* Big Ideas Book	Assess, page T91
herbivore **carnivore** **omnivore**	Extend Learning: Learning Masters 40–41	Assess, page T97
predator **prey**	Science Inquiry and Writing Book: *Life Science*	Assess, page T101
		Assess, page T103
		Assess, page T111

TECHNOLOGY RESOURCES

STUDENT RESOURCES

myNGconnect.com

- Student eEdition
- Big Ideas Book
- Science Inquiry and Writing Book
- Explore on Your Own Books
- Read with Me
- Vocabulary Games
- Digital Library
- Enrichment Activities

National Geographic Kids

National Geographic Explorer!

TEACHER RESOURCES

myNGconnect.com

- Teacher eEdition
- Teacher's Edition
- Science Inquiry and Writing Book
- Explore on Your Own Books
- Online Lesson Planner
- National Geographic Unit Launch Videos
- Assessment Handbook
- Presentation Tool
- Digital Library

NGSP ExamView CD-ROM

IF TIME IS SHORT...

FAST FORWARD.

CHAPTER 3 ▫ How Do Living Things Depend On Their Environment?

LESSON	PACING	OBJECTIVES
7 **Different Environments** pages T112–T119	**35** minutes	Describe forest, prairie, tundra, and ocean environments and the types of plants and animals they support. Classify living things as producers or different kinds of consumers.
8 Guided Inquiry *Investigate Food Chains and Webs* pages T119a–T119d **Science Inquiry and Writing Book** pages 40–43	**40** minutes	Investigate through Guided Inquiry (answer a question; make and compare observations; collect and record data and observations; generate explanations and conclusions based on evidence; share findings; ask questions based on observations to increase understanding; adjust explanations based on findings and new ideas). Trace the flow of energy through various living and nonliving systems. Know that when solving a problem, it is important to plan and get ideas and help from other people.
9 **Conclusion and Review** pages T120–T121	**15** minutes	
10 NATIONAL GEOGRAPHIC LIFE SCIENCE EXPERT *Behavioral Ecologist* pages T122–T123 NATIONAL GEOGRAPHIC BECOME AN EXPERT *Bats: Winged Wonders of the Night* pages T124–T125—T132	**35** minutes	Describe how behavioral ecologists use science and technology in their careers. Explain how so many different kinds of bats can live on Barro Colorado Island.

FAST FORWARD
ACCELERATED PACING GUIDE

DAY 1 40 minutes

Directed Inquiry

Investigate Owl Pellets, page T85e

DAY 2 35 minutes

NATIONAL GEOGRAPHIC LIFE SCIENCE EXPERT *Behavioral Ecologist,* page T122

NATIONAL GEOGRAPHIC BECOME AN EXPERT *Bats: Winged Wonders of the Night,* page T124–T125

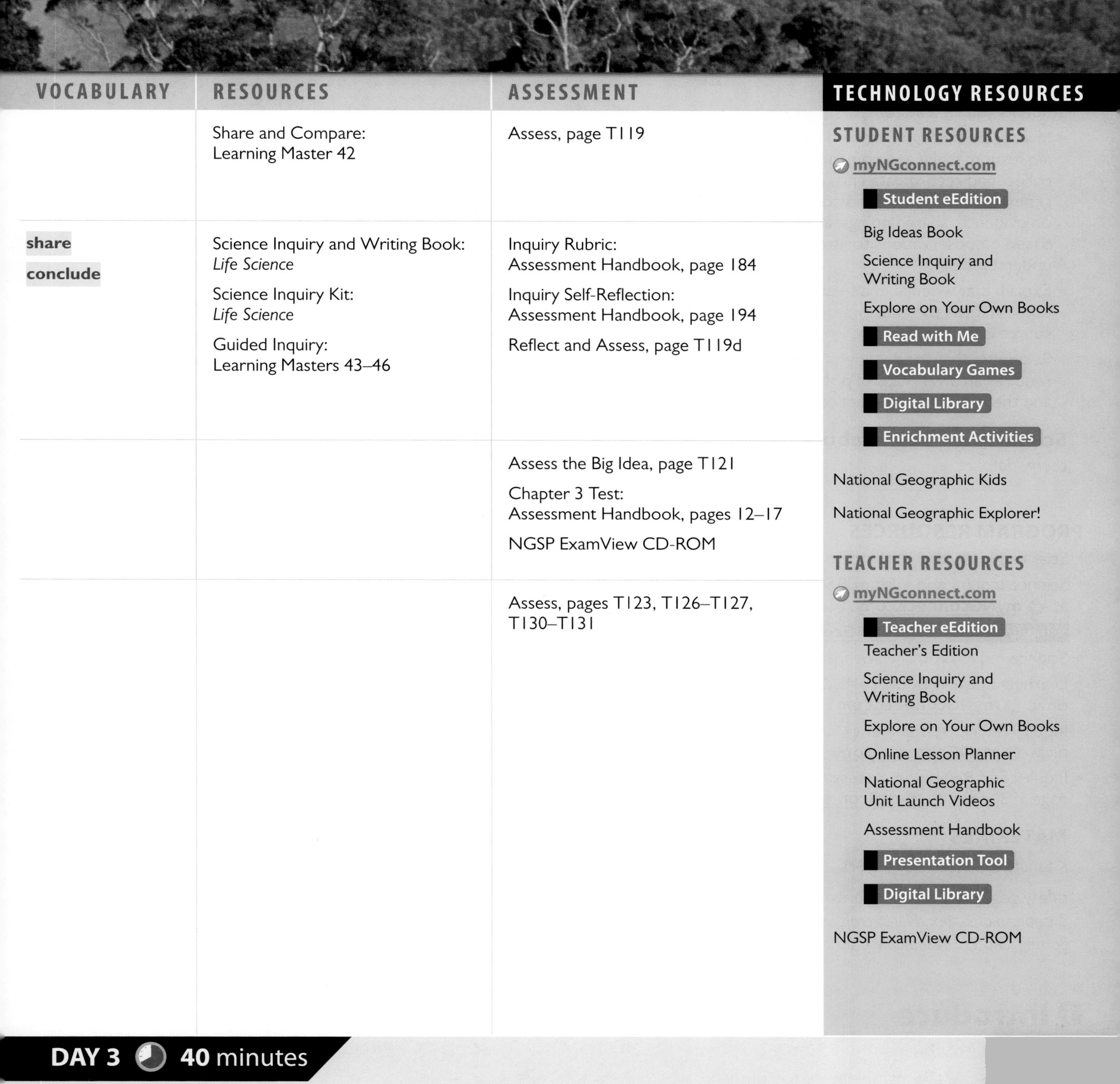

VOCABULARY	RESOURCES	ASSESSMENT
	Share and Compare: Learning Master 42	Assess, page T119
share **conclude**	Science Inquiry and Writing Book: *Life Science* Science Inquiry Kit: *Life Science* Guided Inquiry: Learning Masters 43–46	Inquiry Rubric: Assessment Handbook, page 184 Inquiry Self-Reflection: Assessment Handbook, page 194 Reflect and Assess, page T119d
		Assess the Big Idea, page T121 Chapter 3 Test: Assessment Handbook, pages 12–17 NGSP ExamView CD-ROM
		Assess, pages T123, T126–T127, T130–T131

TECHNOLOGY RESOURCES

STUDENT RESOURCES

myNGconnect.com

- **Student eEdition**
- Big Ideas Book
- Science Inquiry and Writing Book
- Explore on Your Own Books
- **Read with Me**
- **Vocabulary Games**
- **Digital Library**
- **Enrichment Activities**

National Geographic Kids

National Geographic Explorer!

TEACHER RESOURCES

myNGconnect.com

- **Teacher eEdition**
- Teacher's Edition
- Science Inquiry and Writing Book
- Explore on Your Own Books
- Online Lesson Planner
- National Geographic Unit Launch Videos
- Assessment Handbook
- **Presentation Tool**
- **Digital Library**

NGSP ExamView CD-ROM

DAY 3 **40** minutes

Guided Inquiry

Investigate Food Chains and Webs,
page T119a

Objectives

Students will be able to:

- Investigate through Directed Inquiry (answer a question; make and compare observations; collect and record data and observations; generate explanations and conclusions based on evidence; share findings; ask questions based on observations to increase understanding).
- Explain that when animals eat plants or other animals the energy stored in the food source is passed to them.
- Recognize that conducting science activities requires an awareness of potential hazards and the need for safe practices.

Science Process Vocabulary

count, data

PROGRAM RESOURCES

- Science Inquiry and Writing Book: *Life Science*
- Science Inquiry and Writing Book eEdition at myNGconnect.com
- Inquiry eHelp at myNGconnect.com
- Science Inquiry Kit: *Life Science*
- Learning Masters Book, pages 35–38, or at myNGconnect.com
- Inquiry Rubric: Assessment Handbook, page 184, or at myNGconnect.com
- Inquiry Self-Reflection: Assessment Handbook, page 193, or at myNGconnect.com

MATERIALS

Kit materials are listed in italics.

safety goggles; *protective gloves; owl pellet; 2 paper plates; hand lens; craft stick; forceps; Bone Sorting Chart; 6 resealable plastic bags;* masking tape

❶ Introduce

Tap Prior Knowledge

- Point to the owl photograph at the top of the page. Ask students what they know about owls. Say: **Owls are birds that are mostly active at night. They have large, powerful eyes and sharp beaks and talons.** Ask: **What do you think owls eat?** (Accept reasonable answers.)

Directed Inquiry
Investigate Owl Pellets
Question
What objects can you find in an owl pellet?
Science Process Vocabulary
count verb
When you count, you tell the number of something.
I count 4 bones that are alike.
data noun
Data are observations and information that you collect and record in an investigation.
I will record these data in my science notebook.
Materials
safety goggles
protective gloves
owl pellet
2 paper plates
hand lens
craft stick
forceps
Bone Sorting Chart
6 resealable bags
tape
36

MANAGING THE INVESTIGATION

Time

40 minutes

Groups

Small groups of 4

Teaching Tips

- Remind students to be careful with the forceps so they do not poke themselves or each other.
- The owl pellets are densely packed and may be difficult for students to pull apart at first. Advise students to be gentle with the forceps so as not to crush or crack any bones.
- After students complete this activity, store the bones in resealable plastic bags for use in the Chapter 3 Guided Inquiry on page 40.

What to Do

1. Owls usually swallow their food whole. They digest some parts of the food. They eject the other parts through their mouths in the form of pellets. Pellets may contain bones, fur, teeth, and other animal parts.

2. Put on your safety goggles and gloves. Unwrap the owl pellet and place it on a paper plate. Use the hand lens to **observe** the owl pellet. Record your observations in your science notebook. **Predict** how many different animal parts you will find in the pellet.

3. Hold the owl pellet in place with the craft stick while you gently pull apart the pellet with the forceps. Use the forceps to remove pieces of bone, teeth, and other materials.

37

What to Expect

- Students should dissect the owl pellet. The contents of the pellet may include fur, feathers, and different types of bones from rodents, such as mice or voles, and shrews, moles, or birds.
- Students should use the Bone Sorting Chart to determine the type of bone, as well as which animal it came from. Students will graph the number of each type of bone they find in the pellet.

Introduce, continued

Connect to the Big Idea

- Review the Big Idea Question, *How do living things depend on their environment?* Explain to students that this inquiry will focus on how owls get the energy they need to survive.
- Have students open their Science Inquiry and Writing Books to page 36. Read the Question and invite students to share ideas about what they might find in an owl pellet.

❷ Build Vocabulary

Science Process Words: count, data

Use this routine to introduce the words.

1. **Pronounce the Word** Say **data.** Have students repeat it in syllables.
2. **Explain Its Meaning** Choral read the sentence. Ask students for another word or phrase that means the same as **data.** (information and observations)
3. **Encourage Elaboration** Ask: **What data might you collect in this investigation?** (I will collect data about the materials in an owl pellet.)

ELL Use Cognates

Spanish speakers may know the word **dato** in their home language. Use it to help them access the English word **data.**

Repeat for the word **count.** To encourage elaboration, ask: **How will you find the number of each type of bone in the owl pellet?** (I will **count** the bones.)

❸ Guide the Investigation

- Read the inquiry steps on pages 37–38 together with students. Move from group to group and clarify steps if necessary.
- Ask: **Why is it important to wear safety goggles and gloves in this investigation?** (to avoid injury) Discuss the importance of safety in all investigations. Explain that students should evaluate the potential hazards and prepare to prevent them if they occur.
- As students dissect the owl pellet in step 3, they should be aware that there may be more than one set of animal bones in a pellet.

Guide the Investigation, continued

- Students should use the Bone Sorting Chart to identify the bones. Use the chart to point out differences between bones. Focus on the skulls and teeth of rodents, shrews, moles, and birds and discuss how they are alike and different.

⚠ SafetyFirst

Students should not touch the owl pellet with their bare hands. Have students wash their hands after observing the owl pellet.

❹ Explain and Conclude

- Guide students as they record their observations using a chart similar to the one on page 39. Students should list each material they found in the owl pellet and how many they found.
- Have students create a bar graph with the types of bones they found. Make sure they have marked and labeled an *x*-axis and a *y*-axis. Follow the graph on page 39 as an example.
- Students should use the skulls they find as a tool for identifying the animals and estimating how many animal remains were in the pellet.
- Have students share observations and ideas within groups and with the class. Discuss with students the difference between actual observations, such as finding bones in the owl pellet, and inferences they make based on those observations, such as inferring that owls eat other animals.
- Ask: **Why do you think owls eat other animals?** (They eat other animals to get energy from the food.) Discuss how an owl might get energy from a mouse. The mouse would get energy from plants, and the plants would get energy from the sun.
- Have students consider their answers to question 2 in Explain and Conclude on page 39. Ask: **Does your estimate make sense? Is it likely that an owl would have eaten as many different animals as your estimate?** Discuss students' explanations.

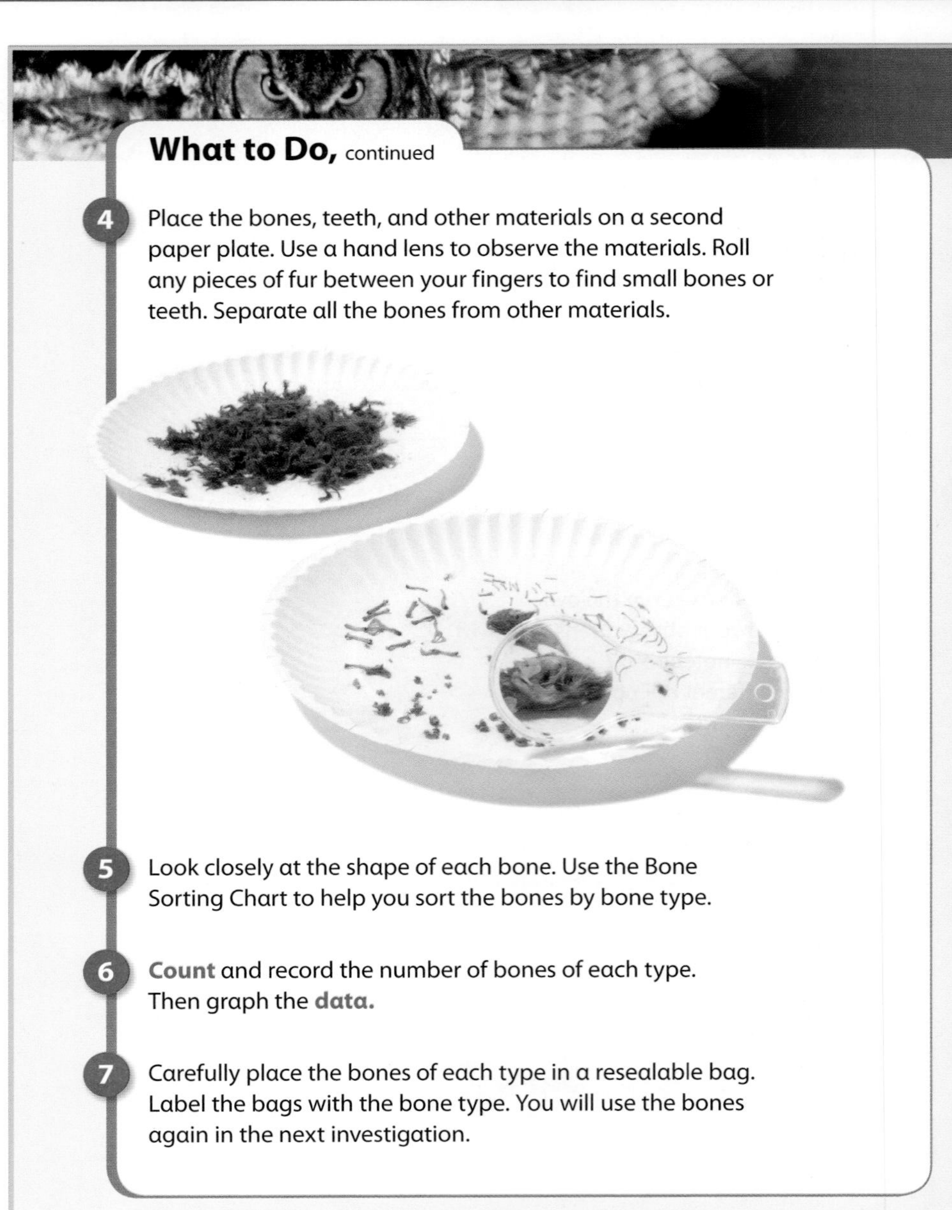

What to Do, continued

4 Place the bones, teeth, and other materials on a second paper plate. Use a hand lens to observe the materials. Roll any pieces of fur between your fingers to find small bones or teeth. Separate all the bones from other materials.

5 Look closely at the shape of each bone. Use the Bone Sorting Chart to help you sort the bones by bone type.

6 **Count** and record the number of bones of each type. Then graph the **data.**

7 Carefully place the bones of each type in a resealable bag. Label the bags with the bone type. You will use the bones again in the next investigation.

38

NATIONAL GEOGRAPHIC Raise Your SciQ!

Owl Pellets Owls do not have teeth and cannot chew their food. Instead, owls often swallow small prey whole. Although the flesh is digested, bones, feathers, and fur are not. Owls regurgitate this material in the form of a pellet. Other birds, such as eagles and hawks, also produce pellets. However, these pellets usually do not contain bones, because these birds are able to digest bones. An owl pellet usually contains all of the bones of the animals an owl has eaten during one night of feeding. The owl produces a pellet about 10–20 hours after eating.

Record

Write in your science notebook.
Use a table and graph like these.

my SCIENCE notebook

Owl Pellet

Object	Observations	How Many?
Whole pellet		1
Skulls		

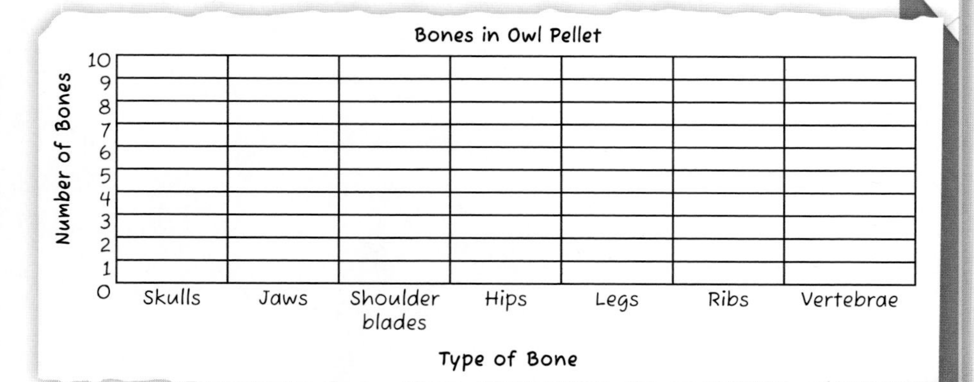

Explain and Conclude

1. What materials did you remove from the owl pellet? How do these materials help you trace the energy the owl gets from its food back to the sun?
2. Based on the **data** in your graph, **estimate** how many different animals are represented by the bones in the pellet. Explain.
3. How might you use owl pellets to make **inferences** about what owls eat?

Think of Another Question

What else would you like to find out about the objects found in owl pellets? How could you find an answer to this new question?

39

Inquiry Rubric at myNGconnect.com	Scale			
The student **observed** and dissected an owl pellet.	4	3	2	1
The student collected **data** by **sorting** bones by type and counting the number of each type.	4	3	2	1
The student created a graph with data about the bones in the pellet.	4	3	2	1
The student **estimated** the number of animal remains in the pellet.	4	3	2	1
The student **shared conclusions** about owl pellets with other students.	4	3	2	1
Overall Score	4	3	2	1

Explain and Conclude, continued

Answers

1. Possible answer: I removed mouse bones and fur from the owl pellet. The owl gets energy from the mouse, the mouse gets energy from plants it eats, and the plants get energy from the sun.
2. Possible answer: I think two different animals are represented in the pellet because I found two skulls.
3. Possible answer: I can infer from the feathers, bones, and fur in the pellet that owls eat other animals.

5 Find Out More

Think of Another Question

Students should use their observations to generate new questions that arise from their investigations.

6 Reflect and Assess

- To assess student work with the Inquiry Rubric shown below, see Assessment Handbook, page 184, or go online at myNGconnect.com
- Have students use the Inquiry Self-Reflection on Assessment Handbook, page 193, or at myNGconnect.com

my SCIENCE notebook

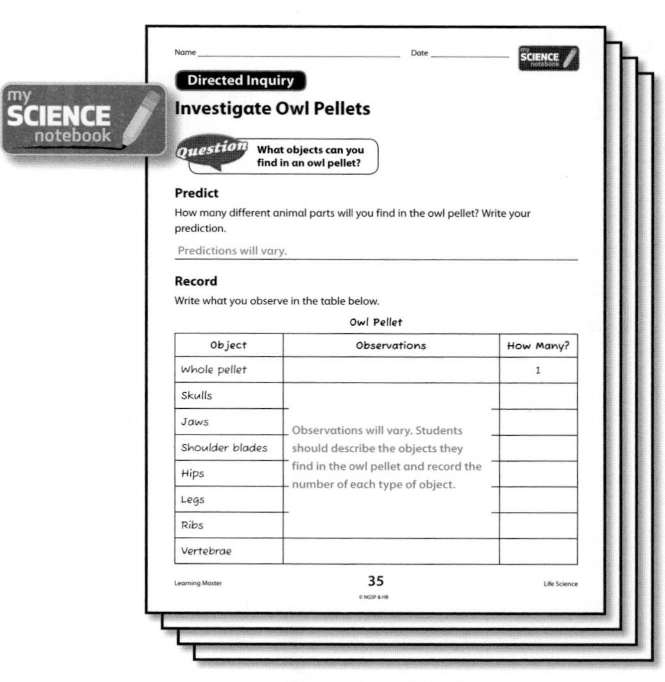
Name ______ Date ______

Directed Inquiry

Investigate Owl Pellets

Question: What objects can you find in an owl pellet?

Predict

How many different animal parts will you find in the owl pellet? Write your prediction.

Predictions will vary.

Record

Write what you observe in the table below.

Owl Pellet

Object	Observations	How Many?
Whole pellet	Observations will vary. Students should describe the objects they find in the owl pellet and record the number of each type of object.	1
Skulls		
Jaws		
Shoulder blades		
Hips		
Legs		
Ribs		
Vertebrae		

Learning Master 35 Life Science

Learning Masters 35–38 or at myNGconnect.com

LESSON 2 ▫ Big Idea Question and Vocabulary

Objectives

Students will be able to:

- Identify needs that living things get from their environment.

Science Academic Vocabulary

herbivore, carnivore, omnivore, predator, prey

PROGRAM RESOURCES

- Big Ideas Book: *Life Science*
- Big Ideas Book: *Life Science* eEdition at myNGconnect.com
- Vocabulary Games at myNGconnect.com
- Digital Library at myNGconnect.com
- Enrichment Activities at myNGconnect.com
- Read with Me at myNGconnect.com
- Learning Masters Book, page 39, or at myNGconnect.com

CHAPTER 3
TECHTREK
myNGconnect.com
HOW DO LIVING THINGS DEPEND ON THEIR ENVIRONMENT?
The African savanna is rich in life. The cape buffalo, cattle egret, and grasses shown here are part of this richness. These living things depend on each other. Cape buffalo graze on the grasses. Cattle egrets and other birds find food by living closely with cape buffalo and other grazers. All of these living things depend on the sun. How?
This cape buffalo and cattle egret depend on each other and the rest of their environment.
86
87

❶ Introduce

Tap Prior Knowledge

- Ask students to name living and nonliving things that they may encounter during the day. Have students discuss why these living and nonliving things are important to their lives.

❷ Focus on the Big Idea

Big Idea Question

- Read the Big Idea Question aloud and have students echo it.
- Preview pages 90–91, 92–97, 98–101, 104–111, and 112–119, linking the headings with the Big Idea Question.

Differentiated Instruction

ELL Vocabulary Support

BEGINNING

Involve students in a chanting rhyme for the word *herbivore.*

I eat plants like grass and more.

No meat for me, I'm an herbivore!

Work with students to develop rhymes for other vocabulary words.

INTERMEDIATE

Provide Academic Language Frames to help students learn vocabulary words:

- *An ____ only eats plants.*
- *A ____ only eats animals.*
- *An ____ eats plants and animals.*

ADVANCED

Have students complete Academic Language Stems to define vocabulary terms:

- *An herbivore eats . . .*
- *A carnivore eats . . .*
- *An omnivore eats . . .*

Focus on the Big Idea, continued

- With student input, post a chart that displays the chapter headings. Have students orally share what they expect to find in each section. Then read page 86 aloud.

How Do Living Things Depend on Their Environment?

What Living Things Need from Their Environment

Communities of Living Things

Food Chains and Food Webs in Communities

Interactions Among Living Things

Different Environments

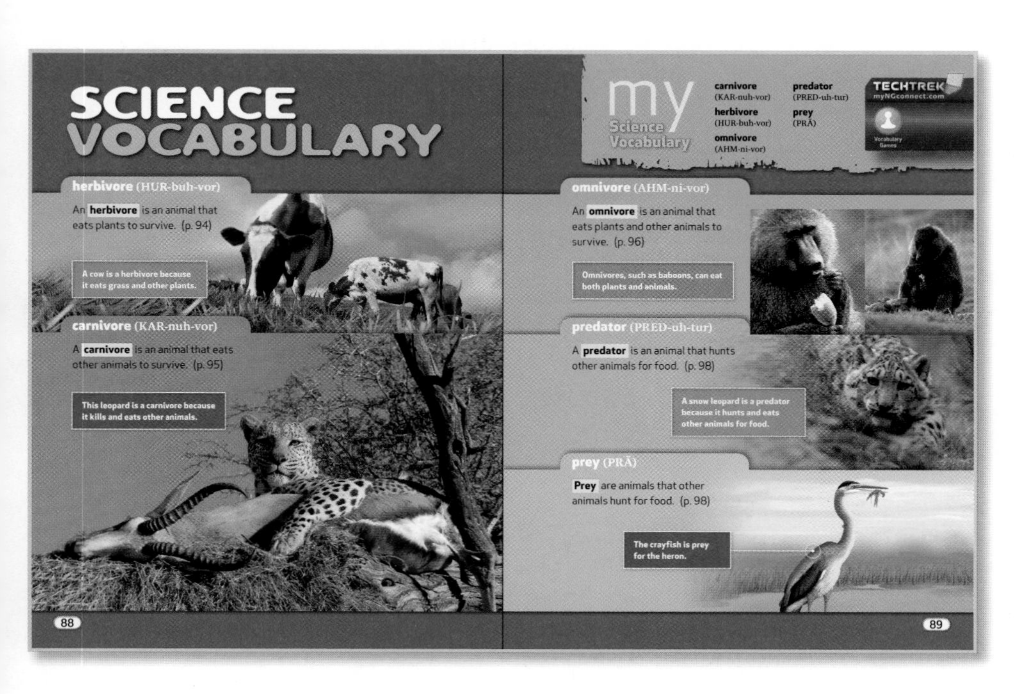

SCIENCE VOCABULARY

herbivore (HUR-buh-vor)

An **herbivore** is an animal that eats plants to survive. (p. 94)

A cow is a herbivore because it eats grass and other plants.

carnivore (KAR-nuh-vor)

A **carnivore** is an animal that eats other animals to survive. (p. 95)

This leopard is a carnivore because it kills and eats other animals.

my Science Vocabulary

carnivore (KAR-nuh-vor)
herbivore (HUR-buh-vor)
omnivore (AHM-ni-vor)
predator (PRED-uh-tur)
prey (PRĀ)

TECHTREK myNGconnect.com
Vocabulary Games

omnivore (AHM-ni-vor)

An **omnivore** is an animal that eats plants and other animals to survive. (p. 96)

Omnivores, such as baboons, can eat both plants and animals.

predator (PRED-uh-tur)

A **predator** is an animal that hunts other animals for food. (p. 98)

A snow leopard is a predator because it hunts and eats other animals for food.

prey (PRĀ)

Prey are animals that other animals hunt for food. (p. 98)

The crayfish is prey for the heron.

88 89

❸ Teach Vocabulary

Have students look at pages 88–89, and use this routine to teach each word. For example:

1. **Pronounce the Word** Say **herbivore** and have students repeat it.
2. **Explain Its Meaning** Read the word, definition, and sample sentence, and use the photo to explain the word's meaning. Point out the cows. Say: **Cows eat grass, so a cow is an herbivore.**
3. **Encourage Elaboration** Ask: **What other animals eat plants? Give some examples.** (Possible answers: Horses, sheep, and goats eat grass and hay; squirrels eat nuts; parrots and finches eat seeds, nectar, and fruit.)

Repeat for the words **carnivore**, **omnivore**, **predator**, and **prey** using the following Elaboration Prompts:

- **Why is the leopard an example of a carnivore?** (It eats other animals only.)
- **What is an omnivore?** (an animal that eats plants and other animals)
- **The snow leopard is a predator. How does it get food?** (by hunting and killing other animals)
- **What is the prey of the heron?** (a crayfish)

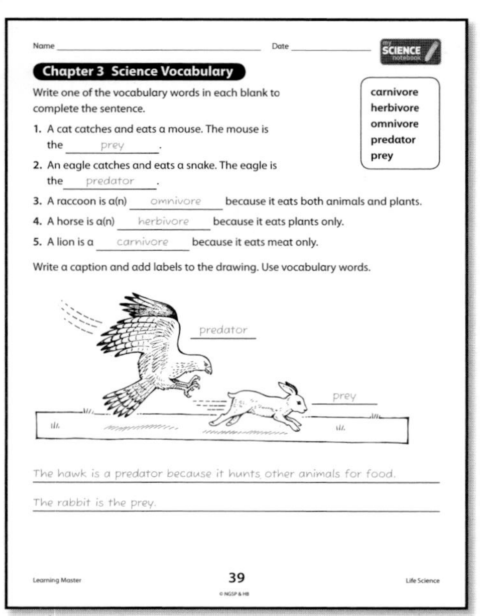

Name ______ Date ______

Chapter 3 Science Vocabulary

Write one of the vocabulary words in each blank to complete the sentence.

carnivore
herbivore
omnivore
predator
prey

1. A cat catches and eats a mouse. The mouse is the prey.
2. An eagle catches and eats a snake. The eagle is the predator.
3. A raccoon is a(n) omnivore because it eats both animals and plants.
4. A horse is a(n) herbivore because it eats plants only.
5. A lion is a carnivore because it eats meat only.

Write a caption and add labels to the drawing. Use vocabulary words.

predator
prey

The hawk is a predator because it hunts other animals for food.

The rabbit is the prey.

Learning Master 39 Life Science

© NGSP & HB

Learning Master 39 or at myNGconnect.com

Objectives

Students will be able to:

- Identify the needs of living things.
- Recognize the way plants and animals get the energy they need for life.

❶ Introduce

Tap Prior Knowledge

- Ask students to name things they need to live, including both living and nonliving things. Have them categorize each thing as living or nonliving.

Set a Purpose and Read

- Read the heading. Tell students that they will read about the needs of living things and how they meet these needs. Then have students read pages 90–91.

❷ Teach

Identify the Needs of Living Things

- Have students study the photo on pages 90–91 and read the labels. Tell them that the photo shows an environment on the grasslands of Africa, which is home to a wide variety of plants and animals. Ask: **What things do all of the plants and animals need from their environment?** (air, water, energy, and space to live)
- Ask: **Are these plants and animals all meeting their needs? How can you tell?** (Yes. The plants and animals are alive and appear healthy. The environment supplies them with air, water, and enough space to live in.)
- Point to the flamingos. Ask: **What if flamingos were the only kind of living thing in this environment? Could they meet all of their needs here? Explain.** (No, they could not meet their need for energy. Like all animals, flamingos need to eat other living things for their food.)

What Living Things Need from Their Environment
Find the living things in this picture. Each kind of living thing is different from the others. But like you, they all need air, water, energy, and space to live. Their environment supplies all of these things. Plants also need soil. It provides some materials plants need to grow.
What are the animals and plants getting from their environment?
PLANTS
ANIMALS
SOIL
90

Science Misconceptions

Food for Plants? Gardeners often use fertilizer to help plants grow. Fertilizers are included in products sold as "plant food," so some students may believe that these products are food for plants. Explain that plants make their own food using energy directly from the sun. Fertilizer contains minerals that help plants grow. However, it does not contain energy that plants use and is not truly food for plants.

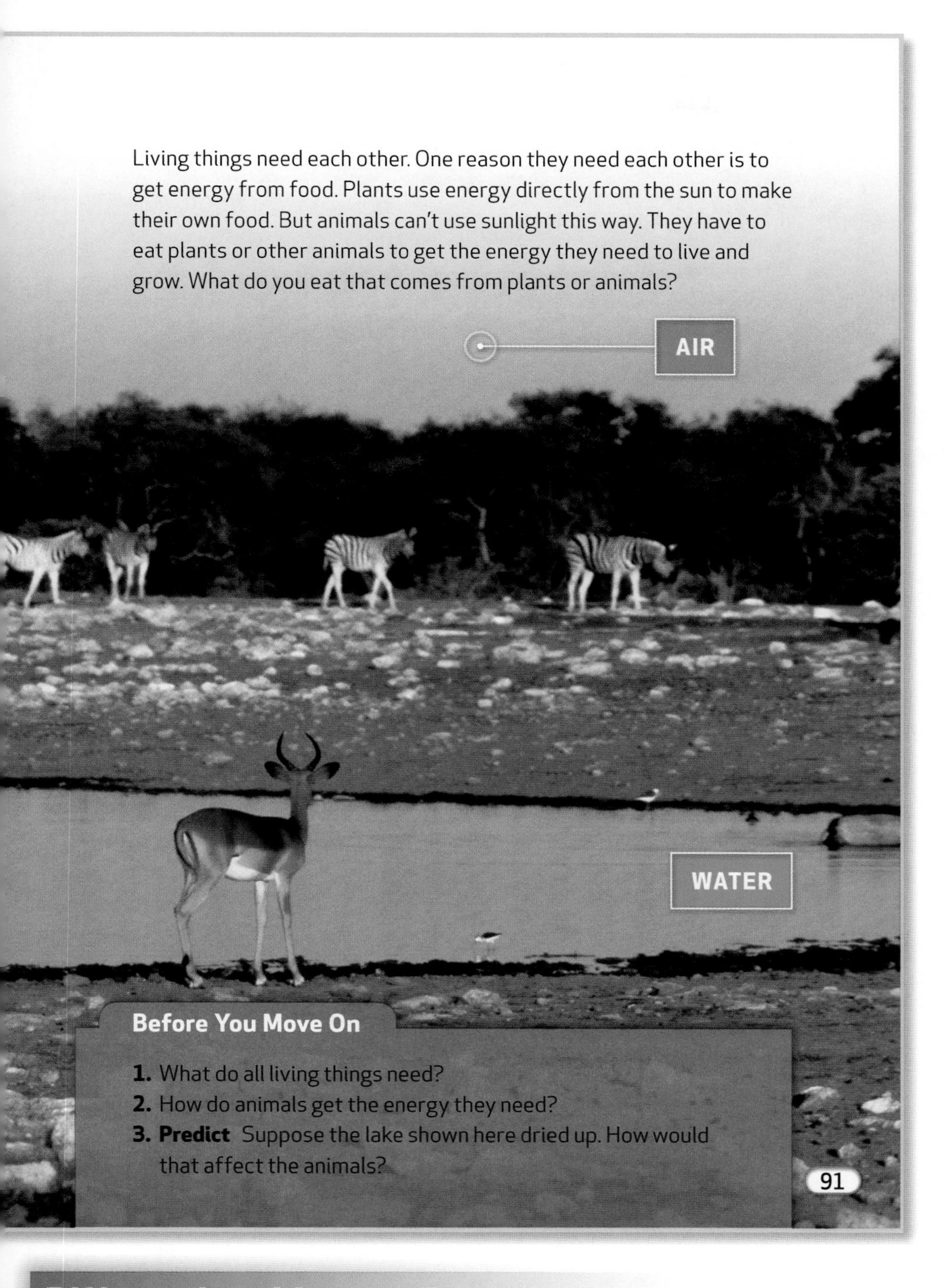
Living things need each other. One reason they need each other is to get energy from food. Plants use energy directly from the sun to make their own food. But animals can't use sunlight this way. They have to eat plants or other animals to get the energy they need to live and grow. What do you eat that comes from plants or animals?

Before You Move On

1. What do all living things need?
2. How do animals get the energy they need?
3. **Predict** Suppose the lake shown here dried up. How would that affect the animals?

91

Teach, continued

Compare Environments

- Have students describe a familiar place where plants and animals live, such as a local park or vacant lot. Ask: **How is the environment in this place different from the environment shown in the photo?** (Possible answer: The plants and animals are different.)
- Ask: **How are the two environments alike?** (Both have plants and animals, as well as nonliving things such as air, water, and soil.) Remind students that all living things have the same basic needs. Help them use these needs as criteria to determine whether an unfamiliar object is alive or not.

How Plants and Animals Get Food

- Review with students the needs of all living things—air, water, energy, and space to live—as listed on page 90. Then ask: **How do plants and animals get energy?** (Plants can use energy directly from sunlight, while animals cannot do this. Animals must eat other living things to meet their need for energy.)

3 Assess

» Before You Move On

1. **List** **What do all living things need?** (All living things need air, water, energy, and space to live.)
2. **Explain** **How do animals get the energy they need?** (Animals get the energy they need by eating plants and other animals.)
3. **Predict** **Suppose the lake shown here dried up. How would that affect the animals?** (They would have to move to another place to meet their need for water and the food that lives in it.)

Differentiated Instruction

ELL Language Support for Identifying Needs of Living Things

BEGINNING

Ask students yes/no questions to discuss needs of living things.

Does a plant need air?

Do animals make their own energy?

Do plants get energy from the sun?

INTERMEDIATE

Provide Academic Language Frames to help students describe needs of living things:

- *Animals get energy from* ___.
- *Plants use energy from the* ___.
- *Both plants and animals need* ___.

ADVANCED

Show students how to repeat questions in their answers. Ask: **What do living things need from their environment?** (Living things need from their environment...)

LESSON 3 ▫ Communities of Living Things

Objectives

Students will be able to:

- Recognize communities as groups of living things that depend on each other for food.
- Recognize that green plants are producers because they make food using sunlight, water, and air.

PROGRAM RESOURCES

- Digital Library at myNGconnect.com

❶ Introduce

Tap Prior Knowledge

- Have students name people who are important in their community. Ask: **How do the members of our community help us live and grow?** (Possible answers: Teachers help us learn; police officers help us stay safe; doctors and nurses help us stay healthy.)

Preview and Read

- Read the heading. Then have students preview the photos on pages 92–97.
- Have students read pages 92–97.

❷ Teach

Describe Interactions in a Community

- Point to the panda, and then ask: **What is one way that the panda interacts with the community in which it lives?** (It eats bamboo.) Remind students that a community includes all the living things in an area, including the plants and animals. Bamboo is an important part of the panda's community.

Communities of Living Things
The plants and animals shown here have something in common. They belong to the same community. A community is all the different kinds of living things in an area. Members of a community interact with and depend on each other. For example, the panda interacts with the bamboo by eating it. The snow leopard interacts with the panda and other animals by eating them. These interactions provide energy for the animals. But how does the bamboo get its energy?
Giant panda bears eat almost nothing but bamboo.
This snow leopard depends on other animals for food. It has its eye on some food right now.
92

NATIONAL GEOGRAPHIC Raise Your SciQ!

Pandas and Bamboo Today, giant pandas live in the bamboo forests of central China. Pandas eat bamboo almost exclusively. However, their digestive systems do not digest bamboo efficiently. For this reason, they eat very large quantities of bamboo—nearly 18 kilograms (40 pounds) per day. Because much of the bamboo forests have been cut down, the number of giant pandas has been decreasing.

Producers Bamboo is a producer. So are the grass, trees, and other plants where you live. Producers use air, water, and energy from the sun to make their own food. They store this food in their leaves, stems, and other green parts. The food is a source of energy for the producers. It's also a source of energy for living things that don't produce their own food, such as pandas and snow leopards.

Bamboo is a kind of grass. Food is stored in its green leaves and stems.

93

Differentiated Instruction

ELL Language Support for Describing Interactions in a Community

BEGINNING

Help students explain how living things interact in a community by asking yes/no questions. For example:

Are plants part of a community? Are animals part of a community? Do the living things in a community need each other?

INTERMEDIATE

Provide Academic Language Frames to help students describe a community.

- *Giant pandas and snow leopards live in the same ____.*
- *Bamboo and other plants are ____.*
- *Plants are a source of ____ for many animals.*

ADVANCED

Have students complete Academic Language Stems about interactions:

- *A panda interacts with plants by . . .*
- *A snow leopard interacts with animals by . . .*

Teach, continued

- Point to the snow leopard, and then ask: **What is one way that the snow leopard interacts with its community?** (It eats other animals that live in the same area.)
- Ask: **Why must the panda and snow leopard each interact with the other living things of their community?** (They must get food to meet their need for energy.) Remind students that all living things need energy to live.

Describe Producers

- Say: **The word *produce* means "make." Why do you think that plants are called *producers*?** (Plants make their own food.) **What materials do they use to make food?** (air, water, and energy from the sun) Point out that all of these materials are nonliving.
- Ask: **What happens to the food that producers make?** (The producers store some of the food, and they use it as a source of energy to live and grow.) Explain that when animals eat plants, the energy stored in the plants is passed to the animals.

Compare Different Producers

- Have students study the photo of bamboo on page 93 and read the caption. Ask: **How is bamboo different from the kinds of grasses we grow on lawns?** (Bamboo is taller, has a thicker stem, and has many leaves.)
- Ask: **How are bamboo and other grasses alike?** (All have green parts, and they are producers.) Explain that the green parts of plants, such as the leaves of most plants, are the parts that produce food.

myNGconnect.com

Have students use the Digital Library to find more photos of producers and consumers.

Computer Presentation Students can assemble the photos and write captions to make a computer presentation on producers and consumers.

Objectives

Students will be able to:

- Identify consumers as living things that must eat food to get energy.
- Identify examples of consumers as either herbivores or carnivores.

Science Academic Vocabulary

herbivore, carnivore

Teach, continued

Academic Vocabulary: *herbivore, carnivore*

- Write the word **herbivore**. Place a vertical line between the *i* and the *v*. Say: **The first part of the word means "grass or plant." The second part of the word means "to eat."** Ask: **What is an herbivore?** (a grass or plant eater)
- Write the word **carnivore**. Place a vertical line between the *i* and the *v*. Ask: **What does the second part of this word mean?** (to eat) **The word part *carni-* means "flesh," so what is a carnivore?** (a flesh or meat eater)

Consumers Living things that cannot make their own food are called consumers. All animals are consumers. They have to eat other living things to get energy. What kinds of living things do consumers eat? The answer depends on the type of consumer.

Cows, horses, goats, and sheep eat plants. An animal that eats plants is an **herbivore**. Large, flat teeth help herbivores grind leaves and stems.

When you eat beef or drink milk, you get some of the energy the cows got by eating grass.

94

Raise Your SciQ!

NATIONAL GEOGRAPHIC

Eating Grass Grass leaves are especially tough and stringy, and many animals cannot eat them. Cows are able to eat grass because of their unusually long and complex digestive tract, which includes a four-chambered stomach. Horses, sheep, goats, and other grazing animals have similar digestive tracts. Bacteria live in the digestive tracts of these animals and help break down grass leaves and other tough plant parts.

Cats, seals, snakes, and hawks eat other animals. An animal that eats other animals is a **carnivore**. Most carnivores have sharp teeth or beaks that can bite and tear meat.

A leopard is a carnivore. This one guards the impala it killed for food.

95

Differentiated Instruction

Extra Support

Have students begin a Venn diagram by drawing two circles that are not touching. They should use the labels *eats plants, eats animals, herbivore,* and *carnivore* for the circles. They can then add an overlapping circle with the labels *eats plants* and *animals* and *omnivore* after they read page 96.

Challenge

Have students begin a similar Venn diagram using the labels *eats plants, eats animals, herbivore,* and *carnivore* for the circles. Then students should list names of animals from the chapter in the appropriate part of the diagram. Students should add the overlapping circle with the labels *eats plants and animals* and *omnivore* after they read page 96 and list names of animals there, too.

Teach, continued

Identify and Compare Consumers

- Have students observe the photo of the cow on page 94 and the leopard on page 95. Ask: **Why are both the cow and leopard classified as consumers?** (Both are animals, and they eat other living things for food.)
- Ask: **How is the cow's food different from the leopard's food?** (The cow eats plants, while the leopard eats other animals.) Review the definitions of **herbivore** and **carnivore** from the text.
- Ask: **Do you think a cow could kill and eat other animals, like a leopard does? Why or why not?** (No. One reason is that a cow's large, flat teeth are suited for eating grass, not biting into meat.) Explain that a cow also has a specialized digestive system for eating grass. Other body features such as the legs and mouth are not suited for catching and eating other animals.
- Ask: **Do you think a leopard could eat a large meal of grass, like a cow does? Why or why not?** (No. One reason is that a leopard's sharp teeth would not chew grass very well.) Explain that grasses and other plants are made mostly of a tough material called cellulose. Cellulose is much more difficult to digest than meat. **Carnivores** have digestive systems specialized for eating meat.
- Say: **The size of an animal does not indicate what it eats. Some herbivores, such as elephants, are larger than many carnivores. Many insects are carnivores, including hornets and tiger beetles.**

LESSON 3 ▫ Communities of Living Things

Objectives

Students will be able to:

- Identify examples of consumers that are omnivores.
- Describe the role of decomposers in a community.

Science Academic Vocabulary

omnivore

PROGRAM RESOURCES

- Digital Library at myNGconnect.com
- Learning Masters Book, pages 40–41, or at myNGconnect.com

Teach, continued

Academic Vocabulary: *omnivore*

- Ask: **What part of the word omnivore do you already know?** (vore) **What does it mean?** (to eat) Say: **The prefix *omni-* means "all." An omnivore eats both plants and animals.**

Recognize Omnivores

- Have students study the photos on pages 96–97 and read the captions. Ask: **What do baboons eat?** (Baboons eat both plants and animals; they are omnivores.)
- Ask: **How might being able to eat both plants and animals help the community?** (An omnivore might be able to shift to another food if many herbivores were in the area.)

Raccoons, skunks, pigs, and bears eat plants as well as other animals. An animal that eats both plants and other animals is an **omnivore**. Humans are omnivores too. Omnivores have some teeth that help with chewing plants and some teeth that bite and tear meat.

Eating certain kinds of foods is one of the roles, or jobs, of a consumer. The different roles help keep enough food for the community. For example, if all animals were herbivores, an area might soon have no plants left to eat.

This baboon is eating the meat of a young impala.

Baboons round out their diet with fruit.

96

Extend Learning

MANAGING THE INVESTIGATION

Time

30 minutes for set-up, then 5 minutes at least once a week

Groups

Whole class

PROGRAM RESOURCES

- Learning Masters 40–41

MATERIALS

- two compost bins, such as plastic shoe boxes with lids or other plastic containers
- dirt or potting soil
- water
- earthworms or red worms
- dead plant matter, such as kitchen scraps, grass clippings, or fallen leaves

Investigate the Actions of Decomposers

Preview

How do worms change their environment and help plant matter decompose?

Students will prepare two compost bins, one with worms and one without. Fill each bin with soil and plant matter, then observe and compare how the plant matter decomposes. Conclude that worms help return dead plants to the soil.

What To Do

1. Mix dead plant matter into the soil. Add water to make the mixture moist. Distribute the mixture evenly into bins.
2. Add the worms to one of the compost bins. Leave the bins uncovered, or punch air holes in the lids.
3. Place the compost bins side by side in a cool, dark place.
4. Observe the compost bins once or twice a week. Add water to the bins if necessary.

Decomposers Consumers that feed on the remains of plants and animals are decomposers. Think of leaves that fall to the ground. After a few months, the leaves might have many holes. Decomposers are at work. As they break down the dead matter, they release some of the materials in that matter back into the soil. Then plants can use these materials to grow. Some decomposers are too small to see with your eyes alone.

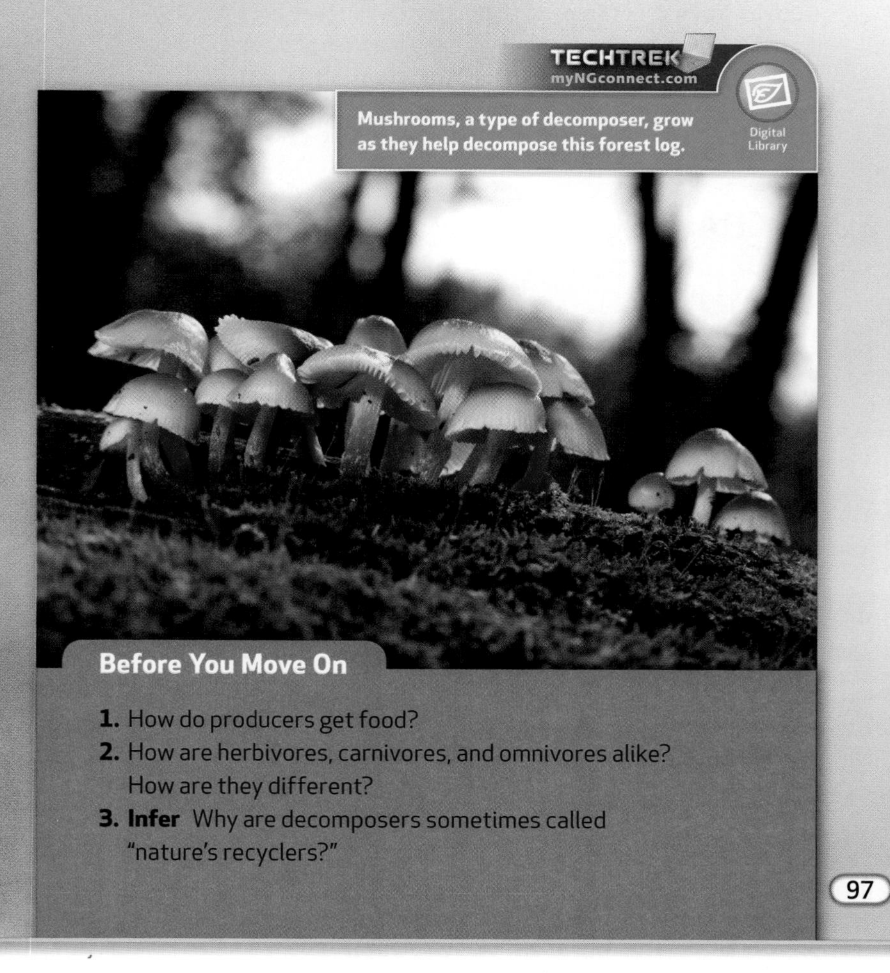

Mushrooms, a type of decomposer, grow as they help decompose this forest log.

Before You Move On

1. How do producers get food?
2. How are herbivores, carnivores, and omnivores alike? How are they different?
3. **Infer** Why are decomposers sometimes called "nature's recyclers?"

97

Teach, continued

Describe the Role of Decomposers

- Say: **Earthworms and mushrooms are decomposers, or detritivores. They eat detritus, or decaying matter.** Ask: **How are decomposers helpful in a community?** (They break down dead matter and release some of the materials in that matter back into the soil.)

myNGconnect.com

Have students use the Digital Library to find more photos of decomposers.

Whiteboard Presentation Help students prepare a computer presentation about decomposers.

3 Assess

» Before You Move On

1. **Recall How do producers get food?** (They use the energy in sunlight to make food.)
2. **Compare and Contrast How are herbivores, carnivores, and omnivores alike? How are they different?** (They are all consumers. Herbivores eat plants, carnivores eat other animals, and omnivores eat both.)
3. **Infer Why are decomposers sometimes called "nature's recyclers"?** (Decomposers recycle materials from dead matter back into the soil. Then plants can use the materials to grow.)

Learning Masters 40–41, or at myNGconnect.com

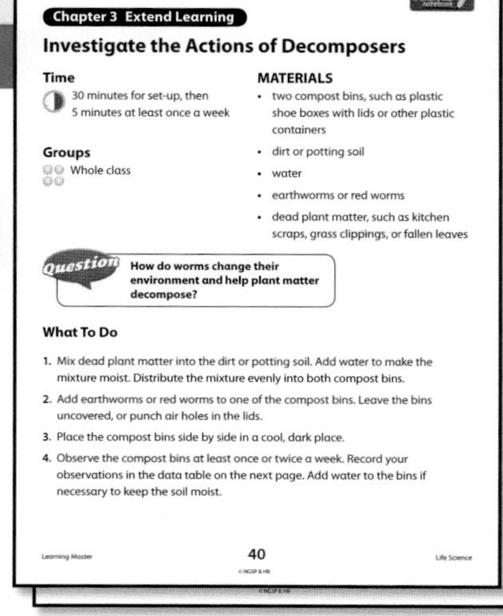

Name ______ Date ______

Chapter 3 Extend Learning

Investigate the Actions of Decomposers

Time
30 minutes for set-up, then 5 minutes at least once a week

Groups
Whole class

MATERIALS
- two compost bins, such as plastic shoe boxes with lids or other plastic containers
- dirt or potting soil
- water
- earthworms or red worms
- dead plant matter, such as kitchen scraps, grass clippings, or fallen leaves

Question How do worms change their environment and help plant matter decompose?

What To Do
1. Mix dead plant matter into the dirt or potting soil. Add water to make the mixture moist. Distribute the mixture evenly into both compost bins.
2. Add earthworms or red worms to one of the compost bins. Leave the bins uncovered, or punch air holes in the lids.
3. Place the compost bins side by side in a cool, dark place.
4. Observe the compost bins at least once or twice a week. Record your observations in the data table on the next page. Add water to the bins if necessary to keep the soil moist.

Learning Master 40 Life Science

Explain Results

Students should explain that:

- Worms help decompose plant matter and help return the matter to the soil.
- Worms change the soil and plant matter into fine pieces, called castings.
- Discuss how worms and other decomposers fill an important role in the soil. They recycle the nutrients of dead organisms so that other plants may use them to grow.

CHAPTER 3
LESSON 4 ▫ Food Chains and Food Webs in Communities

Objectives
Students will be able to:
- Describe how energy passes from one living thing to another in a community.
- Distinguish predators from prey.

Science Academic Vocabulary
predator, prey

❶ Introduce

Tap Prior Knowledge
- Have students describe an example of a chain, such as a chain made of paper or metal links. Ask: **What makes up a chain?** (many links joined together) **Why is every link of the chain important?** (If one link breaks, the chain breaks in two and may no longer be usable.)

Set a Purpose and Read
- Say: **All animals depend on the living things of their community for food. Let's look at the way these animals can affect one another.**
- Have students read pages 98–101.

❷ Teach

Academic Vocabulary: *predator, prey*
- Pronounce and write **predator**. Say: **A predator hunts and eats other animals.**
- Have students look back through previous pages of the chapter. Ask: **Which of the consumers we read about are predators?** (snow leopards, leopards, cats, seals, snakes, hawks)
- Pronounce and write **prey**. Say: **A prey is an animal that a predator hunts and may eat.**
- Have students look at the photo on page 95. Ask: **Which animal is the predator?** (leopard) **Which animal is the prey?** (impala)

Food Chains and Food Webs in Communities

When you see a squirrel eating a nut or the sun shining on a plant, you are seeing a link in a food chain. A food chain is one path that energy takes through a community.

Trace the flow of energy in the food chain shown here. Cattails and other producers use the sun's energy to make food. Consumers such as crayfish get some of that energy when they eat the cattails. Later, a great blue heron might grab the crayfish and eat it. The heron is a **predator** —an animal that hunts and eats other animals. The crayfish is the **prey** —an animal that is hunted and then eaten.

FOOD CHAIN IN A POND

Most food chains begin with the sun. Producers, such as cattails, use sunlight to make food.

Consumers, such as crayfish, eat the cattails and get the plant's stored energy.

Other consumers, such as this heron, eat the crayfish and get the energy stored in its body.

98

Differentiated Instruction

Extra Support
Have students copy the illustrations of the food chain shown and then write new captions for each illustration.

Challenge
Have students choose a familiar animal and research what it eats as well as what eats it. Have students illustrate these animals in a food chain and include labels and captions.

Science in a Snap! What's Your Food Chain?

Think about your favorite meal. Infer what plant or animal provided each kind of food. Include what the animals ate.

Draw and label each plant or animal on a card. Arrange the cards to show different food chains.

Where do you fit into each food chain?

Tortoises get their energy from dandelions and other plants. Where would you place this photo in the food chain shown on the opposite page?

99

Teach, continued

Text Feature: Diagrams and Captions

- Have students study the diagram of the food chain and read the captions. Ask: **What does a food chain show?** (how energy moves from organism to organism in a community; how organisms in a community depend on each other for food)
- Ask: **What do the arrows in the diagram show?** (the direction of the movement of energy from producers to consumers)
- Ask: **What is the source of the energy in the food chain?** (the sun) Explain that the sun is the energy source for nearly all food chains on Earth. The exceptions are certain food chains in the ocean floor, which get their energy from Earth's hot interior.
- Point out the pond plants in the first box of the diagram. Ask: **Why is a producer always the first living thing in a food chain?** (Only producers are able to use energy from the sun.) **What living things come next in a food chain?** (herbivores, omnivores such as crayfish, or decomposers) Explain that a food chain always consists of a producer followed by one or more consumers. Invite students to trace the flow of energy in the diagram from the sun, to producers, to consumers.

Science in a Snap!

What's Your Food Chain?

Materials index cards, crayons

Have students work independently or in pairs. Have them follow the instructions on page 99 of the Big Ideas Book. Students should record their observations and results in their science notebook.

What to Expect Students should observe and record that each food chain they construct begins with a producer and ends with themselves.

Quick Questions Ask students the following questions:

- **Which animals in your food chains are herbivores? carnivores? omnivores?** (Possible answers: Cows and sheep are herbivores; pigs and chickens are omnivores; few common food animals are carnivores.)
- **How would you classify yourself, as an herbivore, carnivore, or omnivore? Explain.** (Possible answers: Omnivore, because I eat both plants and animals; herbivore, because I eat plants only.)

Objectives

Students will be able to:

- Describe how energy passes from one living thing to another in a community.
- Recognize that several interacting food chains form a food web.

PROGRAM RESOURCES

- Enrichment Activities at myNGconnect.com

Teach, continued

Text Feature: Illustrations

- Have students study the food web shown on pages 100–101. Ask: **Why do you think *food web* is an appropriate term for this illustration?** (Like a spider's web, a food web has many paths that connect and overlap. A food web shows how different food chains connect.)
- Have students use their fingers to trace examples of food chains in the food web. Ask: **Why can one animal, such as the fish or raccoon, be part of many different food chains?** (One animal often eats many different foods, or many different animals may eat it.)

Enrichment Activities

myNGconnect.com

Have students use Enrichment Activities to learn more about food chains and food webs.

Integrated Technology

Computer Presentation Students can use the information they learn to make a computer presentation about food chains and food webs.

Food Webs You eat more than one kind of food. Most other consumers do too. A heron's prey includes fish, insects, mice, snakes, and other animals. A raccoon eats just about anything it can get its paws on, from berries and nuts to worms and fish. So a community includes many food chains that are all linked to each other. All of these connected food chains make up a food web. The drawing shows part of a pond food web. What does the heron eat? What eats the fish?
TECHTREK
myNGconnect.com
Enrichment Activities
In this food web, the arrows point from the thing being eaten to the thing that eats it. How many food chains can you find in this food web?
sun
great blue heron
snail
cattails
cricket
bullfrog
algae
100

Raise Your SciQ!

The Importance of Food Webs Scientists often study food webs to explain changes in the populations and health of community members. In the 1960s, biologist Rachel Carson showed that the pesticide DDT was traveling through many food webs to collect in the bodies of birds, where it led to thinner egg shells. Scientists today are studying the effects of mercury, lead, and other toxic metals in lake and ocean food webs, where they are collecting in the bodies of many fish.

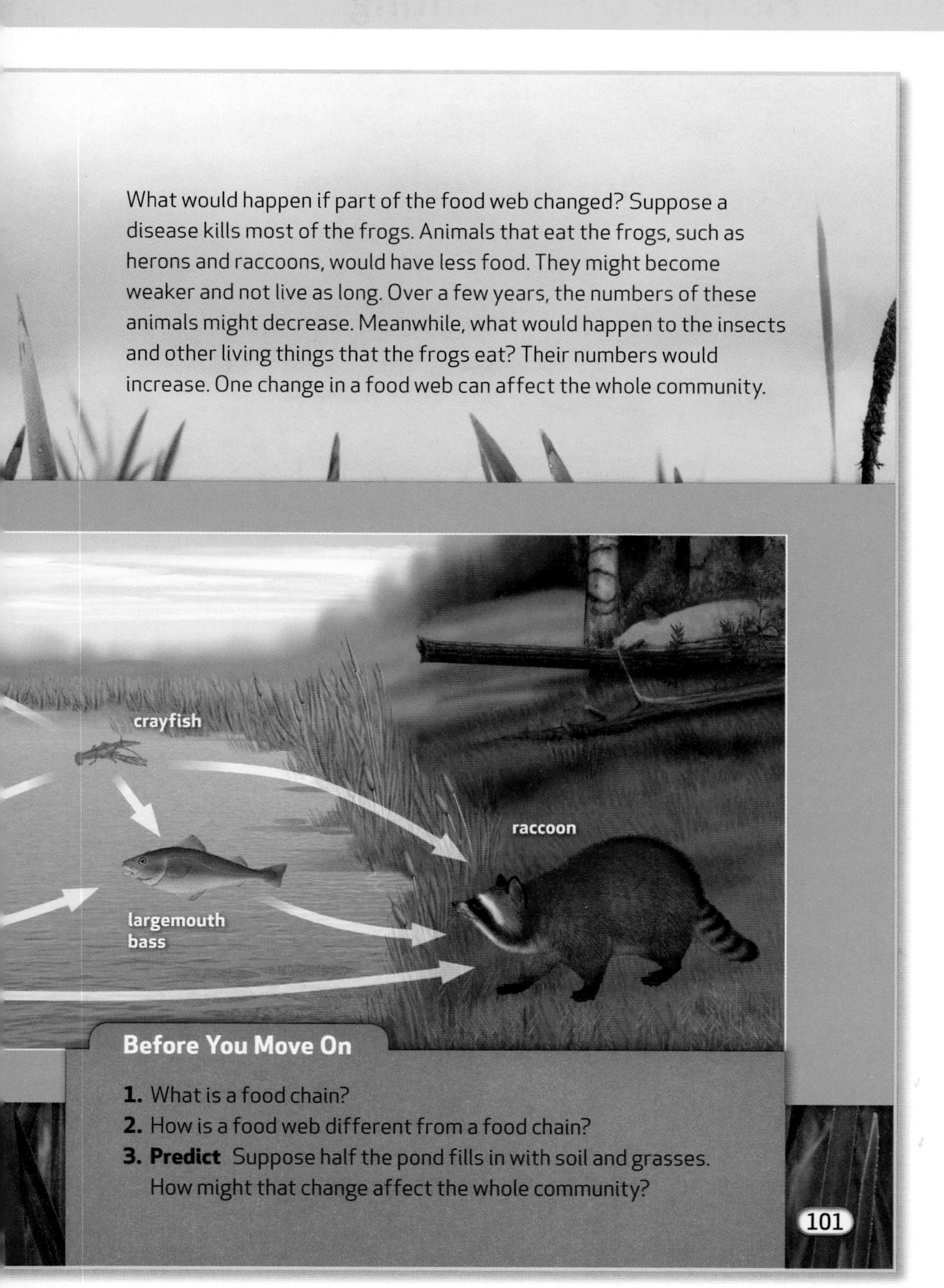

What would happen if part of the food web changed? Suppose a disease kills most of the frogs. Animals that eat the frogs, such as herons and raccoons, would have less food. They might become weaker and not live as long. Over a few years, the numbers of these animals might decrease. Meanwhile, what would happen to the insects and other living things that the frogs eat? Their numbers would increase. One change in a food web can affect the whole community.
crayfish
raccoon
largemouth bass
Before You Move On
1. What is a food chain?
2. How is a food web different from a food chain?
3. Predict Suppose half the pond fills in with soil and grasses. How might that change affect the whole community?
101

Differentiated Instruction

ELL Language Support for Identifying Feeding Relationships in a Pond Food Web

BEGINNING

Ask students either/or questions to help them discuss food webs. For example:

Is a great blue heron a predator or prey?

Is a cattail a predator or a producer?

INTERMEDIATE

Help students match an organism with its role in the pond food web by identifying it as *producer*, *prey*, or *predator*.

- ***crayfish*** *(prey)*
- ***great blue heron*** *(predator)*
- ***cattails*** *(producer)*

ADVANCED

Provide sequence words to help students describe a single food chain. For example:

Most food chains begin with the sun. First, ... Second, ... Then ...

Teach, continued

Identify Feeding Relationships in a Pond Food Web

- Have students use the illustration to answer questions about the feeding relationships in the pond community. For example: **Which organisms are predators?** (great blue heron, bullfrog, largemouth bass, raccoon) **Which organisms are prey?** (snail, cricket, bullfrog, crayfish, largemouth bass)
- Discuss ways in which a community is affected when there is a change in the number of any of the predators or prey of a community. For organisms with only one source of food, for example, a change in the food source could cause tremendous harm.

Identify Plants that People Eat

Science Inquiry and Writing Book: *Life Science*, page 14

Read the instructions together with students. Time students as they make their list. Then have them work in pairs. Students should record their ideas in their science notebook.

What to Expect Students should list foods, such as fruits, vegetables, nuts, and beans. Most plants get their energy from the sun.

3 Assess

1. **Define What is a food chain?** (one path energy takes through a community that starts with a producer and has one or more consumers)
2. **Contrast How is a food web different from a food chain?** (A food web consists of many connected food chains.)
3. **Predict Suppose half the pond fills in with soil and grasses. How might that change affect the whole community?** (Possible answer: Animals that eat the grass, such as insects, might increase in number. Then the number of animals that eat insects, such as frogs, also might increase.)

CHAPTER 3
LESSON 5 ▫ Wolves Choose Fishing Over Hunting

Objectives

Students will be able to:

- Describe how energy passes from one living thing to another in a community.
- Explain why wolves choose to hunt salmon over deer in some places during the fall.

❶ Introduce

Tap Prior Knowledge

- Ask students to discuss what they know about wolves. Encourage them to discuss the ways that wolves are presented in stories and whether they think those stories present wolves correctly or incorrectly.

Preview and Read

- Read the heading. Tell students they will read about ways wolves in British Columbia find food.
- Have students read pages 102–103.

❷ Teach

Identify Wolves' Feeding Behaviors

- Say: **Like other animals, wolves get their energy by eating food.** Ask: **How do wolves usually get their food?** (by hunting deer and other large prey)
- Ask: **Why were scientists surprised about the wolves they studied in British Columbia?** (The scientists observed wolves catching salmon instead of hunting deer and other land animals.)

NATIONAL GEOGRAPHIC

WOLVES CHOOSE FISHING OVER HUNTING

Wolves are large predators. They chase down deer, moose, and other large prey. But scientists have discovered something surprising. When given the choice, wolves would rather fish than hunt.

Scientists have studied wolves in British Columbia, Canada. During most of the year, wolves eat deer and other land animals. But during the fall, thousands of salmon swim upstream to lay their eggs. That's when salmon become the wolves' favorite food.

This wolf dives into the stream to catch a meal.

102

Social Studies in Science

Salmon Run of British Columbia Each year, thousands of salmon return from the Pacific Ocean to the freshwater streams of British Columbia in Canada. The salmon return as adults to spawn and lay eggs. Remarkably, they return to the very streams in which they hatched. Scientists think that salmon can detect chemical markers that indicate their native streams. Animals that prey on the returning salmon include black bears, grizzly bears, bald eagles, golden eagles, and seagulls.

Hunting salmon instead of deer makes a lot of sense. Wolves often get hurt while chasing down deer. The wolf might get gouged by antlers or kicked by powerful legs and sturdy hooves. Plus, the wolf has to use a lot of energy to chase the deer, and sometimes the deer gets away. It's a lot easier to just stand by the riverside and grab a fish as it swims by.
Wolves eat only the head of the salmon. Birds, crayfish, and other animals eat the body. Decomposers break down what's left. So the wolves' choice to go fishing supplies energy for much of the community.
Wolves play an important role in food chains and food webs.
103

Differentiated Instruction

Extra Support

Have students identify the vocabulary words or other unfamiliar words that are used on pages 102–103. These words include *predator, prey, decomposer,* and *community.* Have them explain or define the meanings of these words to a partner.

Challenge

Tell students that salmon eat herring and other small fish, and these fish eat algae and other small producers. Have students draw and label a food chain that includes salmon and wolves. Remind them to include decomposers in the food chain.

Teach, continued

- Ask: **Why do the wolves only fish for salmon in the fall?** (Salmon are plentiful only in the fall, when they come to the streams to lay eggs.)
- Ask: **How do the wolves help other animals that live in or near the stream?** (The wolves only eat the salmon's heads. Crayfish, birds, and other animals eat other parts of the fish.) Explain that not all carnivores kill prey. Some may rely on other carnivores, like wolves, to kill prey for them.
- Point out that fishing for salmon affects communities on both the water and land. Ask: **When wolves catch and eat salmon, how are land animals affected?** (Possible answers: Animals that live on land may eat some of the salmon; more moose, deer, and other animals may survive because the wolves are not hunting them.)

Text Feature: Photos and Captions

- Have students observe the photo sequence of the wolf on pages 102–103. Ask: **How do wolves catch salmon?** (They spot a salmon in the water, catch it with their mouths, and carry it away.)

❸ Assess

1. **List What kinds of animals are prey for wolves?** (deer, moose, other large prey, and salmon)
2. **Explain Why do wolves fish for salmon?** (In the fall, there are a lot of salmon. It is easier and less dangerous for wolves to catch salmon than it is for wolves to hunt larger prey.)
3. **Predict Suppose the salmon population became very low. What might happen to wolves and other members of the land and stream food webs?** (Possible answers: Wolves would have fewer salmon to catch and might have to hunt deer all year long. If they're hunting deer, then the deer population would go down. That would affect the land food web. The stream food web would also be affected because the birds, crayfish, and decomposers depend on the salmon that the wolves catch. Also, some wolves might die because they'll have less salmon to eat.)

LESSON 6 ▫ Interactions Among Living Things

Objectives

Students will be able to:

- Describe ways that organisms interact with one another.
- Classify living things as predators or prey.

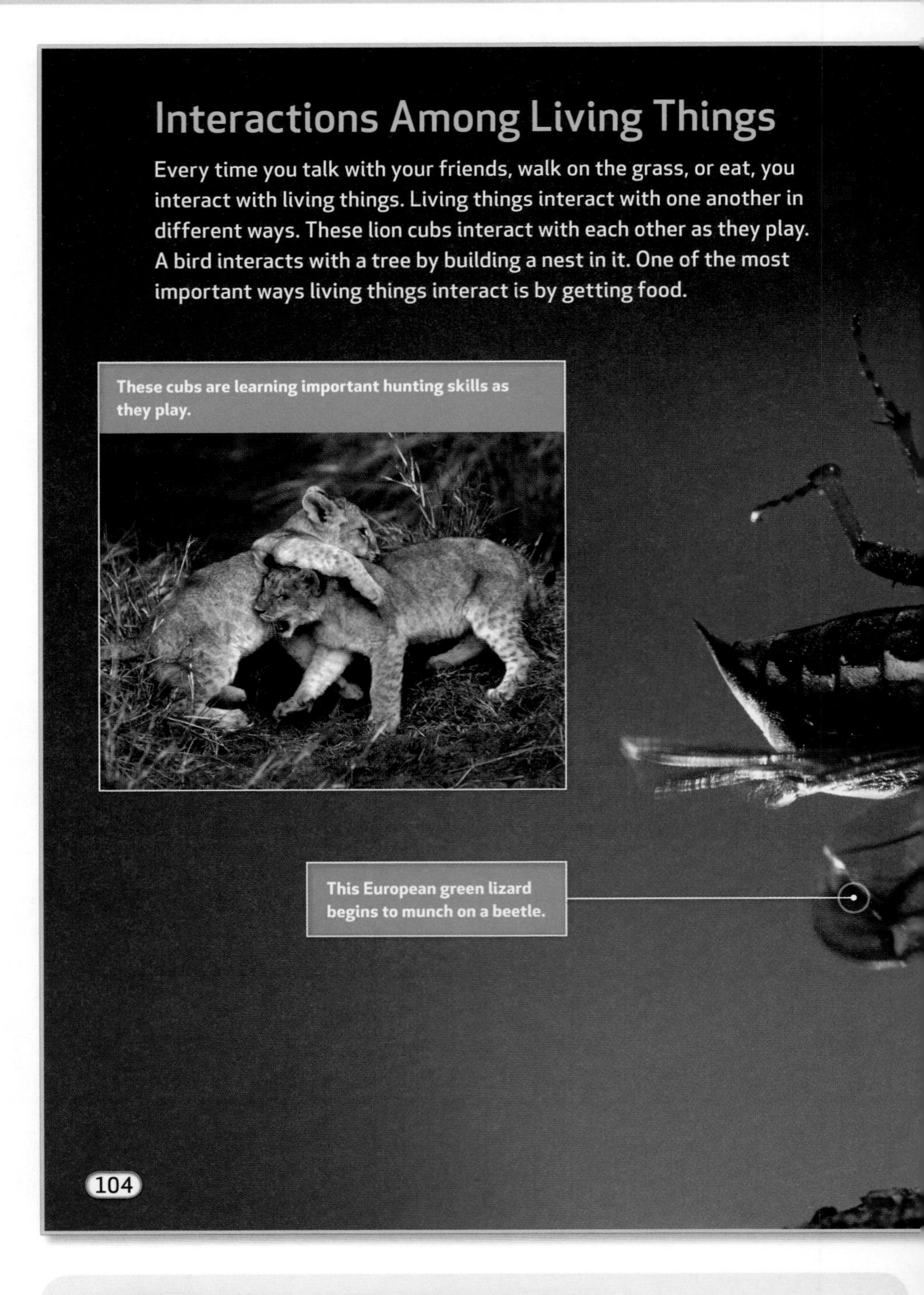

Interactions Among Living Things
Every time you talk with your friends, walk on the grass, or eat, you interact with living things. Living things interact with one another in different ways. These lion cubs interact with each other as they play. A bird interacts with a tree by building a nest in it. One of the most important ways living things interact is by getting food.
These cubs are learning important hunting skills as they play.
This European green lizard begins to munch on a beetle.
104

❶ Introduce

Tap Prior Knowledge

- Ask students to describe examples they have observed of two organisms interacting with one another. Discuss examples in which both organisms benefit, such as puppies playing, and in which one organism is harmed, such as a robin catching a worm or a frog catching a fly.

Set a Purpose

- Read the heading. Say: **Living things depend on one another in many ways. In this lesson we'll read about the ways they interact.**

❷ Teach

Identify Predators and Prey

- Have students read pages 104–105.
- Have students study the photo of the lizard eating the beetle and then read the caption. Ask: **Which animal is predator and which animal is prey? How do you know?** (The lizard hunted the beetle and is eating it, so the lizard is the predator and the beetle is the prey.)
- Ask: **What are some other examples of predators and their prey?** (Possible answers: robin and worm, leopard and impala, bullfrog and cricket) Have students review the photos on pages 92–101 for examples of predators and their prey.

NATIONAL GEOGRAPHIC **Raise Your SciQ!**

Population Shifts Sometimes the population of plants and animals in an environment stays relatively constant. Often, however, their numbers shift significantly from year to year. Sometimes these shifts indicate a serious change or threat to the environment, such as an increase in pollution, the spread of a disease, or the introduction of an invasive species. Yet for many predators and their prey, populations rise and fall as part of a natural cycle. For example, when the population of wolves falls, the population of deer may rise. The increase in the deer population allows the wolf population to recover.

This lizard and beetle show how predators and prey interact. Predators need their prey for food. But what would happen if more and more lizards moved into the area? They may eat all the beetles. Then the lizards will have less food and may die.
The prey also need their predators. If fewer lizards are in the area, more and more beetles may live. But there may not be enough food, water, or space for all of them. Many prey animals may die of hunger or thirst. Predators and prey need each other to keep the community healthy.
105

Differentiated Instruction

ELL Language Support for Identifying Predator/Prey Interactions

BEGINNING

Have students review the illustrations on pages 92, 95, 100–101, and 104–105. In each illustration, have partners identify the predator and prey.

INTERMEDIATE

Have students identify predator/prey interactions by completing Academic Language Frames, such as:

- *A lizard is a ____ of beetles.*
- *Beetles are the ____ of lizards.*

ADVANCED

Have students write a new caption for the photo on pages 104–105. Students should include the words *predator* and *prey* in the caption.

Teach, continued

Describe the Importance of Predator/Prey Interactions

- Have students discuss how the lizards and beetles of an area depend on one another. Ask: **How do lizards depend on beetles?** (Beetles are food for lizards.) **What do you think would happen to the lizards if the number of beetles became very low?** (Possible answers: The lizards would be forced to find other food, or they might starve and die.)
- Ask: **Do lizards help the beetles in any way? How so?** (Yes. By eating beetles, lizards help keep the beetle population in an area from growing too large.) Discuss how a healthy environment has populations of predators and prey that are in balance with one another.

Make Inferences

Ask: **When two living things interact, must one of them always be harmed? Give an example.** (No. When lion cubs play together, as shown in the photo, both learn hunting skills that will help them survive.) Remind students that a feeding relationship, such as between the lizard and beetle, is only one of the ways that living things interact.

Identify the Niches and Habitats of Living Things

- Point out that the lizard and beetle share the same habitat, which is the place where they live. Say: **Every living thing fills a specific niche, or role, in its habitat. Feeding relationships are part of a niche.**
- Have students observe the photo on page 105. Ask: **What does the photo show about the niche of the lizard?** (Its niche involves catching and eating beetles.) **What does it show about the niche of the beetle?** (Its niche includes being food for lizards.)
- Encourage students to identify other habitats and niches for plants and animals as they read the chapter.

Objectives

Students will be able to:

- Explain how parasites get their food.
- Recognize examples of parasites and their hosts.

Teach, continued

Text Feature: Photo and Caption

- Point to the inset photo of the flea on page 106. Explain that a computer added the tan color to the flea. Ask: **How does the size of this picture of a flea compare to the size of a real flea? Is the picture much smaller, much larger, or the same size?** (The picture is much larger.) Point out the white box and leader line, which indicate that the flea could live on a small area of the cat's fur.
- Have students read the caption. Ask: **Why do you think a cat's fur is a good home for a flea?** (Possible answer: The flea can take in nutrients from the cat's blood; the flea can hide from enemies in the fur.)
- Ask: **Does the flea help the cat, harm it, or neither help nor harm it?** (It harms the cat.) **How can you tell?** (The cat is harmed when it loses blood to the flea, and the flea irritates the cat's skin.)

Recognize Parasites and Their Hosts

- Have students read pages 106–107.
- Ask: **Which of these two animals is the parasite, and which is the host?** (The flea is the parasite, and the cat is the host.) **Why is a flea an example of a parasite?** (It lives on another animal's body, and it harms that animal.)
- Ask: **Do you think the flea will kill the cat? Why or why not?** (Possible answer. No. Although the flea harms the cat, it likely will not take enough blood to kill it.) Discuss how parasites may weaken their hosts but usually do not kill them directly. If the parasite did kill its host, it would lose a place to live.

Parasites When a bird lives in a nest in a tree, the bird does not harm the tree. But some consumers do harm other living things by living on or in them. These consumers are called parasites. They take nutrients from other living things. The animal or plant that provides the nutrients is called a host. Parasites harm their hosts but usually do not kill them.

Cats may be the host of several kinds of parasites.

A flea digs through the cat's fur and into the skin.

106

NATIONAL GEOGRAPHIC Raise Your SciQ!

Parasites A parasite is an organism that lives on or inside the body of another organism, called the host, and that harms the host. Examples of parasites include many species of tapeworms and roundworms. These worms feed on blood from their host's intestinal tract or blood vessels. In humans, most parasitic worms are acquired from contaminated water or food, especially undercooked meat, fish, and shellfish. Parasitic infections are a serious health threat in developing countries.

Some parasites live on their hosts. Fleas dig through the skin of pets and other animals to suck their blood. After the meal of blood, the flea jumps off. Ticks, on the other hand, attach themselves to their hosts as they dig through the skin to suck blood.

Not all parasites are consumers. Some are producers. Look at the green clumps on the tree. These are plants called mistletoe. They take water and nutrients that travel through the tree trunk and branches.

These clumps of mistletoe are not part of the tree. They are parasites that live on the tree.

107

Teach, continued

Text Feature: Photos and Captions

- Have students observe the photo of the tree and mistletoe on page 107 and read the caption. Ask: **What are the two kinds of plants that the photo shows?** (a tree and mistletoe)
- Ask: **Why is mistletoe classified as a parasite?** (It lives on the tree trunk and harms the tree. It takes water and nutrients from the tree's trunk and branches.) Remind students that parasites may be plants or animals.
- Point to the photo of the flea on page 106. Ask: **How does mistletoe living on a tree compare to a flea living on a cat?** (Both live as parasites, harming their hosts.) Discuss how cats and other animals may feel discomfort or pain from a parasite, while a tree does not.

Make a Classification Chart

Have students make a two-column chart in their science notebook for classifying interactions as harmful to one organism or helpful to both organisms. Have them make entries in the chart for each of the interactions described in the lesson.

Science Misconceptions

Not All Bad Because parasites harm their hosts, students may think of them as harmful in all ways. Explain that parasites are natural parts of the environment, and while they harm their hosts they may be beneficial to other organisms. Ticks, for example, are food for guinea fowl and other ground-feeding birds. Mistletoe is a food source and nesting area for many animals.

Objectives

Students will be able to:

- Describe examples of how plants and animals interact with one another.

Teach, continued

Describe Helpful Interactions Between Plants and Animals

- Have students read pages 108 and 109.
- Point to the photo of the bee on page 108. Ask: **Where can you observe yellow pollen grains in this photo?** (on the parts of the flower and sticking to the bee's body) Explain that the flower makes a very large number of pollen grains.
- Ask: **How is the bee helping the plant that made this flower?** (The bee is spreading pollen from one flower to another. This helps new plants form and grow.) Explain that flowers make seeds, and seeds can grow into new plants. Before a seed can form, however, a flower needs to receive pollen from another flower.
- Ask: **How do both the bee and plant benefit from the bee's visit to the flower?** (The bee gets food from the flower, while the plant gets its pollen spread to other flowers so that seeds can form.) Explain that bees use both pollen and nectar for food. Nectar is a sweet liquid that flowers make.

Living Things Help Each Other Sometimes living things help each other when they interact. For example, the bee in the photograph is helping the plant. The plant is also helping the bee. The flower of the plant makes yellow grains of pollen. The bee uses the pollen for food. The pollen sticks to the bee as it flies from flower to flower. The pollen then rubs off onto the other flowers. The flowers use the pollen to grow new plants.

Bees carry yellow grains of pollen from flower to flower.

108

NATIONAL GEOGRAPHIC Raise Your SciQ!

Bee Troubles Honey bees helps pollinate a huge number of farm crops, including apples, oranges, lemons, broccoli, cantaloupe, avocados, carrots, and cucumbers. Unfortunately, beekeepers have been reporting huge losses in their colonies, a phenomenon called Colony Collapse Disorder. Scientists have not identified the cause of this disorder. The cause could be related to bee parasites, such as varroa mites, or to environmental stresses, such as pesticides, or to a combination of many factors. Although other insects can and do pollinate farm crops, none perform the task as effectively as honey bees.

Living things help each other in many other ways. In each interaction below, think about how both living things are helped.

HELPFUL INTERACTIONS

SPREADING SEEDS Animals that eat the fruits of plants spread the plants' seeds to a new area. The plants may grow well in the new area.

PROVIDING SHELTER Stinging ants live in the hollow thorns of acacia trees. The thorns keep away ant predators. The ants destroy vines and other plants that might crowd out the acacia trees.

109

Teach, continued

Text Features: Photos and Captions

- Have students study the photos on page 109 and read the captions. Then have them compare the photos with the photo of the bee and flower on page 108. Ask: **What do all three photos have in common?** (All show animals and plants helping each other.)
- Point to the bird eating the berry. Ask: **How are the bird and plant helping each other?** (The bird gets food from the plant, and the plant gets its seeds spread to new places.) Remind students that a bush might make hundreds of tiny berries, such as the one in the photo. Each berry contains one or more seeds. The bird might drop a berry or its seeds in a place where the seeds could grow.
- Point to the ants and acacia tree. Ask: **How are the ants and acacia tree helping each other?** (The ants get shelter and food, and they eat the vines and other plants that would harm the acacia tree.)

Monitor and Fix Up

Students may be confused by the different helpful and harmful interactions among living things. Discuss possible fix-up strategies. Help students identify and learn any unfamiliar words or terms. Students may also review and compare the photos and captions on pages 106–109 or complete the Differentiated Instruction activities on this page.

Differentiated Instruction

Extra Support

Have students review the photos on pages 106–109 and then work with partners to describe one helpful interaction and one harmful interaction among living things. Encourage students to identify the parasites that they observe.

Challenge

Have students select two photos from pages 106–109 and write new captions for them. Students should choose one photo that shows a parasite/host interaction and one photo that describes an interaction in which both organisms benefit.

Objectives

Students will be able to:

- Explain how individuals within a species may compete with each other.
- Identify the kinds of resources for which living things compete.

Teach, continued

Explain Competition Among Animals

- Have students read pages 110 and 111.
- Ask: **What things do animals compete for?** (Possible answers: food, water, and shelter) **Why do they compete for these resources?** (Animals need these resources to survive.) Discuss how food, water, and shelter may be scarce in the place where an animal lives.
- Point to the photo of the two horses on page 110. Ask: **What is the environment like in this place?** (It is dry and dusty; the plants appear brown and cover the ground only in spots) Say: **Like other animals, horses need food, water, and shelter.** Ask: **Which resources might be scarce in this environment?** (food and water)
- Read the caption. Ask: **How are these horses competing?** (by fighting) Discuss how the fighting might end when one horse gives up and walks away.

Living Things Compete with Each Other If two hungry robins see the same worm, what do they do? They compete for it. The faster or stronger robin might end up with the worm. All living things need the same resources, such as food, water, and shelter. How are these horses competing with each other?

These horses fight for water near a water hole.

110

Differentiated Instruction

ELL Language Support for Identifying Interactions Among Living Things

BEGINNING

Help students learn words to identify interactions by asking either/or questions, such as:

Is a flea a parasite or a host?

When horses fight, do they compete or cooperate?

INTERMEDIATE

Help students learn words to identify interactions by using Academic Language Frames, such as:

- *A ____ lives on or inside another organism and harms it.*
- *A ____ is harmed by a parasite.*
- *Living things may ____ for the same resources.*

ADVANCED

Have students complete Academic Language Stems to describe interactions:

- *A parasite harms its host by . . .*
- *Animals compete for . . .*
- *An example of a helpful interaction is . . .*

An area has only a certain amount of resources. Living things have to compete for them, like the robins trying to get the worm and the horses trying to get water. Now look at the picture below. How are the vines competing with the tree?

Strangler figs are plants that compete with trees for sunlight by growing up the trunks of the trees.

Before You Move On

1. What is a parasite?
2. How do some bees and plants help each other?
3. **Infer** Why would a parasite benefit from its host being healthy?

111

NATIONAL GEOGRAPHIC Raise Your SciQ!

Strangler Fig The strangler fig gets its name for its strangling-like effect on the tree it grows around. Every year, the vines of the strangler fig grow thicker and thicker. Eventually, the effect of the vines is so great that the tree can no longer take in the mass of water and nutrients that it needs from the soil. The strangler fig also grows leaves that block the sunlight. After many years, the tree dies from lack of water, nutrients, and sunlight. The strangler fig continues growing around the dead trunk.

Teach, continued

Text Features: Photos and Captions

- Have students look at the photo on page 111 and read the caption. Point to the tree trunk and identify the branching stem of the strangler fig growing around it.
- Ask the question that concludes the text: **How are the vines competing with the tree?** (The vines are living in the same space as the tree. They are taking nutrients from the soil and receiving sunlight that the tree could use.) Remind students that the strangler fig, like most plants, has leaves that make food using the energy of sunlight. The leaves are not shown in the photo.
- Ask: **How are the other plants in the photo competing with one another?** (All are growing closely together in the same area. They compete for space to grow and for sunlight.) Point out that the photo is relatively dark. The leaves on top of the trees block much of the sunlight from the forest floor.

3 Assess

» Before You Move On

1. **Define** **What is a parasite?** (A parasite is a living thing that lives on or inside its host and that harms the host.)
2. **Explain** **How do some bees and plants help each other?** (Bees use pollen from flowers as food. They also spread pollen to other flowers, which lets the plants reproduce.)
3. **Infer** **Why would a parasite benefit from its host being healthy?** (An unhealthy host might die quickly, and the parasite would either have to find a new host or it would die.)

Objectives

Students will be able to:

- Describe forest, prairie, tundra, and ocean environments and the types of plants and animals they support.
- Classify living things as producers or different kinds of consumers.

❶ Introduce

Tap Prior Knowledge

- Have students name and describe plants and animals that live in nature in your region. Then have them name and describe those that live elsewhere, with possible examples including cactuses, banana trees, seaweed, and a wide variety of animals. Lead students to recognize that different plants and animals live in different environments around the world.

Set a Purpose

- Read the heading. Then have students preview the photos on pages 112–119. Tell students that they will learn about several different kinds of environments and some of the interactions of the living and nonliving things in them.

❷ Teach

Describe a Forest Environment

- Have students read pages 112–113. Then have them observe the photos and read the captions. Ask: **What plants live in the forest environment?** (trees, with ferns, mosses, and other small plants growing on the forest floor) Explain that the photo shows only one example of a forest. Other forests are home to different kinds of trees and other plants.
- Point to the photo of the eagle on page 112. Ask: **How do eagles depend on the plants of the forest?** (They build their nests on treetops.) Point to the deer. Ask: **How do deer depend on the plants?** (Deer eat plants.) Remind students that an *herbivore* is a consumer that eats plants.

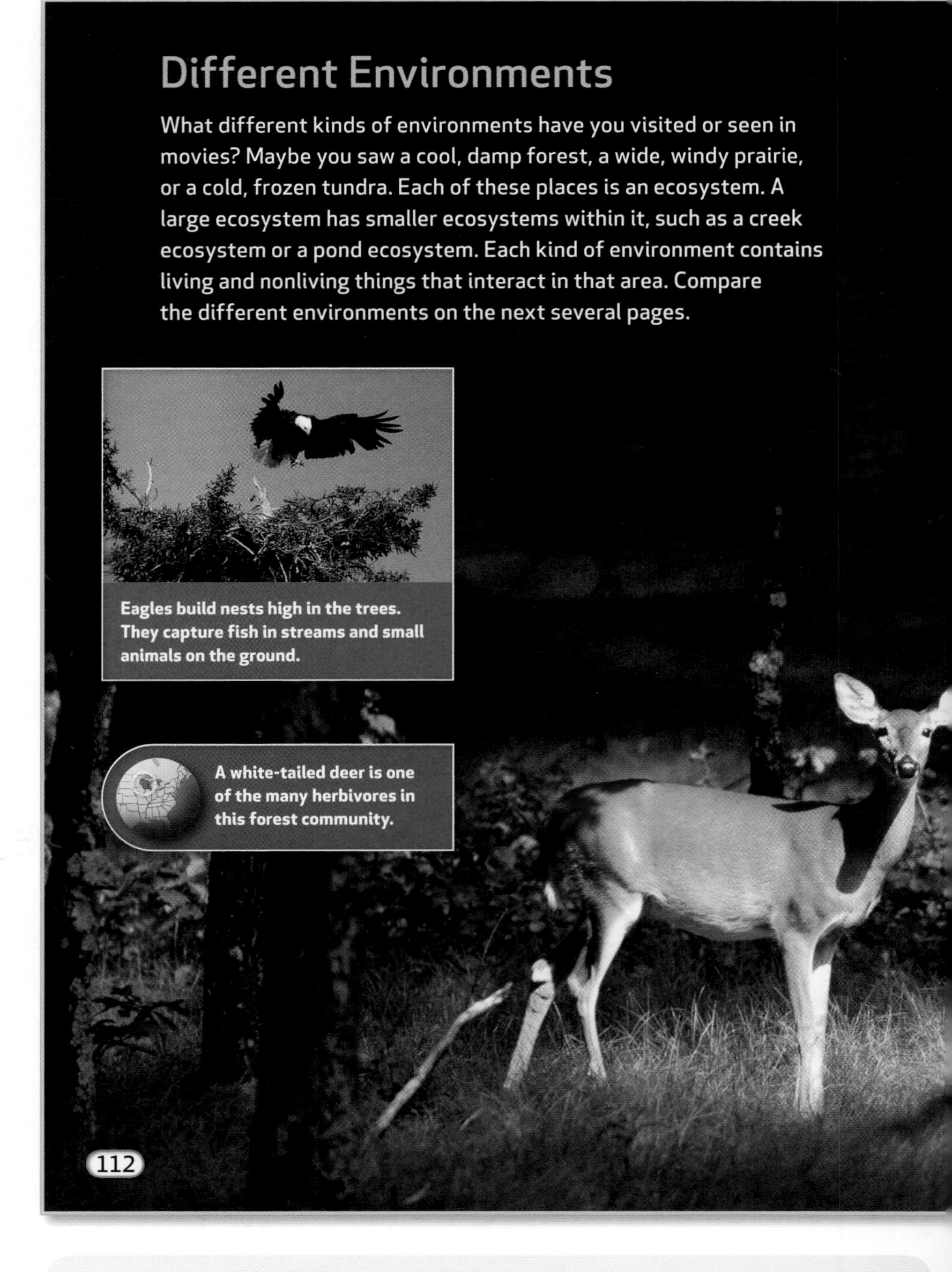

Different Environments
What different kinds of environments have you visited or seen in movies? Maybe you saw a cool, damp forest, a wide, windy prairie, or a cold, frozen tundra. Each of these places is an ecosystem. A large ecosystem has smaller ecosystems within it, such as a creek ecosystem or a pond ecosystem. Each kind of environment contains living and nonliving things that interact in that area. Compare the different environments on the next several pages.
Eagles build nests high in the trees. They capture fish in streams and small animals on the ground.
A white-tailed deer is one of the many herbivores in this forest community.
112

Social Studies in Science

Earth's Forests Show students a world map that identifies biomes (types of environments). Help students read the map's legend to identify lands that have forests. Identify the different kinds of forests, which include temperate rain forests (such as those found in the northwest United States), tropical rain forests (found in tropical regions), and the taiga, or coniferous forest (found in Canada and Alaska and regions of similar latitude).

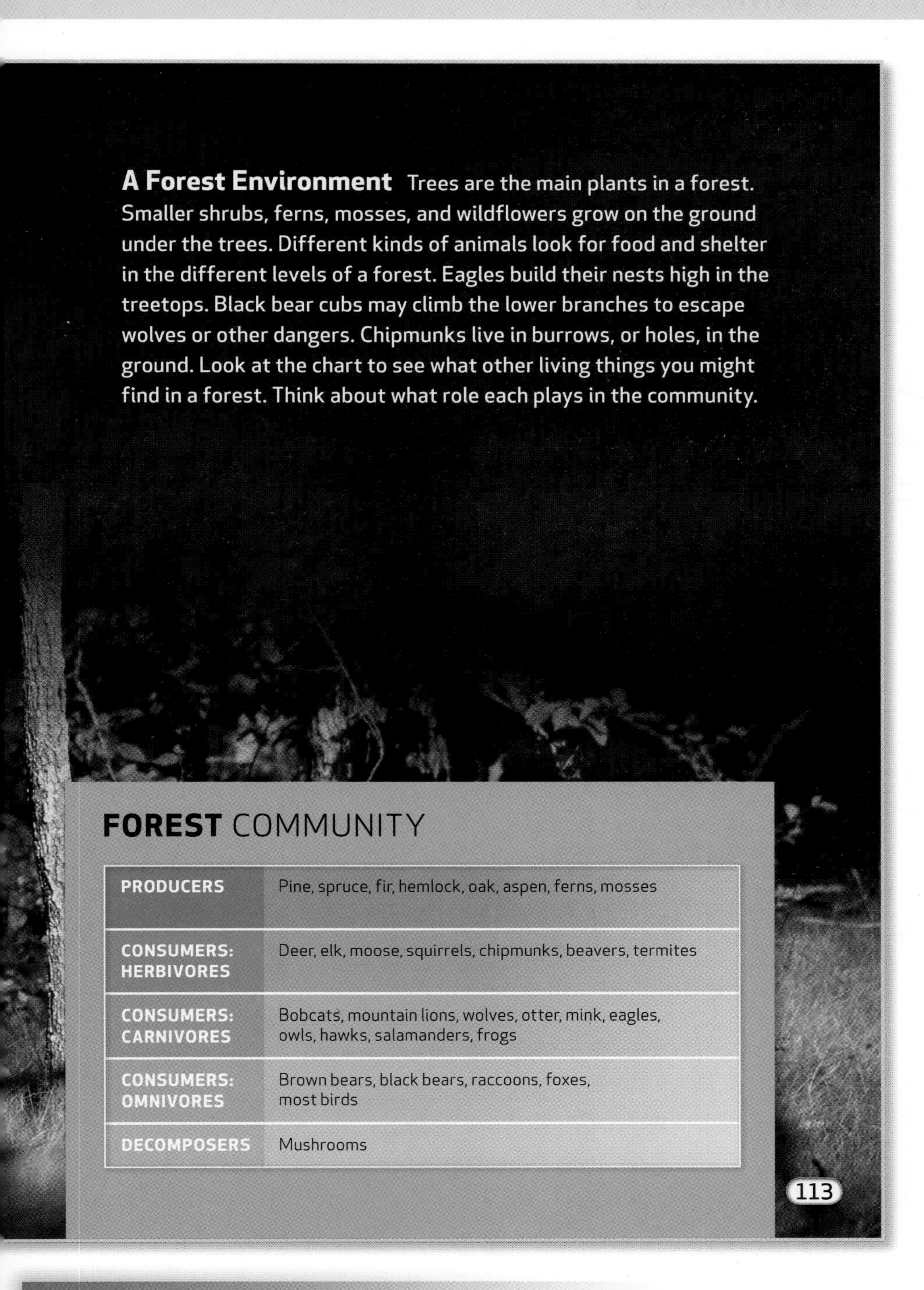

A Forest Environment Trees are the main plants in a forest. Smaller shrubs, ferns, mosses, and wildflowers grow on the ground under the trees. Different kinds of animals look for food and shelter in the different levels of a forest. Eagles build their nests high in the treetops. Black bear cubs may climb the lower branches to escape wolves or other dangers. Chipmunks live in burrows, or holes, in the ground. Look at the chart to see what other living things you might find in a forest. Think about what role each plays in the community.

FOREST COMMUNITY

PRODUCERS	Pine, spruce, fir, hemlock, oak, aspen, ferns, mosses
CONSUMERS: HERBIVORES	Deer, elk, moose, squirrels, chipmunks, beavers, termites
CONSUMERS: CARNIVORES	Bobcats, mountain lions, wolves, otter, mink, eagles, owls, hawks, salamanders, frogs
CONSUMERS: OMNIVORES	Brown bears, black bears, raccoons, foxes, most birds
DECOMPOSERS	Mushrooms

113

Teach, continued

Identify Parts of the Forest Environment

- Have students compare the forest with a house with different levels: a roof and upper floors, lower floors, the ground floor, and the basement. Explain that different animals live in or use different parts of the house. Have students use the chart on page 113 to answer the following questions. Students may also name other animals not listed on the chart. Ask:
  - **Which animals live near or use the roof and upper floors of the forest?** (Possible answers: the eagle, owls, hawks, other birds, and squirrels)
  - **Which animals live in or use the lower floors of the forest?** (Possible answers: brown or black bears and bear cubs, raccoons)
  - **Which organisms live on the ground floor of the forest?** (Possible answers: deer, elk, moose, bobcats, wolves, salamanders, frogs, foxes, mosses, mushrooms)
  - **Which animals live in the basement of the forest?** (termites, chipmunks, ants, earthworms)

Identify Roles of the Forest Community

- If necessary, have students review the definitions of the terms presented in the chart on page 113. Then ask: **How do the producers, consumers, and decomposers of the forest community interact with one another?** (They form food chains, which are feeding relationships.) Review how consumers eat producers or other consumers. Decomposers break down their remains.
- Point out that the bobcat is a carnivore. Ask: **Why do bobcats depend on the plants of the forest, even though bobcats do not eat plants?** (Bobcats hunt and eat other animals, and their prey eat plants.) Remind students that all living things in a food chain depend on one another and that all food chains begin with producers, such as plants.

Differentiated Instruction

ELL Language Support for Identifying Roles of the Forest Community

BEGINNING

Have students copy the first column of the chart on page 113 and then draw one example of an organism for each entry. Have students label their drawings.

INTERMEDIATE

Provide Academic Language Frames, such as these:

- *A forest is an example of a(n) ____.*
- *Pine and spruce trees are ____ of the forest.*
- *Deer and bears are ____ of the forest.*

ADVANCED

Help students create a chart that compares forest community members. Then help students use their charts as aids to describe how forest community members interact.

Objectives

Students will be able to:

- Describe forest, prairie, tundra, and ocean environments and the types of plants and animals they support.
- Classify living things as producers or different kinds of consumers.

Teach, continued

Describe a Prairie Environment

- Read the heading on page 114, and then have students observe the photo of the prairie dog on the prairie. Ask: **What does the prairie environment look like?** (It is mostly flat and somewhat dry, and the land is covered in grasses and dirt.)
- Have students read the caption and study the locator map. Say: **The prairie environment is found in the midwestern United States, which spans from north to south across the western center of the country.** Discuss how today, much of the wild prairie has been taken over for farms, ranches, and cities.
- Have students read pages 114 and 115.

A Prairie Environment A prairie has mostly grasses and wildflowers. The climate affects the kinds of producers that live here. Tall grasses grow in warmer, wetter climates. Short grasses grow in cooler, drier climates. The kinds of animals in a prairie depend on the kinds of plants that grow there. Compare the chart of the prairie community with the forest community. Which living things can be found in both environments?
Black-tailed prairie dogs live in burrows on the short-grass prairies of the midwestern United States.
114

NATIONAL GEOGRAPHIC Raise Your SciQ!

Fires Wilderness fires can be destructive and dangerous. Scientists are also concerned about the apparent rise in their frequency due to drier weather, a possible effect of global warming. Yet fires are also natural events that are a healthy part of some ecosystems. On prairies, wildfires help prevent the growth of trees and keep grasses as the dominant plant. In forests, wildfires clear away old trees and allow new plants to replace them. The cones of the jack pine depend on the heat from fire to release their seeds.

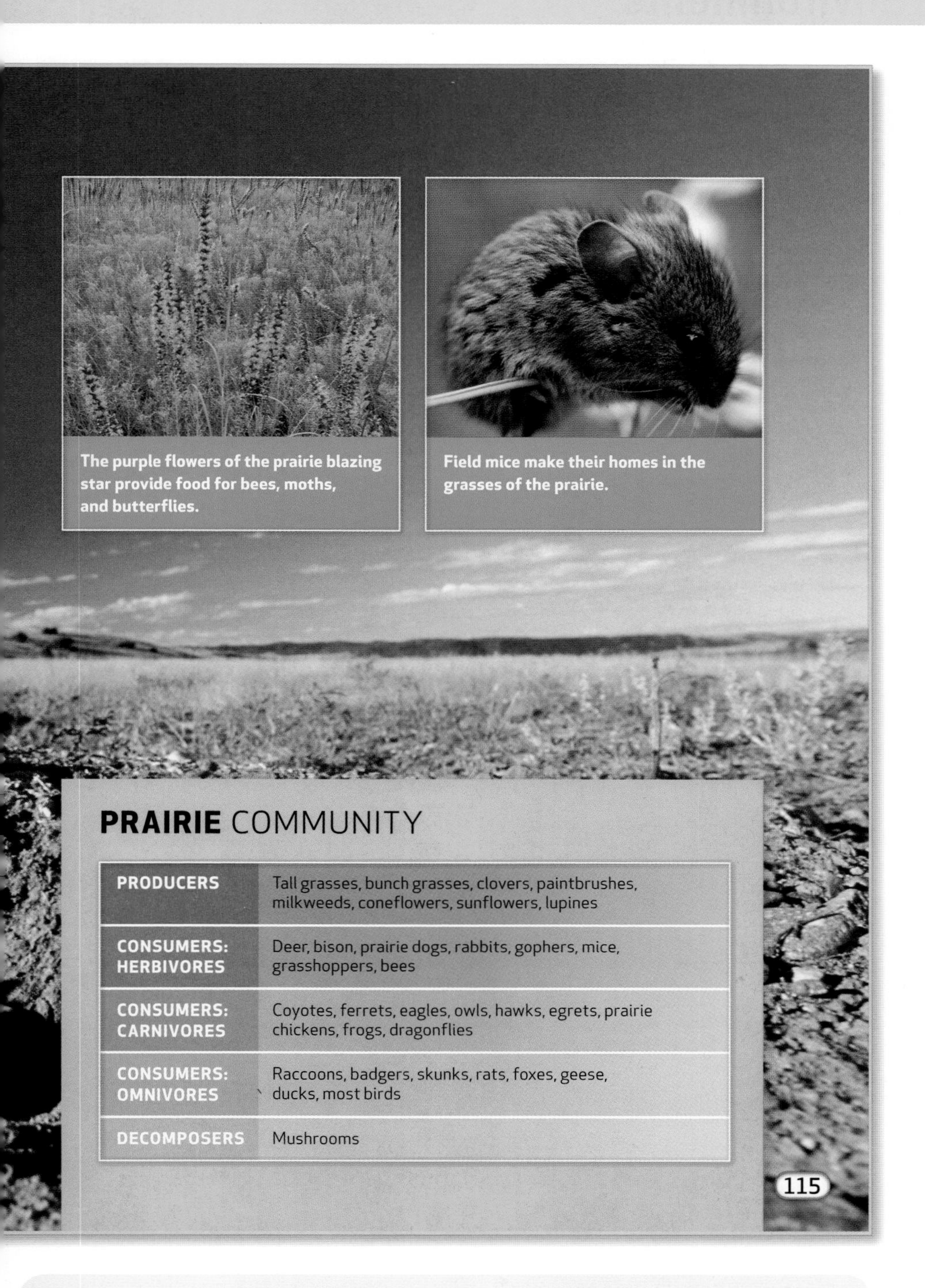

The purple flowers of the prairie blazing star provide food for bees, moths, and butterflies.

Field mice make their homes in the grasses of the prairie.

PRAIRIE COMMUNITY

PRODUCERS	Tall grasses, bunch grasses, clovers, paintbrushes, milkweeds, coneflowers, sunflowers, lupines
CONSUMERS: HERBIVORES	Deer, bison, prairie dogs, rabbits, gophers, mice, grasshoppers, bees
CONSUMERS: CARNIVORES	Coyotes, ferrets, eagles, owls, hawks, egrets, prairie chickens, frogs, dragonflies
CONSUMERS: OMNIVORES	Raccoons, badgers, skunks, rats, foxes, geese, ducks, most birds
DECOMPOSERS	Mushrooms

115

Teach, continued

Describe Interactions Among Members of the Prairie Community

- Point to the photo of the prairie dog on page 114. Then have a volunteer find the prairie dog in the chart. Ask: **How are prairie dogs classified?** (as consumers that are herbivores) Explain that prairie dogs eat the grasses of the prairie.
- Have another volunteer read the list of carnivores in the chart on page 115. Ask: **Which of the carnivores do you think might catch and eat a prairie dog?** (Possible answer: coyotes, ferrets, eagles, owls, hawks) Explain that a ferret is an animal like a weasel. Some ferrets hunt prairie dogs exclusively.
- Ask: **How do you think it helps the prairie dog to live in a burrow?** (Possible answer: The burrow provides shelter and protection from predators and the weather.) Discuss how a prairie has very few places where animals can take shelter. Prairie dogs, gophers, rabbits, and ferrets all find shelter underground. A few large prairie animals, such as bison, live above ground at all times.

Compare the Prairie and Forest Environments

- Have students answer the question that concludes the text on page 114. Ask: **Which living things can be found in both the forest and prairie environments?** (deer, eagles, frogs, raccoons, birds, mushrooms)
- Ask: **What do you think is the biggest difference between the forest and prairie?** (Possible answer: the presence of trees) Discuss how the different levels of trees—treetops, middle branches, floor, and roots—each provide food and shelter for many animals. Without trees, a prairie does not have these levels of habitats.

Make a Compare and Contrast Chart

Have students prepare a chart in their science notebook that compares and contrasts the forest and prairie environments. Encourage students to draw pictures and add labels to illustrate the two environments.

Social Studies in Science

Grasslands Show students a world map that identifies biomes (types of environments.) Help students read the map's legend to identify grasslands, of which prairies are a part. Point out the large regions of grasslands in the midwestern United States and the savannah of eastern Africa. Discuss how many familiar zoo animals, including lions, giraffes, zebras, gazelles, and wildebeests, are native to the African savannah.

Objectives

Students will be able to:

- Describe forest, prairie, tundra, and ocean environments and the types of plants and animals they support.
- Classify living things as producers or different kinds of consumers.

Teach, continued

Describe a Tundra Environment

- Read the heading on page 116, and then have students observe the photo of the caribou on the tundra. Ask: **What does the tundra environment look like?** (It is mostly flat but may include mountains. Only small plants grow. The weather is very cold.) Explain that snow covers the frozen ground in winter and only the top layer thaws during the summer.
- Have students read the caption and study the locator map. Ask: **Where is Alaska?** (in the far northwest corner of North America) Discuss how northern Alaska is near the North Pole. Explain that due to the tilt of Earth, winters experience nearly 24 hours of darkness while summers have nearly 24 hours of sunlight.
- Have students read pages 116 and 117.

A Tundra Environment Tundra environments are found near the Arctic and high in the mountains. The tundra is one of the harshest environments on Earth. The climate is very cold for most of the year. The ground is frozen throughout most of the year. As a result, few large plants grow there. Instead, the ground is covered with lichens, mosses, and small flowering plants. Lichens often form colorful patches on rocks and soil.
Denali National Park in Alaska is tundra country.
116

Differentiated Instruction

Extra Support

Have students use the photos in the Big Ideas Book to describe the forest, prairie, and tundra environments to a partner. Encourage students to describe the climate of each environment as well as its plants and animals.

Challenge

Have students draw a picture to illustrate a forest, prairie, or tundra environment. Encourage students to show and label a variety of the plants and animals listed in the Big Ideas Book. Encourage students to research unfamiliar plants or animals that they would like to include.

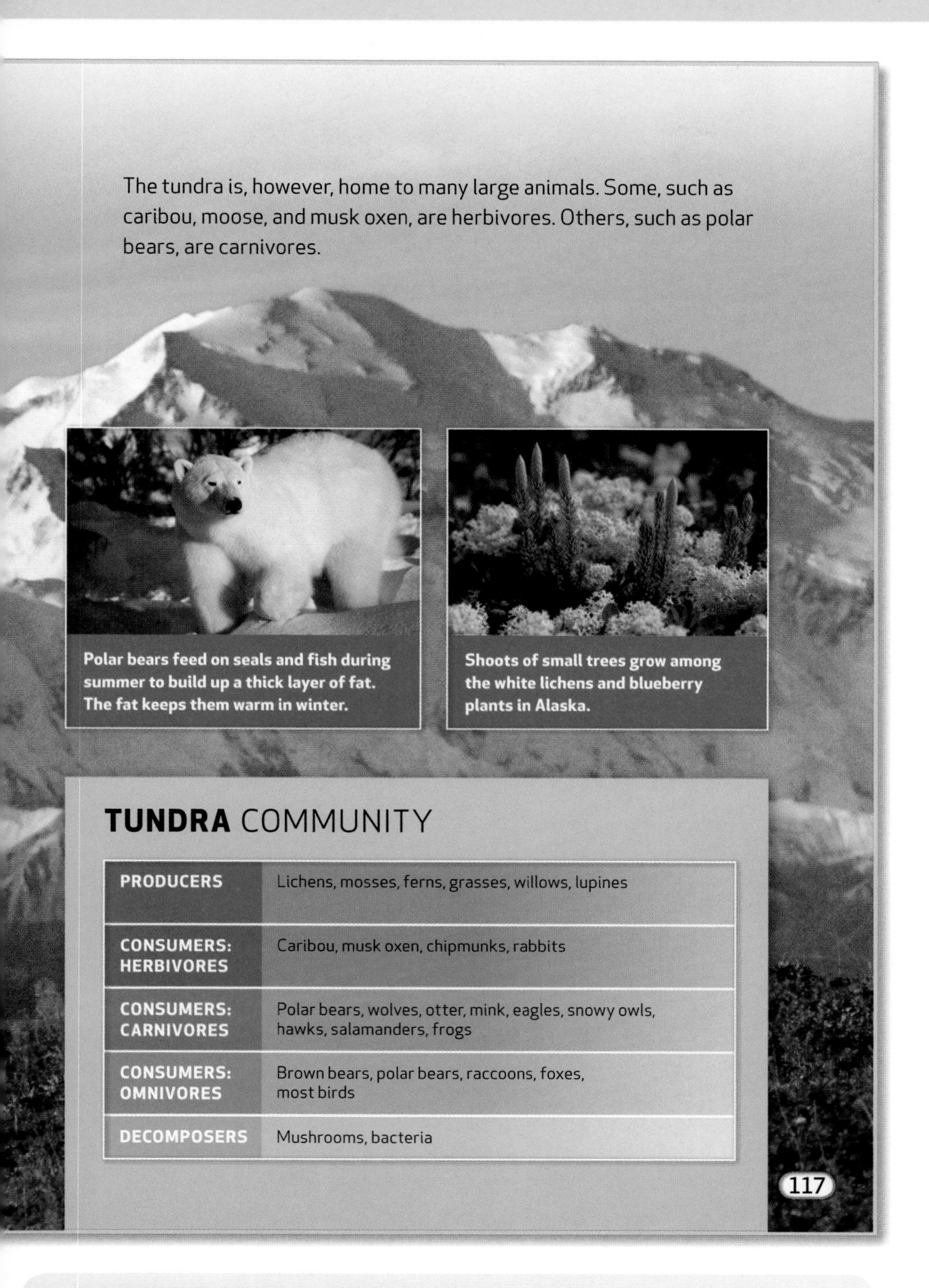

The tundra is, however, home to many large animals. Some, such as caribou, moose, and musk oxen, are herbivores. Others, such as polar bears, are carnivores.

Polar bears feed on seals and fish during summer to build up a thick layer of fat. The fat keeps them warm in winter.

Shoots of small trees grow among the white lichens and blueberry plants in Alaska.

TUNDRA COMMUNITY

PRODUCERS	Lichens, mosses, ferns, grasses, willows, lupines
CONSUMERS: HERBIVORES	Caribou, musk oxen, chipmunks, rabbits
CONSUMERS: CARNIVORES	Polar bears, wolves, otter, mink, eagles, snowy owls, hawks, salamanders, frogs
CONSUMERS: OMNIVORES	Brown bears, polar bears, raccoons, foxes, most birds
DECOMPOSERS	Mushrooms, bacteria

117

NATIONAL GEOGRAPHIC Raise Your SciQ!

Polar Bears Polar bears are born on land, but they spend much of their time hunting for food in the ocean. They mostly eat seals and fish but have been observed to hunt walruses and other larger animals. Although polar bears can swim, they hunt seals from floating patches of ice, called sea ice. Many scientists are concerned that the polar bear population will decline due to global warming, which is predicted to lower the amount and frequency of sea ice. In 2008, the United States government classified polar bears as a threatened species.

Teach, continued

Describe Plants and Animals of the Tundra Community

- Point to the small photo of the tundra plants and lichens on page 117, and have students read the caption. Explain that lichens are not plants but, like plants, are classified as producers. Ask: **How are these plants and lichens alike?** (All are very small.)
- Ask: **Why don't plants grow very large or tall on the tundra?** (The ground is frozen for most of the year.) Discuss how tall plants need long, strong roots to anchor them into the soil. A tree's roots may not be able to penetrate frozen ground, nor could they take in much water or nutrients.
- Have a volunteer read the list of herbivores in the chart on page 117. Then identify the animal on page 116 as a caribou. Ask: **How do the herbivores of the tundra compare to those of the forest and prairie?** (All eat plants. The herbivores of the tundra eat only small plants, however, and they must be able to survive very cold weather.) Point out that chipmunks and rabbits live in the tundra and other environments.
- Point to the photo of the polar bear on page 117, and have a volunteer read the caption. Ask: **Where does the polar bear find food?** (in the ocean) Discuss how some animals, such as the polar bear, may spend their time in two neighboring environments, such as the tundra and the ocean.

Identify Characteristics of Plants and Animals in the Desert

- If possible, show students a photo of cacti in a desert environment. Say: **A desert is a very dry environment that usually is very hot during the day and cold at night. Cacti live well in the desert, but many types of trees and other plants would not survive there.**
- Ask: **How do you predict the roots of a cactus are different from roots of other plants? Explain.** (Possible answer: A cactus has a much greater and wider network of roots than do plants of similar size. This helps the cactus take in as much water as possible.) Have students investigate and identify other characteristics for plant and animal survival in the desert.

Objectives

Students will be able to:

- Describe forest, prairie, tundra, and ocean environments and the types of plants and animals they support.
- Classify living things as producers or different kinds of consumers.

PROGRAM RESOURCES

- Learning Masters Book, page 42, or at myNGconnect.com

Teach, continued

Describe an Ocean Environment

- Read the heading on page 118, and then have students observe the photos on pages 118–119. Ask: **How is the ocean environment different from the other environments we have studied so far?** (It is underwater, and it is much larger.) Explain that oceans cover more of the Earth's surface than all the land environments combined.
- Ask: **Do you think any of the living things shown in the photos could also live in a land environment? Why or why not?** (Possible answer: No. All make or find food in the ocean's watery environment, and they likely could not survive for long on land.) Explain that the sea turtle breathes air and that female turtles come to land to lay eggs. However, sea turtles find food in the ocean and live nearly all their lives there.
- Have students read pages 118 and 119.

Describe the Role of Plankton in Ocean Food Webs

- Ask: **What are plankton?** (tiny living things, such as bacteria and algae, that float in the water)
- **Why are plankton important in ocean food webs?** (They provide the food for many consumers, such as snails, fish, and brine shrimp.) Explain that many plankton are producers. They capture the energy of sunlight and bring it into ocean food webs, just as plants do on land.

An Ocean Environment The ocean is the largest environment on Earth. It also has the largest variety of living things. The surface waters get plenty of sunlight and nutrients. So these waters are full of life. However, you would need a microscope to see much of that life. It is in the form of plankton. These tiny living things float in the water. Plankton include bacteria, algae, and even tiny animals.

This sea turtle makes a meal of soft coral.

118

Think Like a Scientist Math in Science

Earth's Oceans Display a globe, and have students identify the continents and oceans. Have them recognize that oceans cover more of Earth's surface than lands cover. To quantify this fact, draw ten identical squares on the board. Then shade seven of the squares with blue chalk, if available. Explain that if each square stood for an equal section of Earth, seven squares would be covered in water. Write the fraction 7/10 to show this relationship.

Plankton are a major part of ocean food webs. Many plankton are producers, such as algae. Snails, fish, shrimp, and other consumers eat the plankton. Still other consumers eat these consumers. In this way, energy from the sun flows throughout the entire ecosystem.

This cup coral has captured an octopus.

Before You Move On

1. What plants, other than trees, are in a forest?
2. How does the ocean compare with land environments?
3. **Infer** Most consumers in the ocean depend on plankton for food even if they don't eat plankton. How is this true?

119

Share and Compare

Construct a Food Chain Give students the Learning Master. Have them draw pictures of the organisms in the graphic organizer and then label the organisms as producers or consumers. Check their work. Have partners discuss how the organisms of a food web depend on each other.

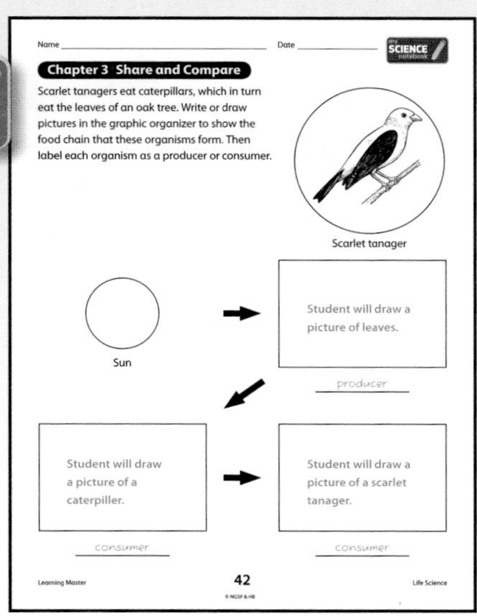

Name ______ Date ______ my SCIENCE notebook

Chapter 3 Share and Compare

Scarlet tanagers eat caterpillars, which in turn eat the leaves of an oak tree. Write or draw pictures in the graphic organizer to show the food chain that these organisms form. Then label each organism as a producer or consumer.

Scarlet tanager

Sun → Student will draw a picture of leaves. (producer)

→ Student will draw a picture of a caterpiller. (consumer) → Student will draw a picture of a scarlet tanager. (consumer)

Learning Master 42 Life Science

© NGSP & HB

Learning Master 42 or at myNGconnect.com

Teach, continued

- Ask: **How is an ocean food chain like food chains on land?** (The sun's energy is captured by a producer and passed from one consumer to the next.) **How is an ocean food chain different?** (The producers are algae, not plants. The consumers are also different types of animals from those that live on land.)

Identify Characteristics of Plants and Animals in a Wetland Environment

- If possible, show students a photo of a marsh, swamp, or other wetlands. Say: **A wetland environment is covered or soaked in water for at least part of the year. The water could be salty, like ocean water, or fresh, like the water in lakes, rivers, and ponds.**
- Ask: **How might the trees and grasses that live well in wetlands be different from those that live on dry land?** (Possible answer: In wetlands, the trees and grasses must be able to stand up in soft, wet soil.) Discuss how cypress trees, which grow in many wetlands, have very wide roots that help support them.
- Have students investigate and identify characteristics for plant and animal survival in wetlands.

❸ Assess

» Before You Move On

1. **List What plants, other than trees, are in a forest?** (shrubs, ferns, mosses, and wildflowers)
2. **Compare How does the ocean compare with land environments?** (The ocean is much larger and has a larger variety of living things than land environments.)
3. **Infer Most consumers in the ocean depend on plankton for food even if they don't eat plankton. How is this true?** (Even if a consumer doesn't eat plankton, it may eat another consumer that eats plankton. So most ocean consumers depend on plankton for food.)

Objectives

Students will be able to:

- Investigate through Guided Inquiry (answer a question; make and compare observations; collect and record data and observations; generate explanations and conclusions based on evidence; share findings; ask questions based on observations to increase understanding; adjust explanations based on findings and new ideas).
- Trace the flow of energy through various living and nonliving systems.
- Know that when solving a problem, it is important to plan and get ideas and help from other people.

Science Process Vocabulary

share, conclude

PROGRAM RESOURCES

- Science Inquiry and Writing Book: *Life Science*
- Science Inquiry and Writing Book eEdition at myNGconnect.com
- Inquiry eHelp at myNGconnect.com
- Science Inquiry Kit: *Life Science*
- Learning Masters Book, pages 43–46, or at myNGconnect.com
- Inquiry Rubric: Assessment Handbook, page 184, or at myNGconnect.com
- Inquiry Self-Reflection: Assessment Handbook, page 194, or at myNGconnect.com

MATERIALS

Kit materials are listed in italics.

protective gloves; bones from *owl pellet; forceps; 4–6 index cards; Bone Sorting Chart; Skeleton diagrams;* 2–3 sheets of white paper; 1 sheet of construction paper; glue; *Food Web diagram*

Guided Inquiry

Investigate Food Chains and Webs

Question How can you use an owl pellet to infer a food chain and food web?

Science Process Vocabulary

share verb

When you **share** results, you tell or show what you have learned.

conclude verb

You **conclude** when you use data from an investigation to come up with a decision or an answer.

I conclude that these two bones came from the same kind of animal.

Materials

40

MANAGING THE INVESTIGATION

Time

40 minutes

Groups

Small groups of 4

Advance Preparation

- Use the bones from the owl pellets students dissected in the Chapter 3 Directed Inquiry, page 36. Each group's bones should be stored in resealable plastic bags.
- Students will also use the Bone Sorting Chart from the Directed Inquiry.

❶ Introduce

Tap Prior Knowledge

- Invite students to draw a simple food chain on the board. Ask: **Which organism can make its own food?** (a plant) **Where does the energy for this food chain begin?** (the sun)

What to Do

1. Put on the plastic gloves. Use the forceps to carefully remove the bones from each resealable bag and place them on an index card. Make sure to keep the bones from each bag together.

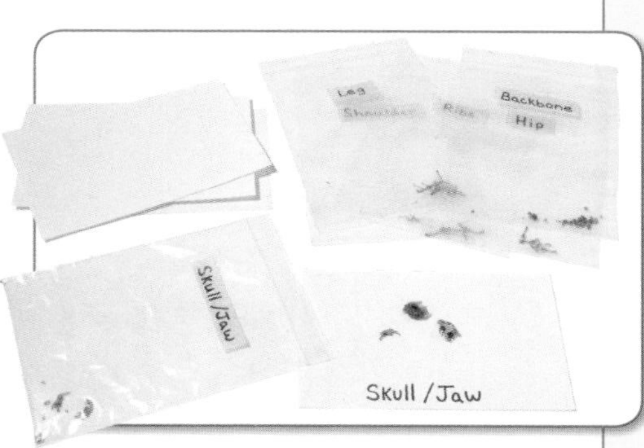

2. Use the Bone Sorting Chart to identify the animal to which each bone belongs. If you have bones from more than one animal, sort all the bones from each animal and place them on separate sheets of paper. Choose the animal for which you have the most bones. Place the bones from the other animals into a resealable bag and set the bag aside.

41

Teaching Tips

- Remind students to handle the bones gently. Ask: **Why is it important to be gentle with the bones?** (They might break.)
- Have students wash their hands after handling the bones.
- The owl pellets are sterile, so they can be thrown in the garbage. Place the pellets in a resealable plastic bag before disposal.

What to Expect

- Students will reconstruct an animal skeleton. They will use the Skeleton and the Food Web diagrams to draw a food chain. Students will then work together to create a food web.

Introduce, continued

Connect to the Big Idea

- Review the Big Idea Question, *How do living things depend on their environment?* Explain that this inquiry will include constructing a food chain and a food web, which will show how energy is transferred from producers to consumers.
- Have students open their Science Inquiry and Writing Books to page 40. Read the Question and invite students to share ideas about how they can use the contents of an owl pellet to help make a food chain and a food web.

2 Build Vocabulary

Science Process Words: share, conclude

Use this routine to introduce the words.

1. **Pronounce the Word** Say **conclude**. Have students repeat it in syllables.
2. **Explain Its Meaning** Choral read the sentence and the example in the speech bubble. Ask students for another word or phrase that means the same as **conclude**. (use information to make a decision or answer a question)
3. **Encourage Elaboration** To encourage elaboration, ask: **What conclusions might you come up with by studying bones?** (You could draw a conclusion about the kind of animal they come from or the type of bone.)

ELL Use Language Frames

Write this sentence on the board: *After studying the data, I will _____ what the owl eats.* Have students read it aloud and fill in the blank to practice using the word **conclude**.

Repeat for the word **share**. To encourage elaboration, say: **Give an example of something you have shared with others.** (Students may mention toys, books, or ideas.)

3 Guide the Investigation

- Read the inquiry steps on pages 41–42. Move from group to group and clarify steps if necessary.
- In step 2, have groups work together to identify the animals to which the bones belong. Remind students that each member of the group can contribute ideas to help with the investigation.

Guide the Investigation, continued

- If students find evidence of more than one skeleton in their owl pellet, you may allow extra time for students to assemble both skeletons on separate pieces of paper. If time is limited, each group should only assemble the skeleton with the most identifiable bones.
- When students create their food chains in step 4, guide them in using the Food Web diagram. Explain that each food chain begins with a producer that receives energy from the sun. Each food chain ends with a consumer. Most food chains will include three to six living things.
- For step 5, construct a class food web on the board. Have each group come up and write the names of the animals in their food chain. They should connect the animals with arrows. Groups should add to what is already on the board.

❹ Explain and Conclude

- Guide students as they share their skeletons and food chains with the class. Ask: **As you constructed your skeleton, why do you think it is important to get ideas from more than one person?** Discuss how sharing ideas with others helps to solve a problem.
- Discuss how the parts of the class food web overlap and connect. Ask: **What do you observe about the different food chains in the food web?** Guide students to understand that the different food chains overlap because they contain some of the same animals and plants.
- Ask: **Do you think the animal you identified is a consumer or a producer?** (consumer) Discuss how the animals fit in the food chain and the food web.
- Have students summarize their observations from the investigation. Ask: **What conclusions can you make based on your observations?** Discuss the differences between observations and conclusions.

What to Do, continued

3. Use the Skeleton diagram for the type of animal you found to make a skeleton on construction paper. Glue the bones to the paper. Draw in any missing bones. Label the skeleton with the name of the animal.

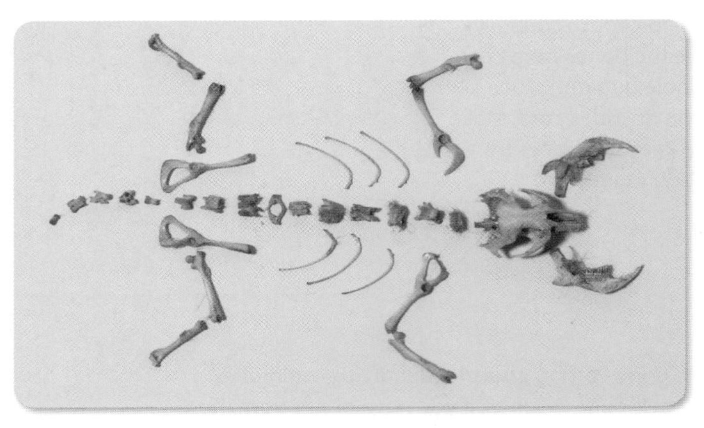

4. Use the results of the investigation and the Food Web diagram to draw a food chain that includes the owl and the animal whose skeleton you put together.

5. **Share** your skeleton and food chain with other groups. **Compare** the food chains that include different animals. Construct a class food web that includes all the animals identified in the owl pellets.

42

How Scientists Work

Descriptive Investigations and Experiments Explain to students that this is a descriptive investigation, not an experiment. They are not identifying, manipulating, or controlling variables in this investigation. However, that does not mean that the data collected have less value than data from an experiment. In fact, scientists learn a great deal by observing objects systematically and using their observations to draw conclusions about the natural world.

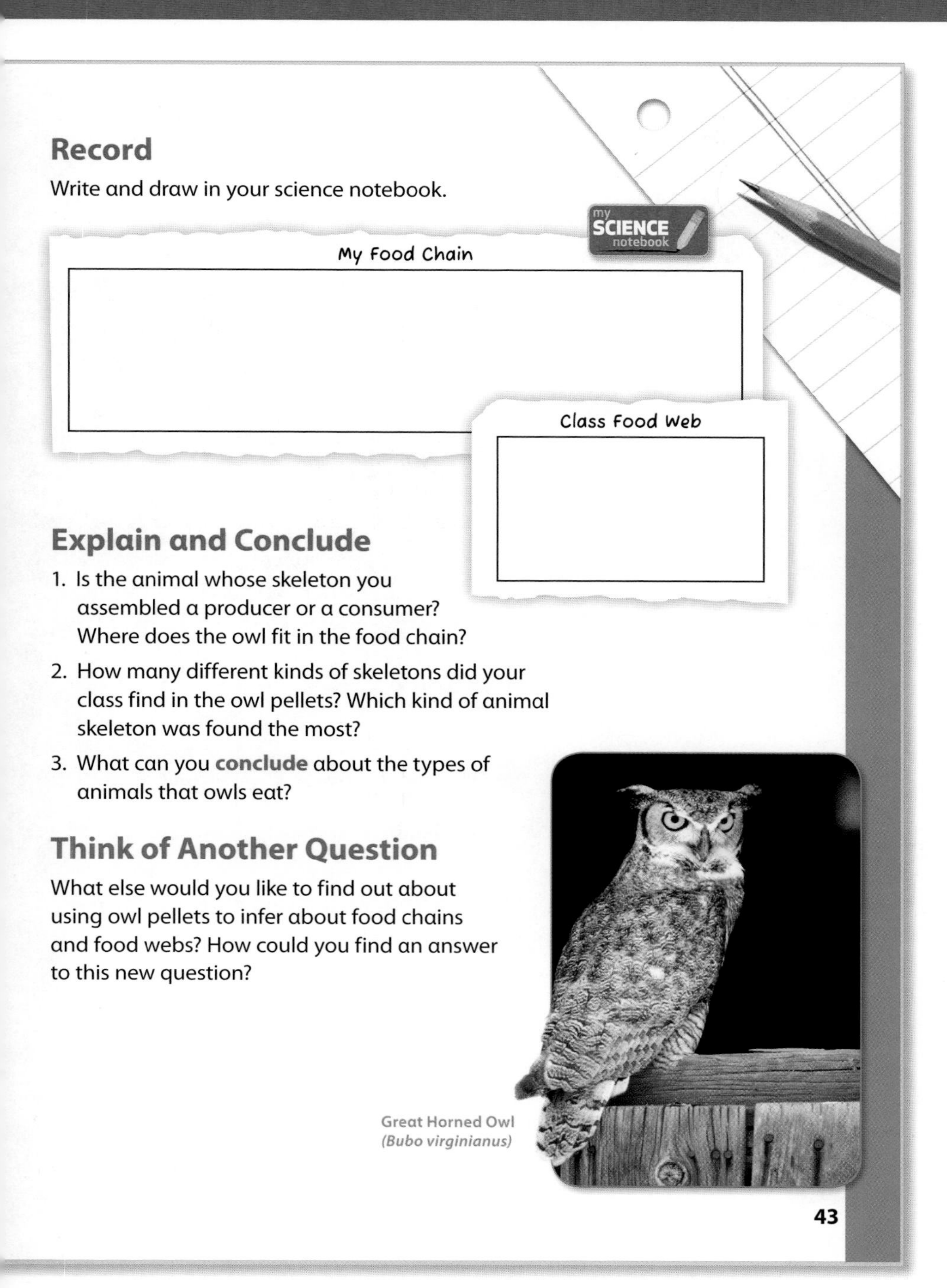

Record

Write and draw in your science notebook.

my SCIENCE notebook

My Food Chain

Class Food Web

Explain and Conclude

1. Is the animal whose skeleton you assembled a producer or a consumer? Where does the owl fit in the food chain?
2. How many different kinds of skeletons did your class find in the owl pellets? Which kind of animal skeleton was found the most?
3. What can you **conclude** about the types of animals that owls eat?

Think of Another Question

What else would you like to find out about using owl pellets to infer about food chains and food webs? How could you find an answer to this new question?

Great Horned Owl
(Bubo virginianus)

43

Inquiry Rubric at myNGconnect.com	Scale			
The student **observed** the bones and identified the animal to which the bones belonged.	4	3	2	1
The student sorted the bones and reconstructed the skeleton of an animal.	4	3	2	1
The student analyzed **data** and made a short food chain.	4	3	2	1
The student drew **conclusions** about the types of animals that owls eat.	4	3	2	1
The student **shared** and **compared** ideas with others to create a larger food web.	4	3	2	1
Overall Score	**4**	**3**	**2**	**1**

Explain and Conclude, continued

Answers

1. Answers will vary, but students should explain that the animal whose remains were in the owl pellet is a consumer. The owl is also a consumer, but it is higher up on the food chain than its prey.
2. Answers will vary depending on students' findings.
3. Possible answer: Owls eat small animals, such as rodents and birds.

5 Find Out More

Think of Another Question

Students should use observations to generate new questions that could be studied using readily available materials. Record student questions and discuss what students could do to find answers.

6 Reflect and Assess

- To assess student work with the Inquiry Rubric shown below, see Assessment Handbook, page 184, or go online at myNGconnect.com
- Have students use the Inquiry Self-Reflection on Assessment Handbook, page 194, or at myNGconnect.com

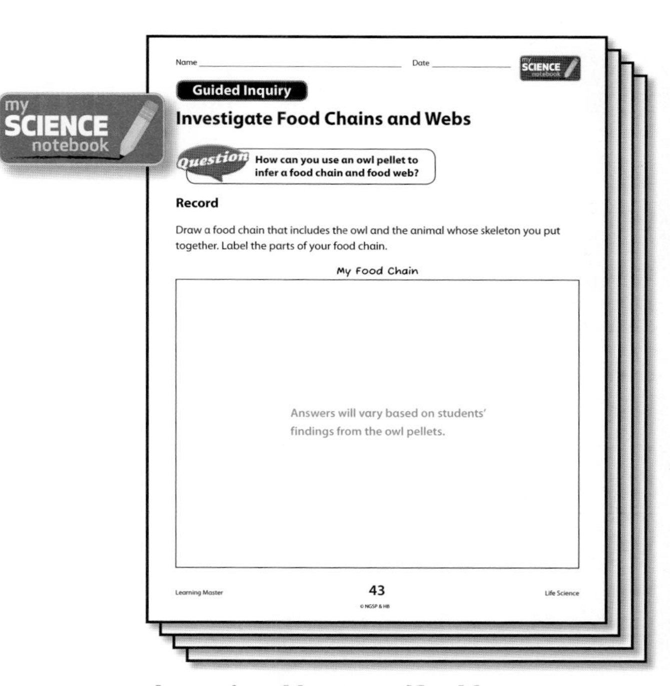

Name ______ Date ______

Guided Inquiry

Investigate Food Chains and Webs

Question: How can you use an owl pellet to infer a food chain and food web?

Record

Draw a food chain that includes the owl and the animal whose skeleton you put together. Label the parts of your food chain.

My Food Chain

Answers will vary based on students' findings from the owl pellets.

Learning Master 43 Life Science

Learning Masters 43–46 or at myNGconnect.com

PROGRAM RESOURCES

- Chapter 3 Test, Assessment Handbook, pages 12–17, or at myNGconnect.com
- NGSP ExamView CD-ROM

❶ Sum Up the Big Idea

- Display the chart from page T88–T89. Read the Big Idea Question. Then add new information to the chart.
- Ask: **What did you find out in each section? Is this what you expected to find?**

How Do Living Things Depend on Their Environment?

What Living Things Need from Their Environment

Brook: Living things get air, water, space, and energy from food where they live.

Communities of Living Things

Jordana: Consumers depend on producers or other consumers to get energy from food.

Food Chains and Food Webs in Communities

Mairead: Food chains and food webs show how energy from food moves from one living thing to the next.

Interactions Among Living Things

Barry: Living things may be harmed when they interact, such as a parasite living off its host, or living things may help one another.

Different Environments

Alexi: Different communities of living things live together in forests, prairies, tundras, and ocean environments.

Conclusion

Energy passes through a community by moving along food chains and food webs. Producers such as plants use energy from the sun to make their food. Consumers, including animals and humans, cannot make their own food. They get energy by eating producers or other consumers. Each kind of environment contains living and nonliving things that interact. These things interact in different ways.

Big Idea Living things depend on each other and their environment for food.

Vocabulary Review

Match each term with its correct definition.

A. predator
B. prey
C. herbivore
D. carnivore
E. omnivore

1. An animal that eats plants to survive
2. An animal that eats both plants and other animals to survive
3. An animal that hunts other animals for food
4. Animals that other animals hunt for food
5. An animal that eats other animals to survive

120

Review Academic Vocabulary

Academic Vocabulary Tell students that scientists use special words when describing their work to others. For example:

herbivore **carnivore** **omnivore**
predator **prey**

Have students write the list in their science notebook and use each word in a sentence that tells how living things depend on their environment. They should then share their sentences with a partner.

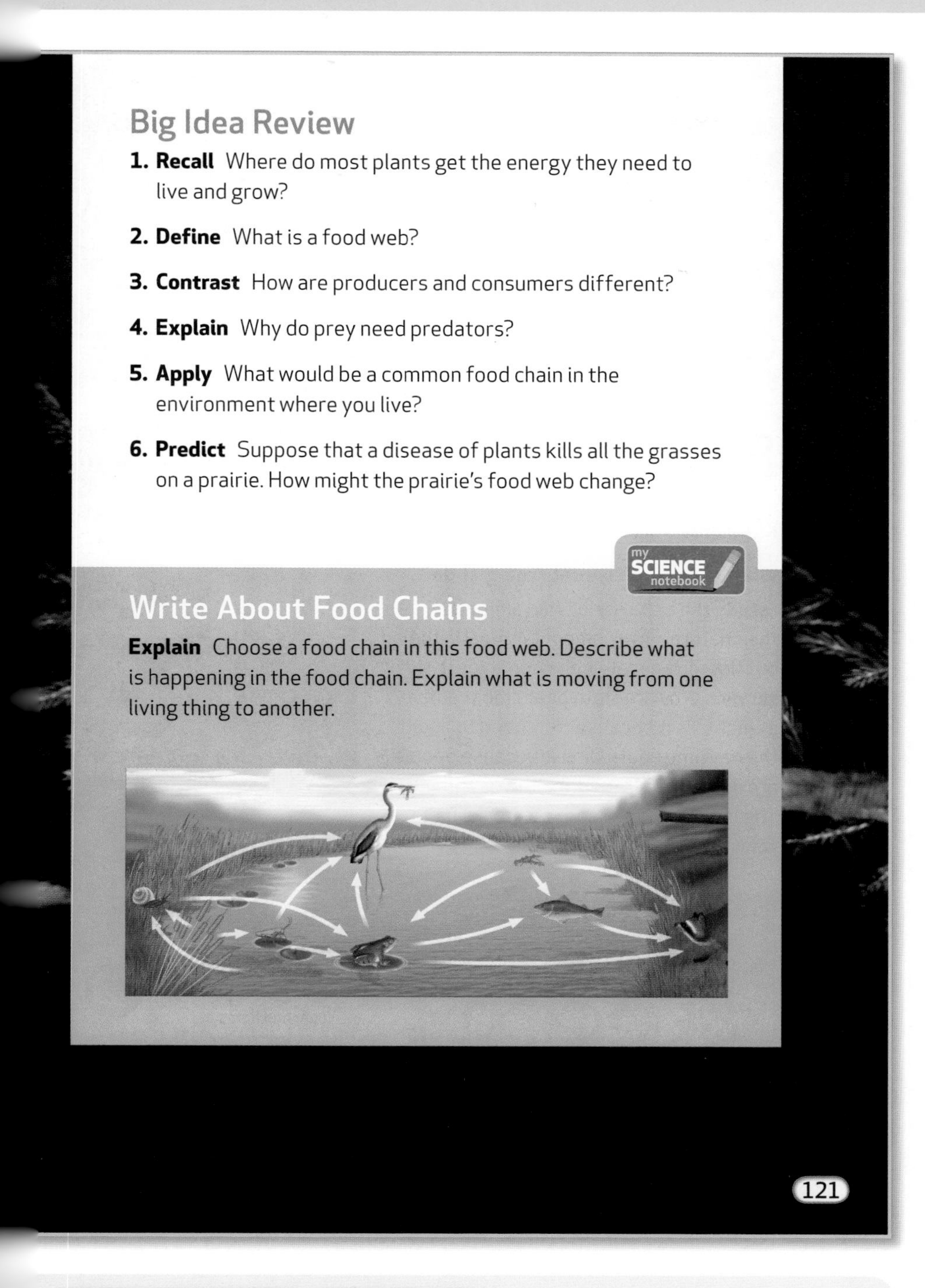

Big Idea Review

1. **Recall** Where do most plants get the energy they need to live and grow?
2. **Define** What is a food web?
3. **Contrast** How are producers and consumers different?
4. **Explain** Why do prey need predators?
5. **Apply** What would be a common food chain in the environment where you live?
6. **Predict** Suppose that a disease of plants kills all the grasses on a prairie. How might the prairie's food web change?

my SCIENCE notebook

Write About Food Chains

Explain Choose a food chain in this food web. Describe what is happening in the food chain. Explain what is moving from one living thing to another.

121

❷ Discuss the Big Idea

Notetaking

Have students write in their science notebook to show what they know about the Big Idea. Have them:

1. Write about what living things get from their environment.
2. Write about producers and the different kinds of consumers found in a community.
3. Write about the way the sun's energy is passed from one living thing to the next in a community.

❸ Assess the Big Idea

Vocabulary Review

1. C **2.** E **3.** A **4.** B **5.** D

Big Idea Review

1. from the sun
2. A food web is a diagram of all of the connected food chains in a community.
3. Producers get energy from the sun. Consumers get energy by eating producers or other consumers.
4. Predators keep the number of prey down. If there are too many prey, the prey may not be able to get enough food and might die.
5. Answers will vary. Accept all reasonable responses that include a producer followed by one or more consumers. Students should include examples of food chains based on where they live.
6. Possible answer: If all the grasses on a prairie were to die from a disease, the consumers that eat grass would have to find a different kind of producer to eat. They might move to a new location. This might mean that some other consumers on the prairie would have to eat less or might have to eat something else.

Write About Food Chains

In this food web, all food chains begin with the sun. In one food chain, algae are eaten by the snail, which is eaten by the bullfrog, which is eaten by the heron. The energy that came originally from the sun is passed from one organism to the next along the food chain.

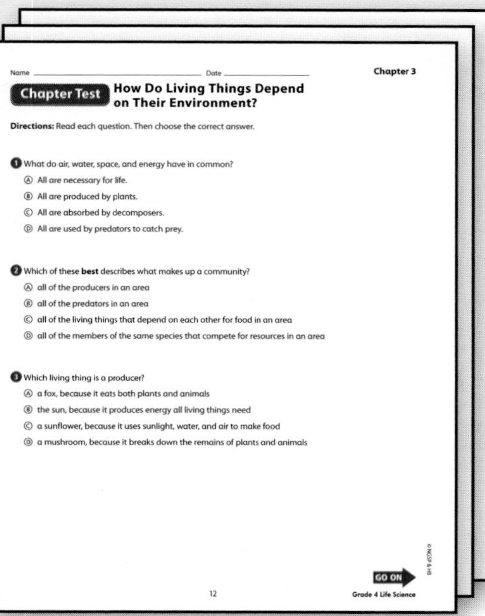

Name ______ Date ______ Chapter 3

Chapter Test How Do Living Things Depend on Their Environment?

Directions: Read each question. Then choose the correct answer.

1. What do air, water, space, and energy have in common?
   Ⓐ All are necessary for life.
   Ⓑ All are produced by plants.
   Ⓒ All are absorbed by decomposers.
   Ⓓ All are used by predators to catch prey.

2. Which of these **best** describes what makes up a community?
   Ⓐ all of the producers in an area
   Ⓑ all of the predators in an area
   Ⓒ all of the living things that depend on each other for food in an area
   Ⓓ all of the members of the same species that compete for resources in an area

3. Which living thing is a producer?
   Ⓐ a fox, because it eats both plants and animals
   Ⓑ the sun, because it produces energy all living things need
   Ⓒ a sunflower, because it uses sunlight, water, and air to make food
   Ⓓ a mushroom, because it breaks down the remains of plants and animals

GO ON

12 Grade 4 Life Science

Assess Student Progress Chapter 3 Test

Have students complete their Chapter 3 Test to assess their progress in this chapter.

Chapter 3 Test, Assessment Handbook, pages 12–17, or at myNGconnect.com

or NGSP ExamView CD-ROM

Objectives

Students will be able to:

- Describe how behavioral ecologists use science and technology in their careers.

PROGRAM RESOURCES

- Big Ideas Book: *Life Science*
- Big Ideas Book: *Life Science* eEdition at myNGconnect.com
- Digital Library at myNGconnect.com

❶ Introduce

Tap Prior Knowledge

- Show students a map of Asia. Ask: **Where is India?** (Asia) Ask a volunteer to point out central India on the map.
- Ask students to observe the photo of the sloth bear on page 123 and to compare it to other animals with which they are familiar.

Preview and Read

- Read the heading. Have students preview the photos on pages 122–123. Ask them to think about what a behavioral ecologist might do.
- Have students read pages 122–123.

❷ Teach

myNGconnect.com

Have students use the Digital Library to find images of animals they have observed.

Computer Presentation Students may present the photos in a computer presentation about the behaviors of these animals.

NATIONAL GEOGRAPHIC

Curious about animals? How about being a behavioral ecologist?

Some people can spend hours watching how animals act. If you're one of them, you might want to be a behavioral ecologist. You could work in some of the wildest places on Earth! That's what behavioral ecologist K. Yoganand does. He talks about tracking wild sloth bears in central India.

What does a behavioral ecologist do?

I study the behavior of wild animals in their natural environment. My goal is to find out how an animal's behavior helps it survive. Why are sloth bears most active at night? Is it because they need to avoid the daytime heat? Or is it because people are in their habitat during the daytime? The more we know about animals in the wild, the better we can protect them.

Yoganand holds a sloth bear cub while examining its health.

Yoganand (left) and an assistant put a radio collar on a sloth bear.

122

NATIONAL GEOGRAPHIC **Raise Your SciQ!**

Sloth Bears Sloth bears are omnivorous, mostly nocturnal mammals of the forests of South Asia. They hunt for ants and termites by using their long, powerful claws to dig through rock-like termite mounds or to dig up ant nests from the ground. They also eat the flowers and fruits of mango, fig, and ebony trees. Sloth bears are threatened by habitat loss and by hunting.

TECHTREK
myNGconnect.com

Student eEdition | Digital Library

What do you like most about your job?

I love being outdoors. Some people pay a lot of money to observe animals in an Asian forest or African savanna. It's a vacation for them. I do it as part of my job. How cool is that?

How did you study wild sloth bears?

I spent a lot of time in the forest observing and collecting data. I wanted to find out what a typical day is like for sloth bears. When do they sleep? What do they eat? Where do they give birth? What do they do when they meet a tiger? To answer these and other questions, I had to observe them in the wild. So I put radio collars on the bears to track them.

Was your research successful?

Yes! We now have a better idea of what people need to do to help sloth bears survive. I hope that my work helps us find ways to protect them and the land they need to live.

How do you think a sloth bear uses its long, sharp claws?
123

Differentiated Instruction

Extra Support

Have students list questions that they think a behavioral ecologist would investigate about sloth bears. Encourage them to include questions that relate to the chapter content on food chains, food webs, and the flow of energy in ecosystems. Examples: *What do sloth bears eat? Are sloth bears herbivores, carnivores, or omnivores? What animals prey on sloth bears?*

Challenge

Have students research and report on the life of sloth bears or another animal that lives in the wild in central India. Encourage students to include the animal's role in food chains and food webs. Students may share their reports orally with the class.

Teach, continued

Describe Behavioral Ecologists

- Ask: **What is K. Yoganand's goal for his work?** (to find out how an animal's behavior helps it survive) **Why does he think this goal is important?** (because the more we learn about wild animals, the better we can protect them)
- Point to the photo of the sloth bear on page 123. Say: **An animal's behaviors are its actions. What are some of the sloth bear's behaviors?** (Possible answers: walking, eating, sleeping, giving birth, responding to danger) **How does K. Yoganand study these behaviors?** (He uses radio collars to track the bears, and he observes their behavior in the wild.) Discuss how scientists depend on the skill of observation to learn about nature. For example, the best way to learn how a sloth bear behaves when a tiger is near is to observe it happening.

Find Out More

- Ask: **Is this a career you would find interesting? Why or why not?** (Accept all answers, positive or negative.) Discuss how behavioral ecologists spend much of their time outdoors, carefully observing animals in their natural environment.
- Encourage interested students to research where besides Asia and Africa behavioral ecologists study animals and which animals they study.

3 Assess

1. **Recall What is the goal of Yoganand's research?** (His goal is to find out how a sloth bear's behavior helps it survive where it lives.)
2. **Explain How does Yoganand study sloth bears?** (He observes them directly and uses radio collars to track their movement.)
3. **Draw Conclusions How is a sloth bear affected by other living things in its environment?** (Possible answers: It eats other living things for food; it may be eaten or threatened by tigers or other animals.)

NATIONAL GEOGRAPHIC
BECOME AN EXPERT
Bats: Winged Wonders of the Night
Have you ever seen a bat flying around at night? The United States has 47 kinds of bats. But one small island in the Panama Canal has more than 70 kinds of bats. The island is Barro Colorado. How did one small place come to have so many different kinds of bats?
Each kind of bat became an expert in finding food a different way. By dividing the resources where they live, the bats of Barro Colorado Island become parts of different food chains and food webs. The bats avoid competing with each other for food. The following pages show some of the many different kinds of bats on Barro Colorado and around the world. Think about what each kind eats and how it finds its food.
TECHTREK
myNGconnect.com
Student eEdition
Digital Library
Barro Colorado Island, Panama
Small disk-winged bats use suction cups at the base of their thumb and foot to climb inside a curled leaf. These bats eat insects.
The chestnut short-tailed bat feeds mostly on fruits.
124
125

PROGRAM RESOURCES

- Big Ideas Book: *Life Science*
- Big Ideas Book: *Life Science* eEdition at myNGconnect.com
- Digital Library at myNGconnect.com

Access Science Content

Compare Bats of Barro Colorado Island

- Have students read the text and captions on pages 124–125. Say: **Barro Colorado Island is a tropical rain forest.** Ask: **What is a tropical rain forest like?** (hot and rainy with tall trees and many other plants)
- Have students observe and compare the photos of bats. Ask: **How are these bats alike?** (All have brown, hairy bodies with wings.) **How are they different?** (They have different sizes and body shapes. Some are small enough to live inside a curled leaf; others have horn-like structures on their faces.) **How do different kinds of bats having different features help them fill their roles in the community?** (They can live in different places, eat different foods, and avoid competing with one another.)

Notetaking

Have students find the Panama Canal on a world map and observe the large photograph on pages 124–125. In their science notebook, have them write a descriptive paragraph about what they think it would feel like if they were standing on the ground here.

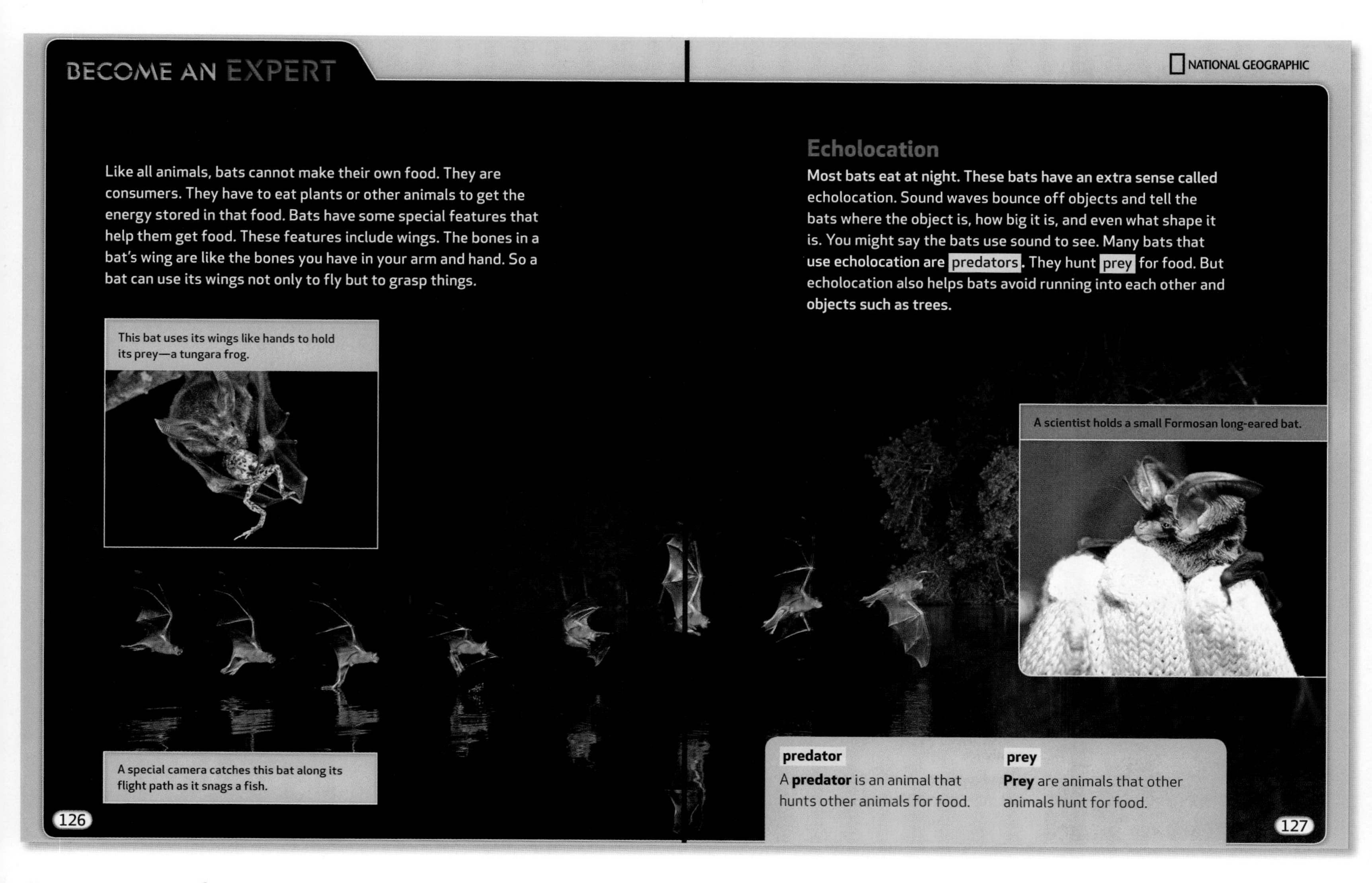
BECOME AN EXPERT

NATIONAL GEOGRAPHIC

Like all animals, bats cannot make their own food. They are consumers. They have to eat plants or other animals to get the energy stored in that food. Bats have some special features that help them get food. These features include wings. The bones in a bat's wing are like the bones you have in your arm and hand. So a bat can use its wings not only to fly but to grasp things.

This bat uses its wings like hands to hold its prey—a tungara frog.

A special camera catches this bat along its flight path as it snags a fish.

Echolocation

Most bats eat at night. These bats have an extra sense called echolocation. Sound waves bounce off objects and tell the bats where the object is, how big it is, and even what shape it is. You might say the bats use sound to see. Many bats that use echolocation are **predators**. They hunt **prey** for food. But echolocation also helps bats avoid running into each other and objects such as trees.

A scientist holds a small Formosan long-eared bat.

predator
A **predator** is an animal that hunts other animals for food.

prey
Prey are animals that other animals hunt for food.

126

127

Access Science Content

Describe How Bats Get Food

- Have students read pages 126–127. Ask: **Why are these bats examples of predators?** (They hunt other animals for food.) **Which animals are their prey?** (the frog and fish) **What helps a bat to find food at night?** (echolocation)

Differentiated Instruction

ELL Language Support for Identifying Predators and Prey

BEGINNING

Have students observe the photos on pages 126-127 and identify the predators and prey. Then have them define both words.

INTERMEDIATE

Provide Academic Language Frames, such as:

- *A ____ hunts other animals for food.*
- *____ are animals that are hunted and eaten.*

ADVANCED

Provide Academic Language Stems, such as:

- *Predators get food by . . .*
- *Prey are animals that . . .*

Assess

1. **Explain How do bats use echolocation?** (to find food at night and to avoid bumping into one another and other objects)
2. **Analyze Why are so many different kinds of bats able to live on Barro Colorado Island?** (Each kind of bat gets food in a different way and lives in a different place in the forest, and thus they do not compete with one another.)

LESSON 10 ▫ Become an Expert

BECOME AN EXPERT

NATIONAL GEOGRAPHIC

Bats as Carnivores

Most bats are carnivores. They get energy by eating insects, fish, frogs, and other small animals. The bats use echolocation to find these animals, even fish under water. Other features, like those shown here, make the bat a successful predator.

Vampire Bats

A few bats, called vampire bats, eat the blood of large animals, such as horses and cows. Unlike the vampires in stories, the bats do not suck blood. A chemical in their saliva keeps the blood from clotting, or hardening. Then the bat simply laps up the blood.

Vampire bats have razor-sharp teeth that they use to cut the skin of animals.

carnivore
A **carnivore** is an animal that eats other animals to survive.

128

Bats as Omnivores

A few bats are omnivores. They eat plant parts, such as fruit, as well as small animals.

The greater spear-nosed bat is an omnivore. Here it feeds on a lizard, but it also eats fruits, insects, and pollen.

Claws on the feet of the greater bulldog bat are perfect for grabbing and holding prey.

omnivore
An **omnivore** is an animal that eats both plants and other animals to survive.

129

Access Science Content

Compare Feeding Habits of Bats

- Have students read the text on page 128, observe the photo, and read its caption. Ask: **Why are most bats classified as carnivores?** (because they eat other animals) Review the definition of **carnivore** at the bottom of the page. Then point to the photo. Ask: **What body parts or abilities help the vampire bat be a successful carnivore?** (sharp teeth, echolocation, and a chemical that stops blood from clotting)
- Have students read page 129. Say: **The greater spear-nosed bat is an omnivore.** Ask: **How is this bat's food different from the food of the vampire bat?** (The greater spear-nosed bat eats plants as well as animals.)

Make Inferences

Have students observe the photo of the vampire bat, and then focus their attention on the size and shape of the bat's ears. Ask: **How might large, wide ears help a bat find food?** (Their large size helps the bat hear sounds, which is how the bat finds food and other objects.)

Notetaking

As students read about different kinds of bats, have them write down new words they encounter in their science notebook. Then have them write a descriptive sentence that uses the word correctly. Students may also include drawings to illustrate their sentences.

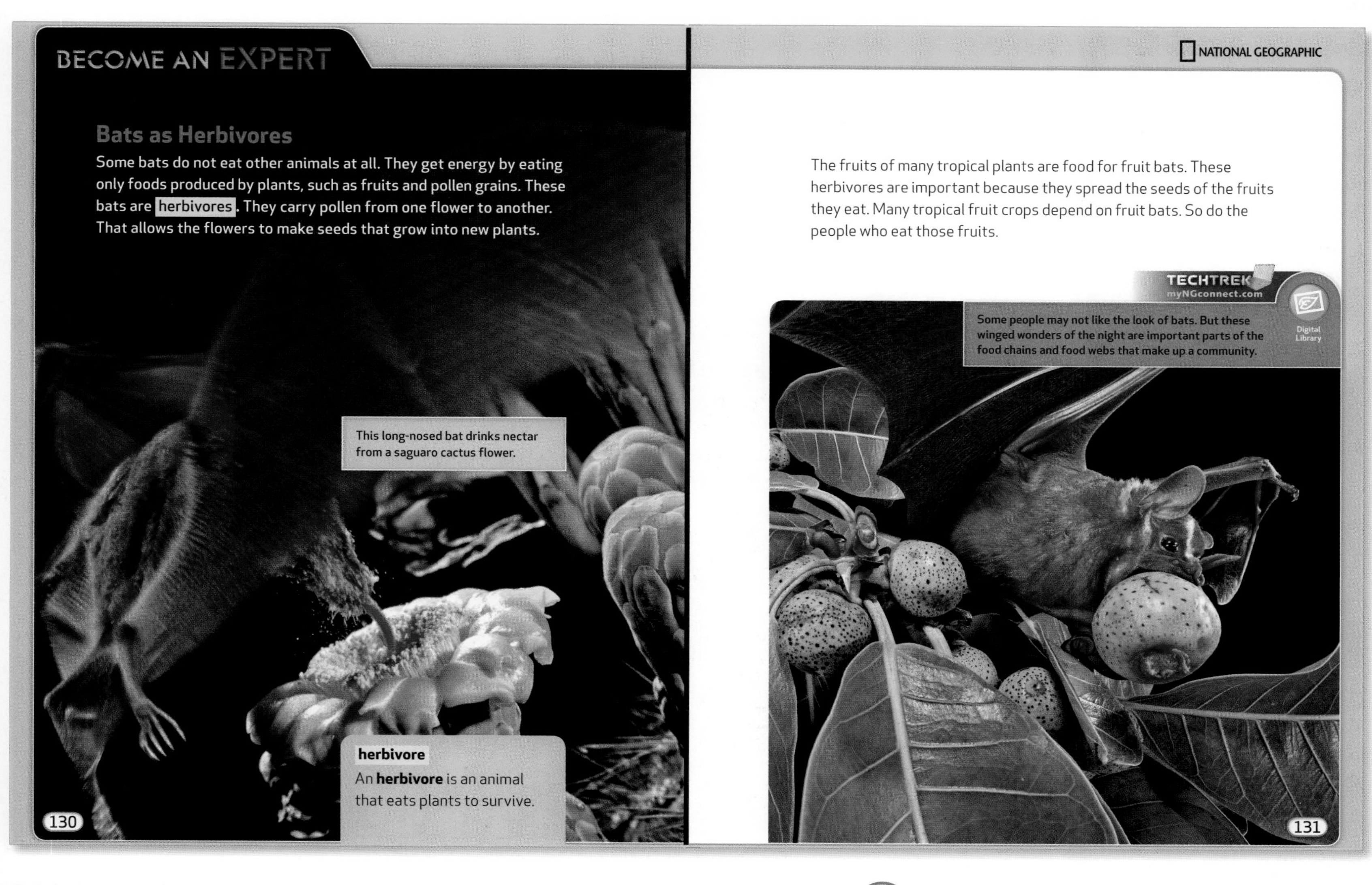

BECOME AN EXPERT
NATIONAL GEOGRAPHIC
Bats as Herbivores
Some bats do not eat other animals at all. They get energy by eating only foods produced by plants, such as fruits and pollen grains. These bats are herbivores. They carry pollen from one flower to another. That allows the flowers to make seeds that grow into new plants.
This long-nosed bat drinks nectar from a saguaro cactus flower.
herbivore
An herbivore is an animal that eats plants to survive.
130
The fruits of many tropical plants are food for fruit bats. These herbivores are important because they spread the seeds of the fruits they eat. Many tropical fruit crops depend on fruit bats. So do the people who eat those fruits.
TECHTREK
myNGconnect.com
Some people may not like the look of bats. But these winged wonders of the night are important parts of the food chains and food webs that make up a community.
Digital Library
131

PROGRAM RESOURCES

- Digital Library at myNGconnect.com

Access Science Content

Describe the Importance of Bats That Are Herbivores

- Have students read pages 130–131 and study the photos. Ask: **What is an herbivore?** (an animal that eats plants only) **What foods from plants are eaten by bats that are herbivores?** (nectar and pollen grains from flowers; fruits, such as from fig trees) **How do these bats help the plants that they visit?** (They may spread pollen from flower to flower, or they may spread seeds.) Discuss how many flowers depend on animals to spread pollen before seeds are formed.
- Have students review and compare the photos of bats on pages 124–129. Ask: **Could the long-nosed bat shown on page 130 catch and eat a lizard, like the greater spear-nosed bat can? Why or why not?** (No. The long-nosed bat is an herbivore and eats plants only.) Discuss how each bat can be classified as a carnivore, omnivore, or **herbivore** according to the food it eats and that each has body parts and behaviors that are suited to taking in this food.

myNGconnect.com

Have students use the Digital Library to find more photos of bats.

Assess

1. **Recall How can some bats help plants reproduce?** (They spread pollen from flower to flower or spread seeds to other places where they might grow.)
2. **Compare and Contrast How are most bats alike? What is one way that they are different?** (All bats are winged mammals, and most use echolocation to fly at night. They differ in that some bats are carnivores that eat other animals, others are omnivores that eat plants and animals, and still others are herbivores that eat plants only.)

Share and Compare

Turn and Talk

Ask students to turn to partners and discuss the adaptations that allow bats to find their way at night. Prompt students by asking:

1. **Recall What kinds of foods do bats eat?** (Some bats eat plants, other bats eat animals, and still others eat both plants and animals.)
2. **Explain What is echolocation?** (In echolocation, bats bounce sound waves off objects.)
3. **Analyze Most bats feed in the dark of night. How are they able to find food or avoid running into trees?** (Most bats use echolocation to find food and identify trees and other objects as they fly.)

Read

Have students practice reading the pages they select at least twice before they read aloud to others. Encourage students to ask questions of one another about their interest in bats and opinions of them.

Write

my
SCIENCE
notebook

Have students write a conclusion that summarizes what they learned about bats. Then have them write the Big Idea in their own words. Then ask them to compare what they wrote with a classmate. Ask if partners included the same ideas in their conclusions.

Draw

my
SCIENCE
notebook

Have students draw a picture of a bat as it looks for or eats food in its environment. Have small groups compare their drawings, and then discuss the kinds of foods that different bats may eat.

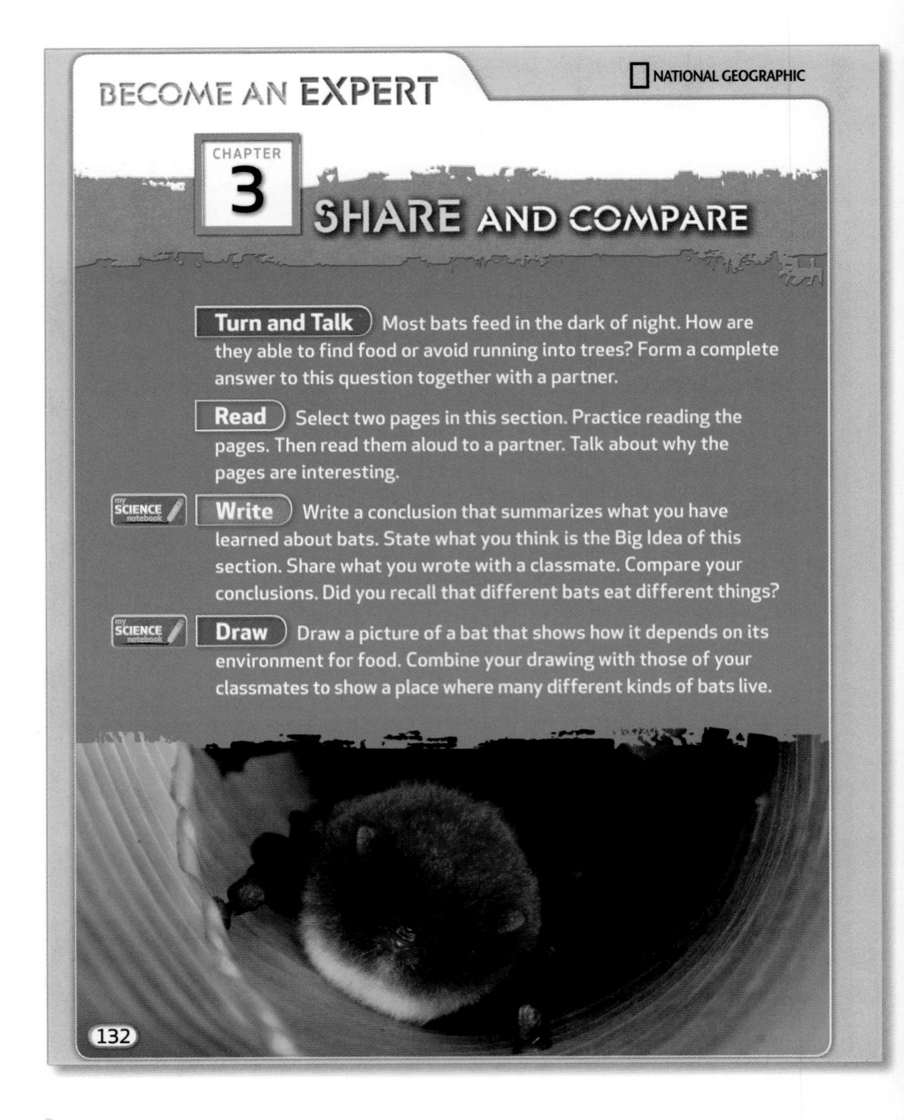
BECOME AN EXPERT
NATIONAL GEOGRAPHIC
CHAPTER 3
SHARE AND COMPARE
Turn and Talk Most bats feed in the dark of night. How are they able to find food or avoid running into trees? Form a complete answer to this question together with a partner.
Read Select two pages in this section. Practice reading the pages. Then read them aloud to a partner. Talk about why the pages are interesting.
Write Write a conclusion that summarizes what you have learned about bats. State what you think is the Big Idea of this section. Share what you wrote with a classmate. Compare your conclusions. Did you recall that different bats eat different things?
Draw Draw a picture of a bat that shows how it depends on its environment for food. Combine your drawing with those of your classmates to show a place where many different kinds of bats live.
132

Sum Up

Tell students that to sum up a text helps them pull together the ideas of the text. Remind students that they summed up the Become an Expert lesson in the Write section of Share and Compare. Have students take turns reading the conclusion they wrote to a partner. Invite them to compare and contrast their conclusions.

Read Informational Text

Explore on Your Own books provide additional opportunities for your students to:

- deepen their science content knowledge even more as they focus on one Big Idea.
- independently apply multiple reading comprehension strategies as they read.

After applying the strategies throughout the chapter, students will independently read Explore on Your Own books. Facsimiles of the Explore on Your Own pages are shown on pages T132a–T132h.

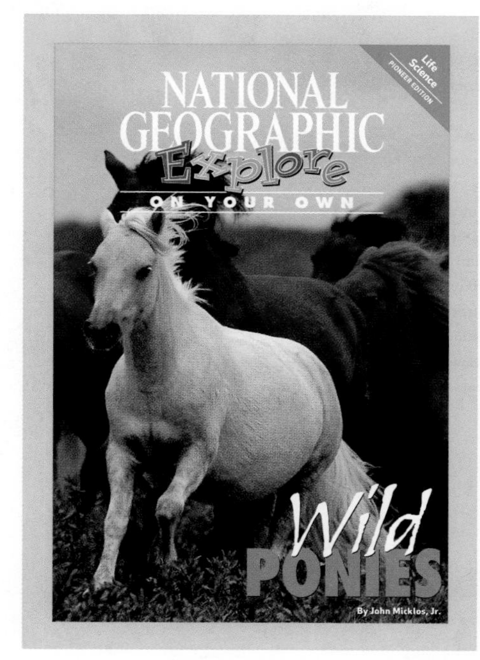

Pioneer

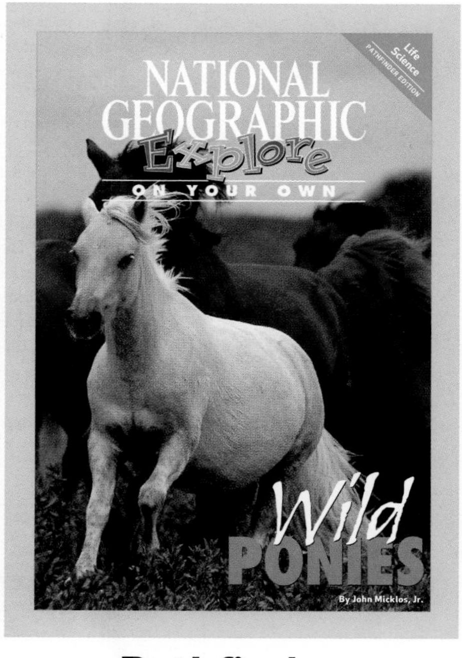

Pathfinder

Life Science
PIONEER EDITION

It's Time to Explore on Your Own!

Good readers use multiple strategies as they read on their own. Use the four key reading comprehension strategies below:

1 PREVIEW AND PREDICT

- Look over the text.
- Form ideas about how the text is organized and what it says.
- Confirm ideas about how the text is organized and what it says.

2 MONITOR AND FIX UP

- Think about whether the text is making sense and how it relates to what you know.
- Identify comprehension problems and clear up the problems.

3 MAKE INFERENCES

- Use what you know to figure out what is not said or shown directly.

4 SUM UP

- Pull together the text's big ideas.

Remember that you can choose different strategies at different times to help you understand what you are reading.

NATIONAL GEOGRAPHIC
School Publishing

PIONEER EDITION

By John Micklos, Jr.

CONTENTS

Wild PONIES

Assateague Island's Mane Attraction

BY JOHN MICKLOS, JR.

The July morning is hot. Yet a crowd of people stands in the sunshine. These people do not mind the heat. They are excited. They are about to see wild ponies.

The ponies live on Assateague Island. Once a year, cowboys round them up. They herd the ponies to another island named Chincoteague. How do the ponies get from one island to the other? They swim!

Splash! The ponies swim across a waterway. It is called a **channel**. Their trip takes about five minutes.

The ponies reach Chincoteague. They step out of the water and onto the beach. People cheer. The ponies have arrived!

In the Swim. Wild ponies swim from Assateague Island to Chinoteague Island.

2

3

Salty Snack. The wild ponies eat a salty plant called cordgrass. Every spring, baby ponies are born on Assateague Island. The young ponies eat cordgrass too.

The Pony Sale

Soon cowboys herd the ponies to town. The next day, the young ponies are sold. Why sell the ponies?

- The pony sale raises money for the local fire department.
- The sale also keeps the number of ponies down. The island can feed only a few ponies.

History and Mystery

Wild ponies have lived on Assateague Island for hundreds of years. How did they get there? No one knows for sure. Most experts think the ponies came from England. People brought them when they moved to America. Later, people turned the animals loose on the island.

Book Link

A Whinny-ing Book

Read *Misty of Chincoteague* by Marguerite Henry (Simon & Schuster). This famous book first appeared in 1947 and has helped children all over the world learn about the ponies.

Small in Size

People call these animals ponies. That is because they are small like ponies. They are less than 147 centimeters (58 inches) tall. Yet they are actually horses. Why are they so small? Life on the island is hard.

The island is a harsh **habitat**, or home. In the summer, the weather is hot. Insects bite. In the winter, the weather is cold and windy. Life on the island is not easy for ponies.

Pony Groups

The ponies live together in small groups called **bands**. Some bands have only two ponies. Others may have a dozen ponies.

Each band usually has one **stallion**. That is an adult male. His job is to protect the band.

A band also has adult females called **mares**. In spring, the mares have babies. The mares care for the new ponies while they grow.

There are many bands of ponies on the island. Together, they form two groups, or **herds**. One lives on the Maryland side of the island. The other lives on the Virginia side. A fence keeps the herds apart.

4

5

On the Fence. It is not always easy to choose a favorite pony at the pony sale.

Pony Numbers

Assateague is not a big island. It has enough food for only about 300 ponies. Yet mares have new babies each spring. How does the number of ponies stay at the right size?

People control the size of one pony herd with a **vaccine**. It is a kind of medicine. The vaccine keeps mares from having babies that year.

The pony sale controls the size of the other herd. A few of the ponies are sold each year. The sale raises money for the fire department. It also helps the ponies.

The pony sale lets some ponies go to new homes. The rest of the herd returns to the island. They will roam free—and have plenty to eat.

Wordwise

band: small group of horses

channel: waterway between two landmasses that lie close to each other

habitat: place where something lives

herd: large group of horses (often contains many bands)

mare: adult female horse

stallion: adult male horse

vaccine: medicine

Pony Parts

The Assateague ponies are small compared to other horses. Yet they have had big success in their island home. What parts help these tough ponies live in the wild.

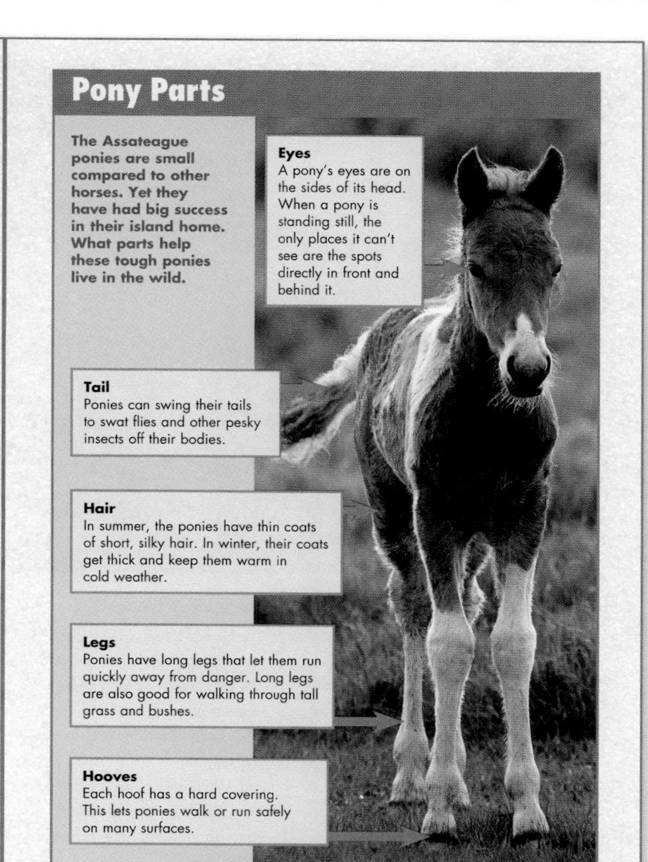

Eyes
A pony's eyes are on the sides of its head. When a pony is standing still, the only places it can't see are the spots directly in front and behind it.

Tail
Ponies can swing their tails to swat flies and other pesky insects off their bodies.

Hair
In summer, the ponies have thin coats of short, silky hair. In winter, their coats get thick and keep them warm in cold weather.

Legs
Ponies have long legs that let them run quickly away from danger. Long legs are also good for walking through tall grass and bushes.

Hooves
Each hoof has a hard covering. This lets ponies walk or run safely on many surfaces.

Island Refuges

Piping plover

Fly By. Many of the island's birds don't live on the islands. They visit on their way to their winter and summer homes.

Chincoteague and Assateague are famous for their ponies. Yet other animals live there too. The islands are home to many kinds of wildlife.

Parts of each island are protected. That means there are rules about what people can do. In these areas, the wildlife is safe.

Safe Spaces

The Chincoteague National Wildlife Refuge is one of the protected areas. It includes land on both islands.

Many animals live in the refuge. Some live in sand dunes. Others like the marshes, or soggy land. Some make homes in the forest.

Flocking to the Islands

The islands also give shelter to traveling animals. Some birds live in the north during the summer. When the weather turns cold, they fly south. They travel many miles to their warm winter homes. In spring, they fly back along the same path.

All that flying gets tiring. Where do these birds rest? They stop at Chincoteague and Assateague Islands.

New Horizons. The refuge is a protected area. It helps animals and plants survive.

Wet and Wild

Life is pretty wild on the islands. Wind whips the sand from place to place. Water washes the sand away. Yet plants and animals find a way to survive.

Tough grasses grow along the beach. Ghost crabs burrow in the sand. Many kinds of shellfish live in marshes. These are areas where ocean water covers the land.

Life in the Forest

The islands also have forestland. Tall pine trees grow from the sandy soil. The pine trees give food and shelter to many animals.

Squirrels leap from tree to tree. Owls hunt from the treetops. Deer and wild ponies roam the forest floor. They nibble on bushes and plants. Foxes and raccoons also find meals in the forest.

Safety in the Wild

The refuge does more than protect the land. It helps animals survive.

Bald eagles and piping plovers are in danger of dying out. The refuge gives them a safe habitat. The birds can find food. They have a place to raise their young.

The islands are full of amazing plants and animals. The refuge gives them safe homes in the wild.

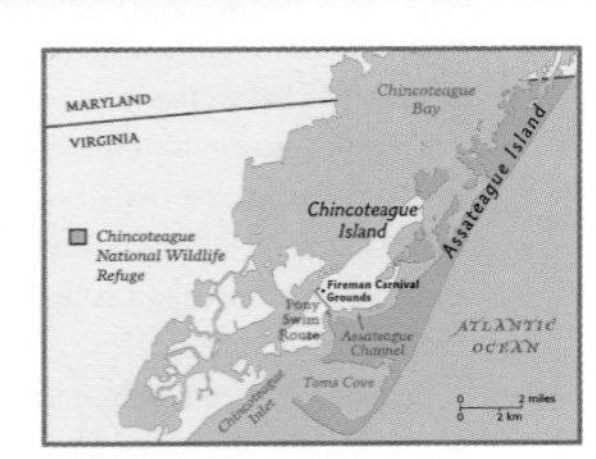

Wet Land. Chincoteague and Assateague are narrow islands. They are between the bay and the Atlantic Ocean.

CONCEPT CHECK

ISLAND LIFE

Take a run at these questions to see what you have learned.

1. Why do ponies swim to Chincoteague Island?
2. Are the ponies on Assateague Island really ponies? Explain.
3. Why is life hard for wild ponies?
4. Why do people control the number of ponies on the islands?
5. Why are the islands good homes?

INDEX

Acknowledgments
Grateful acknowledgment is given to the authors, artists, photographers, museums, publishers, and agents for permission to reprint copyrighted material. Every effort has been made to secure the appropriate permission. If any omissions have been made or if corrections are required, please contact the Publisher.

Photographic Credits
Cover Rich Pomerantz; 2-3 Medford Taylor/National Geographic Image Collection; 4-5 James L. Stanfield; 5 Mark Thiessen; 6 Medford Taylor/National Geographic Image Collection; 6 Rich Pomerantz; 7 James L. Stanfield; 8-9 James L. Amos; 9 James P. Blair; 10 James L. Amos; 10 Raymond Gehman; 11 James P. Blair; 12 James L. Amos.

Illustrator Credits
11 NG Maps

Program Authors
Randy Bell, Ph.D., Associate Professor of Science Education, University of Virginia, Charlottesville, Virginia; Malcolm B. Butler, Ph.D., Associate Professor of Science Education, University of South Florida, St. Petersburg, Florida; Kathy Cabe Trundle, Ph.D., Associate Professor of Early Childhood Science Education, The Ohio State University, Columbus, Ohio; Judith Sweeney Lederman, Ph.D., Director of Teacher Education and Associate Professor of Science Education, Department of Mathematics and Science Education, Illinois Institute of Technology, Chicago, Illinois; David W. Moore, Ph.D., Professor of Education, College of Teacher Education and Leadership, Arizona State University, Tempe, Arizona

National Geographic School Publishing
Hampton-Brown
www.NGSP.com

Printed in the USA.
RR Donnelley, Johnson City, TN

ISBN-13: 978-0-7362-7758-7

10 11 12 13 14 15 16 17 18 19
10 9 8 7 6 5 4 3 2 1

Life Science
PATHFINDER EDITION

It's Time to Explore on Your Own!

Good readers use multiple strategies as they read on their own. Use the four key reading comprehension strategies below:

1 PREVIEW AND PREDICT
- Look over the text.
- Form ideas about how the text is organized and what it says.
- Confirm ideas about how the text is organized and what it says.

2 MONITOR AND FIX UP
- Think about whether the text is making sense and how it relates to what you know.
- Identify comprehension problems and clear up the problems.

3 MAKE INFERENCES
- Use what you know to figure out what is not said or shown directly.

4 SUM UP
- Pull together the text's big ideas.

Remember that you can choose different strategies at different times to help you understand what you are reading.

NATIONAL GEOGRAPHIC
School Publishing

Wild PONIES

PATHFINDER EDITION

By John Micklos, Jr.

CONTENTS

Cheers rise as the **ponies** splash into the water. It's a July morning on Assateague, an island located off the coasts of Virginia and Maryland. Every year at this time, local cowboys round up the wild ponies that live on the southern end of the island. At low tide, they herd the ponies across a narrow waterway called a **channel** to another island named Chincoteague. Thousands of people come to watch.

The ponies swim across the channel in about five minutes. The crowd cheers again as the animals reach the shore. Back on dry land, the ponies shake the water from their manes. Then they start to graze calmly. Some wander right up to the fence that separates them from the onlookers.

In the Swim. Wild ponies cross Assateague Channel to Chincoteague Island during the annual pony swim. Once ashore, cowboys on horseback lead the ponies to the carnival grounds. Some of the foals will be auctioned there. The auction helps control the pony population.

Salty Snack. The wild ponies feed on a salty plant called cordgrass. Every spring, foals are born on Assateague Island. The baby ponies graze on cordgrass too.

Soon cowboys herd the ponies through town to the carnival grounds. The next day most of the young ponies, called **foals**, will be auctioned, or sold to the highest bidder. The pony auction does three things:

- It raises money for the Chincoteague Fire Department.
- It allows some people to take home a foal.
- And most important, it keeps the pony population at the proper size. Resources such as food will support only about 150 ponies on the southern end of Assateague Island. A larger number would hurt the island's **ecology**, or balance of life.

History and Mystery

Assateague is a long, narrow island. It stretches between southern Maryland and northern Virginia. On one side is the Atlantic Ocean. On the other side is a quiet bay.

The ponies have been roaming free on the island for hundreds of years. They are **feral** animals. This means that their ancestors once were tame.

No one knows exactly how the ponies got to the island. Some people believe that long ago the first ponies were transported by ship from Spain. They think the ship wrecked near the island in a storm, and the ponies swam ashore.

Book Link

A Whinny-ing Book

Read *Misty of Chincoteague* by Marguerite Henry (Simon & Schuster). This famous book first appeared in 1947 and has helped children all over the world learn about the ponies.

Most experts, though, think the first settlers of mainland Maryland and Virginia brought the ponies with them from England. Later they turned the animals loose to graze on Assateague Island.

Harsh Habitat

Today's ponies lead a hard life. In the summer they face hot weather and biting insects. In the winter they must grow thick coats to protect themselves from bitter winds.

Spring and fall are the best seasons. The weather on the island is mild, and there is plenty of grass for the ponies to eat. The ponies also eat leaves and twigs. They even munch on poison ivy, which doesn't seem to bother them.

These island grazers may be the size of ponies—less than 147 centimeters (58 inches) tall—but they are actually horses. Experts think that the harsh **habitat**, or place where they live, accounts for their small size. In fact, when some of the auctioned foals leave Assateague and receive better food and shelter, they grow to horse size. But people have been calling them ponies for years, and the name has stuck.

Pony Bands

The ponies live together in small groups called **bands**. Some bands may have as few as two ponies. Others may have a dozen. In most bands there are usually several **mares**, or adult females, some foals, and one adult male.

The adult male pony is called a **stallion**. It is his job to protect the band. Sometimes stallions try to steal ponies from other bands. This can lead to fights between stallions. They bite and kick with their heavy hooves until one stallion backs away.

In the spring mares give birth. Within minutes, their foals begin to walk on wobbly legs. Soon they are running and playing. At first they drink their mother's milk to help them grow. Then they begin to eat grass as the older ponies do.

On the Fence. It's not always easy to choose a favorite foal at the auction.

Managing the Herds

There are two main groups, or **herds**, of wild ponies on Assateague Island. Each herd has 100 to 150 ponies and includes many pony bands. One herd lives on the Maryland side of the island. The other lives on the Virginia side. A fence at the state line keeps the herds apart.

National Park Service rangers manage the herd on the Maryland side of the island. They control the number of ponies there by using a special **vaccine**, or medicine. Each year they inject the vaccine into some of the mares. The vaccine keeps the mares from having babies that year.

The Chincoteague Fire Department manages the herd on the Virginia side of the island. It controls the number of ponies there through the annual pony auction.

Return to the Wild

At the auction, some people bid on ponies to take home. Others just come to watch. In 2001, 85 ponies were sold. One foal sold for $10,500. That set a new record price for the auction. In all, that auction raised $167,000.

The day after the auction, Chincoteague cowboys herd the ponies back to the water's edge. Crowds cheer again as the ponies swim home to Assateague Island. There they will be free to roam again for another year.

Wordwise

band: small group of horses
channel: waterway between two landmasses that lie close to each other
ecology: how plants and animals live in relation to each other
feral: wild animals whose ancestors were once tame
foal: young horse
habitat: place where something lives
herd: large group of horses (often contains many bands)
mare: adult female horse
pony: small horse that is less than 147 centimeters (58 inches) tall when fully grown
stallion: adult male horse
vaccine: medicine swallowed or injected into the body

Pony Parts

The Assateague ponies may be small compared to other horses. Yet they have had big success in their island home. These tough ponies have features, or parts, that help them survive in the wild.

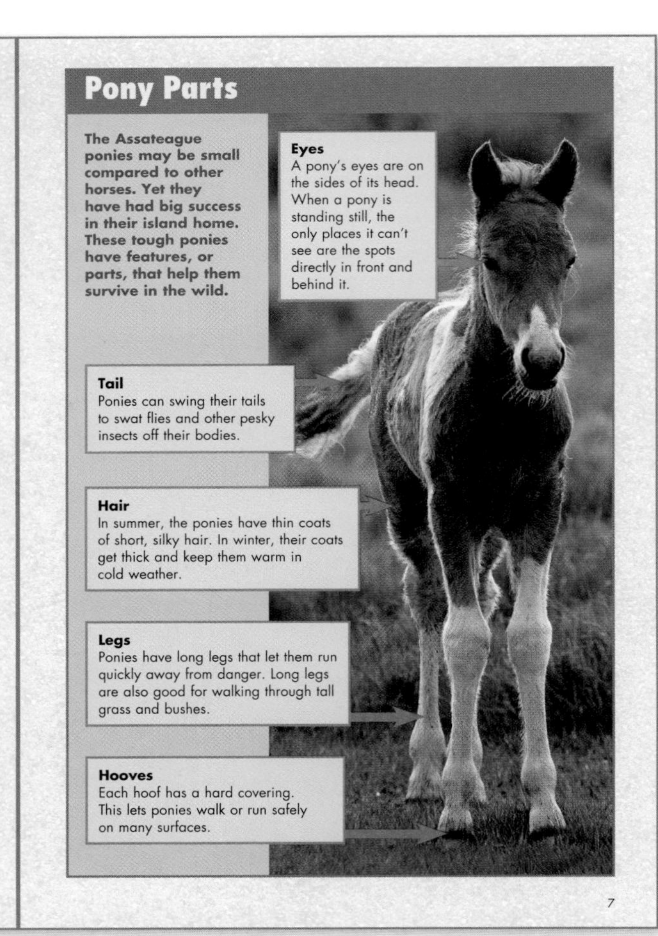

Eyes
A pony's eyes are on the sides of its head. When a pony is standing still, the only places it can't see are the spots directly in front and behind it.

Tail
Ponies can swing their tails to swat flies and other pesky insects off their bodies.

Hair
In summer, the ponies have thin coats of short, silky hair. In winter, their coats get thick and keep them warm in cold weather.

Legs
Ponies have long legs that let them run quickly away from danger. Long legs are also good for walking through tall grass and bushes.

Hooves
Each hoof has a hard covering. This lets ponies walk or run safely on many surfaces.

Island Refuges

Fly By. Many of the island's feathered friends don't live on the islands. They visit Chincoteague and Assateague when they are flying to their seasonal homes.

Chincoteague and Assateague are famous for their ponies. Yet this pair of islands is home to an amazing variety of other animals. Why do so many creatures thrive on these small islands? Certain sections of the island are protected. That means there are rules about what people can do on the land—and what they can't do.

Safe Spaces

The Chincoteague National Wildlife Refuge is one of these protected patches. It includes more than 5,666 hectares (14,000 acres) of land on both Chincoteague and Assateague Islands.

The refuge provides a safe space for plants and animals. It protects their habitats, or the places where they live. The refuge offers plenty of natural shelters too. Animals find homes among the pine trees, in the soggy marshes, and within dry sand dunes.

The islands make nice year-round homes for animals that hunt in the ocean. Many birds nest on the islands. During the day, they catch fish in the ocean. At night, the birds return to the islands to sleep.

Flocking to the Islands

The islands are also popular vacation spots for animals on the move. Every year, flocks of birds migrate, or travel long distances between their winter and summer homes.

These birds live in the north during the summer. When the weather turns cold, they travel south. They fly thousands of miles to warm winter homes. In spring, they fly back along the same path.

So many birds travel along the eastern coast of North America that scientists call it the Atlantic Flyway. It's like a highway for birds. But flying for days on end can wear the birds out. Where do they stop to rest? You guessed it—Chincoteague and Assateague Islands.

8 9

New Horizons. The island refuge is a protected area. It gives new life to animals and plants that are struggling to survive.

Wet and Wild

Life can get pretty wild on the islands. Chincoteague and Assateague are barrier islands. They provide a barrier that protects the mainland from rough ocean waves. Wind whips the sand from place to place. Tides sweep the sand away. Yet even in this harsh habitat, many living things find a way to thrive. Tough grasses grow in the dunes. Ghost crabs burrow in the sand.

Saltwater marshes are another kind of habitat on the islands. These are areas where ocean water covers the land. Many kinds of shellfish and other small animals live in this salty habitat. Black ducks and other migratory birds depend on these critters for food.

Life in the Forest

Chincoteague and Assateague are flat and windswept. Yet on the highest ground, you can also find forests. Here, the land doesn't flood with salty ocean water as it does on the beach or in the marshes. So these areas have a very different look. Tall pine trees tower above the sandy soil.

Instead of water-loving birds and side-stepping crabs, the forests are home to other animals. Squirrels leap from tree to tree. Deer and wild ponies nibble on the bushes and plants. Foxes and raccoons roam the forest floor in search of food. Owls nest in the treetops. At night they swoop down to scoop up mice and other tasty treats.

Saving Species

The Chincoteague Island National Wildlife Refuge does more than just protect the land for animals. It also helps species survive. For example, bald eagles and piping plovers are struggling to survive in the wild. They get a boost by living on these special islands. The refuge gives them a safe habitat. Here the birds can find food and a safe place to raise their young.

Ponies may be the superstars of these two islands in the Atlantic Ocean. But it's the variety of wildlife that makes the islands so amazing. Many kinds of animals live on these islands. Here ponies run wild and the animals outnumber the people.

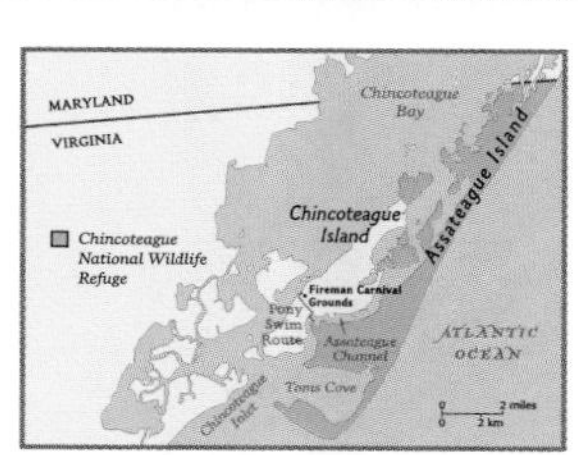

Protecting the Land. Chincoteague and Assateague are barrier islands. They protect the mainland from rough ocean waves.

10 11

CONCEPT CHECK

ISLAND LIFE

Take a run at these questions to see how much you've learned.

1. Why do ponies swim from Assateague Island to Chincoteague Island?
2. Are the ponies on Assateague Island really ponies? Explain.
3. Is life easy for the Assateague ponies? Why or why not?
4. Why do people control the number of ponies on the islands?
5. Why do so many kinds of animals come to the islands?

INDEX

Acknowledgments
Grateful acknowledgment is given to the authors, artists, photographers, museums, publishers, and agents for permission to reprint copyrighted material. Every effort has been made to secure the appropriate permission. If any omissions have been made or if corrections are required, please contact the Publisher.

Photographic Credits
Cover Rich Pomerantz; 2-3 Medford Taylor/National Geographic Image Collection; 4-5 James L. Stanfield; 5 Mark Thiessen; 6 Medford Taylor/National Geographic Image Collection; 6 Rich Pomerantz; 7 James L. Stanfield; 8-9 James L. Amos; 9 James P. Blair; 10 James L. Amos; 10 Raymond Gehman; 11 James P. Blair; 12 James L. Amos.

Illustrator Credits
11 NG Maps

Program Authors
Randy Bell, Ph.D., Associate Professor of Science Education, University of Virginia, Charlottesville, Virginia; Malcolm B. Butler, Ph.D., Associate Professor of Science Education, University of South Florida, St. Petersburg, Florida; Kathy Cabe Trundle, Ph.D., Associate Professor of Early Childhood Science Education, The Ohio State University, Columbus, Ohio; Judith Sweeney Lederman, Ph.D., Director of Teacher Education and Associate Professor of Science Education, Department of Mathematics and Science Education, Illinois Institute of Technology, Chicago, Illinois; David W. Moore, Ph.D., Professor of Education, College of Teacher Education and Leadership, Arizona State University, Tempe, Arizona

National Geographic School Publishing
Hampton-Brown
www.NGSP.com

Printed in the USA.
RR Donnelley, Johnson City, TN

ISBN-13: 978-0-7362-7761-7

10 11 12 13 14 15 16 17 18 19
10 9 8 7 6 5 4 3 2 1

How Do Adaptations Help Living Things Survive?

LIFE SCIENCE

CHAPTER
4
HOW DO ADAPTATIONS HELP LIVING THINGS SURVIVE?
One secret to survival in the wild is to avoid being seen. That's what this peacock butterfly is doing. A hungry bird might never see it. Other animals may be frightened of its wings. How does the peacock butterfly's body covering help it stay alive?
TECHTREK
myNGconnect.com
134
135

Chapter 4 Contents

After reading Chapter 4, you will be able to:

- Recognize how adaptations are important to the survival of living things. ADAPTATIONS, MOVEMENT, GETTING FOOD, DEFENSE AND PROTECTION, SENSING AND COMMUNICATING, PLANT ADAPTATIONS
- Identify examples of animal adaptations that are used for movement, protection, sensing, communication, reproduction, and meeting the animal's needs in its environment. ADAPTATIONS, MOVEMENT, GETTING FOOD, DEFENSE AND PROTECTION, SENSING AND COMMUNICATING
- Identify examples of plant adaptations that are used for protection, reproduction, and meeting the plant's needs in its environment. PLANT ADAPTATIONS
- Recognize that modern animals and plants often resemble fossilized species. FOSSILS AND EXTINCTION
- Science in a Snap! Identify examples of animal adaptations that are used for movement, protection, sensing, communication, reproduction, and meeting the animal's needs in its environment. GETTING FOOD

133

Preview and Predict

Tell students that they can preview a text by looking over it to get an idea of what the text is about and how it is organized. Tell students that predicting is forming ideas about what they will read. Students should confirm their ideas with others.

Have students preview and make predictions about the chapter. Remind them to use section heads, vocabulary words, pictures, and their background knowledge to make predictions.

CHAPTER 4 ▫ How Do Adaptations Help Living Things Survive?

LESSON	PACING	OBJECTIVES
1 **Directed Inquiry** *Investigate Plant Fossils* pages T133g–T133j **Science Inquiry and Writing Book** pages 44–47	**40** minutes	Investigate through Directed Inquiry (answer a question; make and compare observations; collect and record data and observations; generate explanations and conclusions based on evidence; share findings; ask questions based on observations to increase understanding). Explain how fossils provide evidence of the history of Earth. Ask "why" questions based on observations. Differentiate fact from opinion and explain that scientists do not rely on claims or conclusions unless they are backed by observations that can be confirmed.
2 **Big Idea Question and Vocabulary** pages T134–T135—T136–T137 **Adaptations** pages T138–T139	**30** minutes	Describe how an animal's body structure can help it survive. Describe how an animal's coloration can help it survive. Recognize how adaptations are important to the survival of living things. Identify examples of animal structures that are adaptations.
3 **Movement** pages T140–T141	**20** minutes	Identify body parts of animals that are adaptations for moving in water and in air. Explain how adaptations for movement can help a living thing survive.
4 **Getting Food** pages T142–T147	**30** minutes	Recognize that hunger is an internal signal that causes an animal to hunt for food. Explain how an animal uses its adaptations for getting food. Recognize adaptations that animals use to attract prey.
5 **Defense and Protection** pages T148–T149	**20** minutes	Identify and explain how animals use an adaptation for defense and protection. Distinguish between an adaptation and a variation.

VOCABULARY	RESOURCES	ASSESSMENT
compare **infer**	Science Inquiry and Writing Book: *Life Science* Science Inquiry Kit: *Life Science* Directed Inquiry: Learning Masters 47–50	Inquiry Rubric: Assessment Handbook, page 185 Inquiry Self-Reflection: Assessment Handbook, page 195 Reflect and Assess, page T133j
adaptation	Vocabulary: Learning Master 51 *Life Science* Big Ideas Book	Assess, page T139
		Assess, page T141
camouflage		Assess, page T147
variation		Assess, page T149

TECHNOLOGY RESOURCES

STUDENT RESOURCES

myNGconnect.com

- Student eEdition
- Big Ideas Book
- Science Inquiry and Writing Book
- Read with Me
- Vocabulary Games
- Enrichment Activities
- Digital Library

National Geographic Kids

National Geographic Explorer!

TEACHER RESOURCES

myNGconnect.com

- Teacher eEdition
- Teacher's Edition
- Science Inquiry and Writing Book
- Online Lesson Planner
- National Geographic Unit Launch Videos
- Assessment Handbook
- Presentation Tool
- Digital Library

NGSP ExamView CD-ROM

IF TIME IS SHORT...
FAST FORWARD.

CHAPTER 4 ▫ How Do Adaptations Help Living Things Survive?

LESSON	PACING	OBJECTIVES
6 Sensing and Communicating pages T150–T157	**35** minutes	Identify the importance of adaptations for sensing. Describe adaptations animals have for seeing and hearing. Describe adaptations animals have for touch and smell. Explain advantages of animals using sound to communicate. Recognize ways that smell is used for communication. Identify examples of adaptations of animals for reproduction.
7 Plant Adaptations pages T158–T163	**30** minutes	Describe how a carnivorous plant has adaptations to survive in its environment. Identify adaptations of plants that reduce water loss. Describe an adaptation that plants use for defense. Explain how flowers, seeds, and fruits are adaptations for plant reproduction.
8 Fossils and Extinction pages T164–T167	**25** minutes	Explain how scientists use fossils in the study of Earth's history. Identify examples of modern plants that resemble a fossilized species. Identify examples of modern animals that resemble a fossilized species. Explain why some living things were able to survive for a long time while others became extinct.
9 NATIONAL GEOGRAPHIC **Camels and People: Surviving in the Sahara** pages T168–T169	**20** minutes	Explain how adaptations can help a living thing survive in its environment.

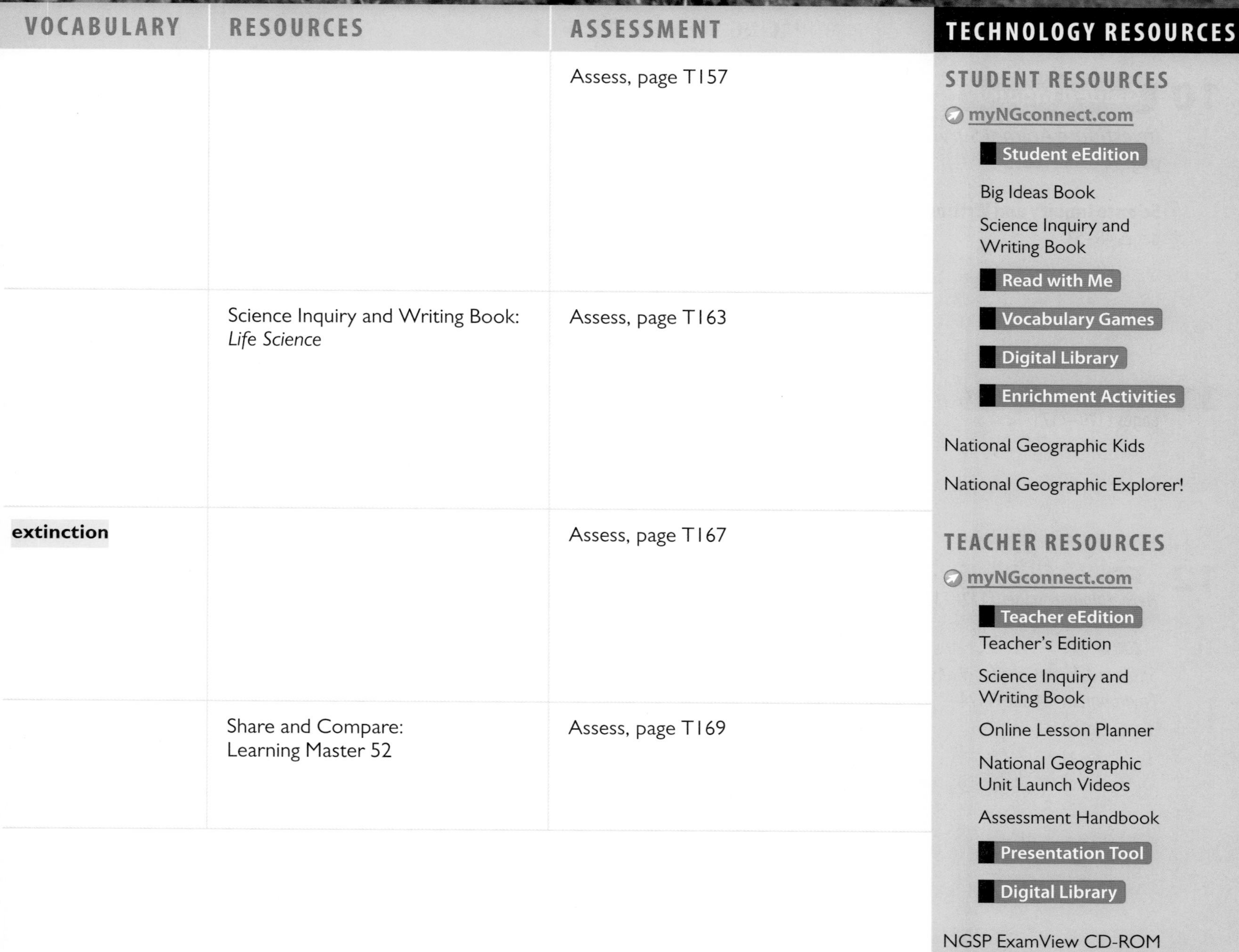

VOCABULARY	RESOURCES	ASSESSMENT
		Assess, page T157
	Science Inquiry and Writing Book: *Life Science*	Assess, page T163
extinction		Assess, page T167
	Share and Compare: Learning Master 52	Assess, page T169

TECHNOLOGY RESOURCES

STUDENT RESOURCES

myNGconnect.com

- Student eEdition
- Big Ideas Book
- Science Inquiry and Writing Book
- Read with Me
- Vocabulary Games
- Digital Library
- Enrichment Activities

National Geographic Kids

National Geographic Explorer!

TEACHER RESOURCES

myNGconnect.com

- Teacher eEdition
- Teacher's Edition
- Science Inquiry and Writing Book
- Online Lesson Planner
- National Geographic Unit Launch Videos
- Assessment Handbook
- Presentation Tool
- Digital Library

NGSP ExamView CD-ROM

IF TIME IS SHORT...

FAST FORWARD.

CHAPTER 4 ▫ How Do Adaptations Help Living Things Survive?

LESSON	PACING	OBJECTIVES
10 Guided Inquiry *Investigate Animal Survival* pages T169a–T169d **Science Inquiry and Writing Book** pages 48–51	**30** minutes	Investigate through Guided Inquiry (answer a question; make and compare observations; collect and record data and observations; generate explanations and conclusions based on evidence; share findings; ask questions based on observations to increase understanding; adjust explanations based on findings and new ideas). Predict which model animal will survive in a specific environment based on its special coloration. Understand how adaptations such as camouflage can help animals survive in their environment. Organize and analyze data by making tables and graphs.
11 **Conclusion and Review** pages T170–T171	**15** minutes	
12 NATIONAL GEOGRAPHIC LIFE SCIENCE EXPERT *Herpetologist* pages T172–T173 NATIONAL GEOGRAPHIC BECOME AN EXPERT *Snakes: Adaptations for Many Environments* pages T174–T175—T180	**35** minutes	Recognize how different scientists' work has contributed to general scientific understanding. Explain how adaptations can help a living thing survive in its environment.

FAST FORWARD
ACCELERATED PACING GUIDE

DAY 1 40 minutes

Directed Inquiry

Investigate Plant Fossils, page T133g

DAY 2 35 minutes

NATIONAL GEOGRAPHIC **LIFE SCIENCE EXPERT** *Herpetologist,* page T172

NATIONAL GEOGRAPHIC **BECOME AN EXPERT** *Snakes: Adaptations for Many Environments,* page T174–T175

VOCABULARY	RESOURCES	ASSESSMENT
compare **infer**	Science Inquiry and Writing Book: *Life Science* Science Inquiry Kit: *Life Science* Guided Inquiry: Learning Masters 53–55	Inquiry Rubric: Assessment Handbook, page 185 Inquiry Self-Reflection: Assessment Handbook, page 196 Reflect and Assess, page T169d
		Assess the Big Idea, page T171 Chapter 4 Test: Assessment Handbook, pages 19–25 NGSP ExamView CD-ROM
		Assess, pages T173, T178–T179

TECHNOLOGY RESOURCES

STUDENT RESOURCES

myNGconnect.com

- **Student eEdition**
- Big Ideas Book
- Science Inquiry and Writing Book
- **Read with Me**
- **Vocabulary Games**
- **Digital Library**
- **Enrichment Activities**

National Geographic Kids

National Geographic Explorer!

TEACHER RESOURCES

myNGconnect.com

- **Teacher eEdition**
- Teacher's Edition
- Science Inquiry and Writing Book
- Online Lesson Planner
- National Geographic Unit Launch Videos
- Assessment Handbook
- **Presentation Tool**
- **Digital Library**

NGSP ExamView CD-ROM

DAY 3 30 minutes

Guided Inquiry

Investigate Animal Survival, page T169a

Objectives

Students will be able to:

- Investigate through Directed Inquiry (answer a question; make and compare observations; collect and record data and observations; generate explanations and conclusions based on evidence; share findings; ask questions based on observations to increase understanding).
- Explain how fossils provide evidence of the history of Earth.
- Ask "why" questions based on observations.
- Differentiate fact from opinion and explain that scientists do not rely on claims or conclusions unless they are backed by observations that can be confirmed.

Science Process Vocabulary

compare, infer

PROGRAM RESOURCES

- Science Inquiry and Writing Book: *Life Science*
- Science Inquiry and Writing Book eEdition at myNGconnect.com
- Inquiry eHelp at myNGconnect.com
- Science Inquiry Kit: *Life Science*
- Learning Masters Book, pages 47–50, or at myNGconnect.com
- Inquiry Rubric: Assessment Handbook, page 185, or at myNGconnect.com
- Inquiry Self-Reflection: Assessment Handbook, page 195, or at myNGconnect.com

MATERIALS

Kit materials are listed in italics.

fern, elm, or sycamore fossil; hand lens; tissue paper; pencil; Plant Information Learning Master; construction paper; markers; glue; *chenille stems; clay;* scissors

1 Introduce

Tap Prior Knowledge

- Ask: **What can we use to study plants and animals that lived long ago?** (fossils) Discuss what scientists can learn from fossils.

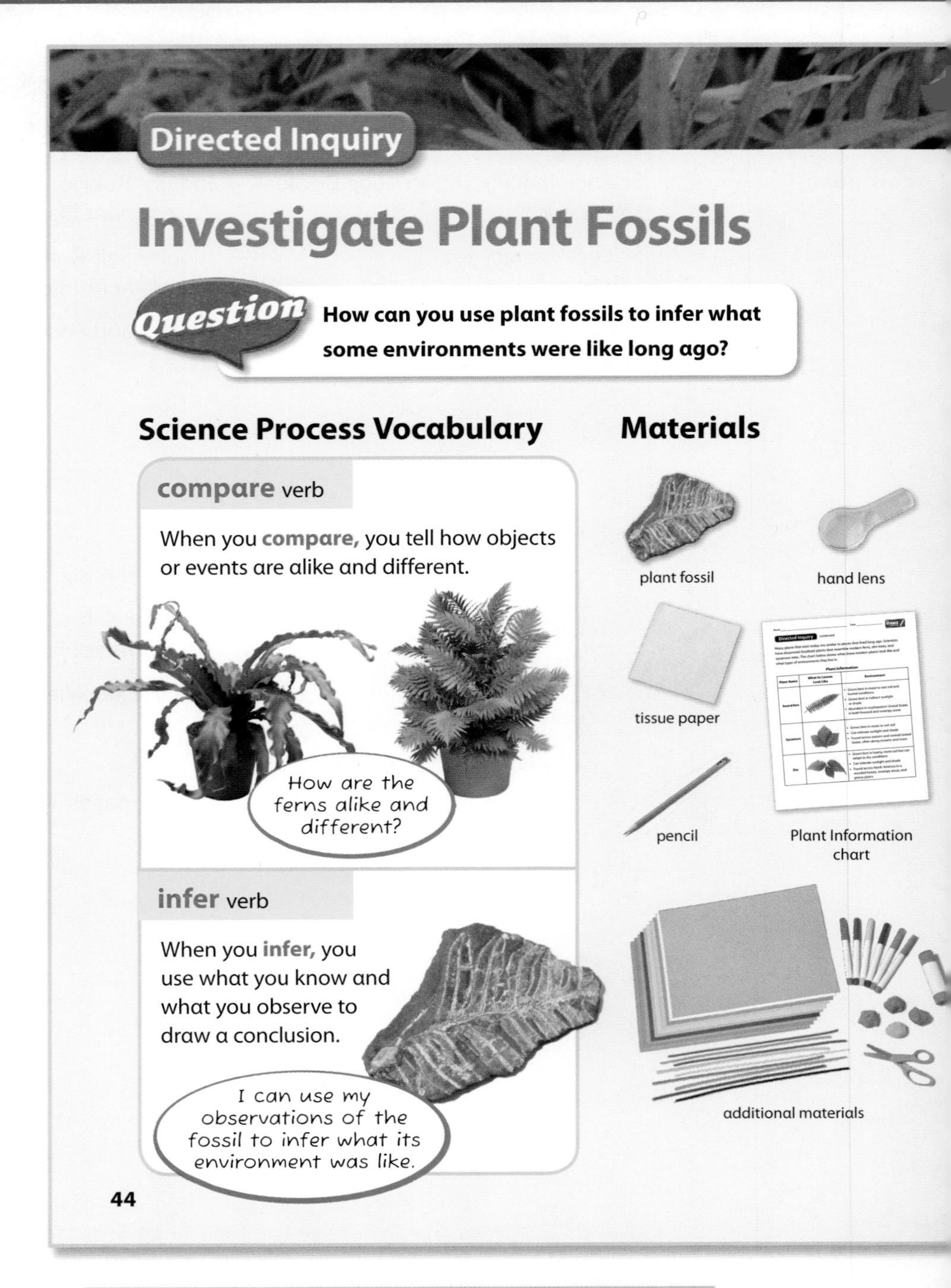

MANAGING THE INVESTIGATION

Time

40 minutes

Groups

Small groups of 4

Advance Preparation

- Copy the Plant Information Learning Masters for the class.
- Sharpen enough pencils for each group. If available, pencils with a softer lead might pick up more detail than a #2 pencil.
- Cut the sheets of tissue paper in half and distribute to students.
- Collect art materials such as markers, construction paper, scissors, chenille stems, glue, and modeling clay for students to use while making their displays.

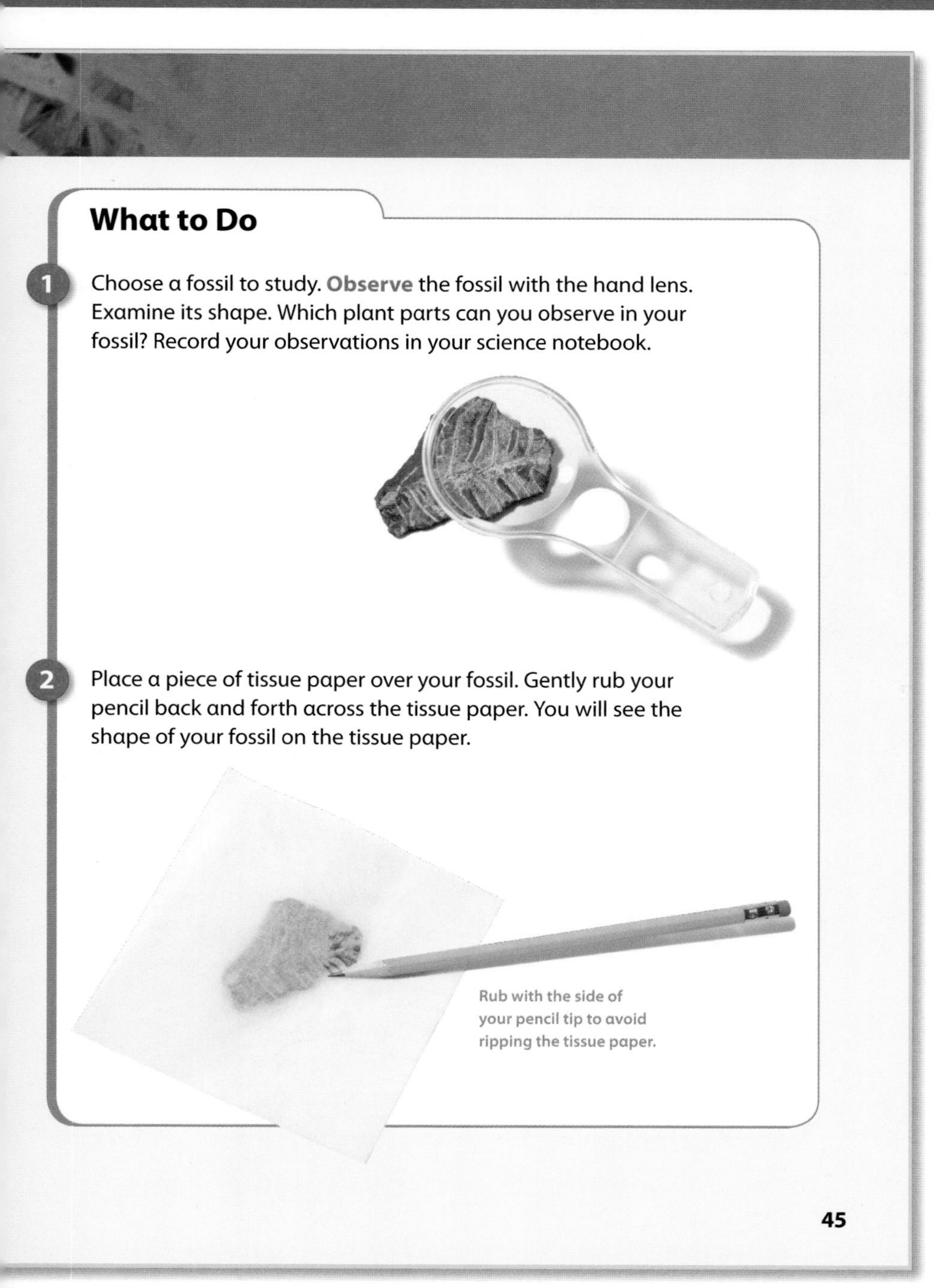

What to Do

1. Choose a fossil to study. **Observe** the fossil with the hand lens. Examine its shape. Which plant parts can you observe in your fossil? Record your observations in your science notebook.

2. Place a piece of tissue paper over your fossil. Gently rub your pencil back and forth across the tissue paper. You will see the shape of your fossil on the tissue paper.

Rub with the side of your pencil tip to avoid ripping the tissue paper.

45

Teaching Tips

- The fossils from the kit are samples of ancient fern, sycamore, and elm species.
- In step 2, remind students to rub gently and with the side of the pencil tip. They may need to rub the pencil back and forth several times across the same area to produce a clear image of the fossil.

What to Expect

- Students will observe a fossil and decide which modern plant the fossil resembles. They will then use information about a similar modern plant to make inferences about the fossilized plant's environment.

Introduce, continued

Connect to the Big Idea

- Review the Big Idea Question, *How do adaptations help living things survive?* Explain to students that this inquiry will help them learn about ancient plants and their environments.
- Have students open their Science Inquiry and Writing Books to page 44. Read the Question and invite students to share ideas about what some plant environments were like long ago.

❷ Build Vocabulary

Science Process Words: compare, infer

Use this routine to introduce the words.

1. **Pronounce the Word** Say **compare**. Have students repeat it in syllables.
2. **Explain Its Meaning** Choral read the sentence. Ask students to define **compare**. (tell how things are alike and different)
3. **Encourage Elaboration** Ask: **What senses might you use to compare a fossilized plant to a living plant?** (sight and touch)

ELL Use Cognates

Spanish speakers may know the verb **comparar** in their home language. Use it to help them understand the English word **compare**.

Repeat for the word **infer**. To encourage elaboration, ask: **What can you use to infer about plant fossils?** (observations of the fossil; information about similar plants alive today)

❸ Guide the Investigation

- Distribute materials. Read the steps on pages 45–46. Clarify steps if necessary.
- Review the different ways plants and animals can become fossilized. Explain to students that sometimes only part of the plant will be preserved, such as a leaf, so the fossil will only show part of the plant.
- In step 1, students may have difficulty distinguishing between the sycamore and elm leaf fossils if their samples only show part of the leaf. Guide students to compare the leaf edges, shapes, and the vein patterns on page 46 with their fossils.

Guide the Investigation, continued

- If students are still unable to identify their fossils, they can respond that their fossil was "either an elm or a sycamore." Explain that scientists who study fossils often find it a challenge to identify them because the fossils are hard to see or have only been partially preserved.

❹ Explain and Conclude

- Students may create a poster, a diorama, a model, or another visual display. Have students share their displays with the class.
- Guide students to evaluate the displays. Ask: **Which information in this display is fact and which is opinion? How do you know?** Discuss how scientists do not rely on claims or conclusions unless they can be confirmed by observations.
- Tell students to further evaluate the displays by asking "why" questions, such as: **Why do you think your fossilized plant grew in a wet, humid, environment?** (because it looks like a modern fern, and ferns grow best in moist soil and humid conditions)
- Guide students as they make inferences about what the environment was like when the fossilized plants were alive. Students should understand that ancient elm trees may have grown in forests, swamps, and plains; sycamores probably grew near water; and ferns grew in wet, shady areas such as forests and swamps. Explain that the environment where the fossils were found may be very different now than it was when the ancient plants were alive.
- Ask: **What information helped you make these inferences?** (information about the environments of similar plants alive today)
- Ask: **Why is making inferences such an important part of scientists' work?** Students should understand that scientists use inferences to form conclusions.

Answers

1. Answers will vary, but students should note similarities in physical characteristics such as leaf shape, pattern of veins, or shape of leaf edge.

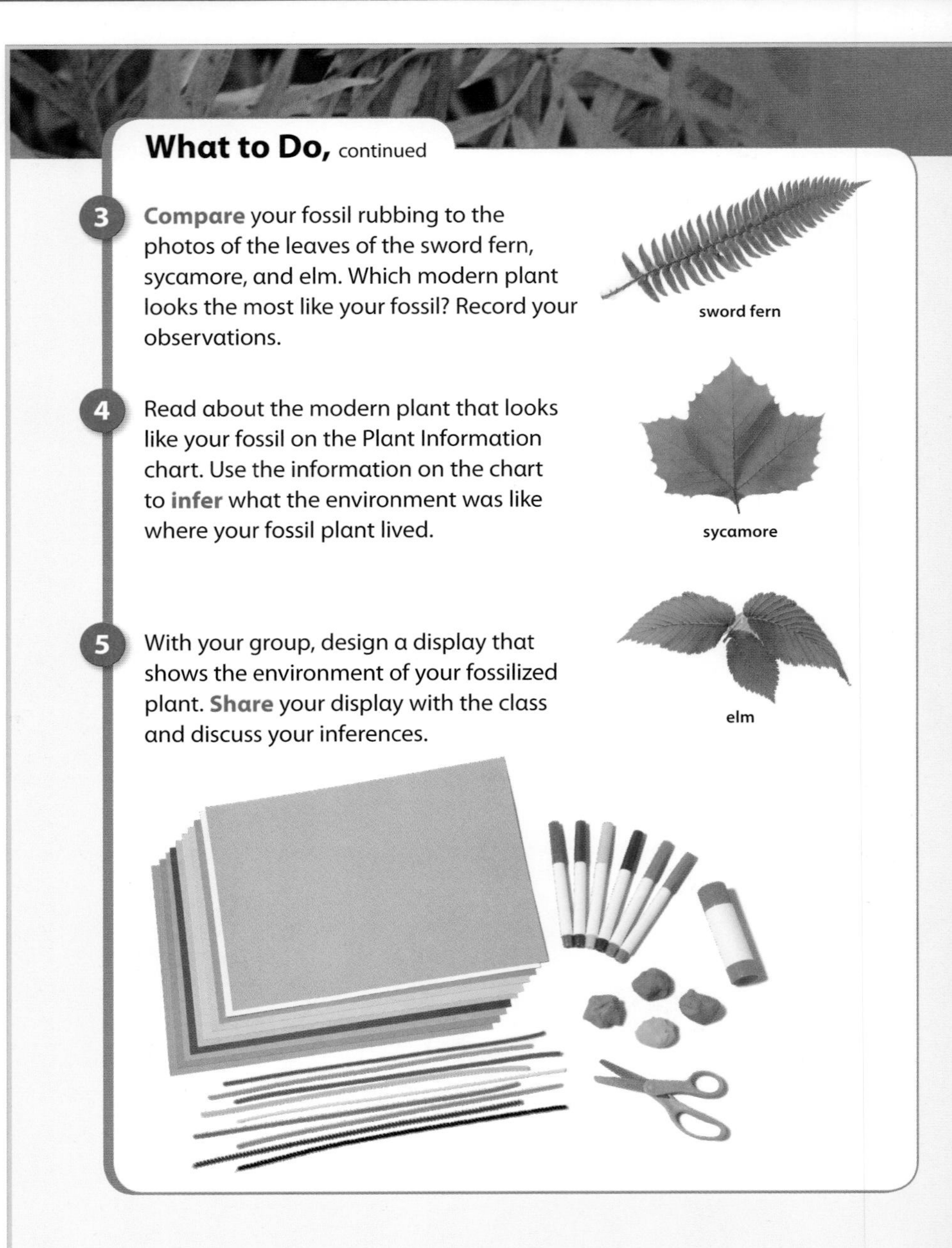

What to Do, continued
3 Compare your fossil rubbing to the photos of the leaves of the sword fern, sycamore, and elm. Which modern plant looks the most like your fossil? Record your observations.
sword fern
4 Read about the modern plant that looks like your fossil on the Plant Information chart. Use the information on the chart to infer what the environment was like where your fossil plant lived.
sycamore
5 With your group, design a display that shows the environment of your fossilized plant. Share your display with the class and discuss your inferences.
elm
46

NATIONAL GEOGRAPHIC Raise Your SciQ!

Sword Fern Species Sword fern is the common name for two different species of ferns that look very similar. *Polystichum munitum,* also known as the western sword fern, only grows in the western United States and Pennsylvania. *Nephrolepis exaltata*, also known as the Boston fern, grows only in the southern and southeastern parts of the United States.

Record

Write and draw in your science notebook.
Use a table like this one.

my SCIENCE notebook

Fossil Observations

Shape	
Plant parts	
Other observations	
Which modern plant looks the most like your fossil?	

Explain and Conclude

1. How was your fossil plant similar to the modern plant on the chart? How was it different?
2. **Compare** your fossil with the fossils of other groups. How are they alike and different?
3. **Infer** what the environment was like at the time your fossil plant lived. Explain your inference.

Think of Another Question

What else would you like to find out about using plant fossils to learn about environments from long ago? How could you find an answer to this new question?

Many different types of ferns live in this forest.

47

Inquiry Rubric at myNGconnect.com	Scale			
The student **observed** a plant fossil, made rubbings, and recorded observations and information in a table.	4	3	2	1
The student **compared** the fossilized plant to a similar modern plant and with other fossils.	4	3	2	1
The student made **inferences** about the fossilized plant's environment.	4	3	2	1
The student made a display to **share** results and **conclusions** with others.	4	3	2	1
The student asked **questions** about the displays of others.	4	3	2	1
Overall Score	4	3	2	1

Explain and Conclude, continued

2. Answers will vary depending on the fossils students choose to observe. Students may note that all the fossils were from leaves, but the shapes of the fossils were different. The part of the leaf that could be seen was also different.

3. Answers will vary depending on the fossils students choose. Students should explain that they used the photos to compare the fossilized plant to a modern plant. They then could infer that the fossilized plant lived in an environment similar to the modern species.

5 Find Out More

Think of Another Question

- Students should use their observations to generate new questions. Have them use these question stems: *What happens when ... ? How can I/we ... ?*

6 Reflect and Assess

- To assess student work with the Inquiry Rubric shown below, see Assessment Handbook, page 185, or go online at myNGconnect.com
- Have students use the Inquiry Self-Reflection on Assessment Handbook, page 195, or at myNGconnect.com

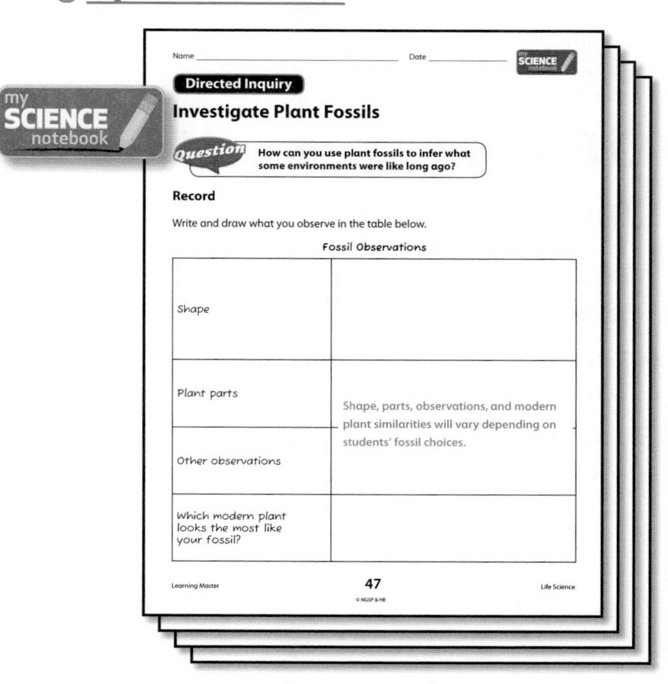
my SCIENCE notebook

Name ______ Date ______

Directed Inquiry

Investigate Plant Fossils

Question: How can you use plant fossils to infer what some environments were like long ago?

Record

Write and draw what you observe in the table below.

Fossil Observations

Shape	
Plant parts	Shape, parts, observations, and modern plant similarities will vary depending on students' fossil choices.
Other observations	
Which modern plant looks the most like your fossil?	

Learning Master 47 Life Science

Learning Masters 47–50
or at myNGconnect.com

CHAPTER 4
LESSON 2 ▫ Big Idea Question and Vocabulary

Objectives

Students will be able to:

- Describe how an animal's body structure can help it survive.
- Describe how an animal's coloration can help it survive.

Science Academic Vocabulary

adaptation, camouflage, variation, extinction

CHAPTER
4
HOW DO ADAPTATIONS HELP LIVING THINGS SURVIVE?
One secret to survival in the wild is to avoid being seen. That's what this peacock butterfly is doing. A hungry bird might never see it. Other animals may be frightened of its wings. How does the peacock butterfly's body covering help it stay alive?
The peacock butterfly is hard to see among the colors of the autumn leaves.
TECHTREK
myNGconnect.com
134
135

PROGRAM RESOURCES

- Big Ideas Book: *Life Science*
- Big Ideas Book: *Life Science* eEdition at myNGconnect.com
- Vocabulary Games at myNGconnect.com
- Digital Library at myNGconnect.com
- Enrichment Activities at myNGconnect.com
- Read with Me at myNGconnect.com
- Learning Masters Book, page 51, or at myNGconnect.com

❶ Introduce

Tap Prior Knowledge

- Briefly review the *Life Science* video. Ask students to discuss the differences among the animals presented in the video. Lead them to recognize that different animals survive in their environments in different ways.

❷ Focus on the Big Idea

Big Idea Question

- Read the Big Idea Question aloud and have students echo it.
- Preview pages 138–139, 140–141, 142–147, 148–149, 150–157, 158–163, and 164–167, linking the headings with the Big Idea Question.

Differentiated Instruction

ELL Vocabulary Support

BEGINNING

Have students work in pairs to describe the vocabulary words to one another. Encourage them to use the words to discuss the photos on pages 136–137.

INTERMEDIATE

Have students use Academic Language Frames to learn the vocabulary words.

- *An ____ helps a living thing survive in its environment.*
- *____ allows an animal to hide.*
- *Different colors among toads are a kind of ____.*

ADVANCED

Have students prepare flash cards for each vocabulary word. Each flash card should include the word, its definition, and an illustration. Partners may use the flash cards to quiz one another.

Focus on the Big Idea, continued

- With student input, post a chart that displays the chapter headings. Have students orally share what they expect to find in each section. Then read page 134 aloud.

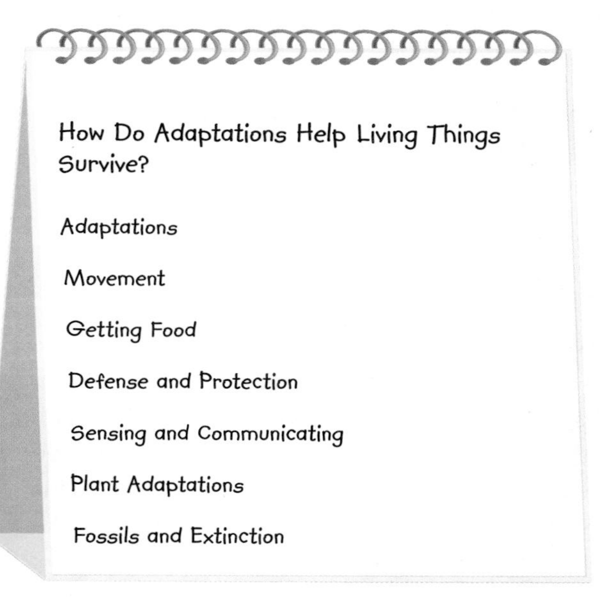

How Do Adaptations Help Living Things Survive?
Adaptations
Movement
Getting Food
Defense and Protection
Sensing and Communicating
Plant Adaptations
Fossils and Extinction

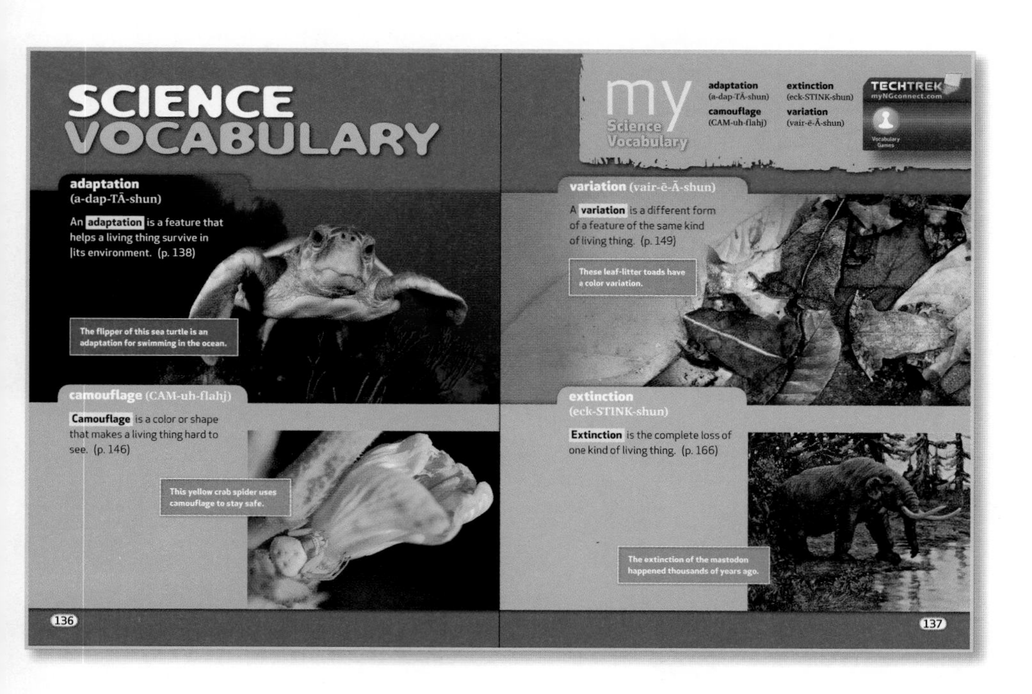

SCIENCE VOCABULARY
adaptation
(a-dap-TĀ-shun)
An adaptation is a feature that helps a living thing survive in its environment. (p. 138)
The flipper of this sea turtle is an adaptation for swimming in the ocean.
camouflage (CAM-uh-flahj)
Camouflage is a color or shape that makes a living thing hard to see. (p. 146)
This yellow crab spider uses camouflage to stay safe.
136
my Science Vocabulary
adaptation
camouflage
extinction
variation
TECHTREK
myNGconnect.com
variation (vair-ē-Ā-shun)
A variation is a different form of a feature of the same kind of living thing. (p. 149)
These leaf-litter toads have a color variation.
extinction
(eck-STINK-shun)
Extinction is the complete loss of one kind of living thing. (p. 166)
The extinction of the mastodon happened thousands of years ago.
137

❸ Teach Vocabulary

Have students look at pages 136–137, and use this routine to teach each word. For example:

1. **Pronounce the Word** Say **adaptation** and have students repeat it.
2. **Explain Its Meaning** Read the word, definition, and sample sentence, and use the photo to explain the word's meaning. Point out the flipper. Say: **Long, wide flippers are an adaptation that helps a sea turtle swim.**
3. **Encourage Elaboration** Ask: **What other adaptations do you see on pages 136 and 137?** (Examples include camouflage and body parts such as a trunk or tusks.)

Repeat for the words **camouflage**, **variation**, and **extinction** using these Elaboration Prompts:

- **How does camouflage help the yellow crab spider?** (Blending into the background helps the spider hide from enemies and surprise prey.)
- **What does the photo show about the color variation of leaf-litter toads?** (Some toads are tan; others are dark brown. Both colors provide camouflage among fallen leaves.)
- **Why can't we observe living mastodons or other extinct animals?** (They have all died out.)

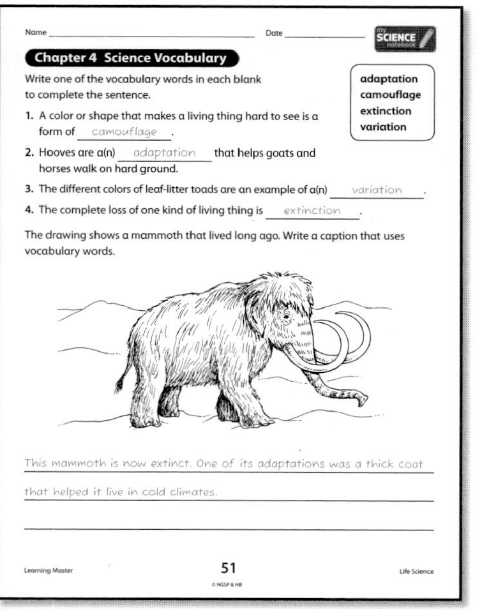

Name ______________________ Date ____________

Chapter 4 Science Vocabulary

Write one of the vocabulary words in each blank to complete the sentence.

adaptation
camouflage
extinction
variation

1. A color or shape that makes a living thing hard to see is a form of ___camouflage___.
2. Hooves are a(n) ___adaptation___ that helps goats and horses walk on hard ground.
3. The different colors of leaf-litter toads are an example of a(n) ___variation___.
4. The complete loss of one kind of living thing is ___extinction___.

The drawing shows a mammoth that lived long ago. Write a caption that uses vocabulary words.

This mammoth is now extinct. One of its adaptations was a thick coat that helped it live in cold climates.

Learning Master 51 Life Science

Learning Master 51 or at myNGconnect.com

Objectives

Students will be able to:

- Recognize how adaptations are important to the survival of living things.
- Identify examples of animal structures that are adaptations.

Science Academic Vocabulary

adaptation

❶ Introduce

Tap Prior Knowledge

- Have students name some familiar animals and describe body parts, skills, or behaviors that help the animals survive. Possible examples include an elephant's trunk, an eagle's talons, and a turtle's shell. Discuss how animals have a variety of features that help them survive.

Set a Purpose and Read

- Ask students if they are familiar with the word **adaptations**. Explain that living things have **adaptations**, and in the lesson they will learn the meaning of the word and observe examples.
- Have students read pages 138–139.

❷ Teach

Academic Vocabulary: *adaptation*

- Reread the last sentence on page 138 that defines the word **adaptation**.
- Point to the flipper of the sea turtle pictured on page 138. Ask: **Why are flippers a useful adaptation for sea turtles?** (The flippers help the sea turtle swim in the water.) Explain that sea turtles have many **adaptations** for life in the water. Among them are powerful lungs that help the sea turtle stay underwater.

Adaptations
Living things come in many sizes and shapes, and their appearance is often quite unique. Look at the turtles in the photos. These animals have scaly skin, a shell, and four limbs. They are alike in more ways than they are different. Now observe the animals more carefully. The sea turtle has flippers and the tortoise has legs with feet. These limbs are features that help the animals move. A feature that enables a living thing to survive in its environment is an adaptation.
A sea turtle uses its flippers to swim gracefully in the ocean.
138

Science Misconceptions

Adaptations Some students may think a plant or animal has only a few physical adaptations, or that adaptations are always unusual and easy to observe, such as the flippers of a sea turtle. Explain that all parts of an organism are adapted to its environment. A sea turtle's adaptations for life underwater include a streamlined body shape, powerful lungs, and large, wide eyes. The mouth and jaw of a sea turtle are adapted for the specific type of food it eats. Like other turtles and tortoises, different types of sea turtles are adapted to eat plants only, animals only, or both plants and animals.

Tortoises can walk to find food. The tortoise's legs help it walk in mud or through a shallow pond. A sea turtle, though, has flippers. The flippers enable it to swim in the ocean. By swimming, the sea turtle is able to catch food and escape predators.

This tortoise uses its legs and feet to move in its environment.

Before You Move On

1. Which part of the tortoise's body is an adaptation for walking?
2. How do sea turtles use their flippers to live in the ocean?
3. **Infer** How does a tortoise's hard shell help the tortoise survive in its environment?

139

Teach, continued

Compare and Contrast Adaptations

- Have students observe and compare the sea turtle and tortoise pictured on pages 138–139. Ask: **How are the sea turtle and tortoise alike?** (Both have the same general body shape, including a head, four legs, and a shell.)
- Ask: **How are they different?** (The sea turtle has flippers for front legs, while the tortoise has short, thin legs. Their shells, heads, and other body parts are also different.)
- Ask: **How is each animal adapted to its environment?** (Each has specific body parts that allow it to get food, move, and meet other needs in its environment.) Explain that **adaptations** help animals survive in many ways. Examples include body parts for getting food, protection, and defense, as well as chemicals that animals make, such as the foul spray of a skunk.

3 Assess

» Before You Move On

1. **Recall Which part of the tortoise's body is an adaptation for walking?** (the legs and feet)
2. **Explain How do sea turtles use their flippers to live in the ocean?** (Flippers help a sea turtle swim through the water so it can find food and escape predators.)
3. **Infer How does a tortoise's hard shell help the tortoise survive in its environment?** (The hardness of the shell protects the tortoise from predators, and its coloration is a form of camouflage.)

Differentiated Instruction

ELL Language Support for Comparing and Contrasting Adaptations

BEGINNING

Help students illustrate a sea turtle and label its features (body shape, eyes, mouth, flippers) that are adaptations. Then help students use their illustrations as aids to describe adaptations.

INTERMEDIATE

Provide Academic Language Frames to help students compare and contrast adaptations.

- *The flippers of the sea turtle are an ____ that help it ____ in water.*
- *The legs of the tortoise are an ____ that help it ____ on land.*

ADVANCED

Help students write new captions for the photos of the sea turtle and tortoise. Tell them that each caption should include the word *adaptation* and identify examples of adaptations in the turtles' bodies.

Objectives

Students will be able to:

- Identify body parts of animals that are adaptations for moving in water and in air.
- Explain how adaptations for movement can help a living thing survive.

❶ Introduce

Tap Prior Knowledge

- Have students describe different ways that they can move, such as walking, running, crawling, and climbing. Then have them describe ways that they cannot move but that other animals are able to, such as flying like a bird.

Set a Purpose and Read

- Point to the photos on pages 140–141 and discuss the different ways that animals can move, including climbing and flying. Explain that students will learn about the adaptations for movement.
- Have students read pages 140–141.

❷ Teach

Identify Adaptations for Movement

- Point to the photo of the kinkajou on page 140 and have students read the caption. Say: **Kinkajous live in forests. What adaptations help them move from branch to branch of a tree?** (The adaptations include four legs with sharp claws, and a grasping tail)
- Ask: **How does a grasping tail help the kinkajou get food?** (By holding on to a branch with its tail, the kinkajou can reach for food with its arms and hands.) Remind students that tree branches may be high above ground. An animal could injure itself if it fell from a tall tree branch.

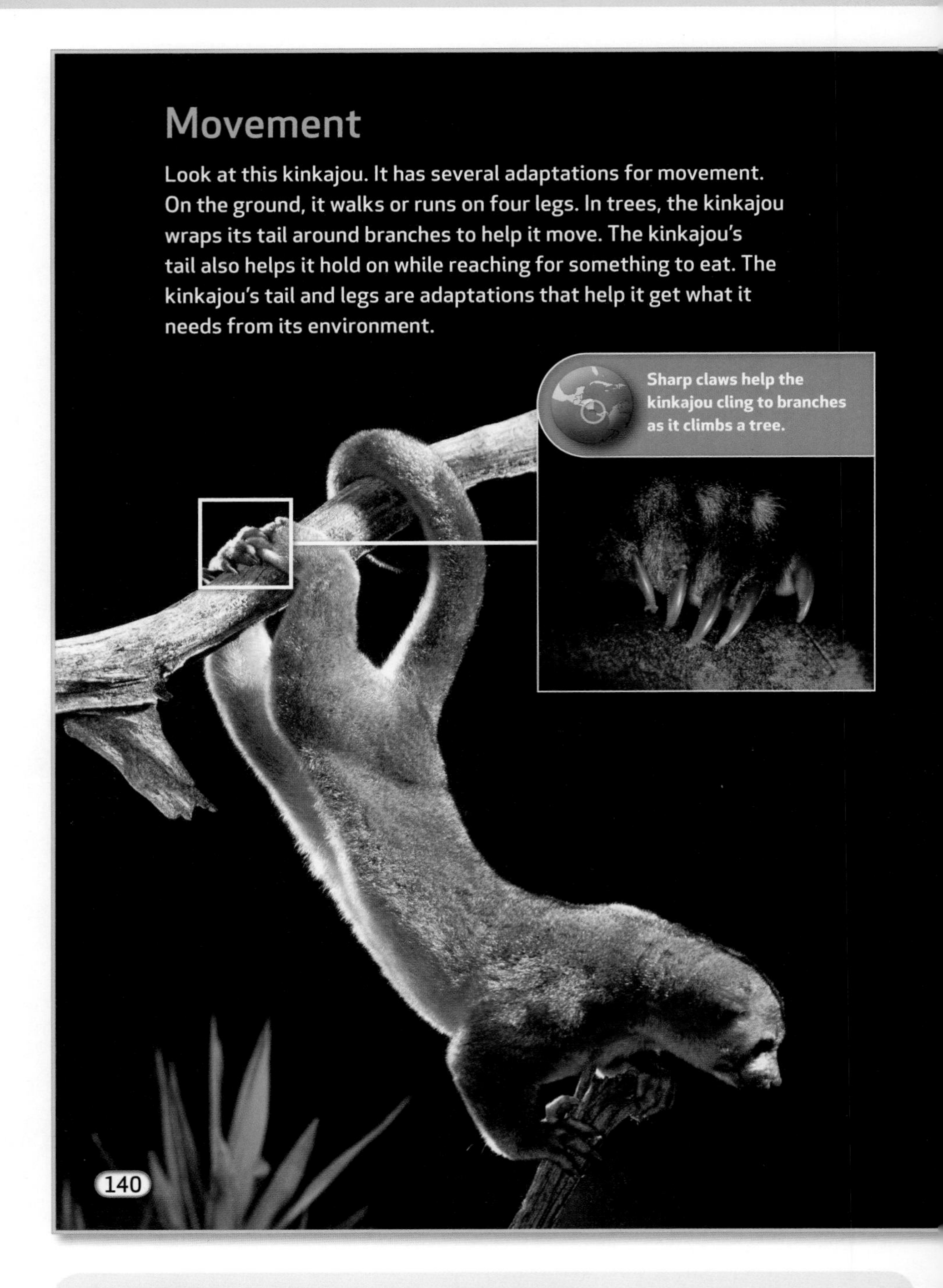

Movement
Look at this kinkajou. It has several adaptations for movement. On the ground, it walks or runs on four legs. In trees, the kinkajou wraps its tail around branches to help it move. The kinkajou's tail also helps it hold on while reaching for something to eat. The kinkajou's tail and legs are adaptations that help it get what it needs from its environment.
Sharp claws help the kinkajou cling to branches as it climbs a tree.
140

NATIONAL GEOGRAPHIC **Raise Your SciQ!**

Adaptations for Flight Most birds fly. Their adaptations for flight include the shapes of their wings. Wings provide the lift, or upward force, that keeps a bird's body aloft. The tail provides stability, while the body's streamlined shape reduces friction with the air. Birds also use a variety of feathers that help them control and stabilize their flying. Other adaptations for flight include lightweight bones and muscles that use oxygen very efficiently.

Animals have many different body parts that help them get what they need. Some body parts are adaptations for movement. Look at the photos. What adaptations help each of these animals move in the air?

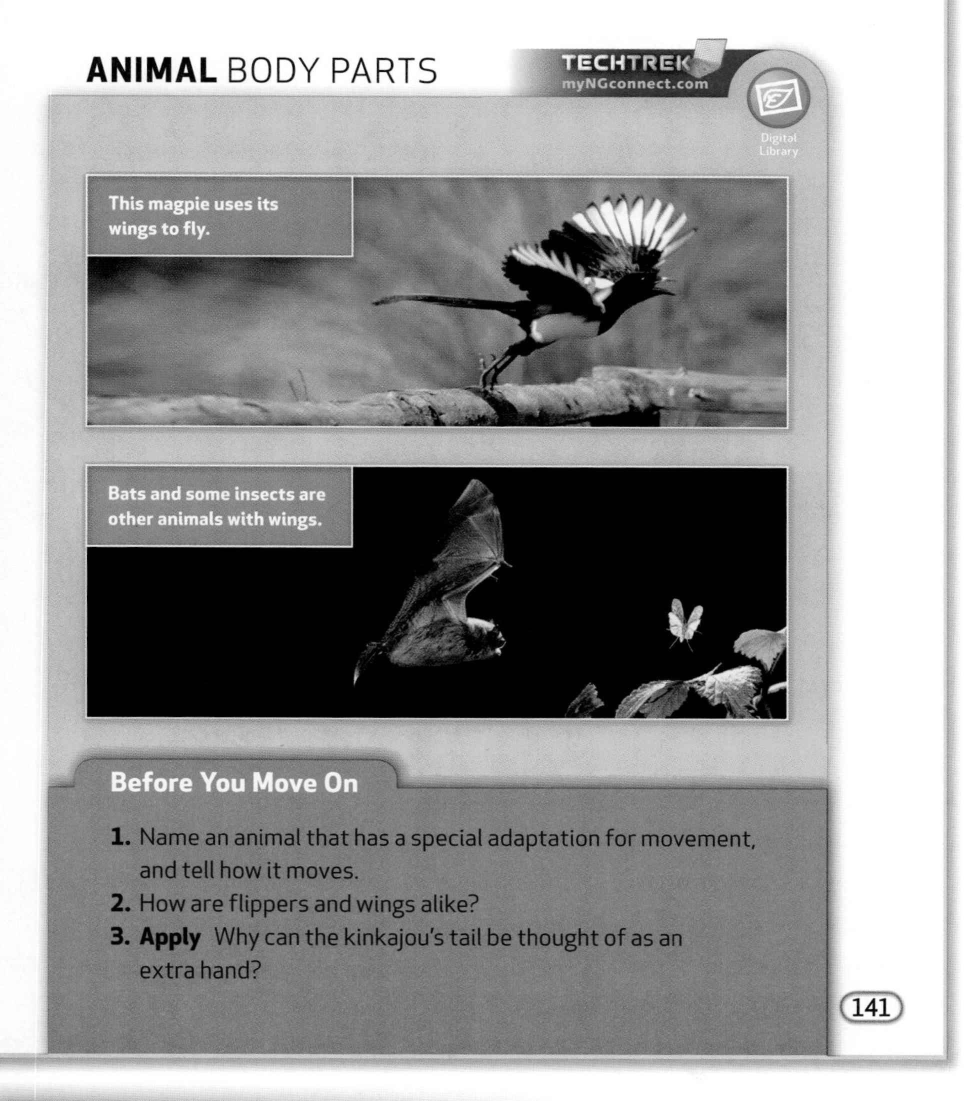

ANIMAL BODY PARTS

This magpie uses its wings to fly.

Bats and some insects are other animals with wings.

Before You Move On

1. Name an animal that has a special adaptation for movement, and tell how it moves.
2. How are flippers and wings alike?
3. **Apply** Why can the kinkajou's tail be thought of as an extra hand?

141

Differentiated Instruction

Extra Support

Have students search through an encyclopedia or old magazines for photos of animals that move on land, in the air, and in water. For each photo, have them identify the body parts that help the animal move.

Challenge

Have students research an animal that moves in an unusual or interesting way. Possible research subjects include snakes, dolphins, eels, moles, spiders, and ostriches. Have them identify the adaptations that help the animal move and report their findings to the class.

Teach, continued

Text Feature: Photos and Captions

- Have students observe the photos on page 141 and read the titles and captions. Ask: **Why can't a kinkajou fly like a magpie?** (It does not have wings.) Explain that even if the kinkajou had wings, it likely would be too heavy to fly. A kinkajou is a mammal. Bats are the only mammals that have adaptations that allow them to fly.
- Point to the magpie. Ask: **How are a bird's wings different from the arms of other animals?** (Possible answers: They are covered in feathers; they end in tips instead of hands and fingers.) Point to the bat. Ask: **How are a bat's wings different from the wings of birds?** (They are covered in skin, not feathers.) Discuss how a bird's feathers and a bat's taut skin give a wing its special shape, which allows these animals to fly.

myNGconnect.com

Have students use the Digital Library to find images of animal movement.

Computer Presentation Students can use the information to make a computer presentation about animal movement.

3 Assess

» Before You Move On

1. **Recall Name an animal that has a special adaptation for movement, and tell how it moves.** (Possible answers: a kinkajou, which has a grasping tail for climbing trees; magpies, bats, and other animals that have wings for flying; sea turtles, which have flippers for moving through water.)
2. **Compare How are flippers and wings alike?** (Both are parts of animals that have special shapes that allow movement through the air or water.)
3. **Apply Why can the kinkajou's tail be thought of as an extra hand?** (Like a hand, the kinkajou's tail can grasp and hold a tree branch. It allows the kinkajou to move easily through the trees of its environment.)

LESSON 4 ▫ Getting Food

Objectives

Students will be able to:

- Recognize that hunger is an internal signal that causes an animal to hunt for food.
- Explain how an animal uses its adaptations for getting food.

❶ Introduce

Tap Prior Knowledge

- Have students describe the body parts and tools they use to eat food. Then discuss how animals do not use forks, knives, and spoons—and many do not have hands!

Preview and Read

- Read the headings and preview the photos on pages 142–147.
- Have students read pages 142–147.

❷ Teach

Recognize Hunger as the Need for Food

- Ask: **When do you feel hungry?** (Possible answers: before lunch or dinnertime, when my stomach feels empty) **What do you want to do when you feel hungry?** (eat food) Have a volunteer read the first sentence on page 142. Discuss how eating food satisfies the feeling of hunger. Animals also eat to satisfy the feeling of hunger.

Getting Food

Hunger is a signal that tells you it's time to eat. Usually, the next step is to find something to satisfy the hunger. For you, that might mean a trip to the pantry for some peanut butter. But for an animal in the wild, it often means hunting for its food, or prey.

A hungry eagle uses powerful wings to hunt for food. Strong feet with claws help catch and hold its prey. Once the animal is caught, the eagle uses its beak to tear the meat and eat it.

The talons, or claws, of an eagle are an adaptation for grasping prey.

The eagle spreads its talons to catch a fish.

142

NATIONAL GEOGRAPHIC Raise Your SciQ!

Adaptations of Teeth An animal's teeth are adapted to the type of food it eats. Sharp, pointed teeth, such as those of lions, wolves, and other meat-eaters, are adapted for cutting and tearing meat. Flat-topped teeth, such as those of zebras, cows, and other plant-eaters, are adapted for grinding and chewing tough plant tissues. Animals that have both kinds of teeth, such as bears, are often able to eat both plants and meat. Especially long teeth, such as the tusks of walruses and elephants, are used not for eating but for grasping or for defense.

Gathering and Eating Food Most animals take in food with help from specialized body parts in their mouths. Each animal's mouth is an adaptation for the kind of food the animal eats. The giraffe has a unique way of gathering and eating its food. The giraffe uses its long neck and long tongue to grab and pull leaves off of high trees. This adaptation helps the giraffe get what it needs from its environment.

Other animals have different adaptations in their mouths that help them eat. Look at the chart below. What adaptations help each of these animals eat?

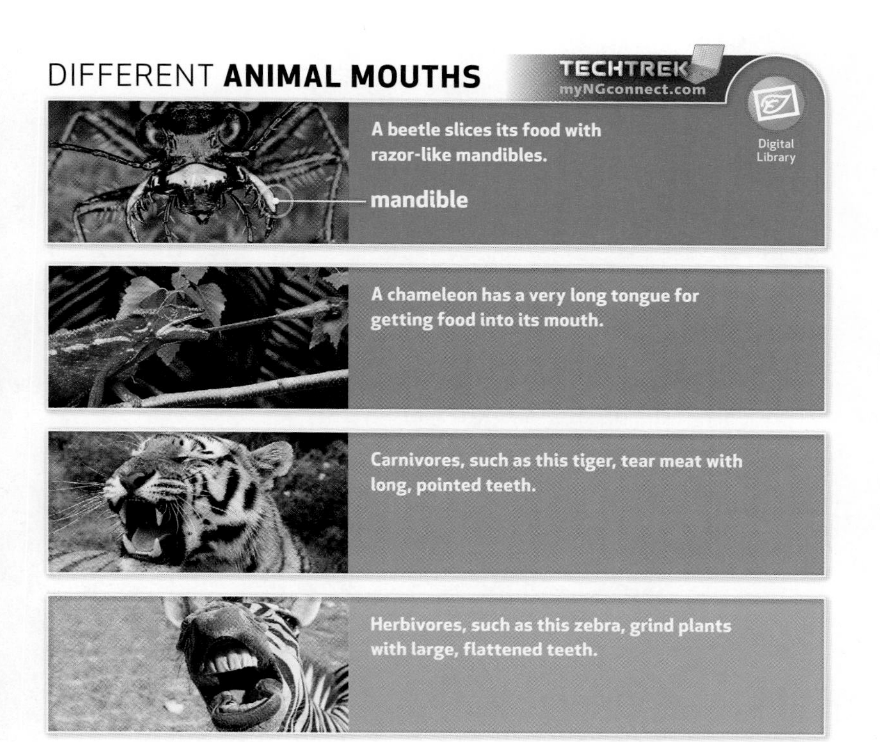

DIFFERENT ANIMAL MOUTHS
TECHTREK
myNGconnect.com
Digital Library
A beetle slices its food with razor-like mandibles.
mandible
A chameleon has a very long tongue for getting food into its mouth.
Carnivores, such as this tiger, tear meat with long, pointed teeth.
Herbivores, such as this zebra, grind plants with large, flattened teeth.

143

Science Misconceptions

Adaptations for Eating Because humans are able to eat a wide variety of foods, some students may mistakenly believe that animals could eat these same foods if they had the opportunity to do so. In fact, many animals have a variety of adaptations that limits their diet. These adaptations include the shapes and parts of the mouth, the length and complexity of the digestive tract, and the ability to digest certain materials in food. Grazing animals, for example, are able to break down cellulose, the tough material that makes up grasses. But many other animals cannot digest cellulose. Although humans can eat plant products, the cellulose in the plants moves through the digestive tract and is removed as waste.

Teach, continued

Text Feature: Chart

- Have students study the photos in the chart on page 143 and read the captions. Ask: **Why can't these animals all eat the same food?** (Possible answer: They all have different mouths. Each type of mouth is adapted for eating only certain foods.)
- Ask: **Which animal is adapted for eating grass?** (the zebra) **Why?** (Its mouth is filled with large, flattened teeth that are ideal for grinding plants.) Have students compare the zebra's teeth with the teeth of the tiger, which are adapted for tearing and cutting meat.
- Ask: **Which animal is adapted for catching flies in mid-air?** (the chameleon) Why? (It has a long, sticky tongue.) Explain that a chameleon can aim and stick out its tongue very quickly.

Record Observations

Have students sketch the animals in the chart on page 143 in their science notebook. Then have them add words, phrases, or labels to describe the adaptations that help animals get food. Students may also add other examples to their charts.

myNGconnect.com

Have students use the Digital Library to find images of animal mouths.

Whiteboard Presentation Students can use the information to make a computer presentation about animal mouths. You can help them show the presentation on a whiteboard.

Objectives

Students will be able to:

- Explain how an animal uses its adaptations for getting food.

Teach, continued

Identify Beak Shape as an Adaptation

- Have students observe the photo of the flamingo on page 144 and read the caption. Say: **Some birds eat seeds or earthworms, but flamingos eat animals called krill.**
- Ask: **Why don't flamingos eat seeds from nearby palms or worms they dig from the soil?** (They are not adapted to eat these foods.) Say: **Suppose small worms lived in the water.** Ask: **Do you think flamingos might eat them? Why?** (Possible answer: Yes. Their bills are curved to scoop small animals floating in the water.)
- Ask: **What adaptation helps flamingos eat krill?** (their long, curved bill or beak) Have students bend one hand into a curved shape. Explain that this shape is ideal for scooping water and small things that live in the water.
- Point to the other photos of birds on pages 144–145. Ask: **Which of these birds has a mouth most like the mouth of the flamingo?** (the duck) **Why?** (Unlike the other birds, the duck also has a wide beak that can scoop water.) Discuss how each bird bill or beak is adapted for eating certain kinds of foods. Because birds have different beaks, they eat different kinds of food.

A bird uses its bill or beak to get food. A bird's bill is very strong. The size and shape of the bill is an adaptation for the food the bird eats.
For example, flamingos wade in shallow water. Their long bills are curved. This shape is an adaptation for scooping small animals out of the water to eat.
This duck's bill is shaped to collect food from the water.
This flamingo uses its beak to eat tiny animals called krill.
144

NATIONAL GEOGRAPHIC Raise Your SciQ!

Birds and Their Food The mouths of all birds are adapted into hard structures called bills or beaks. Most birds catch food using only their beaks, but eagles, hawks, owls, and other birds of prey are able to catch small animals with their talons, or grasping feet. Birds of prey also have keen eyesight for spotting prey from far away. Birds do not have teeth but instead grind food in a gizzard, an organ that is like a second stomach. Many birds swallow small stones, which they store in their gizzard. The movement of the stones helps break apart the food.

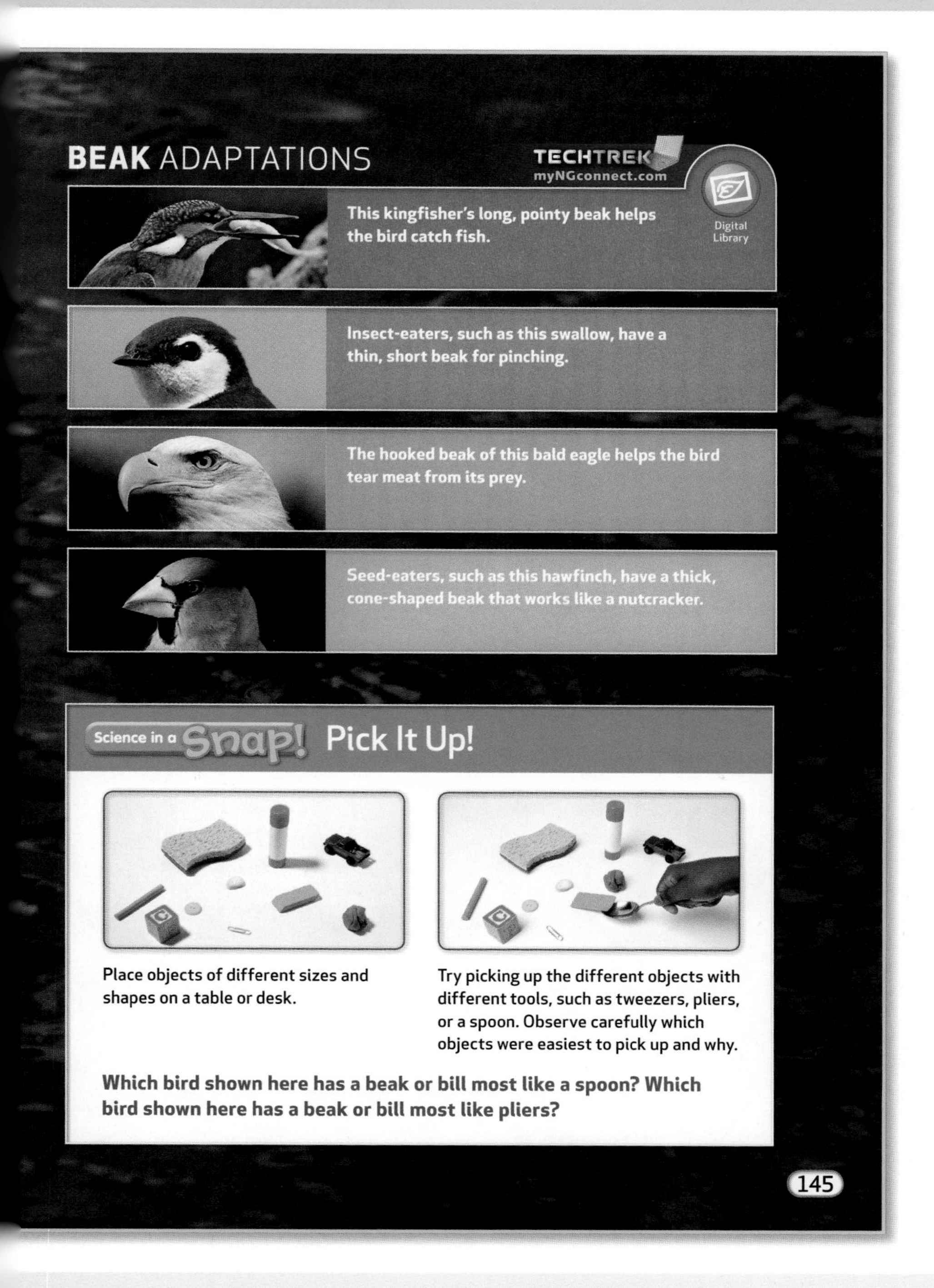

BEAK ADAPTATIONS

TECHTREK myNGconnect.com

Digital Library

This kingfisher's long, pointy beak helps the bird catch fish.

Insect-eaters, such as this swallow, have a thin, short beak for pinching.

The hooked beak of this bald eagle helps the bird tear meat from its prey.

Seed-eaters, such as this hawfinch, have a thick, cone-shaped beak that works like a nutcracker.

Science in a Snap! Pick It Up!

Place objects of different sizes and shapes on a table or desk.

Try picking up the different objects with different tools, such as tweezers, pliers, or a spoon. Observe carefully which objects were easiest to pick up and why.

Which bird shown here has a beak or bill most like a spoon? Which bird shown here has a beak or bill most like pliers?

145

Teach, continued

Text Feature: Chart

- Have students observe and compare the photos in the chart on page 145 and read the captions. Ask: **How are the beaks different?** (They each have different length, thickness, and shape.)
- Point to the kingfisher. Ask: **How is the kingfisher's beak an adaptation for eating fish?** (Possible answer: The beak is long enough and can open wide enough for a fish to fit inside.) Point out that even a medium-sized fish could not fit inside the beaks of the tree swallow and hawfinch.
- Point to the eagle. Ask: **How is the hooked beak an adaptation?** (The hook helps it tear meat.) Explain that unlike the other birds shown in the chart, eagles catch prey with their talons, or grasping feet.
- Point to the hawfinch. Ask: **How is the cone-shaped beak an adaptation for eating?** (Many seeds are very tough. The strength of this kind of beak helps break the seeds apart.)

myNGconnect.com

Have students use the Digital Library to find images of bird beaks.

Digital Booklet Students can use the information to make a digital booklet with captions that describes observations or inferences about bird beaks.

Pick it Up! 

Materials a variety of small objects, such as those pictured on page 145; various tools such as tweezers, pliers, and spoons.

Have students work in pairs, follow the instructions, and record their observations and conclusions in their science notebook.

What to Expect Students should discover that each object is most easily picked up by a specific tool. The flamingo has a beak that is most like a spoon. The hawfinch has a beak that is most like pliers.

Quick Questions Ask students the following questions:

- **In this activity, what do the different objects represent? What do the tools for picking up the objects represent?** (The objects represent kinds of food, and the tools represent different kinds of beaks.)
- **How does different birds who live in the same area having different kinds of beaks aid survival?** (They can eat different kinds of food, so there is less competition for food.)

Objectives

Students will be able to:

- Recognize adaptations that animals use to attract prey.
- Explain how an animal uses its adaptations for getting food.

Science Academic Vocabulary

camouflage

Teach, continued

Academic Vocabulary: *camouflage*

- Point out the word **camouflage**. Have students propose a definition.
- Have students observe the inset photo on page 146 and identify the hidden flounder. Ask: **Why is the flounder hard to see?** (It is using camouflage; its shape and coloring blend into the background of the ocean floor.) Explain that **camouflage** can help an animal hide from both predators and prey.

Identify Adaptations for Attracting Prey

- Have students observe the photo of the anglerfish and its prey on page 146 and then read the caption. Have students identify the anglerfish and the projection on its face that acts as a lure.
- Ask: **Why is the prey attracted to the anglerfish?** (The tip of the projection glows in the dark, and the light attracts the prey.) Explain how deep ocean waters are always dark, even during the daytime. The light made by anglerfish and other animals may be the only light in this environment.

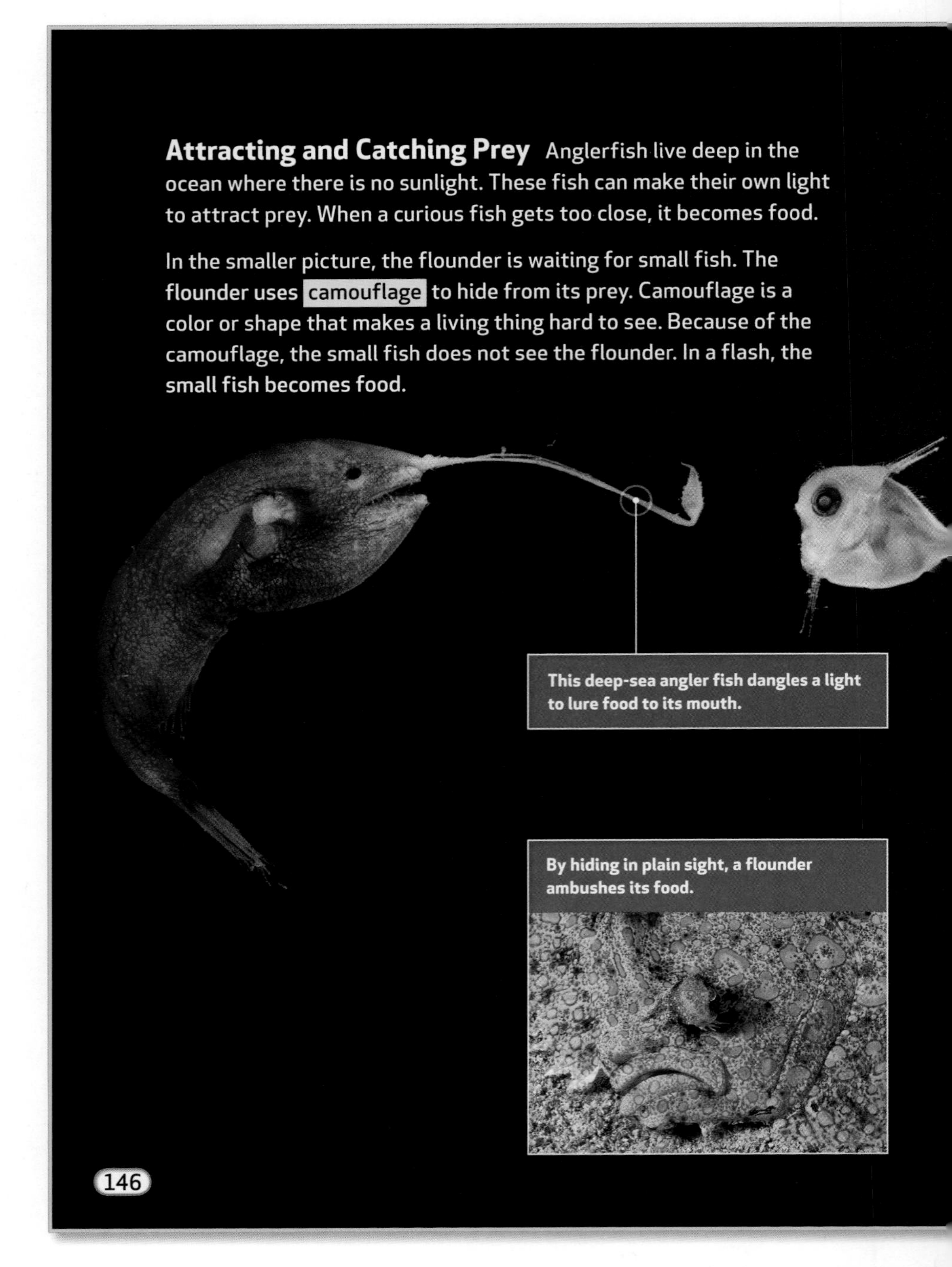

Attracting and Catching Prey Anglerfish live deep in the ocean where there is no sunlight. These fish can make their own light to attract prey. When a curious fish gets too close, it becomes food.

In the smaller picture, the flounder is waiting for small fish. The flounder uses camouflage to hide from its prey. Camouflage is a color or shape that makes a living thing hard to see. Because of the camouflage, the small fish does not see the flounder. In a flash, the small fish becomes food.

This deep-sea angler fish dangles a light to lure food to its mouth.

By hiding in plain sight, a flounder ambushes its food.

146

NATIONAL GEOGRAPHIC Raise Your SciQ!

Examples of Camouflage Animals have a variety of camouflage that helps them blend into the backgrounds of their environment. Zebras' stripes are one example. Lions, the main predator of zebras, are colorblind. They have difficulty distinguishing zebra stripes from the tall grasses that surround them. The stripes also help a dense herd of zebras blend into one another, making it difficult for lions to attack them. Penguins also have camouflage. When they swim in the water, their white fronts blend into the sky above, and their black backs blend into the water below.

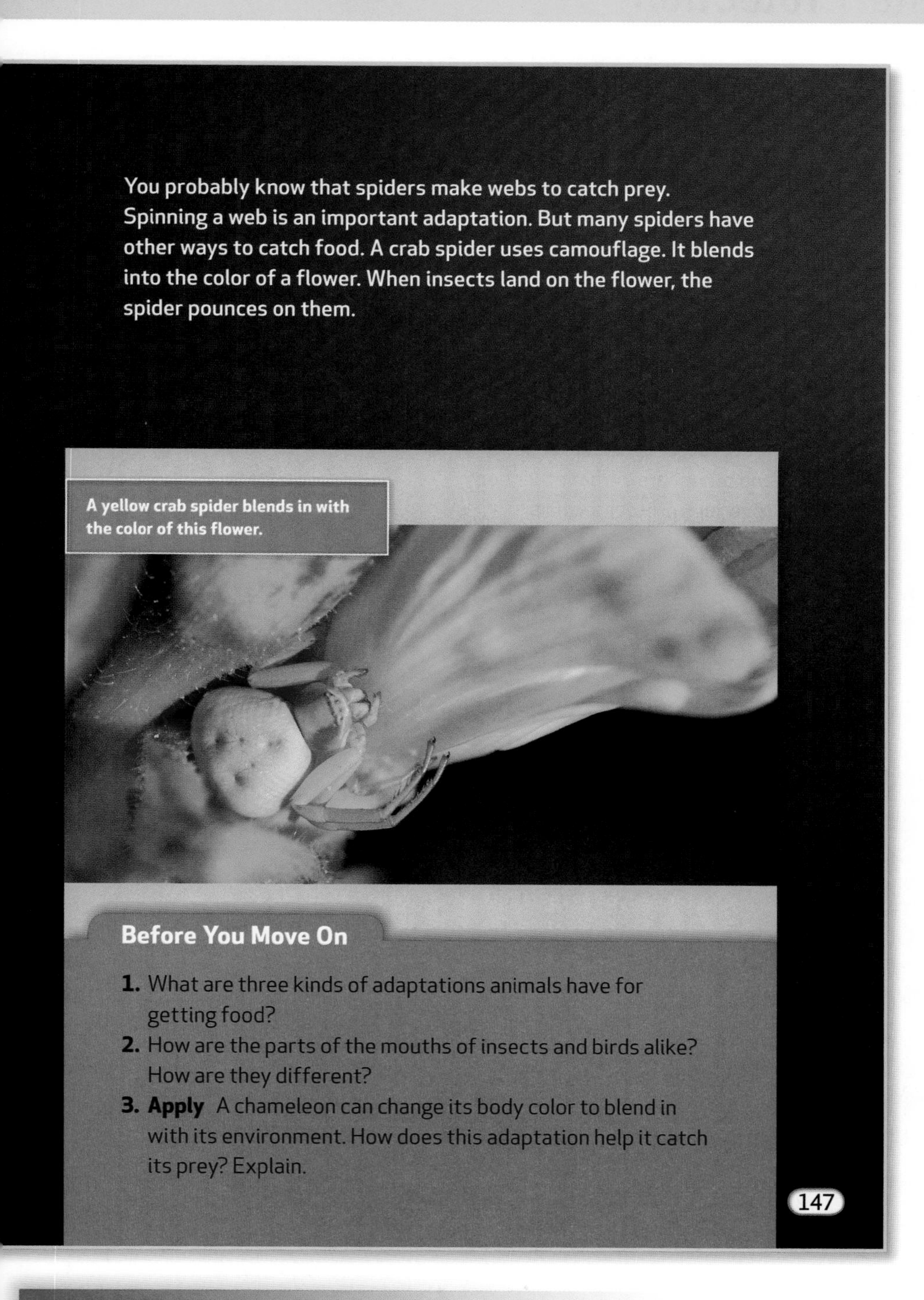

You probably know that spiders make webs to catch prey. Spinning a web is an important adaptation. But many spiders have other ways to catch food. A crab spider uses camouflage. It blends into the color of a flower. When insects land on the flower, the spider pounces on them.

A yellow crab spider blends in with the color of this flower.

Before You Move On

1. What are three kinds of adaptations animals have for getting food?
2. How are the parts of the mouths of insects and birds alike? How are they different?
3. **Apply** A chameleon can change its body color to blend in with its environment. How does this adaptation help it catch its prey? Explain.

147

Teach, continued

Describe Camouflage

- Have students observe the photo of the yellow crab spider on page 147 and then read the caption. Ask: **How is the spider camouflaged?** (Its yellow color blends into the background of the flower.)
- Ask: **How does camouflage help the spider get food?** (Insects visit the flower and do not see the spider. The spider can pounce on the insects and eat them). Explain that insects visit flowers to drink nectar, a sweet liquid that flowers make. Most flowers do not have spiders lurking by them.
- Ask: **Why is camouflage an example of an adaptation?** (Camouflage is a feature that helps an animal survive.) Discuss how **camouflage** can help an animal get food or avoid being food, both of which help the animal survive.

3 Assess

» Before You Move On

1. **List What are three kinds of adaptations animals have for getting food?** (Animals have specialized appendages for moving toward or attracting prey, different kinds of coloring for attracting prey or hiding from it, and behaviors for attracting and catching prey by surprise.)
2. **Compare and Contrast How are the parts of the mouths of insects and birds alike? How are they different?** (The mouthparts of insects and birds are adaptations for grabbing, holding, and breaking up their food into smaller pieces. Insects have pairs of mouthparts that extend out of their mouths, while birds have one bill or beak attached to the front of their head.)
3. **Apply A chameleon can change its body color to blend in with its environment. How does this adaptation help it catch its prey? Explain.** (Because chameleons are camouflaged, their prey may not see them. Chameleons can wait for prey to approach and then catch and eat them by surprise.)

Differentiated Instruction

ELL Language Support for Describing Camouflage

BEGINNING

Have partners identify the camouflaged animals in the photos on pages 146–147 and then use the word *camouflage* to describe the photos to each other.

INTERMEDIATE

Help students describe camouflage using Academic Language Frames.

- *A flounder has ____ because it blends into the ____ of the ocean floor.*

ADVANCED

Have students write new captions for the photos of the flounder and spider that illustrate camouflage. Invite students to read their captions to the class.

LESSON 5 ▫ Defense and Protection

Objectives

Students will be able to:

- Identify and explain how animals use an adaptation for defense and protection.
- Distinguish between an adaptation and a variation.

Science Academic Vocabulary

variation

❶ Introduce

Tap Prior Knowledge

- Have students describe how they protect themselves from rain or cold weather and why this protection is important.

Set a Purpose and Read

- Read the heading. Tell students that animals face many threats in their environment, and many adaptations help defend and protect them. Have students read pages 148–149.

❷ Teach

Academic Vocabulary: *variation*

- Point out the word **variation** on page 149. Write the word *vary* on the board. Say: **The word *vary* means "change."** Then point to the photo of leaf-litter toads on page 149. Ask: **What varies among the shoes that we are wearing today?** (Possible answers: colors, sizes, shapes)

Identify Adaptations for Defense

- Have students observe the photos on page 148 and read the captions. Then point to the porcupine. Ask: **Why do you think the lion is not attacking the porcupine?** (The lion recognizes that the porcupine's quills could hurt it.)

Defense and Protection

What is happening between the lion and the porcupine in the picture below? The lion is probably trying to kill the porcupine. The porcupine naturally defends itself. To do so, it raises its stiff, sharp quills. If the lion gets quills stuck in its face and paws, the porcupine will get away. The quills are an adaptation to defend the porcupine against predators.

Many animals use body coverings to protect themselves. A fish's scales and a snail's shell are body coverings that protect against predators and other harm. What other body coverings protect animals?

The lion does not attack the porcupine because of its sharp quills.

The African crested porcupine uses long, needle-sharp quills to defend itself.

148

Raise Your SciQ!

Warning Coloration Some small animals are brightly colored and very easy to see. Often, this is an adaptation called warning coloration. Many of these animals are poisonous, and their bright colors or markings tell predators to leave them alone. Examples of animals with warning coloration include poison-arrow frogs, tomato frogs, and harlequin bugs. Other animals with bright colors are mimics. They are not poisonous, but predators cannot tell them apart from the poisonous animals they resemble. So the mimics are protected too.

Find the leaf-litter toads among the leaves in the picture below. These toads are all the same kind, but they are not the same color or size. The leaves around the toads are not the same color or size, either. A **variation** is a difference in a feature of one kind of living thing. The variation of the toads help them survive on the forest floor.

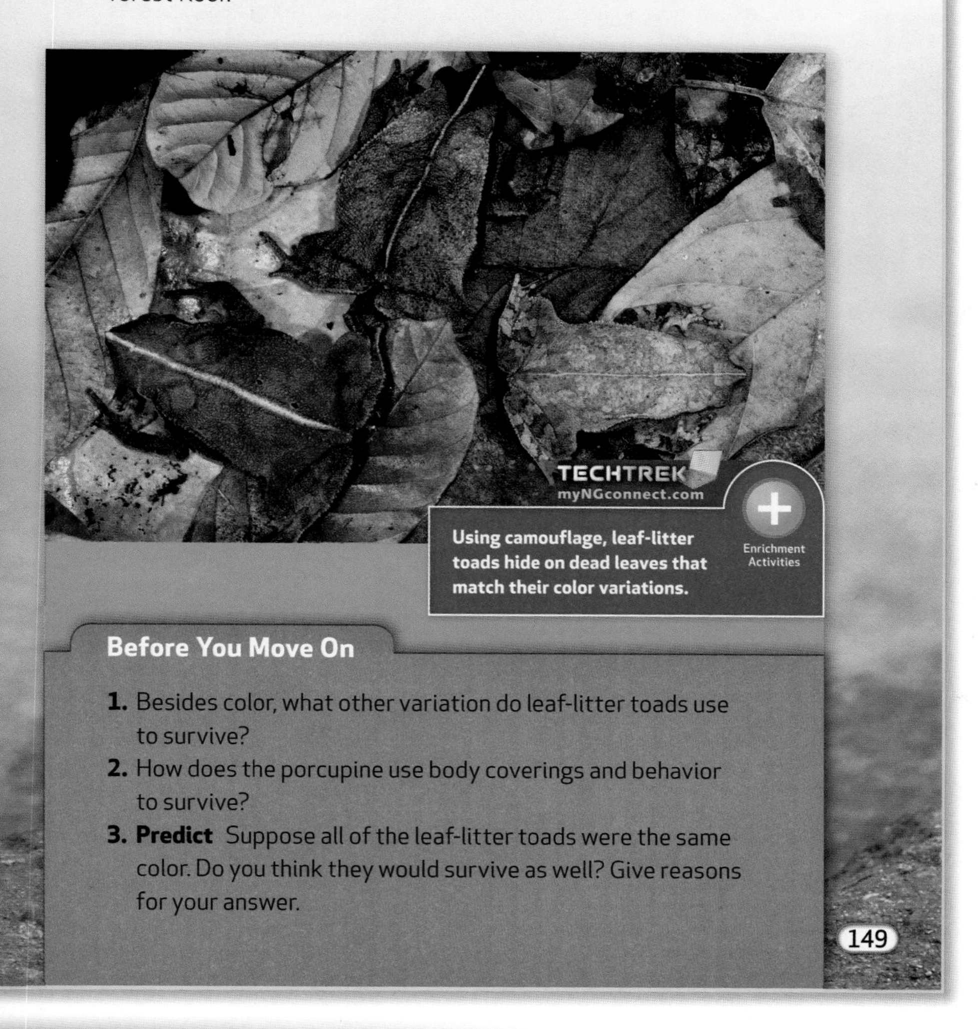

Using camouflage, leaf-litter toads hide on dead leaves that match their color variations.

Before You Move On

1. Besides color, what other variation do leaf-litter toads use to survive?
2. How does the porcupine use body coverings and behavior to survive?
3. **Predict** Suppose all of the leaf-litter toads were the same color. Do you think they would survive as well? Give reasons for your answer.

149

Teach, continued

Identify Adaptations for Protection

- Have students observe the photo of leaf-litter toads on page 149 and read the caption. Ask: **What adaptation helps protect these leaf-litter toads?** (camouflage; the toads blend into the background of fallen leaves.) If necessary, have students review the definition of camouflage on page 146.
- Ask: **What is the variation of the toads?** (their color, which is either tan or dark brown) **How does the variation help protect the toads?** (The variation of the toads' colors matches the colors of the fallen leaves, and this helps the toads blend in.)

myNGconnect.com

Have students use **Enrichment Activities** to find out more about camouflage and variation.

Computer Presentation Students can use the information to make a computer presentation about camouflage and variation.

❸ Assess

» Before You Move On

1. **Name Besides color, what other variation do leaf-litter toads use to survive?** (Variations in size help the toads blend in with different kinds of fallen leaves.)
2. **Summarize How does the porcupine use body coverings and behavior to survive?** (Raising its quills is a behavior that protects the porcupine when it is attacked.)
3. **Predict Suppose all of the leaf-litter toads were the same color. Do you think they would survive as well? Give reasons for your answer.** (They probably would not survive as well because more could be seen among the different colors of fallen leaves.)

Differentiated Instruction

ELL Language Support for Identifying Adaptations

BEGINNING

Write the words *adaptation*, *camouflage*, and *variation* on the board. Have students match each word with its definition.

Feature that helps a living thing survive *(adaptation)*

Blending into the background *(camouflage)*

Differences in a feature *(variation)*

INTERMEDIATE

Help students identify adaptations by completing Academic Language Frames.

- *Blending into the background is called* ____.
- *The different colors of toads is an example of* ____.

ADVANCED

Help students identify adaptations by completing Academic Language Stems.

- *Camouflage helps leaf-litter toads because . . .*
- *Variation helps leaf-litter toads because . . .*

LESSON 6 ▫ Sensing and Communicating

Objectives

Students will be able to:

- Identify the importance of adaptations for sensing.
- Describe adaptations animals have for seeing and hearing.

❶ Introduce

Tap Prior Knowledge

- Have students describe the noises that they have heard animals make. Examples include noises made by pet dogs and cats, songbirds, crickets, and horses. Ask: **Why do you think animals might make sounds?** (Possible answers: to attract a mate, to scare other animals away, to alert other animals of danger)

Set a Purpose

- Read the headings on pages 150–157. Tell students that they will learn about the way animals use their senses, which include the same senses that humans use. They will also learn how animals communicate, or send and receive messages.

❷ Teach

Describe Adaptations for Seeing

- Have students read pages 150–151, observe the photos, and read the captions.
- Point to the loris. Ask: **What are the eyes of the loris like?** (They are huge.) Have students compare the size of the loris's eye and fingers and then compare the size of their own eyes and fingers.
- Explain that the loris hunts at night and sleeps during the day. Ask: **How do huge eyes help the loris?** (The huge size of the eyes helps take in light, thus helping the loris see well at night.)
- Point to the chameleon on page 151. Ask: **What are the chameleon's eyes like?** (They bulge away from its face.) Discuss how a chameleon can move its eyes in the sockets to point in a wide variety of directions. This helps the chameleon see food or enemies approach, even from behind.

Sensing and Communicating

How do you sense what's around you? Eyes for seeing, ears for hearing, and a nose for smelling are some of the body parts you use to gather information. Animals use senses to find food, avoid danger, and communicate.

The slender loris looks out into the dark night with huge eyes. Larger eyes take in more light for hunting prey at night. The loris also uses its ears to hear the sounds of any prey that might be in the area.

This loris uses sight and hearing to hunt for insects, lizards, and birds at night.

150

NATIONAL GEOGRAPHIC Raise Your SciQ!

Eyes and Ears Eyes and ears are complex, sensitive organs. In most animals, muscles narrow or widen the pupil, or opening of the eye. This allows optimal vision in either bright or dark conditions. Many animals perceive colors; others see only in black and white. In a few animals, including primates and owls, both eyes face forward. This helps the brain to judge the distance of objects.

Although the inner ears of most animals are quite similar, different animals are able to hear different ranges of sounds. Dogs, mice, and porpoises hear higher-frequency sounds than humans can hear. Elephants hear lower-frequency sounds.

A great horned owl has large eyes and keen eyesight. But this owl can also use its hearing to find prey in total darkness. You might think its "horns" look like ears. But its ears are on either side of its face, near its eyes. Smooth feathers sweep out from the center of its face. They funnel sounds to the owl's ears, making its hearing even sharper.

The great horned owl has a keen sense of hearing that helps the owl locate its prey.

This chameleon's eyes help it search for prey all around its body.

151

Teach, continued

Describe Adaptations for Hearing

- Ask: **How does keen hearing help the owl survive?** (The owl hunts other animals for its food. Keen hearing helps it find prey.) Discuss how keen vision and hearing are useful adaptations for owls and other predators. When a great horned owl hears a noise, it looks for a small animal that might be making that noise, such as a mouse or rabbit scurrying across the ground.
- Ask: **How do you think a strong sense of hearing helps the prey of owls, such as mice, rabbits, and squirrels?** (Prey use their hearing to listen for approaching predators.) Explain that eyes can see in a forward direction only, but ears can hear sounds that come from all directions. A mouse or rabbit might not see an enemy approaching from behind, but it might hear any noise the enemy makes.

Make a Chart 

Have students make a four-column chart in their science notebook. Have them label the columns *sight, hearing, touch*, and *smell*. Then have them enter examples of each sense as they encounter it on pages 150–153.

Differentiated Instruction

ELL Language Support for Describing Adaptations for the Senses

BEGINNING

Have students match the sense organ with the sense it provides

eyes *(sight)*

ears *(hearing)*

nose *(smell)*

INTERMEDIATE

Help students describe adaptations for the senses by completing Academic Language Frames, such as.

- *The large eyes of the loris provide keen ____.*
- *The owl uses its senses of ____ and ____ to find prey.*

ADVANCED

Help students write new captions for the photos on pages 150–151. Encourage them to use the word *adaptation* in at least one of the captions.

Objectives

Students will be able to:

- Describe adaptations animals have for touch and smell.

Teach, continued

Describe Adaptations for Touch and Smell

- Have students read pages 152–153.
- Ask: **What body part, or organ, allows you to touch objects in the environment?** (the skin, especially on the hands) **What organs do other animals use to sense touch?** (Answers include the whiskers of bearcats, antennae of crabs and insects, and tentacles of star-nosed moles.) Discuss how animals with paws, such as dogs and cats, sense the ground as they walk over it.
- Ask: **What organ allows you to sense smells in the environment?** (the nose) **How do you think the sense of smell is useful to humans and other animals?** (Possible answer: Smell helps identify and locate food.) Discuss how many animals, including dogs, cats, bears, and the animals pictured on pages 152–153, have strong senses of smell.

Touch and Smell Touch and smell also help animals find what they need to survive. This bearcat smells with its nose to find food at night. Its whiskers act as organs of touch.
Many animals have antennae. Antennae are organs of touch for animals such as crayfish, crabs, and insects.
The two long stalks on this crayfish's head are antennae.
antennae
The bearcat uses its senses to find food and stay out of danger.
152

NATIONAL GEOGRAPHIC Raise Your SciQ!

The Star-nosed Mole The star-nosed mole has possibly the most sophisticated sense of touch of any animal. Like other moles, the star-nosed mole lives mostly underground, although it also is an excellent swimmer and moves above ground at times. When it digs or moves through tunnels, it moves its tentacles rapidly and constantly. When the tentacles contact an object, in less than a second the mole can sense and determine whether or not the object is a suitable prey, such as an insect. Scientists also suspect that when the mole hunts underwater, the tentacles may detect electrical signals from prey.

The star-nosed mole is an animal with a multipurpose sense organ. A ring of tentacles form a star at the tip of its nose. Some of the tentacles keep soil out of its nose as the mole digs tunnels underground. The mole uses the shortest tentacles to identify prey. The longer tentacles are the most sensitive.

A star-nosed mole finds its way with its nose instead of its eyes.

153

Teach, continued

Text Feature: Photos and Captions

- Have students observe the photos on pages 152–153 and read the captions. Then point to the bearcat. Ask: **What are the bearcat's whiskers like?** (Possible answer: They look like long, stiff hairs or bristles that stick out from the bearcat's nose and face.)
- Ask: **How are the whiskers useful to the bearcat?** (They are organs of touch.) Explain that the whisker tips are sensitive and allow the bearcat to navigate through tight spaces as its face passes through them.
- Point to the star-nosed mole. Ask: **What body parts of the star-nosed mole act like the whiskers of the bearcat?** (the tentacles around its nose) Discuss how the tentacles are even more sensitive than whiskers. Star-nosed moles use their tentacles to identify and catch prey, most of which are insects and other small animals.
- Say: **Star-nosed moles hunt small animals in underground tunnels. Why do you think touch and smell are more important to the moles than sight?** (It's dark underground, and the small animals that the mole hunts would be very difficult to see.) Point out the small eyes of the mole. Have students compare them to the eyes of the loris on page 150.
- Ask: **Why are sense organs examples of adaptations?** (because they are specialized structures that help an animal survive in its environment)

Differentiated Instruction

Extra Support

Have students draw pictures of different animals and a human and then label the sense organs in each drawing. Encourage students to draw at least four animals, one each for the senses described on pages 150–153. Then have students describe their drawings and the animal senses to a partner.

Challenge

Have students write a short paragraph about the ways animals use their senses. Encourage students to use their own words to describe how sight, hearing, touch, and smell help animals survive.

Objectives

Students will be able to:

- Explain advantages of animals using sound to communicate.
- Recognize ways that smell is used for communication.

Teach, continued

Explain How Animals Benefit From Communication

- Have students read pages 154–155.
- Explain that communication is the sharing of information. Ask: **How do people communicate with one another?** (Examples include talking and listening, sending and reading written messages, drawing pictures, and making facial gestures and other body movements.) Discuss how people communicate with one another every day in many ways, including in the classroom.
- Ask: **How does communicating help animals?** (Possible answers: to show they are hungry, to attract mates, to warn or frighten other animals) Ask: **What are some ways animals communicate?** (Many birds and other animals communicate by making and listening to sounds. Skunks communicate by moving their tails and releasing foul-smelling sprays.) Discuss other examples, such as dogs wagging their tails when they are friendly and cats arching their backs and hissing to keep others away.

Communicating Step outside on a spring day and you will likely hear birds singing. They are not really "singing." They are communicating, or sharing information. Young birds make sounds to express hunger and compete for food. Adult birds make sounds to attract a mate or to tell other birds to stay away from their nest. Many animals make sounds to warn of danger, and to mark territories.

These waved albatross use sound to communicate.

154

Raise Your SciQ!

Animal Communication Communication among animals may occur among parents and offspring, members of the same social group (such as a wolf pack or elephant herd), and predators and potential prey. Scientists have identified the meanings of many animal sounds. For example, the different trumpet-like calls that elephants make may communicate messages about food, danger, or mating. Many birds "sing" to attract mates, while crickets do this by chirping and fireflies by flashing the lights on their abdomens. Cats, dogs, and monkeys make hissing or growling noises to ward off predators or other unwelcome animals. Cats also arch their backs and raise their fur, which helps them appear larger and more threatening.

Animals communicate using odors too. Animals have a very good sense of smell. A skunk gives off a foul odor to warn predators to stay away. Skunks and other animals also give off odors to attract mates and mark territories.

Animals also communicate by moving body parts. A skunk raises its tail to show that it is going to spray. A raised tail of a white-tailed deer warns other deer that danger may be near.

When a skunk raises its tail, it is about to release a very strong odor.

Science Misconceptions

The Importance of Communication Students may believe that communication is of minimal or no benefit to animals. In fact, animals may communicate to meet a variety of needs, and some would not survive without communication. Animals that live in close societies, such as ants and bees, communicate almost constantly about finding food, defending the colony, and raising offspring. The young of many birds and mammals call to their parents for food or protection. Meerkats live in a social group called a mob. A few mob members will act as lookouts while the others hunt for food. The lookouts make a warning call if a dangerous animal approaches.

Teach, continued

Recognize Communication by Sound

- Point to the photo of the albatrosses on page 154. Ask: **What messages do you think these albatross might be communicating?** (Possible answers: The messages might concern food or danger.)
- Ask: **How are birds' voices different from human voices?** (Birds can make a variety of calls and sounds. Although parrots and a few other birds can mimic human speech, most birds cannot speak words as humans do.) Explain that humans speak with several body parts, including the vocal cords and tongue, and that speech requires complex control from the brain.

Recognize Communication by Smell

- Point to the photo of the skunk on page 155. Ask: **Why is the foul spray of a skunk a form of communication?** (It sends a message to other animals. The message is "Stay away from me!") Explain that the skunk's spray is both a form of communication and defense. Because animals have learned how a skunk moves before releasing its spray, this movement is also a form of communication.

Make a Chart

Have students make a three-column chart in their science notebook for three type of communication: sounds, movement, and smell. As they read about communication in the Big Ideas Book, have them record entries for each column.

Objectives

Students will be able to:

- Identify examples of adaptations of animals for reproduction.

Teach, continued

Identify Examples of Adaptations for Reproduction

- Have students read pages 156–157, observe the photos, and read the captions.
- Ask: **What is reproduction?** (making more of one's own kind) Discuss how all types of living things, including those shown on pages 156–157, have young or offspring that grow up to be much like their parents.
- Point to the peacock. Explain that the peacock is the male and the peahen is the female. Ask: **Why is the peacock's large, colorful tail an example of an adaptation?** (The tail attracts a peahen, which is the peacock's mate.) Discuss how the male's strutting and displaying of its tail is also an adaptation for attracting a peahen.
- Point to the fireflies. Ask: **How are the flashing lights of fireflies like the tail of the peacock?** (Both serve the same purpose, which is attracting mates.)
- Discuss how adaptations for reproduction are different from other adaptations because they do not necessarily help an individual survive. Instead, they help the survival of a species, or specific kind of living thing.

Reproduction This peacock is doing his best to attract a peahen. His strutting back and forth and his colorful tail feathers are adaptations. The color of the peacock's tail is an adaptation that helps it survive and reproduce. When living things reproduce, they pass along many of the adaptations that help the offspring survive.
A peacock displays his tail feathers to attract a mate.
Two male bighorn sheep fight to decide which is stronger. This type of contest is an adaptation for reproduction in many animal species.
156

NATIONAL GEOGRAPHIC Raise Your SciQ!

Adaptations for Reproduction Animals use a wide variety of structures and behaviors to attract mates. The large, bushy mane of a male lion attracts a lioness. The brightly colored feathers of many male birds attract the females of their species. Behaviors for attracting mates may involve sounds and movements. In a type of sea bird called the blue-footed booby, the males perform an elaborate dance in which they spread their wings and stomp on the ground. In cardinals, robins, and many other small tree-dwelling birds, the males make complex, warbling calls, or songs. Some birds repeat the same song many times, while others, such as the brown thrasher, may sing over a thousand different songs.

In some places, fireflies glowing at night are a common sight in summer. These insects give off light to attract mates. Many animals that live deep in the ocean, such as small shrimp and marine worms, also have this adaptation.

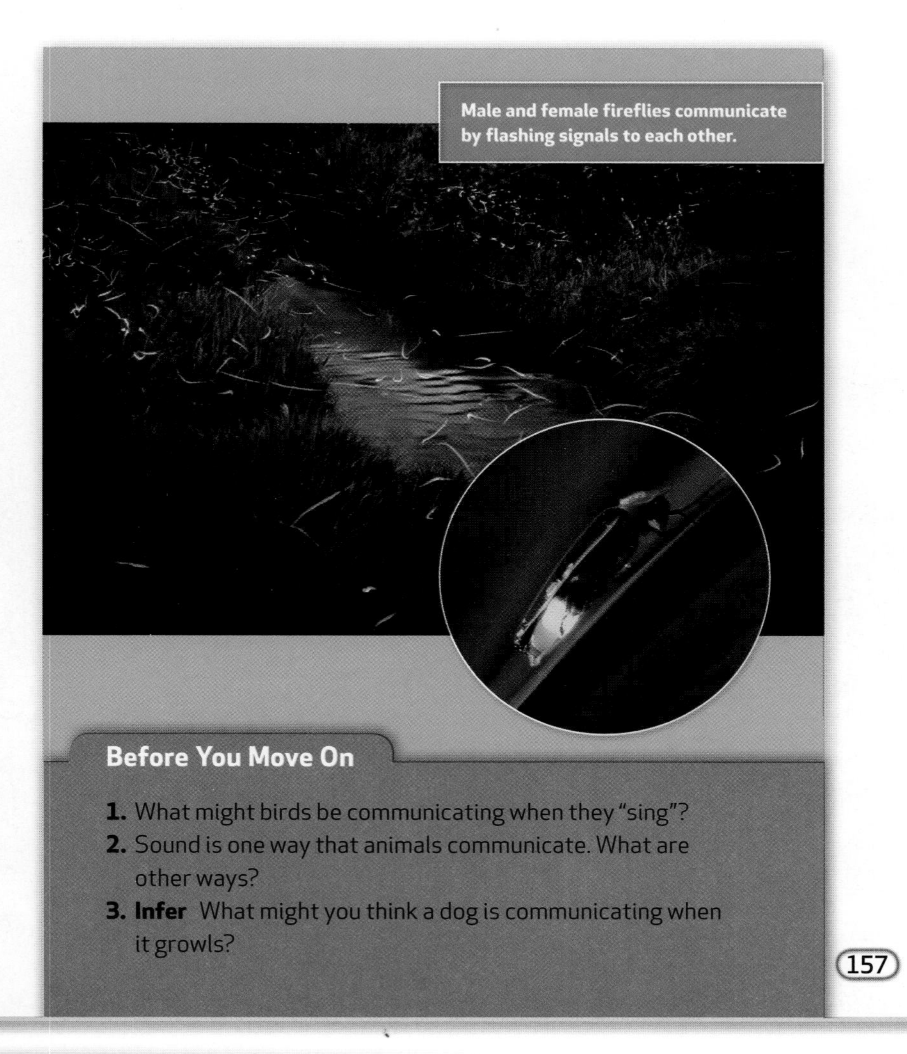

Male and female fireflies communicate by flashing signals to each other.

Before You Move On

1. What might birds be communicating when they "sing"?
2. Sound is one way that animals communicate. What are other ways?
3. **Infer** What might you think a dog is communicating when it growls?

157

Teach, continued

Monitor and Fix-Up

Ask students if anything they read was confusing. Students may be confused by the different adaptations that animals use for sensing, communication, and reproduction. Discuss possible fix-up strategies. For example, students may use their own words to describe the photos and examples of adaptations presented on pages 150–157. They may also compare these adaptations to those in humans or familiar animals. Students may also complete the Differentiated Instruction activity presented on this page.

3 Assess

Before You Move On

1. **Recall** **What might birds be communicating when they "sing"?** (Possible answer: Birds sing to attract mates.)
2. **Explain** **Sound is one way that animals communicate. What are other ways?** (Animals also communicate with smells, such as the foul-smelling spray of a skunk; and by moving their bodies, such as a skunk lifting its tail before it releases its spray.)
3. **Infer** **What might you think a dog is communicating when it growls?** (Possible answer: A dog growls to tell humans or other animals to stay away.)

Differentiated Instruction

ELL Language Support for Identifying Adaptations for Reproduction

BEGINNING

Help students illustrate a peacock or a firefly and label its features (colorful tail, feathers, light) that are adaptations. Then help students use their illustrations and own words to explain how the adaptation attracts mates.

INTERMEDIATE

Have students use Academic Language Frames to describe adaptations for reproduction.

- *Animals ____ when they make offspring.*
- *Bright feathers are an ____ that helps peacocks attract mates.*

ADVANCED

Have students use Academic Language Stems to write sentences about adaptations for reproduction.

- *Peacocks attract mates by . . .*
- *Fireflies attract mates by . . .*
- *Male bighorn sheep fight to . . .*

LESSON 7 ▫ Plant Adaptations

Objectives

Students will be able to:

- Describe how a carnivorous plant has adaptations to survive in its environment.

❶ Introduce

Tap Prior Knowledge

- Ask students to describe their observations of animals visiting or interacting with plants. Examples include animals eating a plant's leaves, stems, or fruits; bees visiting a flower; and birds, squirrels, or other animals living in a tree. Have students describe whether they think the plant benefited from or was harmed by the animal.

Preview and Read

- Read the headings and preview the photos on pages 158–163. Tell students that plants have adaptations that help them survive in their environment, just like animals do.

❷ Teach

Identify Plant Adaptations

- Have students read pages 158–159.
- Point to the inset photo of pitcher plants on page 158. Have a volunteer read the caption. Ask: **What plant parts do you observe in these pitcher plants?** (leaves) Explain that the leaves only appear to be flowers. This fools insects into visiting the leaves.
- Ask: **Why do you think an insect can't escape the leaves of the pitcher plant?** (Possible answers: The sides of the leaves are too slippery to climb, the tube is too narrow to fly out of.)

Plant Adaptations

You may think you're just looking at a patch of yellow flowers, but you are very wrong. Each of these "flowers" is a specialized leaf for trapping insects. Pitcher plants and Venus flytraps capture and digest small animals such as insects and absorb nutrients from them. As you can see, plants have adaptations for survival, too.

The leaves of the pitcher plant form a tube. Insects fall into the tube and cannot get out.

Pitcher plants grow in soil that lacks nitrogen. The plants get this nutrient from animals.

158

Science Misconceptions

Carnivorous Plants Some students may think that flies are food for pitcher plants and Venus flytraps, much like the prey that are food for carnivorous animals. Although carnivorous plants get important nutrients from the insects they catch, they use the energy of sunlight to make their own food and thus are properly classified as producers. Pitcher plants and Venus flytraps are often described as carnivorous plants, but they are not true carnivores.

When an insect lands on a Venus fly trap, it touches tiny hairs that trigger the leaves to snap shut. The plant absorbs nitrogen as the insect decays.

To an insect, the leaves of this Venus flytrap look like a flower.

The Venus flytrap captures the insect and digests it.

159

Differentiated Instruction

Extra Support

Have students identify any new or unfamiliar words in the text and captions on pages 158–159. Examples may include *specialized, digest, nitrogen, decay*, and *nutrients*. Help students use context clues to figure out the meanings of these words and explain them to a partner. Or students may consult the glossary or a dictionary.

Challenge

Have students write two or three sentences that explain how the Venus flytrap catches insects and the benefit the plant gains from insects. Encourage students to include the word *adaptation* in their sentences.

Teach, continued

- Ask: **What happens to insects that land on the leaves?** (They fall into a tube and cannot get out. Over time, their bodies break apart and provide the plant with nutrients.) Explain that the pitcher-like tube is filled with a liquid that drowns the insects.
- Ask: **Why are the pitcher-shaped leaves of pitcher plants examples of adaptations?** (Like other adaptations, they are specialized structures that help a living thing survive.) Discuss how most living things get the nutrients they need from the soil. Explain that pitcher plants and Venus flytraps, however, grow in bogs and marshes where nitrogen, an important plant nutrient, is scarce. By trapping insects, these plants get the nitrogen they need.

Text Feature: Photos and Captions

- Have students observe the photos of the insect and Venus flytrap on page 159 and read the captions. Ask: **What happened to the insect?** (It landed on the leaf of the Venus flytrap, triggered hairs on the leaf, and was quickly trapped inside as the leaf closed.) Point out that the small bar-like structures of the leaf act like a cage when the leaf closes.
- Point to the bottom photo. Ask: **What will happen to the insect over time?** (Its body breaks down, supplying nutrients to the plant.) Explain that this process is similar in pitcher plants.
- Ask: **Why is the leaf of the Venus flytrap an example of an adaptation?** (Its structure and action help the plant survive in its environment.)

Compare and Contrast Adaptations

- Have students observe and compare the pitcher plant and Venus flytrap shown on pages 158–159. Ask: **How are the two leaf adaptations alike?** (Both are adapted for trapping insects.)
- Ask: **How are the two leaf adaptations different?** (The shapes of the leaves are different, and thus they trap insects in different ways.) Explain that adaptations among living things can sometimes look very different from one another, yet serve very similar purposes.

Objectives

Students will be able to:

- Identify adaptations of plants that reduce water loss.
- Describe an adaptation that plants use for defense.

Teach, continued

Identify Adaptations That Reduce Water Loss

- Have students read pages 160–161.
- If possible, show students a photo of a pine tree and its needles, or display a real example. Ask: **What kind of plant parts are the needles of pine trees?** (leaves) Explain that needles, like the leaves of most other plants, have green parts that make food for the plant. Ask: **How are needles different from other kinds of leaves?** (They are thin, pointy, and very tough. Other leaves are broad and flat.)
- Ask: **How are pine trees helped by having needles instead of flat leaves?** (Needles reduce water loss.) Explain that water readily evaporates, or escapes as a gas, through the small openings in broad, flat leaves. Needles lose much less water. They help pine trees and similar trees survive in relatively dry environments. The spines of cacti also lose very little water.
- Have students observe the jade plant on page 161 and then read the caption. Ask: **How are the leaves of the jade plant unusual?** (They are thicker and waxier than other leaves.) **What makes the leaves thick?** (stored water) Discuss that leaves and stems storing water is another adaptation. This helps the jade plant survive in a desert environment.

Needles and Thorns You might be surprised to learn that the needles of pines and firs and the spines of a cactus are leaves. These kinds of leaves are adaptations for reducing water loss. The small surface area of each needle reduces the amount of water that evaporates from inside the needle.

The colorful plant shown here grows in tropical climates. Look closely. The plant is armed with thorns. These structures are adaptations to protect the plant against animals that might eat the plant.

The thorns of a bougainvillea help protect the plant.

160

NATIONAL GEOGRAPHIC Raise Your SciQ!

Leaves, Water, and Water Loss Plants use water along with carbon dioxide, a gas in the air, to make sugars, or food. To get water, most plants take in water through the roots, and it moves through the stem to the leaves. But nearly all of this water is transpired, or evaporated into the air through tiny pores in the leaves. Most broad, flat leaves have pores that can be opened or closed. The pores help conserve water. A leaf's waxy coating also helps keep water inside it.

Leaves and Stems A jade plant lives in a hot, dry climate. The plant's thick leaves and stems store water. They have a waxy covering that prevents water loss. The jade plant can survive long periods without water.

Some plants live in wet places. These plants often have large, thin leaves, which help them lose water through evaporation. The size and shape of the leaves are adaptations for surviving in the environment.

The jade plant's leaves and stems help it survive in very hot, sunny places.

161

Science Misconceptions

Many Adaptations Some students may think that adaptations include only unusual or structural traits of a plant, such as the spines of a cactus or the thick leaves of a jade plant. In fact, all plants have adaptations that help them survive, and many adaptations involve small plant parts or chemicals that plants make. For example, many plants make poisons that help protect them from being eaten. Chemicals called auxins regulate how plants grow and change in response to their environment.

Teach, continued

Describe Adaptations That Protect Plants

- Have students observe the photos on page 160 and read the caption. Ask: **Where are the thorns on this plant?** (on the stem beneath the flowers) Help students identify the thorns.
- Ask: **What are thorns like?** (Thorns are short and tough and have a sharp tip.) **How do thorns protect a plant?** (They poke animals that try to eat the plant. Animals may learn to stay away from plants with thorns.) Have students discuss their experiences with thorns, such as those on rose bushes.

Infer Adaptations in Roots

- Explain that all plant parts—including leaves, stems, and roots—may be adapted for a specific environment. Point again to the jade plant. Ask: **Do you predict that the roots of a jade plant are the same as the roots of a plant that grows in very wet soil? Explain.** (Possible answer: No. Jade plants grow in dry soil, so they need different kinds of roots to survive.)
- Explain that the roots of jade plants, cacti, and other desert plants soak up water very quickly after a rain storm. These roots may rot in soil that stays too wet for too long.

Observe Different Kinds of Leaves

Materials jade plant, ivy plant, scissors

Science Inquiry and Writing Book: *Life Science,* page 14

Read the instructions together with students. Help students cut the leaves. Students should record their observations and ideas in their science notebook.

What to Expect The plant with thick leaves with a waxy covering would survive better in a dry environment because the other plant has thin leaves with no waxy covering. Some plants can survive better in dry conditions because their leaves and stems store more water.

Objectives

Students will be able to:

- Explain how flowers, seeds, and fruits are adaptations for plant reproduction.

Teach, continued

Explain How Adaptations Help Plants Reproduce

- Have students read pages 162–163.
- Have students observe the photo on page 162 and read the caption. Ask: **Why do hummingbirds visit bird-of-paradise flowers?** (to sip nectar that the flowers make) Explain that nectar is a sweet liquid that is food for hummingbirds and other animals.
- Ask: **Where is the nectar in this photo?** (inside the tube-shaped flower part in which the bird is dipping its bill) **How is the tube-shaped part an adaptation for attracting hummingbirds and not other animals?** (The hummingbird's bill is adapted to fit inside the tube.)
- Say: **Hummingbirds get nectar from the flower.** Ask: **How is the plant helped by the hummingbird's visit?** (The hummingbird spreads pollen from one flower to another. This helps the flowers make seeds, which is how the plants reproduce.)

Text Feature: Chart

- Have students study the photos in the chart on page 163 and read the title and captions. Ask: **What does the word *dispersal* mean?** (spreading to new places)
- Ask: **Why is it important for seeds to be dispersed from the parent plant?** (Plants and seeds cannot move from place to place on their own. When the seeds are dispersed, they might land in a place to grow away from the parent plant.)

Reproduction The flowers of a colorful bird-of-paradise plant attract hummingbirds looking for food. The long, tubelike bill of the hummingbird is adapted for reaching the flower's nectar. While it eats, the bird picks up some of the flower's pollen. When the bird flies to another flower of the same kind, the pollen rubs off the bird onto that flower. This pollination starts a series of steps that makes seeds and new plants. In this way, flowers are adaptations for reproduction.
The bird-of-paradise flower and the hummingbird's bill are adaptations for pollination.
162

NATIONAL GEOGRAPHIC Raise Your SciQ!

Adaptations of Pollination Flowering plants have a variety of adaptations for pollination. Many flowers have brightly colored petals and sweet odors that attract insects, birds, or other animal pollinators. Hummingbirds are attracted to red flowers, while bees are more attracted to yellow, blue, and violet flowers. Bats fly at night and often pollinate plants with large, white flowers that open at night. Grasses have flowers that are relatively small and plain, and their pollen is spread by wind.

After pollination, plants produce seeds. A seed contains an offspring of a plant. Most seeds develop inside of a fruit. A seed's best chance to grow into a healthy plant is to be spread, or dispersed, away from the parent plant. Fruits have structures that help do that. When seeds disperse, young plants will not have to compete with their parents for resources such as water, light, and nutrients.

ADAPTATIONS FOR SEED DISPERSAL

BY WIND
These wings carry away these four maple seeds on the wind.

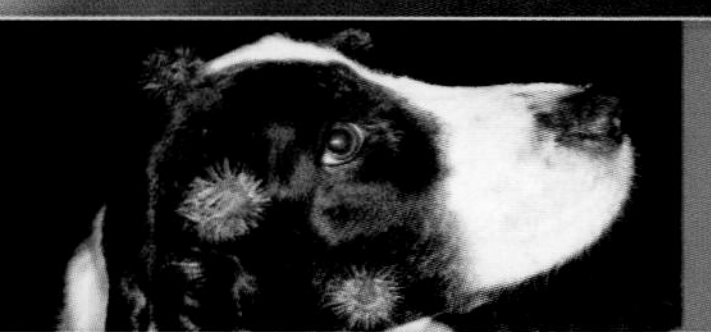

BY ANIMALS
The fruit of a burdock is covered with hooks that cause them to stick to animal fur.

BY WATER
A coconut palm's seeds are partly hollow and float away on water.

Before You Move On

1. What are three adaptations of leaves and stems that help plants survive?
2. Why is the dispersal of seeds important for the survival of plants?
3. **Predict** Could a pitcher plant survive without trapping insects in its leaves? Why or why not?

163

Differentiated Instruction

ELL Language Support for Adaptations for Reproduction

BEGINNING

Have students match new or unfamiliar words with their meaning, for example:

Making more of one's own kind *(reproduction)*

The spreading of pollen from flower to flower *(pollination)*

Spreading to new places *(dispersal)*

INTERMEDIATE

Help students describe adaptations for reproduction using Academic Language Frames, such as these:

- *Flowers are adaptations for ____.*
- *Hummingbirds visit flowers to get ____.*
- *Seed ____ may involve the wind, animals, or water.*

ADVANCED

Have students describe adaptations for reproduction using Academic Language Stems, such as these:

- *Flowers are adaptations that help plants by . . .*
- *Hummingbirds visit flowers to . . .*
- *The wind, animals, or water may help plants by . . .*

Teach, continued

Discuss Seed Dispersal

- Point to the maple seeds. Ask: **Why are maple seeds and fruits adapted for dispersal by the wind, not animals?** (The seeds are attached to wings that the wind can blow far away.) Explain that the wings can also float on water. Streams and rivers sometimes help disperse maple seeds, too.
- Point to the burdock fruits sticking to the dog. Ask: **How do you think these fruits became stuck to the dog's fur?** (Possible answer: The fur brushed against the fruit while the dog was walking by the plant.) Explain that when the burdock fruit falls or is brushed off the dog's fur, it might land in a good place to grow.
- Point to the coconut. Ask: **Why can coconuts be dispersed by ocean water?** (They float.) Discuss how the coconut's hard, thick covering protects the seed inside it, which in this photo has already sprouted.

3 Assess

» Before You Move On

1. **List What are three adaptations of leaves and stems that help plants survive?** (Thorns provide defense against animals that might eat the plant; waxy coverings on leaves and a needle-like shape help prevent water loss; a thin leaf shape helps remove excess water.)
2. **Explain Why is the dispersal of seeds important for the survival of plants?** (By growing away from their parents, young plants do not have to compete with their parents and thus have a better chance to survive.)
3. **Predict Could a pitcher plant survive without trapping insects in its leaves? Why or why not?** (The pitcher plant would not survive. It needs to absorb nutrients from the trapped insects that decay inside it.)

LESSON 8 ▫ Fossils and Extinction

Objectives

Students will be able to:

- Explain how scientists use fossils in the study of Earth's history.
- Identify examples of modern plants that resemble a fossilized species.

❶ Introduce

Tap Prior Knowledge

- Have students name and describe very old things that they have observed. Examples could include old trees, buildings, and antique furniture. Discuss how old things can be clues to the past.

Preview and Read

- Read the headings and preview the photos on pages 164–167. Tell students that plants and animals have lived for a very long time on Earth, including long before humans appeared.
- Have students read pages 164–167.

❷ Teach

Explain Fossil Evidence of Earth's History

- Ask: **What is a fossil?** (the remains or trace of a living thing from long ago) Explain that fossils are evidence of ancient life on Earth.
- Have students observe the trilobite fossil on page 164 and read the caption. Ask: **What does this fossil show about the animal?** (It shows the animal's shape and suggests it had body sections.) Discuss how the fossil trilobite resembles insects or crustaceans, such as certain crabs.

Fossils and Extinction

If you look closely at some rocks, you might discover fossils known as trilobites. A fossil is the remains or trace of a living thing from long ago. Most fossils form in layers of rock, though some fossils form in ice and even in hardened tree sap.

Fossils help us learn about Earth's history. They show how Earth has changed over time and how it may continue to change. For example, the trilobite fossil shown here is from South Dakota. From this find and other evidence, scientists are able to tell that a shallow ocean once covered South Dakota.

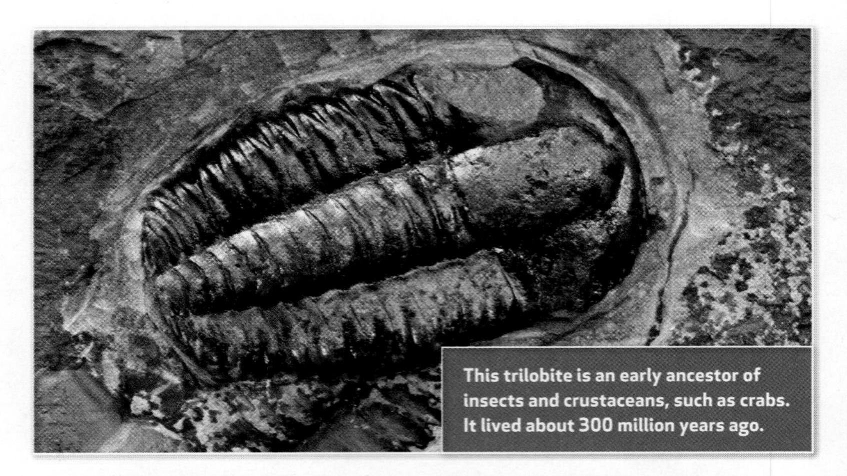

This trilobite is an early ancestor of insects and crustaceans, such as crabs. It lived about 300 million years ago.

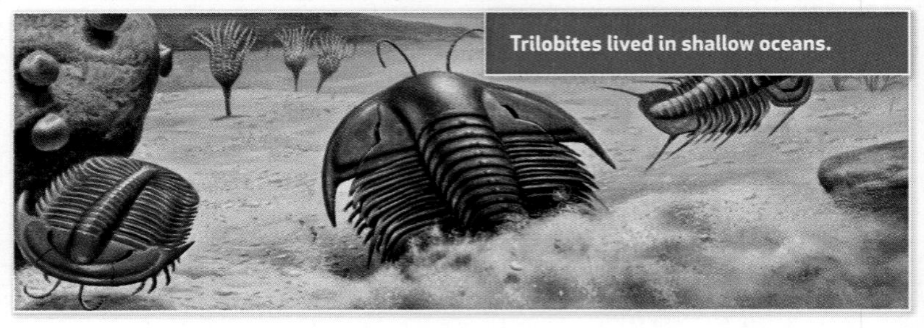

Trilobites lived in shallow oceans.

164

NATIONAL GEOGRAPHIC Raise Your SciQ!

Fossil Evidence Scientists have used fossils to explain many events in Earth's history. Fern fossils have been discovered in Antarctica, which is evidence that this continent was once much warmer. Fish fossils have been discovered inside mountains, evidence that the mountain rocks formed underwater. Fossils of the same species have been discovered on landmasses now separated by oceans, which was key evidence for the theory that Earth's crust moves slowly about the planet. Fossils from different geologic eras show how living things have changed and developed adaptations over time.

Do you recognize this fossil as a leaf from a fern? The fossil looks much like the leaves of modern ferns. When this plant was growing millions of years ago, ferns lived in swamps covering very large areas. Today, ferns still grow in swamps. But other ferns survive in dry places. These ferns developed adaptations that helped them survive as conditions changed.
TECHTREK
myNGconnect.com
Digital Library
This fern leaf fossil has been in this rock for millions of years.
The leaf of a modern fern resembles the fossil shown above.
165

Teach, continued

- Point to the illustration of trilobites below the fossil, and have students read the caption. Ask: **Where did trilobites live?** (in shallow oceans) **Why can trilobite fossils be found in rocks beneath dry land, like the land in South Dakota?** (The oceans dried up as the environment changed.) Explain that the age of fossils may be many millions of years. Earth's land, oceans, and climate have changed greatly during these years, as have its plants and animals.
- Ask: **If we found a trilobite fossil in the rocks beneath the school yard, what might that mean about the land here?** (It might mean that the land in our region was once covered in water.)

Compare Fossil Plants and Modern Plants

- Have students observe and compare the photos on page 165 and read the captions. Ask: **How can we tell that the fossil was a type of fern?** (The shape of the fossil looks just like a fern frond, as shown by the photo of the modern Boston fern.)
- Ask: **Do you think the ancient fern was exactly the same as a modern Boston fern, similar but not the same, or very different? Explain.** (Possible answer: Similar but not the same. Over millions of years, fern plants developed new adaptations.) Explain that plants and animals have lived on Earth for many millions of years. They have changed greatly during this time.

myNGconnect.com

Have students use the Digital Library to find images of fossils

Digital Booklet Students can use the information to make a digital booklet with captions that describes observations or inferences about fossils.

Differentiated Instruction

ELL Language Support for Describing Fossils

BEGINNING

Help students describe fossils by presenting questions with choices. For example:

- Are fossils made from things that were living or nonliving?
- Are fossils made from things that were alive in the past or in the present?

INTERMEDIATE

Help students define fossils using an Academic Language Frame, such as:

- *A ____ is the remains or trace of a ____ living thing.*

ADVANCED

Help students write new captions for the photos on pages 164–165. Tell students that the captions should explain what fossils show about the plants or animals that lived long ago.

LESSON 8 ▫ Fossils and Extinction

Objectives

Students will be able to:

- Identify examples of modern animals that resemble a fossilized species.
- Explain why some living things were able to survive for a long time while others became extinct.

Science Academic Vocabulary

extinction

Teach, continued

Academic Vocabulary: *extinction*

- Point out the term **extinct**. Have a volunteer read the sentence that defines it.
- Point to the illustration of the mastodon on page 167. Say: **Mastodons went extinct thousands of years ago.** Ask: **What does this mean about them?** (Every mastodon on Earth died. There were no more to reproduce baby mastodons.) Explain that **extinction** is permanent. Mastodons and other **extinct** living things will never return to Earth again.

Identify Modern Animals That Resemble Extinct Species

- Have students observe the elephants on page 166 and compare them to the mastodon on page 167. Explain the differences between the animals. The mastodon was heavier than the elephant, and its body was covered in thick hair. Ask: **What happened to mastodons?** (They went extinct over 10,000 years ago when the climate became warmer.) Explain that mastodons could not survive the climate change.

Compare Fossils with Living Things Alive Today in Your State

- Have students list some familiar plants and animals that live in your state or region. Record their responses on the board. Ask: **Do you think these same living things lived in our state thousands of years ago? Or millions of years ago?** (Accept all answers.)

Comparing Living and Extinct Organisms You won't find trilobites swimming in the ocean today. They are extinct. **Extinction** is the complete loss of one kind of living thing. Although trilobites are extinct, many of their relatives, such as crabs, crayfish, and lobsters, remain. These animals have adapted to the present conditions of their environment, but they share some of the same traits as trilobites.

Many other extinct organisms are closely related to organisms living today. For example, if you compare the modern elephant to the American mastodon, you will see similarities. Mastodons lived during the last ice age. They became extinct over 10,000 years ago when the climate warmed.

The elephant is a distant relative of the extinct mastodon.

166

NATIONAL GEOGRAPHIC Raise Your SciQ!

Mastodons and Mammoths An especially cold climate favored mastodons and their close relatives, wooly mammoths. One reason might be that large animals can out-compete smaller animals when food is scarce. Mastodons stood as high as 3 meters (10 feet) tall, and they weighed over 3,500 kilograms (4 tons). Mammoths were even larger than mastodons.

Scientists have discovered mammoth fossils preserved in ice. The fossils sometimes include soft body parts, such as hair, muscles, and eyeballs.

We know about mastodons because of their fossilized bones. Teams of scientists have dug up many remaining bones. These scientists then have assembled the bones back together into a skeleton. They compare the skeleton of the extinct animal with its living relatives. Comparing fossils with living organisms helps scientists learn why one kind of living thing survived and another kind did not.

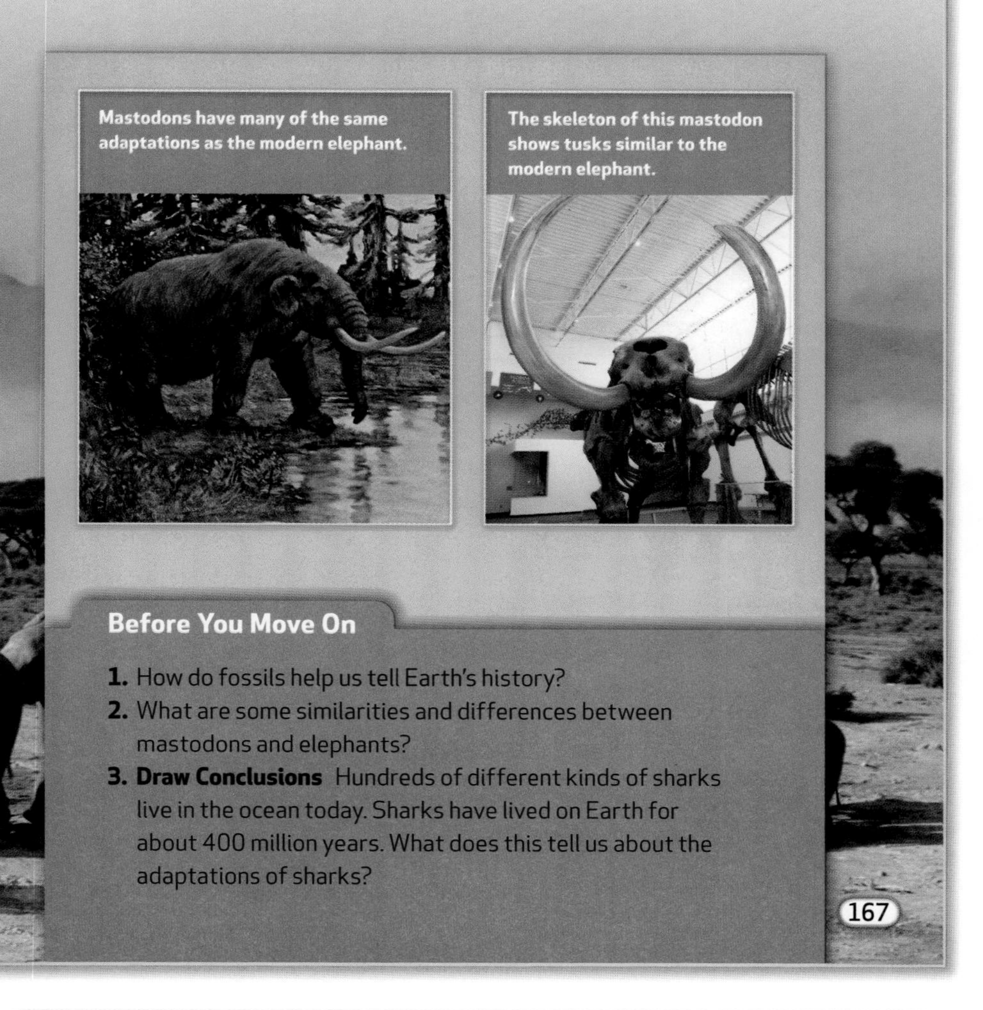

Mastodons have many of the same adaptations as the modern elephant.

The skeleton of this mastodon shows tusks similar to the modern elephant.

Before You Move On

1. How do fossils help us tell Earth's history?
2. What are some similarities and differences between mastodons and elephants?
3. **Draw Conclusions** Hundreds of different kinds of sharks live in the ocean today. Sharks have lived on Earth for about 400 million years. What does this tell us about the adaptations of sharks?

167

Teach, continued

- Ask: **How could we learn about the living things that lived here long ago?** (Possible answer: We could study fossils found here.) Have students research the fossils that were unearthed in your state. Specific fossils to research include ferns and mastodons, as well as water-dwelling organisms such as trilobites, crinoids (relatives of sea stars), gastropods (snails and their relatives), bivalves (clams, oysters, and their relatives), and fish. Have students compare the fossils with organisms alive today.

❸ Assess

» Before You Move On

1. **Summarize How do fossils help us tell Earth's history?** (Fossils show the kinds of living things that have been on Earth, and they also show what the environment was like at different times.)
2. **Analyze What are some similarities and differences between mastodons and elephants?** (They are similar in that both are large animals with bodies of similar parts, including four legs, trunks, and tusks. They differ in the shapes of their heads, sizes of their ears and tusks, and other characteristics.)
3. **Draw Conclusions Hundreds of different kinds of sharks live in the ocean today. Sharks have lived on Earth for about 400 million years. What does this tell us about the adaptations of sharks?** (These facts suggest that sharks have a wide variety of adaptations that have helped them survive many changes to their environment.)

Science Misconceptions

Family Relationships Some students may think that mastodons were the direct ancestors of modern elephants, meaning that mastodons gradually became elephants after many generations of offspring. In fact, mastodons and elephants are more like close cousins. Both share an ancient common ancestor that gave rise to mastodons and mammoths (which went extinct) and modern elephants.

LESSON 9 ▫ Camels and People: Surviving in the Sahara

Objectives

Students will be able to:

- Explain how adaptations can help a living thing survive in its environment.

PROGRAM RESOURCES

- Learning Masters Book, page 52, or at myNGconnect.com

❶ Introduce

Tap Prior Knowledge

- Have students describe how they feel when they play outdoors on very hot days. Lead students to recognize that they have ways to get water and to keep their bodies cool. Have students discuss the needs of animals that live in hot, dry places.

Set a Purpose and Read

- Remind students that they have read about adaptations that help animals survive in different environments. Explain that now they will read about camels, which have adaptations for surviving in the desert.
- Have students read pages 168–169.

❷ Teach

Describe the Desert Environment

- Have students observe the photos on pages 168–169 and read the captions. Ask: **What do the photos show about the environment of the Sahara?** (It is extremely hot and dry. The ground is covered in sand, and no plants are growing in it. Camels and people are the only animals.) Explain that water flows out of the ground in oases, or small areas of the desert where some plants and animals can live.

NATIONAL GEOGRAPHIC
CAMELS AND PEOPLE
SURVIVING IN THE SAHARA
The native people of the Sahara are nomads. They move from one place to another, carrying all that they have with them. Camels are a large part of the nomads' way of life. A camel is a large, grazing mammal of the desert. For more than 4,000 years, camels have carried the Saharan people and their goods across the desert. The people use the milk and meat of camels. They also weave camel hair into cloth, rope, and rugs.
A nomadic herder rides a camel that also carries the herder's belongings.
168

Differentiated Instruction

ELL Language Support for Identifying Adaptations of Camels

BEGINNING

As students read page 169, have them list the different adaptations that help camels survive in the desert. Then have them compare lists with a partner.

INTERMEDIATE

Have students use Academic Language Frames to identify the adaptations of camels.

- *A hump is an ____ that helps camels store ____.*
- *Wide, soft feet are an ____ that helps camels ____.*

ADVANCED

Help partners write sentences to identify the adaptations of camels. Provide an Academic Language Frame:

- *One adaptation that helps camels survive is their . . .*

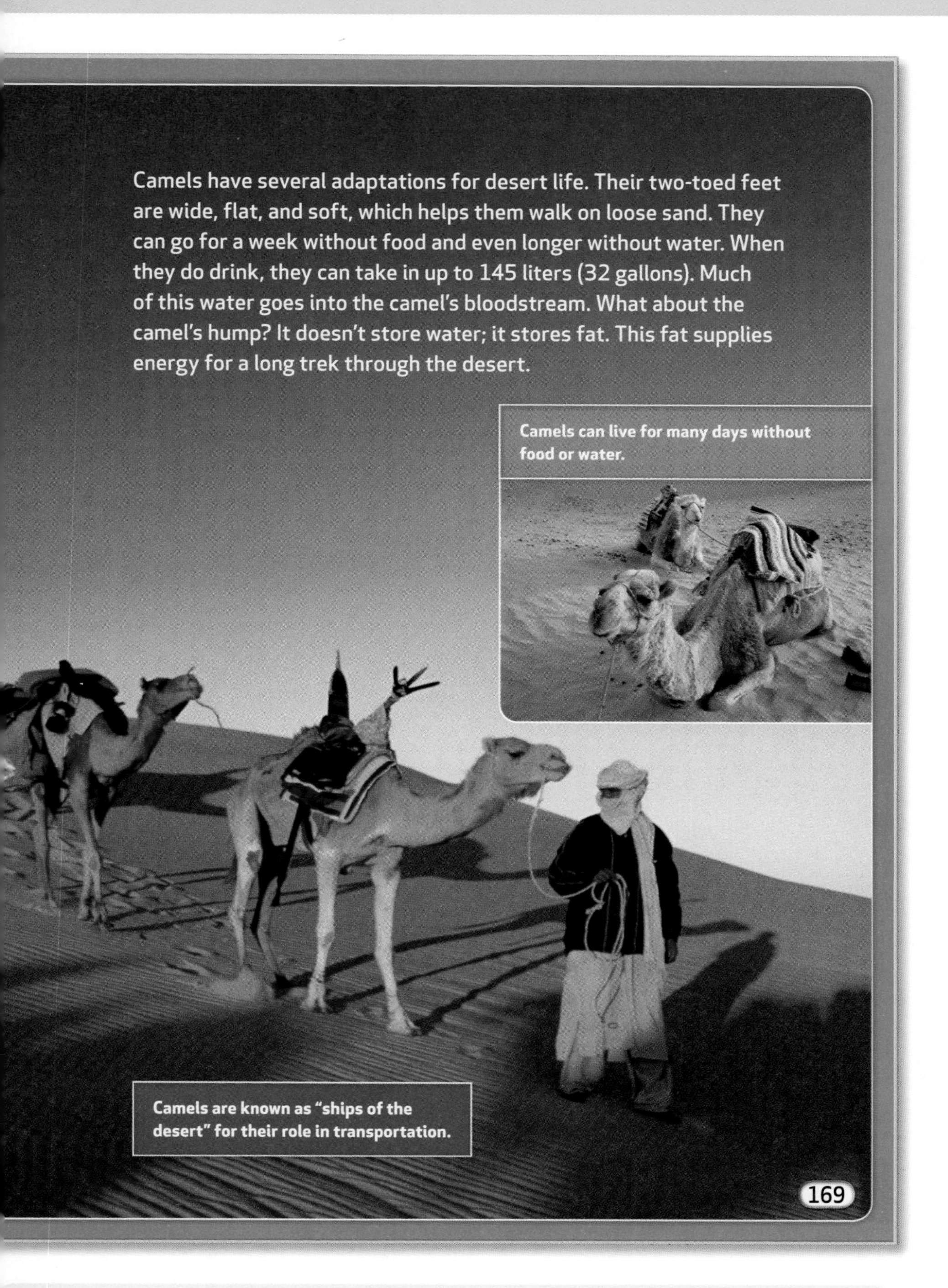

Camels have several adaptations for desert life. Their two-toed feet are wide, flat, and soft, which helps them walk on loose sand. They can go for a week without food and even longer without water. When they do drink, they can take in up to 145 liters (32 gallons). Much of this water goes into the camel's bloodstream. What about the camel's hump? It doesn't store water; it stores fat. This fat supplies energy for a long trek through the desert.

Camels can live for many days without food or water.

Camels are known as "ships of the desert" for their role in transportation.

169

Teach, continued

Explain the Adaptations That Help Camels Survive

- Discuss the many adaptations that help camels survive in the desert. Ask: **How does it help the camel to drink a huge amount of water—up to 145 liters—at one time?** (Water is scarce in the desert. By drinking a huge amount of water at once and storing it in their blood, camels can survive a week or longer without drinking again.)
- Ask: **How are a camel's feet adapted to the desert?** (They are wide, flat, and soft, which helps the camel walk well on sand.)

3 Assess

1. **Recall What is a camel?** (A camel is a large, grazing mammal. It has many adaptations for life in the desert.)
2. **Explain Why are camels called "ships of the desert"?** (Because they are useful for crossing the desert, much like ships can cross the ocean.)
3. **Evaluate Why are camels more useful in the desert than horses, oxen, and other animals that people ride or use to carry things?** (Camels have adaptations for desert life that are lacking on horses, oxen, and most other animals. Horses and oxen need food and water that is not readily available in the desert, and they would suffer in the heat.)

Share and Compare

Examples of Adaptations

my SCIENCE notebook

Give students the Learning Master. Have them find examples of adaptation, variation, camouflage, and extinction and then record their examples in the chart. Check their work, and then have partners share and compare their examples.

Name ______ Date ______ SCIENCE notebook

Chapter 4 Share and Compare

Draw and color illustrations for each word in the table. Then share and compare your examples with a partner.

Student will draw a picture illustrating an adaptation. Adaptation	Student will draw a picture illustrating variation. Variation
Student will draw a picture illustrating camouflage. Camouflage	Student will draw a picture illustrating extinction. Extinction

Learning Master 52 Life Science

Learning Master 52 or at myNGconnect.com

Objectives

Students will be able to:

- Investigate through Guided Inquiry (answer a question; make and compare observations; collect and record data and observations; generate explanations and conclusions based on evidence; share findings; ask questions based on observations to increase understanding; adjust explanations based on findings and new ideas).
- Predict which model animal will survive in a specific environment based on its special coloration.
- Understand how adaptations such as camouflage can help animals survive in their environment.
- Organize and analyze data by making tables and graphs.

Science Process Vocabulary

compare, infer

PROGRAM RESOURCES

- Science Inquiry and Writing Book: *Life Science*
- Science Inquiry and Writing Book eEdition at myNGconnect.com
- Inquiry eHelp at myNGconnect.com
- Science Inquiry Kit: *Life Science*
- Learning Masters Book, pages 53–55, or at myNGconnect.com
- Inquiry Rubric: Assessment Handbook, page 185, or at myNGconnect.com
- Inquiry Self-Reflection: Assessment Handbook, page 196, or at myNGconnect.com

MATERIALS

Kit materials are listed in italics.

6 pieces of construction paper; *hole punch; paper cup (3 oz); stopwatch; white, patterned, and green cloth*

1 Introduce

Tap Prior Knowledge

- Point out the photograph of the octopus on page 51. Ask: **How can this octopus use its color and behavior to hide from predators?** (It blends in with its environment and stays still, making it difficult to see.)

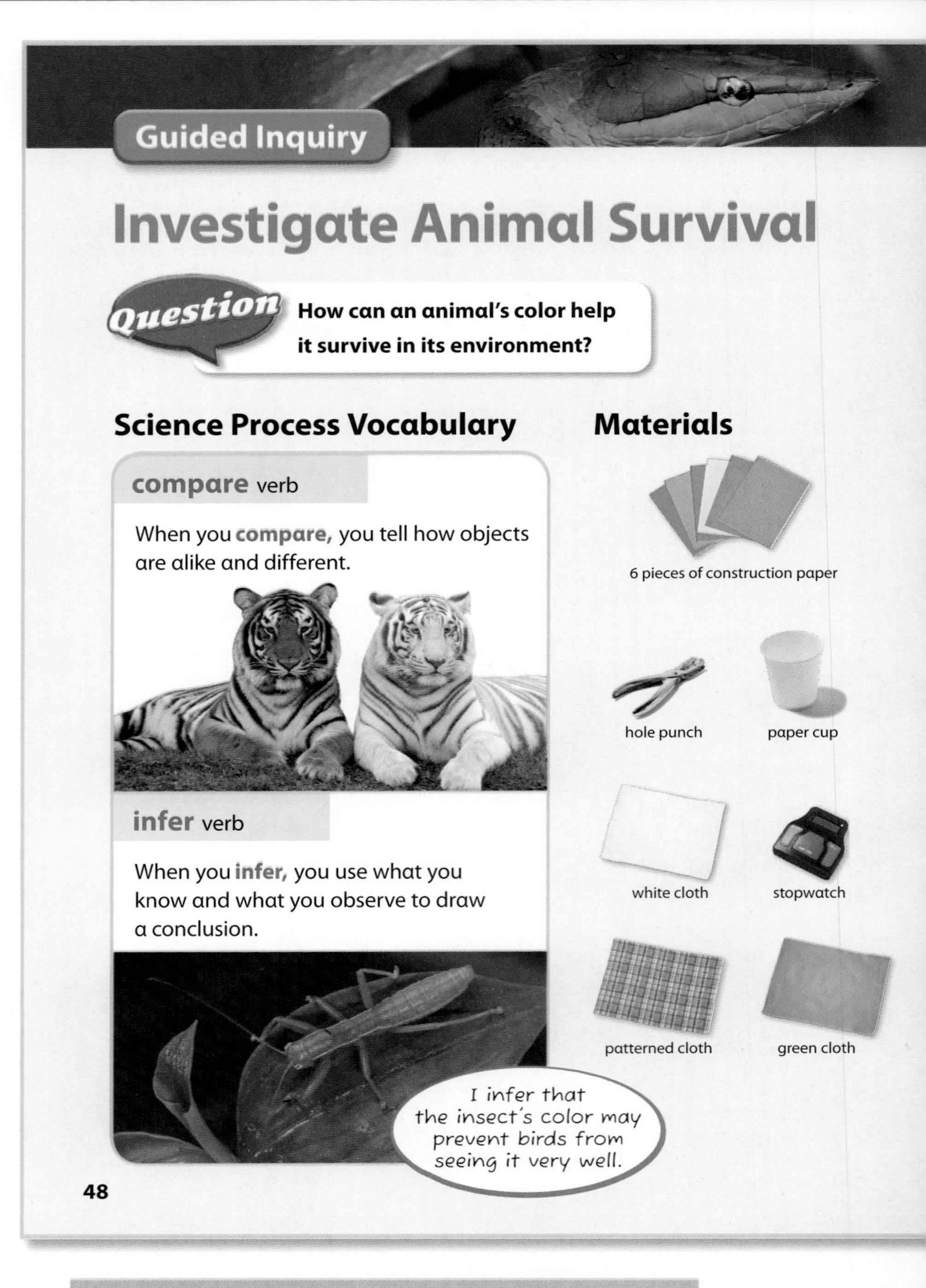

Guided Inquiry
Investigate Animal Survival
Question
How can an animal's color help it survive in its environment?
Science Process Vocabulary
compare verb
When you compare, you tell how objects are alike and different.
infer verb
When you infer, you use what you know and what you observe to draw a conclusion.
I infer that the insect's color may prevent birds from seeing it very well.
Materials
6 pieces of construction paper
hole punch
paper cup
white cloth
stopwatch
patterned cloth
green cloth
48

MANAGING THE INVESTIGATION

Time

30 minutes

Groups

Small groups of 4

Advance Preparation

- Cut sheets of construction paper so that each group receives ¼ sheet. Use white construction paper, 2 colors of construction paper that are found in the pattern of the cloth, and 2 other, non-related colors.
- Cut 22 x 30 cm pieces of white, green, and patterned cloth for each group.

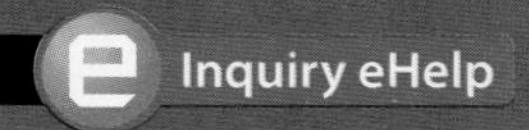

What to Do

1. Use a hole punch to punch out 10 paper dots from each piece of construction paper. Each dot stands for one animal. Place the dots in the paper cup.

2. Place the white cloth on the table. The cloth is a **model** habitat where the animals live. Without letting your partner see, spread out the paper dots on the white cloth.

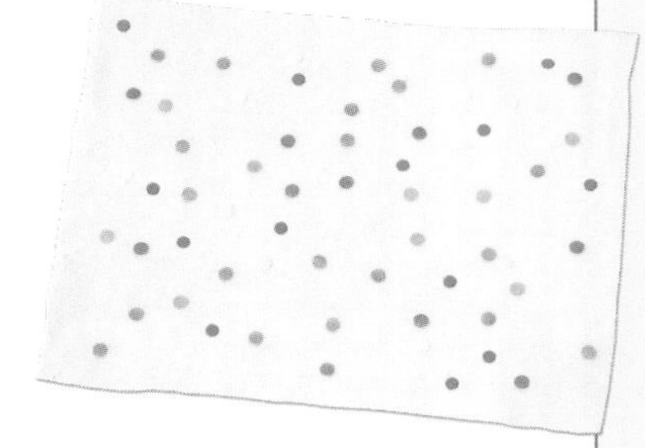

3. Your partner will represent a predator. When you say "Go," your partner should begin removing as many of the dots from the cloth as possible. He or she should pick up the dots one at a time. Use the stopwatch to time your partner for 15 seconds. Then tell him or her to "Stop."

49

Teaching Tips

- Remind students that the word *camouflage* refers to coloring that helps an animal blend in with its environment.
- Show students some photograph examples of camouflage. Explain that camouflage is an example of an adaptation.
- Tell students to use tally marks to record their data. Have students practice using tally marks. Review how to use tally marks, if necessary.

What to Expect

- Students will notice that it is more difficult to see and pick up colored dots if they blend in with the background color of the cloth.

Introduce, continued

Connect to the Big Idea

- Review the Big Idea Question, *How do adaptations help living things survive?* Explain to students that this inquiry will help them learn how camouflage helps animals avoid predation.
- Have students open their Science Inquiry and Writing Books to page 48. Read the Question and invite students to share ideas about how an animal's color helps it survive in its environment.

❷ Build Vocabulary

Science Process Words: compare, infer

Use this routine to introduce the words.

1. **Pronounce the Word** Say **compare.** Have students repeat it in syllables.
2. **Explain Its Meaning** Choral read the sentence. Ask students to define **compare.** (tell how things are alike and how they are different)
3. **Encourage Elaboration** Ask: **How can you compare data from different trials in an investigation?** (I can take detailed notes and then compare notes to see how the data from different trials are alike or different.)

Repeat for the word **infer.** To encourage elaboration, ask: **What can you infer about the octopus on page 51?** (Its color makes it difficult for predators to see it.)

❸ Guide the Investigation

- Distribute materials. Read the steps on pages 49–50 with students. Clarify steps if necessary.
- Tell students that they will model how animals blend in with their habitats. Explain that each cloth represents a different habitat. The white cloth represents a snowy habitat. The green cloth represents a green forest or green grass. The patterned cloth represents a habitat with colorful flowers.
- The colored dots represent the animal's special coloring, which helps it blend in with its environment. Before students act as predators, ask: **Which dot or dots do you predict will be hardest to pick up? Why?**

Guide the Investigation, continued

- After step 6, ask: **What did you notice when you picked up the dots on the patterned cloth?** (The dots that matched the colors on the pattern were harder to see and pick up because they blended in.)

❹ Explain and Conclude

- Guide students as they enter the colors of the cloth and dots in their tables. Then help students create a bar graph on the board using the data from each table. Remind students to label the axes and title the graph.
- Ask: **How do tables and graphs help you interpret your data?** (In tables and graphs, the data are organized in a way that makes it easier to compare trials and find patterns.)
- Have students compare data from different trials with the same color cloth. Ask: **Were your results similar?** If results vary, tell students not to change their data. Instead, they should come up with an explanation for their results.
- Next, have students compare data from trials with different-colored cloth. Ask: **What can you infer about camouflage based on these results?** (A well-camouflaged animal is harder for predators to see. Camouflage is an adaptation that helps animals survive.)
- Ask: **What parts of this model are like the real world? What parts are different?** (The model shows how some animals' coloring makes it harder for them to be seen by predators. However, animals with camouflage often have more complex coloring, such as veining that looks like leaves. Sometimes the way animals move or their shapes can also help them blend in.)
- Invite students to give examples of animals that use camouflage to blend in, such as squirrels, birds, Arctic foxes, etc. Ask: **What other coloring or behaviors might help animals survive?** Guide students to discuss behaviors such as hibernation, migration, defense mechanisms, and movement, as well as colors that warn or attract.

What to Do, continued

4. **Count** the total number of dots of each color that your partner collected. Record the **data** in your science notebook.
5. Place all the dots on the white cloth again. Do steps 3 and 4 two more times.
6. Replace the white cloth with either the patterned cloth or the green cloth. Repeat steps 2–4 three times with the new cloth.

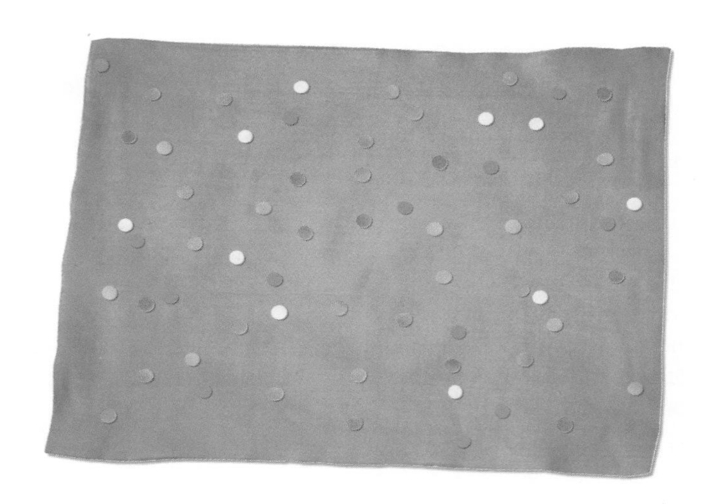

50

NATIONAL GEOGRAPHIC Raise Your SciQ!

Camouflaged Predators Camouflage can be a helpful adaptation for prey to avoid being seen by predators. However, predators can use camouflage, too. Some predators use camouflage to sneak up on their prey. For example, a lion can blend in with the color of dried grasses on the savannah. Other predators use camouflage to blend in with their surroundings as they wait quietly to ambush prey. A flounder remains still and camouflaged on the ocean floor as it waits for an unsuspecting fish to swim close enough to snatch it.

Record

Write in your science notebook. Use tables like these. Write the colors of the dots and the cloth you used.

my SCIENCE notebook

Dots Removed From White Cloth

Trial	____ Dots	____ Dots	____ Dots	____ Dots	____ Dots	____ Dots
1						

Dots Removed From ____________ Cloth

Trial	____ Dots	____ Dots	____ Dots	____ Dots	____ Dots	____ Dots
1						

Explain and Conclude

1. **Compare** the number of dots of each color picked up on the white cloth and on the other cloth. Explain any differences.
2. Use the results of this **investigation** to **infer** how blending in with its environment might help an animal survive.

Think of Another Question

What else would you like to find out about how an animal's color helps it survive? How could you find an answer to this new question?

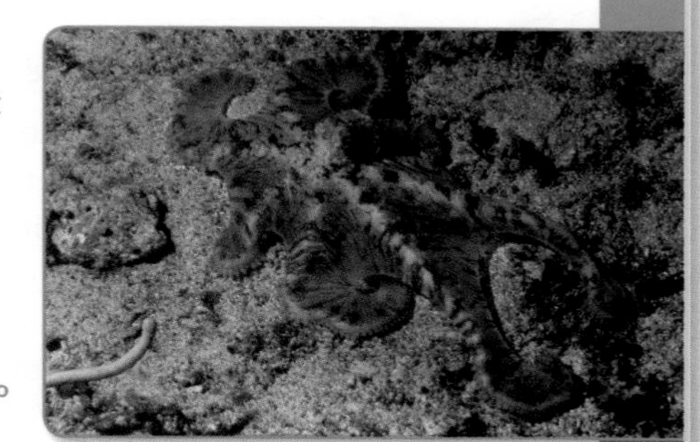

This octopus is difficult to see because of its color.

51

Inquiry Rubric at myNGconnect.com	Scale			
The student **investigated** how camouflage helps prey avoid being seen by predators.	4	3	2	1
The student used a **model** to make qualitative and quantitative **observations.**	4	3	2	1
The student recorded **data** in tables and graphs.	4	3	2	1
The student **shared** and **compared** data from different trials and explained any differences.	4	3	2	1
The student **inferred** how the adaptation of camouflage helps animals survive.	4	3	2	1
Overall Score	4	3	2	1

Explain and Conclude, continued

Answers

1. Possible answer: My partner picked up fewer white dots from the white cloth than other dot colors. The dots that did not blend in with the cloth were easier to see so they were picked up more often. On the other cloth, my partner picked up fewer of the dots that blended in with the color of the cloth.
2. Blending in with its environment can help an animal avoid being seen by a predator. A predator may be able to surprise prey by being well camouflaged.

5 Find Out More

Think of Another Question

- Students should use observations made in this investigation to generate other questions. Record student questions and discuss how students could find answers to the new questions.

6 Reflect and Assess

- To assess student work with the Inquiry Rubric below, see Assessment Handbook, page 185, or go online at myNGconnect.com
- Have students use the Inquiry Self-Reflection on Assessment Handbook, page 196, or at myNGconnect.com

my SCIENCE notebook

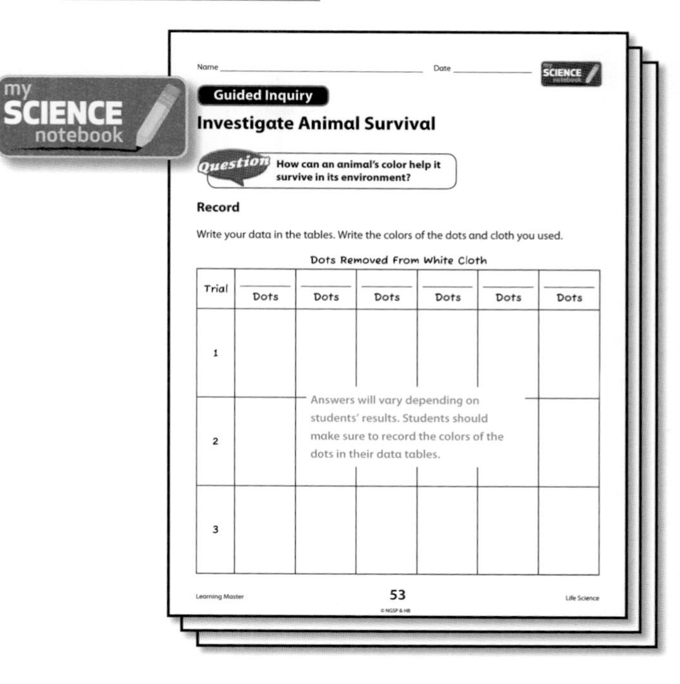

Name ____________ Date ________

Guided Inquiry

Investigate Animal Survival

Question: How can an animal's color help it survive in its environment?

Record

Write your data in the tables. Write the colors of the dots and cloth you used.

Dots Removed From White Cloth

Trial	____ Dots	____ Dots	____ Dots	____ Dots	____ Dots	____ Dots
1						
2						
3						

Answers will vary depending on students' results. Students should make sure to record the colors of the dots in their data tables.

Learning Master 53 Life Science

Learning Masters 53–55 or at myNGconnect.com

PROGRAM RESOURCES

- Chapter 4 Test, Assessment Handbook, pages 19–25, or at myNGconnect.com
- NGSP ExamView CD-ROM

❶ Sum Up the Big Idea

- Display the chart from page T136–T137. Read the Big Idea Question. Then add new information to the chart.
- Ask: **What did you find out in each section? Is this what you expected to find?**

How Do Adaptations Help Living Things Survive?

Adaptations

Frances: Living things have adaptations that help them survive in their environment.

Movement

Evelyn: Adaptations such as wings, feet, claws, and tentacles help animals move in their environment.

Getting Food

Henry: Mouth parts, such as teeth, tongues, and beaks, are adaptations for getting food.

Defense and Protection

Orthea: Tough or sharp body parts, such as a porcupine's quills, are adaptations that help defend and protect an animal from enemies.

Sensing and Communication

José: Animals use ears, eyes, and other sense organs to receive information, and they may communicate with sound, movement, or flashing light.

Plant Adaptations

Hosken: Adaptations help plants in many ways, including water storage, protection, reproduction, and seed dispersal.

Fossils and Extinction

Moira: Fossils show how extinct living things compare with those alive today.

Conclusion

All living things have adaptations. Some adaptations help animals move. Others help them get food. Camouflage and other adaptations protect animals from predators and help them catch prey. Animals and plants also have adaptations that help them reproduce.

Big Idea Adaptations are features that help living things survive in their environment.

Owls have adaptations that help them survive and thrive in their environment.

Vocabulary Review

Match each of the following terms with the correct definition.

A. extinction
B. camouflage
C. variation
D. adaptation

1. A color or shape that makes a living thing hard to see
2. A feature that helps a living thing survive in its environment
3. A different form of a feature of the same kind of living thing
4. The complete loss of one kind of living thing

170

Review Academic Vocabulary

Academic Vocabulary Tell students that scientists use special words when describing their work to others. For example:

adaptation **camouflage** **variation** **extinction**

Have students write the list in their science notebook and use each word in a sentence that tells about how plants and animals live in their environment. They should then share their sentences with a partner.

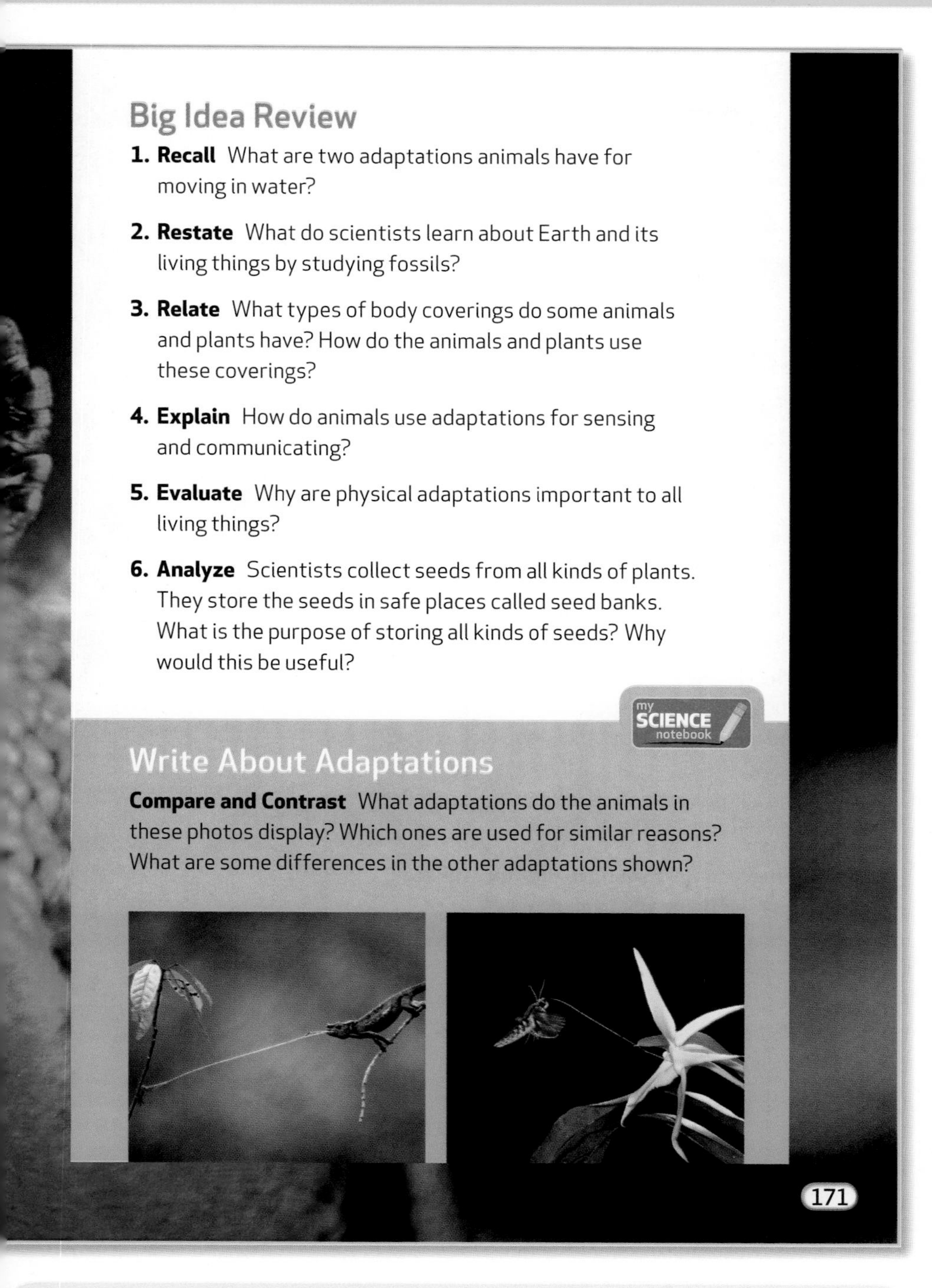

Big Idea Review

1. **Recall** What are two adaptations animals have for moving in water?
2. **Restate** What do scientists learn about Earth and its living things by studying fossils?
3. **Relate** What types of body coverings do some animals and plants have? How do the animals and plants use these coverings?
4. **Explain** How do animals use adaptations for sensing and communicating?
5. **Evaluate** Why are physical adaptations important to all living things?
6. **Analyze** Scientists collect seeds from all kinds of plants. They store the seeds in safe places called seed banks. What is the purpose of storing all kinds of seeds? Why would this be useful?

my SCIENCE notebook

Write About Adaptations

Compare and Contrast What adaptations do the animals in these photos display? Which ones are used for similar reasons? What are some differences in the other adaptations shown?

171

❷ Discuss the Big Idea

Notetaking

Have students write in their science notebook to show what they know about the Big Idea. Have them:

1. Explain how adaptations help animals survive in their environment.
2. Identify adaptations in plants and how they help plants survive in their environment.
3. Compare fossils of plants and animals that lived long ago to plants and animals alive today.

❸ Assess the Big Idea

Vocabulary Review

1. B **2.** D **3.** C **4.** A

Big Idea Review

1. Possible answers include fins, flippers, webbed feet, and tentacles.
2. Fossils show information about the living things that lived long ago, their environment, and how Earth has changed over time.
3. Examples include the quills, scales, and shells of some animals, and the thorns and spines of some plants. All are used for protection.
4. Animals use their senses, such as sight, hearing, and smell, to find food and water, identify enemies or other dangers, and find mates.
5. Physical adaptations help living things move, get food, defend and protect themselves, and send and receive information. They help living things survive and reproduce in their environment.
6. Seeds can grow into adult plants that are similar to their parents. If a type of plant stops growing in the wild, perhaps because its environment changed, then seeds from a seed bank can save the plant from extinction.

Write About Adaptations

Both animals have a very long mouthpart for getting food, and both have legs for moving and eyes for seeing. The moth has wings for moving and antennae for smell and touch. The lizard has four legs with grasping feet and a grasping tail.

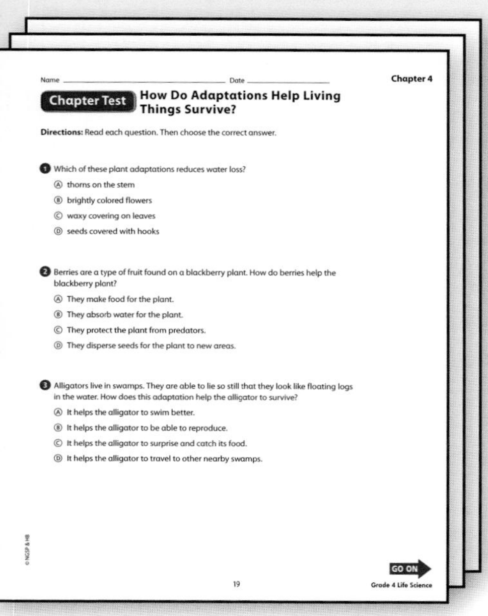

Name ______ Date ______ Chapter 4

Chapter Test **How Do Adaptations Help Living Things Survive?**

Directions: Read each question. Then choose the correct answer.

1. Which of these plant adaptations reduces water loss?
   - Ⓐ thorns on the stem
   - Ⓑ brightly colored flowers
   - Ⓒ waxy covering on leaves
   - Ⓓ seeds covered with hooks
2. Berries are a type of fruit found on a blackberry plant. How do berries help the blackberry plant?
   - Ⓐ They make food for the plant.
   - Ⓑ They absorb water for the plant.
   - Ⓒ They protect the plant from predators.
   - Ⓓ They disperse seeds for the plant to new areas.
3. Alligators live in swamps. They are able to lie so still that they look like floating logs in the water. How does this adaptation help the alligator to survive?
   - Ⓐ It helps the alligator to swim better.
   - Ⓑ It helps the alligator to be able to reproduce.
   - Ⓒ It helps the alligator to surprise and catch its food.
   - Ⓓ It helps the alligator to travel to other nearby swamps.

© NGSP & HB

GO ON

19 Grade 4 Life Science

Assess Student Progress Chapter 4 Test

Have students complete their Chapter 4 Test to assess their progress in this chapter.

Chapter 4 Test, Assessment Handbook, pages 19–25, or at myNGconnect.com

or NGSP ExamView CD-ROM

Objectives

Students will be able to:

- Recognize how different scientists' work has contributed to general scientific understanding.

PROGRAM RESOURCES

- Big Ideas Book: *Life Science*
- Big Ideas Book: *Life Science* eEdition at myNGconnect.com
- Digital Library at myNGconnect.com

1 Introduce

Tap Prior Knowledge

- Ask students to describe their opinions of snakes and frogs, or their observations of these animals if they have encountered them. Encourage all honest and reasonable responses.

Preview and Read

- Read the heading and have students preview the photos. Tell students that they will read about a herpetologist, a scientist who studies reptiles and amphibians.
- Have students read pages 172–173.

2 Teach

Describe the Work of a Herpetologist

- Write the word *herpetologist* on the board, and have a volunteer read the first sentence of the first question on page 172, which defines this word. Ask: **What are some examples of reptiles and amphibians, the kinds of animals that herpetologists study?** (Reptiles include snakes, crocodiles, alligators, and turtles. Amphibians include frogs, toads, and salamanders.) Point out that each photo on pages 172–173 shows a reptile.
- Ask: **Where might herpetologists work, and what would they do there?** (Answers include zoos, where they would help take care of the reptiles and amphibians; in the wild, where they study and identify reptiles and amphibians in their natural environments; and in universities and other schools, where they teach about herpetology.)

NATIONAL GEOGRAPHIC

Kate Jackson, Herpetologist

What does a herpetologist do?

A herpetologist is an expert on amphibians and reptiles, but herpetologists do many different jobs. A herpetologist could take care of reptiles in a zoo, or work outdoors learning about amphibians in a particular pond.

What do you remember liking about science when you were in elementary school?

When I was in elementary school, I knew that I loved animals—especially ones like snakes and toads that most other people don't like. However, I didn't realize that this fascination had anything to do with science, or a possible career studying these animals.

When you were younger, did you ever see yourself doing what you do now?

When I was a little girl growing up in Toronto, Canada, I thought I was the only person in the world who was passionate about snakes. Then I found out about the Ontario Herpetological Society. I learned a huge amount about herpetology from being part of that group.

TECHTREK myNGconnect.com Digital Library

Kate Jackson

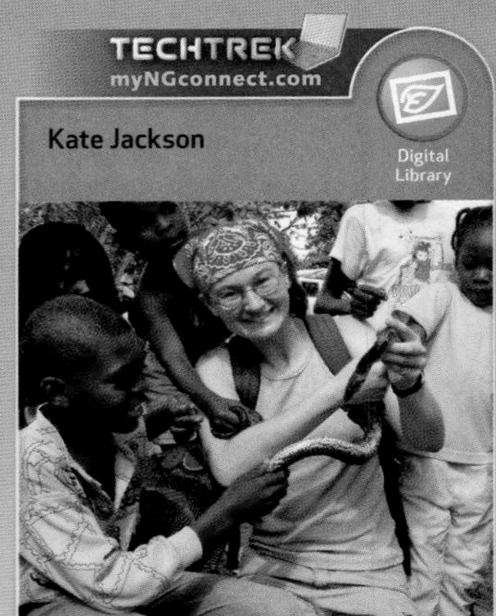

A harmless gopher snake about to attack!

172

Science Misconceptions

Snakes Students may have many misconceptions about snakes and fear them needlessly. Most snakes are harmless to humans. Although some snakes have a venomous bite, humans are not their natural prey. And contrary to a common misconception, the skin of a snake is dry and leathery, not damp and slimy like the skin of many frogs.

Snakes are carnivores that eat mice and other rodents, small birds, eggs, insects, and other foods that fit into their mouths, which can open very widely. Snakes have teeth but cannot chew, so instead they swallow their prey whole. Snakes typically stay in a dormant state for several hours while large prey digest inside their bodies.

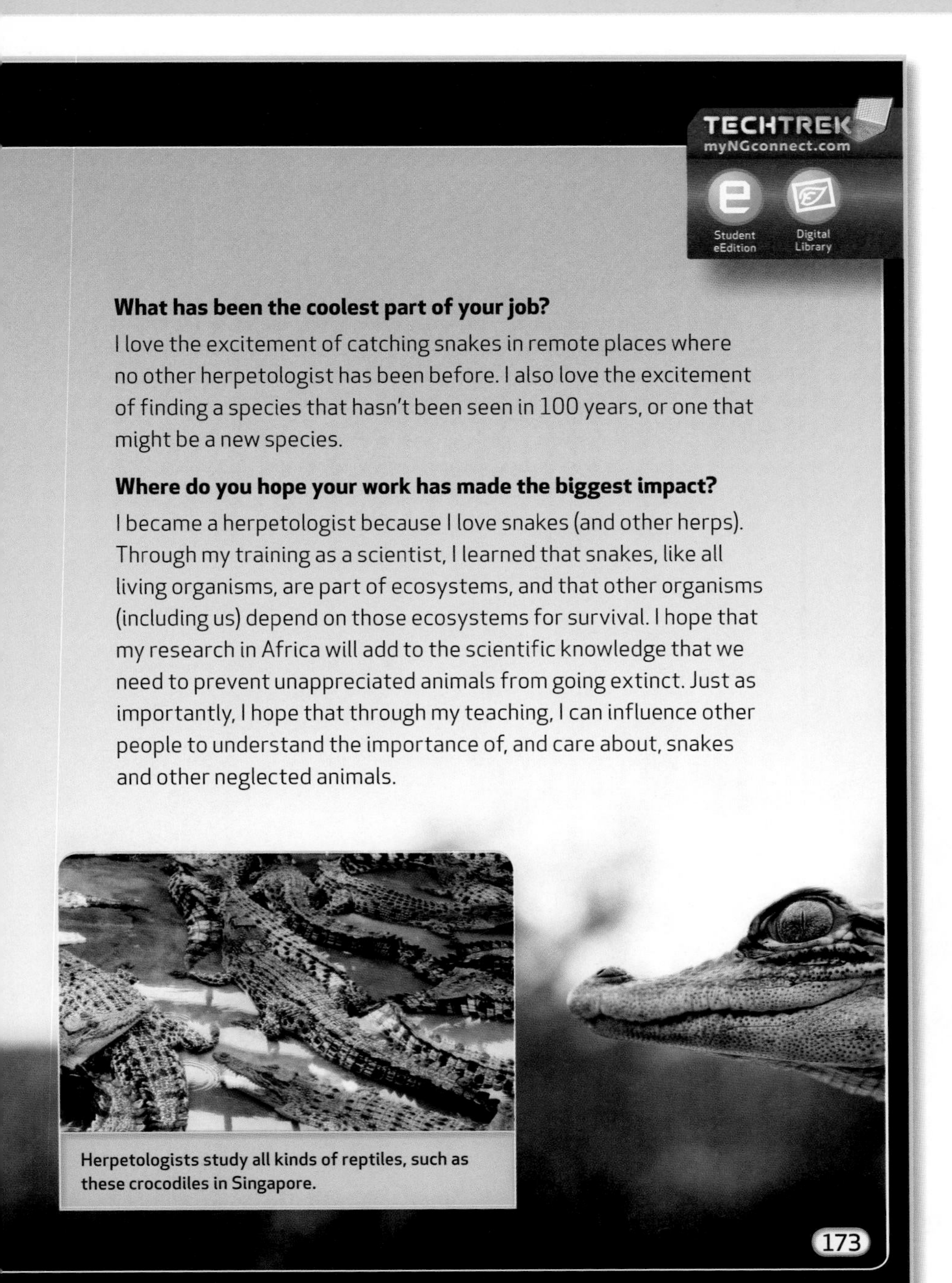

What has been the coolest part of your job?

I love the excitement of catching snakes in remote places where no other herpetologist has been before. I also love the excitement of finding a species that hasn't been seen in 100 years, or one that might be a new species.

Where do you hope your work has made the biggest impact?

I became a herpetologist because I love snakes (and other herps). Through my training as a scientist, I learned that snakes, like all living organisms, are part of ecosystems, and that other organisms (including us) depend on those ecosystems for survival. I hope that my research in Africa will add to the scientific knowledge that we need to prevent unappreciated animals from going extinct. Just as importantly, I hope that through my teaching, I can influence other people to understand the importance of, and care about, snakes and other neglected animals.

Herpetologists study all kinds of reptiles, such as these crocodiles in Singapore.

173

Differentiated Instruction

Extra Support

Have students identify new or unfamiliar words as they encounter them on pages 172–173. Examples may include *herpetologist*, *ecosystem*, and *extinct*. Have partners work to define these words from context clues or by consulting the glossary or a dictionary.

Challenge

Have partners prepare a mock interview with a herpetologist. Have one partner play the role of the reporter and the other partner play the herpetologist. Encourage students to use Kate Jackson as an example for the kinds of work that herpetologists do. Have partners perform their mock interviews for the class.

Teach, continued

- Ask: **Why does Kate Jackson enjoy her job?** (She likes catching snakes and identifying new and rare species of snakes.) Explain that a species is a kind of living thing. Scientists have identified about 2,700 species of snakes.
- Ask: **What does Jackson hope to teach people about snakes?** (that snakes are important animals that should be prevented from going extinct) Discuss how many species of reptiles and amphibians are threatened with extinction around the world because humans are using their land for cities, farms, and other purposes.

Find Out More

- Ask: **Is this a career you would find interesting? Why or why not?** (Accept all answers, positive or negative.) Encourage interested students to research a career in herpetology or other animal specialties.

myNGconnect.com

Have students use the Digital Library to find more photos of reptiles, amphibians, and people who study them.

Whiteboard Presentation Students can use the information to make a computer presentation about reptiles and amphibians. You can help them show the presentation on a whiteboard.

3 Assess

1. **Recall What do herpetologists study?** (reptiles and amphibians)
2. **Explain Why is Kate Jackson interested in protecting the environment where snakes live?** (Snakes depend on their environment to survive. She wants to protect the ecosystems where snakes live so that snakes and other neglected animals can continue living in nature.)
3. **Draw Conclusions Do you think it is important for snakes to be protected? Explain your answer.** (Possible answer: Yes. Snakes fill important roles in ecosystems. If a snake species goes extinct, it will be gone forever, and the ecosystems will be permanently changed.)

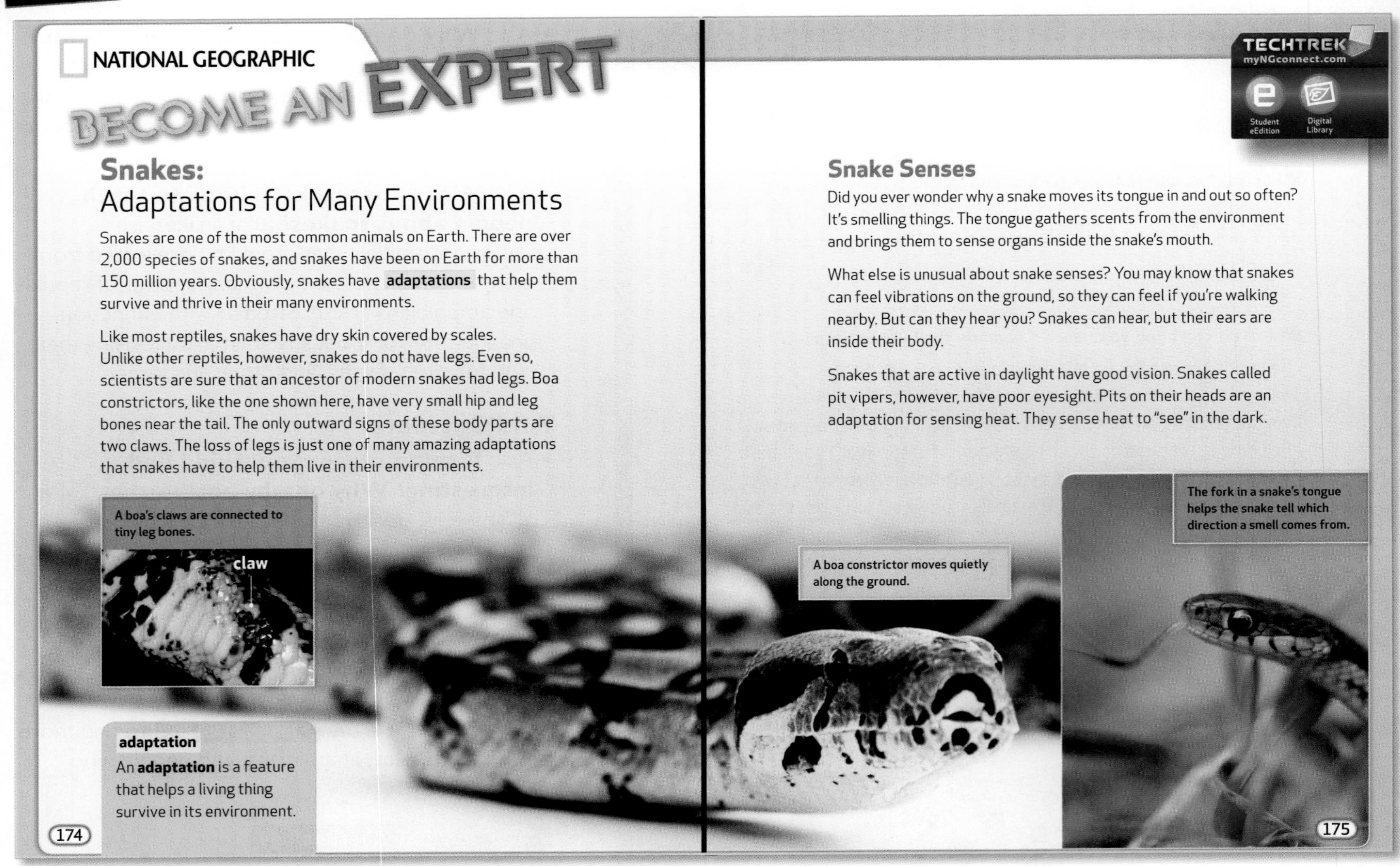

NATIONAL GEOGRAPHIC

BECOME AN EXPERT

Snakes: Adaptations for Many Environments

Snakes are one of the most common animals on Earth. There are over 2,000 species of snakes, and snakes have been on Earth for more than 150 million years. Obviously, snakes have **adaptations** that help them survive and thrive in their many environments.

Like most reptiles, snakes have dry skin covered by scales. Unlike other reptiles, however, snakes do not have legs. Even so, scientists are sure that an ancestor of modern snakes had legs. Boa constrictors, like the one shown here, have very small hip and leg bones near the tail. The only outward signs of these body parts are two claws. The loss of legs is just one of many amazing adaptations that snakes have to help them live in their environments.

A boa's claws are connected to tiny leg bones.

adaptation
An **adaptation** is a feature that helps a living thing survive in its environment.

174

TECHTREK
myNGconnect.com
Student eEdition
Digital Library

Snake Senses

Did you ever wonder why a snake moves its tongue in and out so often? It's smelling things. The tongue gathers scents from the environment and brings them to sense organs inside the snake's mouth.

What else is unusual about snake senses? You may know that snakes can feel vibrations on the ground, so they can feel if you're walking nearby. But can they hear you? Snakes can hear, but their ears are inside their body.

Snakes that are active in daylight have good vision. Snakes called pit vipers, however, have poor eyesight. Pits on their heads are an adaptation for sensing heat. They sense heat to "see" in the dark.

A boa constrictor moves quietly along the ground.

The fork in a snake's tongue helps the snake tell which direction a smell comes from.

175

PROGRAM RESOURCES

- Big Ideas Book: *Life Science*
- Big Ideas Book: *Life Science* eEdition at myNGconnect.com
- Digital Library at myNGconnect.com

Access Science Content

Discuss the Adaptations of Snakes

- Have students read pages 174–175, observe the photos, and read the captions. Then have them read the definition of **adaptation**. Ask: **What are some of the adaptations of snakes?** (Answers include movement without legs, a tongue that can sense odors from the environment, and the pit viper's sense of heat.) Explain that these are only a few of the many adaptations that help snakes survive.
- Point to the inset photo on page 174. Ask: **What do claws and tiny hip and leg bones show about the ancestors of snakes?** (They show that the ancestors had legs.) Explain that snakes, like most kinds of animals, have changed greatly over time.

Notetaking

As students read about snakes, have them write down new words they encounter in their science notebook. Then have them write a descriptive sentence that uses the word correctly. Students may also include drawings to illustrate their sentences.

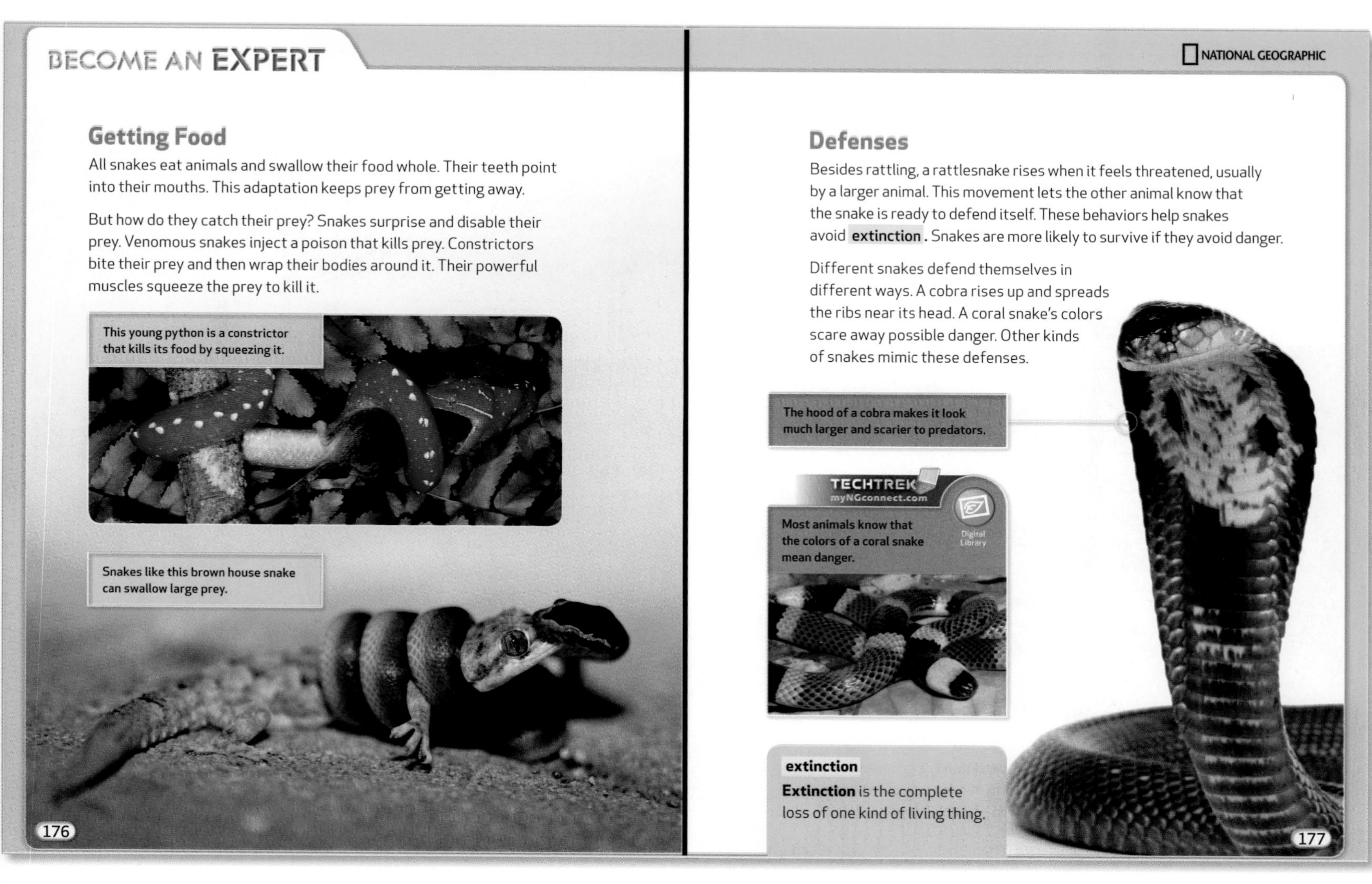

BECOME AN EXPERT

NATIONAL GEOGRAPHIC

Getting Food

All snakes eat animals and swallow their food whole. Their teeth point into their mouths. This adaptation keeps prey from getting away.

But how do they catch their prey? Snakes surprise and disable their prey. Venomous snakes inject a poison that kills prey. Constrictors bite their prey and then wrap their bodies around it. Their powerful muscles squeeze the prey to kill it.

This young python is a constrictor that kills its food by squeezing it.

Snakes like this brown house snake can swallow large prey.

176

Defenses

Besides rattling, a rattlesnake rises when it feels threatened, usually by a larger animal. This movement lets the other animal know that the snake is ready to defend itself. These behaviors help snakes avoid **extinction**. Snakes are more likely to survive if they avoid danger.

Different snakes defend themselves in different ways. A cobra rises up and spreads the ribs near its head. A coral snake's colors scare away possible danger. Other kinds of snakes mimic these defenses.

The hood of a cobra makes it look much larger and scarier to predators.

TECHTREK myNGconnect.com Digital Library

Most animals know that the colors of a coral snake mean danger.

extinction
Extinction is the complete loss of one kind of living thing.

177

PROGRAM RESOURCES

- Digital Library at myNGconnect.com

Access Science Content

Explain How Snakes Get Food and Defend Themselves

- Have students read the text on pages 176–177, observe the photos, and read the captions. Ask: **What is the difference between a venomous snake and a constrictor?** (A venomous snake makes a poison that kills prey, while a constrictor squeezes prey to death.) Then have students review the definition of **extinction** on page 177. Ask: **How do adaptations for defense help snakes avoid extinction?** (By defending themselves, snakes can stay alive and reproduce.)

Make Inferences

Point to the photo of the coral snake on page 177. Say: **The coral snake is poisonous. Other snakes have bands that make them look just like the coral snake, but they are harmless.** Ask: **Why are the mimics protected, too?** (Although the mimics are harmless, they fool predators into thinking they are poisonous, and so they are left alone.)

Digital Library

myNGconnect.com

Have students use the Digital Library to find more photos of snakes.

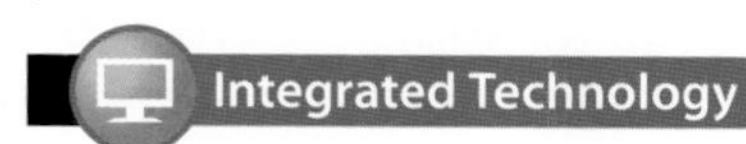

Integrated Technology

Computer Presentation Students can use the information to make a computer presentation about snakes.

LESSON 12 ▫ Become an Expert

BECOME AN EXPERT

NATIONAL GEOGRAPHIC

Camouflage

Can you see the deadly gaboon viper in the photo? It is hidden in the leaves on the ground. Many snakes have colors that help them blend in with their background. **Camouflage** helps them surprise prey and avoid predators. Many other types of snakes use camouflage too. They have many different patterns and colors. **Variations** help them to stay hidden in different environments.

The gaboon viper is one of the most deadly venomous snakes.

viper

A smooth green snake finds grasshoppers by hiding in grass.

camouflage
Camouflage is a color or shape that makes a living thing hard to see.

variation
A **variation** is a different form of a feature of the same kind of living thing.

178

Living in Their Environment

Snakes are adapted to the environment in which they live. A snake's adaptations help it find food and protect itself. Snakes have adapted very well to a variety of environments, and can be found on every continent on Earth except Antarctica.

These pythons are adapted to the jungles of Papua New Guinea.

179

Access Science Content

Discuss Camouflage and Variation in Snakes

- Have students read pages 178–179 and study the photos. Ask: **How does camouflage help some snakes survive?** (By blending into their background, the snakes can surprise prey and hide from predators.) Ask: **How does variation in patterns and colors help snakes survive?** (The variation helps different snakes hide in different environments.)

Differentiated Instruction

ELL Language Support for Discussing Adaptations

BEGINNING

Have students describe how camouflage protects the green snake shown on page 178. Then have them define *camouflage* in their own words.

INTERMEDIATE

Provide Academic Language Frames, such as:

- *An animal with ____ blends into the background.*
- *____ help animals survive in different environments.*

ADVANCED

Provide Academic Language Stems, such as:

- *Camouflage is an adaptation because . . .*
- *Variations are adaptations because . . .*

Assess

1. **Describe What clue shows that snakes' ancestors had legs?** (Some snakes have tiny hip and leg bones with claws attached.)
2. **Analyze What adaptations of snakes make them different from other predators of small animals?** (Possible answers: movement without legs, injecting venom into prey, and the pit viper's ability to sense heat)

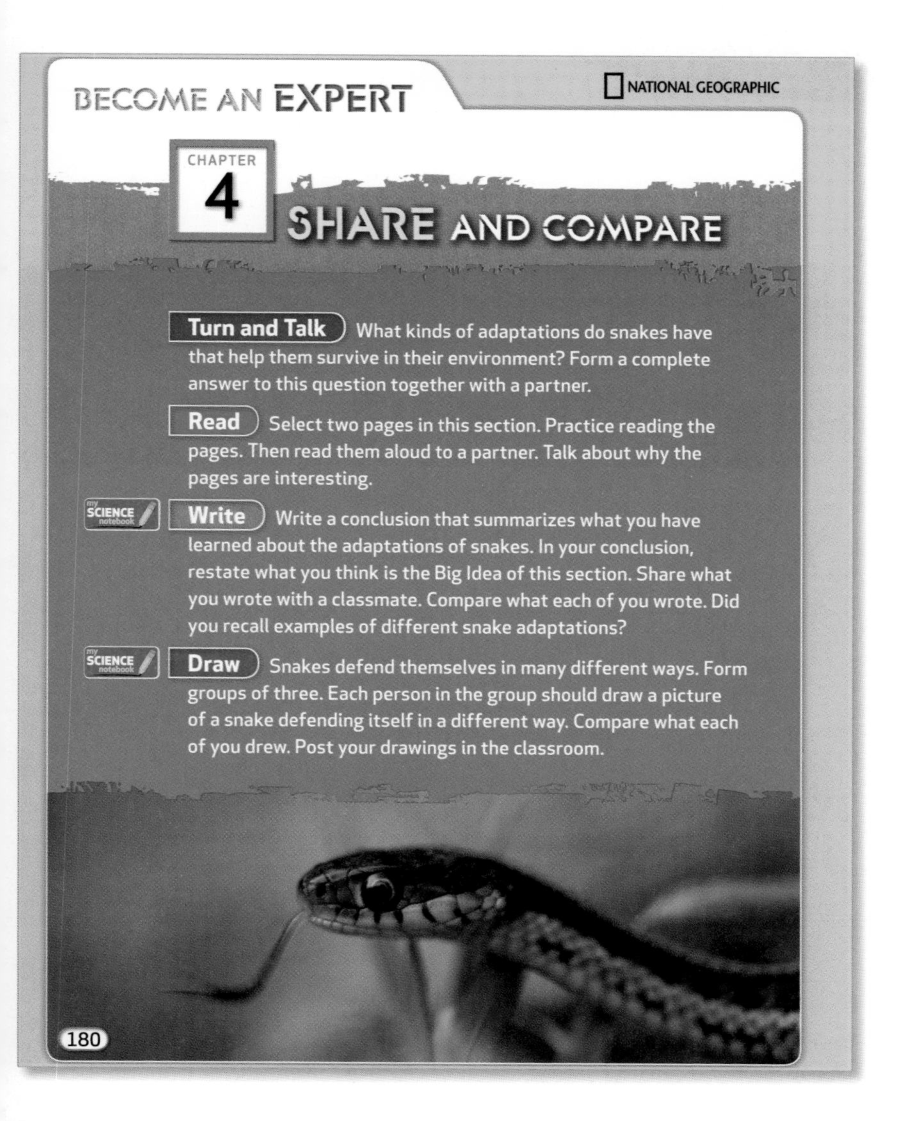
BECOME AN EXPERT
NATIONAL GEOGRAPHIC
CHAPTER 4
SHARE AND COMPARE
Turn and Talk What kinds of adaptations do snakes have that help them survive in their environment? Form a complete answer to this question together with a partner.
Read Select two pages in this section. Practice reading the pages. Then read them aloud to a partner. Talk about why the pages are interesting.
Write Write a conclusion that summarizes what you have learned about the adaptations of snakes. In your conclusion, restate what you think is the Big Idea of this section. Share what you wrote with a classmate. Compare what each of you wrote. Did you recall examples of different snake adaptations?
Draw Snakes defend themselves in many different ways. Form groups of three. Each person in the group should draw a picture of a snake defending itself in a different way. Compare what each of you drew. Post your drawings in the classroom.
180

Sum Up

Tell students that to sum up a text helps them pull together the ideas of the text. Remind students that they summed up the Become an Expert lesson in the Write section of Share and Compare. Have students take turns reading the conclusions they wrote to a partner. Invite them to compare and contrast their conclusions.

Share and Compare

Turn and Talk

Ask students to turn to partners and discuss the adaptations that help snakes survive in their environment. Prompt students by asking:

1. **Recall How do venomous snakes and constrictors kill prey?** (Venomous snakes inject a poison into their prey. Constrictors squeeze prey to kill them.)
2. **Explain How do some snakes benefit from camouflage?** (Camouflage, or blending into the background, helps some snakes surprise prey and hide from predators.)
3. **Analyze What kinds of adaptations do snakes have that help them survive in their environment?** (Answers include adaptations for movement, for sensing the environment, for hunting and killing prey, and camouflage and variation that help them surprise prey and hide from predators.)

Read

As partners read and discuss their selections, encourage them to ask questions to one another about details that interest them, points they find confusing or unclear, or subjects they would like to investigate further.

Write

my SCIENCE notebook

Possible answer: Snakes have a variety of unique or unusual adaptations that help them catch and eat prey, hide from enemies, and survive in their environments. As students share and compare their responses, encourage them to review additional adaptations from the Big Ideas Book.

Draw

my SCIENCE notebook

Tell student groups to discuss their examples before drawing them so that the group presents a variety of snake defenses. Encourage each student to describe his or her own illustration to the other group members.

Notes

LIFE SCIENCE

CHAPTER
5
HOW DO LIVING THINGS
INTERACT WITH THEIR
ENVIRONMENT?
TECHTREK
myNGconnect.com
Everything on Earth is connected, like the threads of a web. Change one thing, and you affect something else. For example, as winter changes to spring, these birds gather on high cliffs to mate and raise their young. How do the birds interact with their environment to live? How do these interactions change the environment?
182
183

Chapter 5 Contents

After reading Chapter 5, you will be able to:

- Give examples of how plants change based on the seasons. **CHANGES IN SEASONS**
- Explain how plants and animals adapt to the changing of the seasons in their environment. **CHANGES IN SEASONS**
- Give examples of how animals adapt to the changing of the seasons in their environment. **CHANGES IN SEASONS**
- Explain how plants can change the environment in many ways. **PLANTS CHANGE THE ENVIRONMENT**
- Describe how plants can harm an environment. **PLANTS CHANGE THE ENVIRONMENT**
- Explain how animals can change the environment by making homes, changing soil, grazing, and feeding. **ANIMALS CHANGE THE ENVIRONMENT**
- Describe how animals can hurt the environment. **ANIMALS CHANGE THE ENVIRONMENT**
- Observe and explain that humans can change the environment in beneficial or harmful ways. **PEOPLE CHANGE THE ENVIRONMENT**
- Recognize and explain that humans depend on their environments to meet their needs. **PEOPLE CHANGE THE ENVIRONMENT**
- Science in a Snap! Observe and explain that humans can change the environment in beneficial or harmful ways. **PEOPLE CHANGE THE ENVIRONMENT**

181

Preview and Predict

Tell students that they can preview a text by looking over it to get an idea of what the text is about and how it is organized. Tell students that predicting is forming ideas about what they will read. Students should confirm their ideas with others.

Have students preview and make predictions about the chapter. Remind them to use section heads, vocabulary words, pictures, and their background knowledge to make predictions.

CHAPTER 5 ▫ How Do Living Things Interact with Their Environment?

LESSON	PACING	OBJECTIVES
1 **Directed Inquiry** *Investigate Plants and Water* pages T181e–T181h **Science Inquiry and Writing Book** pages 52–55	**20** minutes	Investigate through Directed Inquiry (answer a question; make and compare observations; collect and record data and observations; generate explanations and conclusions based on evidence; share findings; ask questions based on observations to increase understanding). Recognize ways plants can impact the environment. Keep accurate records and use evidence to communicate and support reasonable explanations. Make purposeful observation of the natural world using the appropriate senses.
2 **Big Idea Question and Vocabulary** pages T182–T183—T184–T185	**25** minutes	Recognize ways that plants and animals can change the environment.
3 **Changes in Seasons** pages T186–T191	**30** minutes	Give examples of how plants change based on the seasons. Explain how plants adapt to the changing of the seasons in their environment. Give some examples of how animals adapt to the changing of the seasons in their environment. Explain how animals adapt to the changing of the seasons in their environment.
4 **Plants Change the Environment** pages T192–T195	**25** minutes	Explain how plants can change the environment in many ways. Describe how plants can harm an environment.
5 **Animals Change the Environment** pages T196–T201	**30** minutes	Explain how animals can change the environment by making homes and changing soil. Explain how animals can change the environment by grazing and feeding. Describe how animals can hurt the environment.
6 **People Change the Environment** pages T202–T209	**35** minutes	Explain that humans can change the environment in beneficial or harmful ways. Recognize that humans depend on their environments to meet their needs. Observe that humans can change the environment in beneficial or harmful ways. Explain that humans depend on their environments to meet their needs.
7 NATIONAL GEOGRAPHIC **Using Invasive Organisms to Help the Environment** pages T210–T213	**25** minutes	Explain that humans can change the environment in beneficial or harmful ways.

VOCABULARY	RESOURCES	ASSESSMENT
observe **infer**	Science Inquiry and Writing Book: *Life Science* Science Inquiry Kit: *Life Science* Directed Inquiry: Learning Masters 56–58	Inquiry Rubric: Assessment Handbook, page 186 Inquiry Self-Reflection: Assessment Handbook, page 197 Reflect and Assess, page T181h
	Vocabulary: Learning Master 59 *Life Science* Big Ideas Book	
		Assess, page T191
invasive organism		Assess, page T195
		Assess, page T201
recycle **conserve** **pollution**	Science Inquiry and Writing Book: *Life Science*	Assess, page T209
	Share and Compare: Learning Master 60	Assess, page T213

TECHNOLOGY RESOURCES

STUDENT RESOURCES

myNGconnect.com

- **Student eEdition**
- Big Ideas Book
- Science Inquiry and Writing Book
- **Read with Me**
- **Vocabulary Games**
- **Enrichment Activities**
- **Digital Library**

National Geographic Kids

National Geographic Explorer!

TEACHER RESOURCES

myNGconnect.com

- **Teacher eEdition**
- Teacher's Edition
- Science Inquiry and Writing Book
- Online Lesson Planner
- National Geographic Unit Launch Videos
- Assessment Handbook
- **Presentation Tool**
- **Digital Library**

NGSP ExamView CD-ROM

IF TIME IS SHORT...

FAST FORWARD.

CHAPTER 5 ▫ How Do Living Things Interact with Their Environment?

LESSON	PACING	OBJECTIVES
8 Guided Inquiry *Investigate Colors in Green Leaves* pages T213a–T213d **Science Inquiry and Writing Book** pages 56–59	**40** minutes	Investigate through Guided Inquiry (answer a question; make and compare observations; collect and record data and observations; generate explanations and conclusions based on evidence; share findings; ask questions based on observations to increase understanding; adjust explanations based on findings and new ideas). Recognize the characteristics of a fair and unbiased test and the importance of keeping conditions the same. State, orally and in writing, any inferences or generalizations indicated by the data collected.
9 **Conclusion and Review** pages T214–T215	**15** minutes	
10 NATIONAL GEOGRAPHIC LIFE SCIENCE EXPERT *Conservationist* pages T216–T217 NATIONAL GEOGRAPHIC BECOME AN EXPERT *Green Movements Around the World* pages T218–T219—T228	**35** minutes	Recognize that scientists from many different gender and ethnic backgrounds make contributions to science. Explain that humans can change the environment in beneficial ways.

FAST FORWARD
ACCELERATED PACING GUIDE

DAY 1  **20** minutes

Directed Inquiry

Investigate Plants and Water, page T181e

DAY 2 **35** minutes

NATIONAL GEOGRAPHIC **LIFE SCIENCE EXPERT** *Conservationist,* page T216

NATIONAL GEOGRAPHIC **BECOME AN EXPERT** *Green Movements Around the World,* page T218–T219

VOCABULARY	RESOURCES	ASSESSMENT
investigate **infer**	Science Inquiry and Writing Book: *Life Science* Science Inquiry Kit: *Life Science* Guided Inquiry: Learning Masters 61–64	Inquiry Rubric: Assessment Handbook, page 186 Inquiry Self-Reflection: Assessment Handbook, page 198 Reflect and Assess, page T213d
		Assess the Big Idea, page T215 Chapter 5 Test: Assessment Handbook, pages 27–30 NGSP ExamView CD-ROM
		Assess, pages T217, T220–T221, T224–T225, T226–T227

TECHNOLOGY RESOURCES

STUDENT RESOURCES

myNGconnect.com

- Student eEdition
- Big Ideas Book
- Science Inquiry and Writing Book
- Read with Me
- Vocabulary Games
- Enrichment Activities
- Digital Library

National Geographic Kids

National Geographic Explorer!

TEACHER RESOURCES

myNGconnect.com

- Teacher eEdition
- Teacher's Edition
- Science Inquiry and Writing Book
- Online Lesson Planner
- National Geographic Unit Launch Videos
- Assessment Handbook
- Presentation Tool
- Digital Library

NGSP ExamView CD-ROM

DAY 3 45 minutes

Guided Inquiry

Investigate Colors in Green Leaves, page T213a

Objectives

Students will be able to:

- Investigate through Directed Inquiry (answer a question; make and compare observations; collect and record data and observations; generate explanations and conclusions based on evidence; share findings; ask questions based on observations to increase understanding).
- Recognize ways plants can impact the environment.
- Keep accurate records and use evidence to communicate and support reasonable explanations.
- Make purposeful observation of the natural world using the appropriate senses.

Science Process Vocabulary

observe, infer

PROGRAM RESOURCES

- Science Inquiry and Writing Book: *Life Science*
- Science Inquiry and Writing Book eEdition at myNGconnect.com
- Inquiry eHelp at myNGconnect.com
- Science Inquiry Kit: *Life Science*
- Learning Masters Book, pages 56–58, or at myNGconnect.com
- Inquiry Rubric: Assessment Handbook, page 186, or at myNGconnect.com
- Inquiry Self-Reflection: Assessment Handbook, page 197, or at myNGconnect.com

MATERIALS

Kit materials are listed in italics.

safety goggles; plant cutting; *hand lens; index card (4 x 6 in.); waxed paper;* masking tape; pencil; *clay; 2 clear plastic cups (9 oz);* water (100 mL); black marker

For teacher use: scissors

❶ Introduce

Tap Prior Knowledge

- Ask: **How does a plant get water from its environment?** (It absorbs water from the soil through its roots.) **How might a plant give off water to its environment?** (Accept all ideas.)

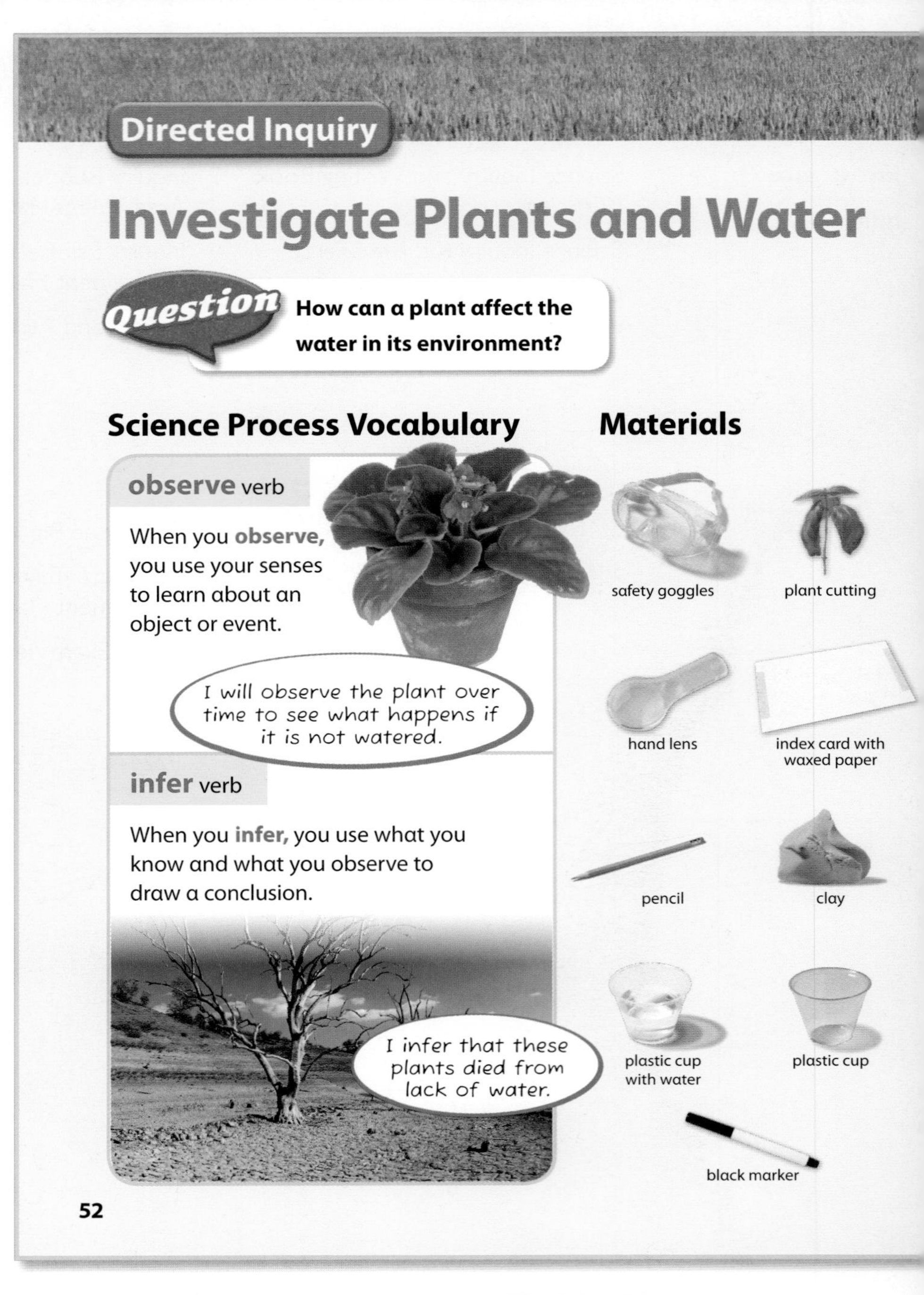

Directed Inquiry
Investigate Plants and Water
Question
How can a plant affect the water in its environment?
Science Process Vocabulary
observe verb
When you observe, you use your senses to learn about an object or event.
I will observe the plant over time to see what happens if it is not watered.
infer verb
When you infer, you use what you know and what you observe to draw a conclusion.
I infer that these plants died from lack of water.
Materials
safety goggles
plant cutting
hand lens
index card with waxed paper
pencil
clay
plastic cup with water
plastic cup
black marker
52

MANAGING THE INVESTIGATION

Time

20 minutes on the first day, 5 minutes on the second day, and 20 minutes on the third day

Groups

Small groups of 4

Advance Preparation

- Fill one 9-oz plastic cup with about 100 mL of water for each group.
- Prepare an index card for each group by taping a 4-by-6-inch piece of waxed paper to one side of each index card.
- Obtain and prepare a plant cutting for each group from a non-poisonous houseplant. Each cutting should have several leaves and be about 6 inches long. A basil plant is pictured.

What to Do

1. Put on your safety goggles. Use the hand lens to **observe** the leaves of the plant cutting. Record your observations in your science notebook.

2. Use the pencil to make a small hole in the center of the index card and waxed paper. Hold the index card so that the waxed paper is on the bottom. Very carefully pull the stem of the plant cutting through the hole. Seal around the opening with clay.

53

Teaching Tips

- Tell students to be careful not to crush the stem with the clay.
- The stem below the index card should sit in the water. The leaves and top of the plant should fit inside the top cup.
- Students should wash their hands after handling the plants.

What to Expect

- After 24 hours, students should see water droplets that have condensed in the top cup. After 48 hours, more droplets should appear in the top cup and the water level in the bottom cup may decrease.

Introduce, continued

Connect to the Big Idea

- Review the Big Idea Question, *How do living things interact with their environment?* Explain to students that this inquiry will help them learn about how plants interact with their environment.
- Have students open their Science Inquiry and Writing Books to page 52. Read the Question and invite students to share ideas about how plants affect the water in their environment.

❷ Build Vocabulary

Science Process Words: observe, infer

Use this routine to introduce the words.

1. **Pronounce the Word** Say **observe.** Have students repeat it in syllables.
2. **Explain Its Meaning** Choral read the sentence and the speech bubble. Ask students for another word or phrase that means the same as **observe.** (watch; notice certain things)
3. **Encourage Elaboration** Ask: **What are some ways you can observe plants?** (by looking carefully; measuring; counting parts, such as leaves or flowers)

ELL Use Cognates

Students may know the verb **observar** in their home language. Use it to help them understand the English word **observe**.

Repeat for the word **infer**. To encourage elaboration, ask: **When you infer, what two things do you use to draw a conclusion?** (your observations and what you already know)

❸ Guide the Investigation

- Read the inquiry steps on pages 53–54 together with students. Move from group to group and clarify steps if necessary.
- In step 1, have students share their recorded observations. Point out that making and recording careful and detailed observations are important so that scientists can accurately compare observations with those made later in the investigation.

Guide the Investigation, continued

- Remind students to observe and record the water level on each day. Ask: **How did the water level change over time?** (The water level became lower each day.) **Do you think water can leave the bottom cup by evaporation? Why or why not?** (No. The cup is not open to the air.)
- Guide students to make a connection about why water might move from the bottom cup to the top cup. After ruling out evaporation, ask: **What else connects the top cup and bottom cup?** (the plant's stem) **What is the purpose of a plant's stem?** (to transport nutrients and water)

❹ Explain and Conclude

- Guide students as they record their observations in the table. Have students compare their results with other groups.
- Students should then make inferences about what happened to the water. Ask: **What do you think happened to water in the bottom cup?** (It traveled up through the plant stem.) Ask: **How do you think the water droplets got to the top cup?** (The plant's leaves released water vapor, which cooled and condensed into droplets.)
- Encourage students to give explanations about how plants interact with their environment. Ask students to respond to the Question on page 52. Students may say that plants affect their environments by releasing water into the air.
- Students should support their ideas by telling about what they observed. Ask: **Why is it important to use evidence to support your ideas?** Remind students that scientists base their scientific explanations on evidence.

Answers

1. Possible answer: I observed that water droplets formed in the top cup after 24 hours. After 48 hours, more droplets had formed in the top cup, and the water level in the bottom cup had gone down.

What to Do, continued

3. Place the plant cutting in the cup of water so the index card rests on top. Cover the top of the plant with the other plastic cup. Observe the plant and the water in the bottom cup. Record your observations. Mark the water level by drawing a line on the bottom cup.

Make sure that the stem of the plant is in the water.

4. Place the plant in a sunny spot.

5. Wait 24 hours and observe the plant and cups. Be sure to observe the water level in the bottom cup. Record your observations.
6. Observe the plant again after 48 hours. Record your observations and **compare** them to your observations from the previous day.

54

NATIONAL GEOGRAPHIC Raise Your SciQ!

Transpiration and the Environment The process by which plants lose water vapor from their leaves is called transpiration. In areas covered with plants, such as forests, transpiration can change local weather conditions. Water vapor from plants increases the humidity of the air. Warm water vapor rises into the atmosphere. There, it cools and condenses. If enough water vapor is present, clouds may form. In some areas, such as tropical rainforests, large amounts of water vapor may rise and cool so quickly that thunderstorms can form.

Record

Write and draw in your science notebook. Use a table like this one.

my SCIENCE notebook

Plant Cutting and Cups

Time	Observations
Plant only	
Plant in water: start	
Plant in water: 24 hours	

Explain and Conclude

1. What did you **observe** in the cups after 24 hours? What did you observe after 48 hours?
2. **Infer** what happens to some of the water that plants take in.
3. Use the results of this **investigation** to infer how plants might affect groundwater. How might they affect water in the air?

Think of Another Question

What else would you like to find out about how plants affect water in the environment? How could you find an answer to this new question?

Trail through forest, Potsdam, Germany

In a year, a tall tree can release over 100,000 liters of water into the air.

55

Inquiry Rubric at myNGconnect.com	Scale			
The student **observed** how water vapor released by the leaves of a plant cutting condensed on a cup.	4	3	2	1
The student collected **data** and recorded observations of the cup over several days.	4	3	2	1
The student **compared** observations with other students.	4	3	2	1
The student **inferred** what happened to the water that the plant cutting took in.	4	3	2	1
The student **shared conclusions** about how plants affect the environment.	4	3	2	1
Overall Score	**4**	**3**	**2**	**1**

Explain and Conclude, continued

2. Possible answer: Some water that plants take in is released as water vapor by their leaves.
3. Answers will vary, but students should explain that plants absorb water from the ground and release it into the atmosphere. Plants may reduce the amount of groundwater and increase the amount of water vapor in the air.

5 Find Out More

Think of Another Question

- Students should use their observations to generate new questions. Students may ask: *What would happen if we used a different plant? How can we make the plant take up water faster?* Discuss what students could do to find answers to the new questions.

6 Reflect and Assess

- To assess student work with the Inquiry Rubric shown below, see Assessment Handbook, page 186, or go online at myNGconnect.com
- Have students use the Inquiry Self-Reflection on Assessment Handbook, page 197, or at myNGconnect.com

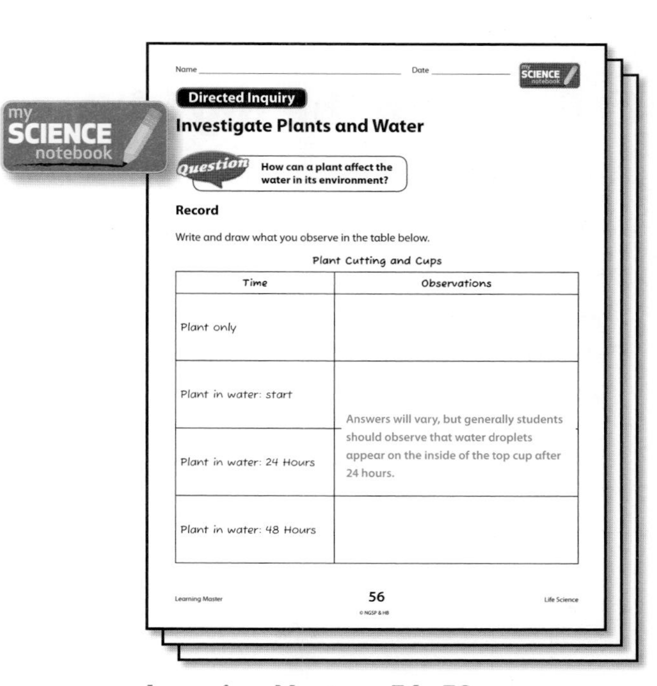

my SCIENCE notebook

Name ______ Date ______

Directed Inquiry

Investigate Plants and Water

Question: How can a plant affect the water in its environment?

Record

Write and draw what you observe in the table below.

Plant Cutting and Cups

Time	Observations
Plant only	
Plant in water: start	Answers will vary, but generally students should observe that water droplets appear on the inside of the top cup after 24 hours.
Plant in water: 24 Hours	
Plant in water: 48 Hours	

Learning Master 56 Life Science

Learning Masters 56–58 or at myNGconnect.com

Objectives

Students will be able to:

- Recognize ways that plants and animals can change the environment.

Science Academic Vocabulary

invasive organism, conserve, recycle, pollution

CHAPTER
5
HOW DO LIVING THINGS
INTERACT WITH THEIR
ENVIRONMENT?
TECHTREK
myNGconnect.com
Everything on Earth is connected, like the threads of a web. Change one thing, and you affect something else. For example, as winter changes to spring, these birds gather on high cliffs to mate and raise their young. How do the birds interact with their environment to live? How do these interactions change the environment?
These seabirds are northern gannets. Thousands of them gather on cliffs on Bonaventure Island, Quebec in Canada each spring.
182
183

PROGRAM RESOURCES

- Big Ideas Book: *Life Science*
- Big Ideas Book: *Life Science* eEdition at myNGconnect.com
- Vocabulary Games at myNGconnect.com
- Digital Library at myNGconnect.com
- Enrichment Activities at myNGconnect.com
- Read with Me at myNGconnect.com
- Learning Masters Book, page 59, or at myNGconnect.com

❶ Introduce

Tap Prior Knowledge

- Ask students if they have heard or used the word *environment* and to use their own words to define it. Remind them that an environment is all the living and nonliving things in an area. Have them observe the photo on pages 182–183 and list the living and nonliving things that they can identify.

❷ Focus on the Big Idea

Big Idea Question

- Read the Big Idea Question aloud and have students echo it.
- Preview pages 186–191, 192–195, 196–201, and 202–209, linking the headings with the Big Idea Question.

Differentiated Instruction

ELL Vocabulary Support

BEGINNING

Help students describe conservation by using Academic Language Frames:

- *To ____ is to use a resource wisely.*
- *One way to conserve resources is to ____.*

INTERMEDIATE

Have students draw a picture that shows a resource being conserved. Have them write a sentence about the picture that uses the word *conserve.*

ADVANCED

Present important words that appear in each definition. For example, *invasive organism* and *place.* Show students how to restate the definitions using the important words. ***Invasive organisms*** **are living things that harm a** ***place*** **because they do not belong there.**

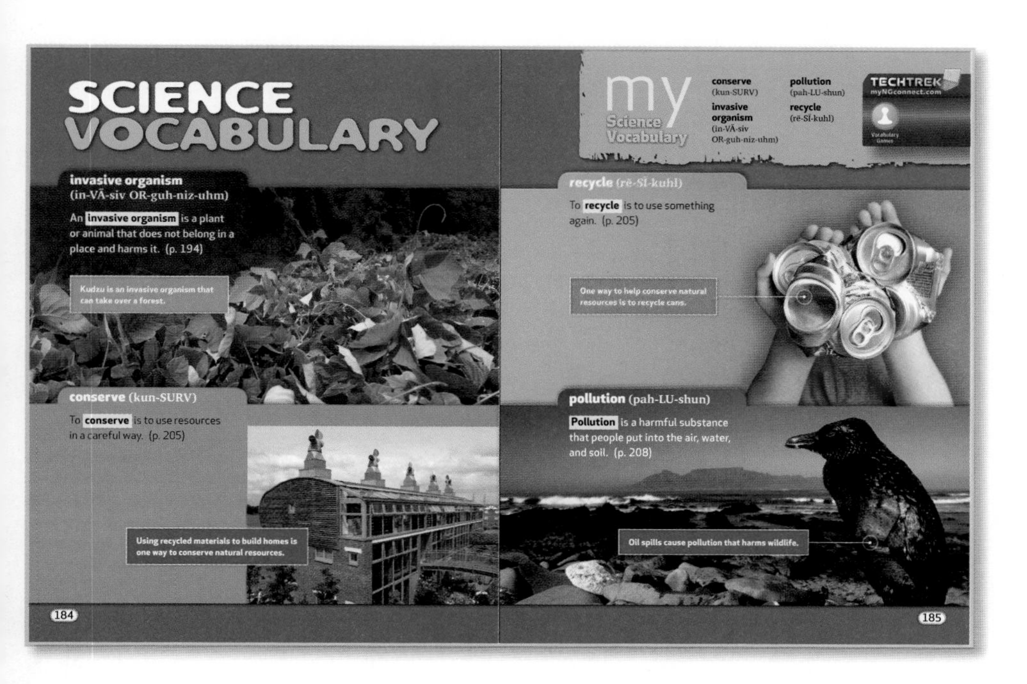

Focus on the Big Idea, continued

- With student input, post a chart that displays the headings. Have students orally share what they expect to find in each section. Then read page 182 aloud.

How Do Living Things Interact with Their Environment?

Changes in Seasons

Plants Change the Environment

Animals Change the Environment

People Change the Environment

❸ Teach Vocabulary

Have students look at pages 184–185, and use this routine to teach each word. For example:

1. **Pronounce the Word** Say **invasive organism** and have students repeat it.
2. **Explain Its Meaning** Read the word, definition, and sample sentence, and use the photo to explain the word's meaning. Say: **To *invade* means "to take over a space." Where would you find an invasive organism?** (in a place where it does not belong, or is not normally found)
3. **Encourage Elaboration** Have students look carefully at the photo. Ask: **What kind of organism is kudzu?** (a plant) **What kind of environment is this plant invading?** (a forest)

Repeat for the words **conserve**, **recycle**, and **pollution** using the following Elaboration Prompts:

- **How can gasoline be conserved?** (Possible answers: by using public transportation or by walking instead of using an automobile; by using fuel-efficient automobiles)
- **How can people recycle glass?** (by taking it to a recycling center)
- **How does pollution enter the air?** (Possible answers: the burning of coal and gasoline, soot from factories)

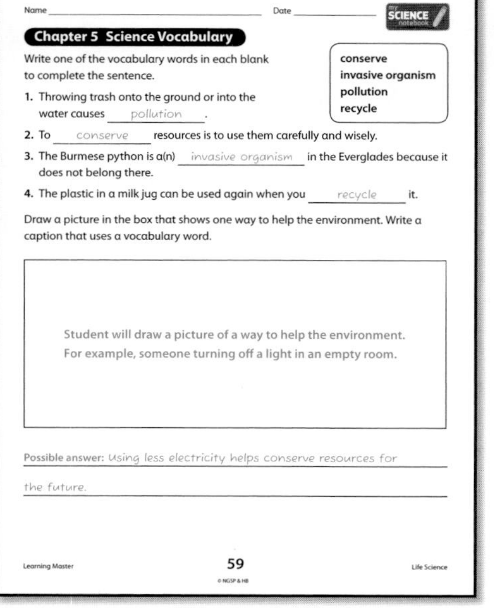

Name ______ Date ______ my SCIENCE notebook

Chapter 5 Science Vocabulary

Write one of the vocabulary words in each blank to complete the sentence.

conserve
invasive organism
pollution
recycle

1. Throwing trash onto the ground or into the water causes pollution.
2. To conserve resources is to use them carefully and wisely.
3. The Burmese python is a(n) invasive organism in the Everglades because it does not belong there.
4. The plastic in a milk jug can be used again when you recycle it.

Draw a picture in the box that shows one way to help the environment. Write a caption that uses a vocabulary word.

Student will draw a picture of a way to help the environment. For example, someone turning off a light in an empty room.

Possible answer: Using less electricity helps conserve resources for the future.

Learning Master 59 Life Science

Learning Master 59 or at myNGconnect.com

Objectives

Students will be able to:

- Give examples of how plants change based on the seasons.
- Explain how plants adapt to the changing of the seasons in their environment.

PROGRAM RESOURCES

- Digital Library at myNGconnect.com

❶ Introduce

Tap Prior Knowledge

- List the seasons on the board: winter, spring, summer, fall. Then ask students to describe the weather in your region during each season and ways that plants and animals change.

Set a Purpose and Read

- Read the headings on pages 186–191. Tell students that they will learn how plants and animals change during the seasons.
- Have students read pages 186–191.

❷ Teach

Recognize Seasonal Changes in Plants

- Point out the photo on pages 186–187. Ask: **What season is shown in this photo?** (fall) **How do you know?** (The leaves of the trees have changed colors from green to orange and yellow. Some leaves have fallen to the ground.)
- Ask: **What will the trees be like in spring?** (New leaves will grow, and flowers may bloom.) **In summer?** (The branches will be full of green leaves, and fruits and seeds may be growing.)
- Explain: **Some trees, like maples and elms, have leaves that change color and drop to the ground in fall. These trees are bare in winter. Then, in spring, new leaves grow in their place.**

Changes in Seasons

Changes in Plants Have you ever heard the saying, "April showers bring May flowers"? At different times of the year, you can see how changes in the environment have changed plants. During the spring, flowers bloom and leaves grow on trees. Some plants grow fruit during the summer. In the fall, leaves on some trees change color. Then the leaves drop. The trees stay bare during the winter.

Some trees drop their leaves in the fall. New leaves will grow in the spring.

186

Science Misconceptions

Changing Leaves Some students may think that leaves change color in fall because the weather becomes colder or drier. Explain that the main reason leaves change color is the changing length of day. Leaves have different colored pigments, the most abundant of which is a green pigment called chlorophyll. In fall, the days grow shorter and the nights grow longer. In response, biochemical processes in the leaf stop maintaining chlorophyll. Other leaf pigments, such as yellow and brown, are revealed instead.

Trees may not change this way where you live. Trees and other plants change more during the seasons in the northern part of the United States than in the southern part. However, even some southern plants show seasonal changes. Some grasses in Florida, for example, turn brown or purple in the fall. The green leaves of dogwood trees and red maples turn red and purple.

During the summer, green grapes grow on the sea grape plant along the Florida coast.

The grapes can turn red and orange in late summer or early fall.

Teach, continued

Compare Seasonal Changes in Plants

- Have students observe the two photos on page 187 and read their captions. Ask: **How do sea grapes in Florida change between summer and fall?** (In summer, the grapes are green and plump. In fall, they can change to a red or orange color.)
- If available, display a map of the United States. Ask: **Where in the United States is Florida?** (in the southeast corner) **How is the weather in the southeast different from the weather in northern states?** (Winters are milder, while summers are hotter.) Ask: **How do the changes in the plants that grow here compare with the changes in plants of northern states?** (They change less with the seasons than northern plants.)

Describe Dormancy in Plants

- Point to the trees shown on page 186. Ask: **What will these trees be like in winter?** (All the leaves will have fallen, and the branches will be bare. The trees will not grow.)
- Define the word *dormancy* as a state of rest or inactivity. Ask: **How long will the trees be dormant?** (until spring arrives)

myNGconnect.com

Have students use the Digital Library to find images of plants changing in different seasons.

Whiteboard Presentation Students can use the information to make a computer presentation about plants changing in different seasons. You can help them show the presentation on a whiteboard.

Differentiated Instruction

ELL Language Support for Describing Seasonal Changes in Plants

BEGINNING

Have students sketch and label a tree in winter, spring, summer, and fall.

Use either/or questions to help them describe their sketch, for example: **Is the winter cold or warm? Does the tree have leaves or is it bare?**

INTERMEDIATE

Provide Academic Language Frames about the four seasons.

- *Flowers bloom and leaves grow during ___.*
- *Fruits grow during ___.*
- *Leaves drop during ___.*

ADVANCED

Help students create a chart that describes changes in plants during the four seasons. For example, title the chart *Seasons* and write headings *Spring, Summer, Fall,* and *Winter.*

Then have students use their charts as aids to describe the changes.

Objectives

Students will be able to:

- Give some examples of how animals adapt to the changing of the seasons in their environment.

Teach, continued

Recognize Seasonal Changes in Animals

- Ask: **What are some ways that animals change with the seasons?** (They grow thick coats of fur in winter and shed their fur in summer. The fur of some animals changes color with the seasons.)
- Have students observe the two photos of the snowshoe hare on page 188. Ask: **How does white fur help the hare in winter?** (The whiteness helps the hare blend in with the snow and hide from enemies.) **How does brown fur help in spring and summer?** (It helps the hare blend in with wooded areas.)
- Explain: **Colors or markings that help an animal blend in with its surroundings are called *camouflage*.** Write the word *camouflage* on the board and have students repeat it.
- Ask: **Why might a snowshoe hare struggle to live in the southern United States, where the weather usually is warm all year long?** (The hare's changing color would provide it with no advantage. The hare might stand out from its surroundings, rather than be camouflaged.) Discuss how the snowshoe hare lives in the north, where snow may cover the ground for much of the year.

Changes in Animals Plants are not the only living things that change with the season. Animals also must adapt to seasonal changes in order to survive. Some animals, such as deer, grow thick coats of fur to keep warm in the winter. During the summer, they shed some of their fur to stay cool. The fur of some animals changes color with the seasons. The change of color helps the animal blend in with its surroundings.
In the spring and summer, the fur of a snowshoe hare is brown.
In the winter, the fur of the snowshoe hare turns white. This helps it blend in with the snow and hide from enemies.
188

Raise Your SciQ! (National Geographic)

Changing Fur Color Like the changing colors of leaves, the changing fur color of snowshoe hares is caused mostly by changing day length. Both in fall and spring, the color change takes about 10 weeks as old fur is shed and new fur is grown in its place. Snowshoe hares will change fur color regardless of the temperature and ground cover of their environment.

Animals change in other ways to survive changes in seasons. During the late summer, some animals eat large amounts of food. Their bodies store the extra energy. They use that energy during the cold winter while they hibernate, or go into a deep sleep.

Other animals migrate, or move, to warmer places during the fall. Many northern birds migrate because there is not enough food in their summer homes during winter.

Woodchucks hibernate during the winter to use less energy.

Woodchucks become active again in late February and March.

189

Teach, continued

Compare Seasonal Changes in Animals

- Write the word *hibernate* on the board and have students pronounce it. Ask: **What does an animal do when it hibernates?** (It goes into a period of inactivity, in some ways like a very deep, extended sleep.) Explain: **Less food is available during a cold, snowy winter. By hibernating, an animal can live off the energy in the food stored in its body.**
- Write the word *migrate* on the board and have students pronounce it. Ask: **What does an animal do when it migrates?** (It travels between different environments, sometimes across great distances.)
- Ask: **How does migrating help many birds that live in northern places?** (By migrating in the fall and spring, the birds can avoid cold winters.) Discuss how many birds migrate from the northern United States to Florida, Mexico, Central America, and South America.

Differentiated Instruction

ELL Language Support for Describing Seasonal Changes in Animals

BEGINNING

Provide Academic Language Frames for students to describe changes in animals.

- *To sleep deeply in winter is to ____.*
- *Many birds ____ to warmer places in fall.*
- *Deer grow ____ coats of fur in winter.*

INTERMEDIATE

Provide Academic Language Stems, such as those below, for students to describe changes in animals during the seasons:

- *In winter, woodchucks . . .*
- *In winter, the snowshoe hare . . .*

ADVANCED

Have students make before and after sketches of an animal that changes fur color from winter to summer, an animal that hibernates, or an animal that migrates. Help them write labels or captions for their drawings using the terms *migrate, hibernate,* or *fur.*

Objectives
Students will be able to:
- Explain how animals adapt to the changing of the seasons in their environment.

PROGRAM RESOURCES
- Digital Library at myNGconnect.com

Teach, continued

Recognize Seasonal Changes in Places with Warm Climates

- Ask students to observe the photo on page 190 and to study the bird's beak. Ask: **What does the beak of this bird look like?** (Possible answers: a spoon, a shovel)
- Point to the word *roseate* in the first sentence and write it on the board. Explain: **The word *roseate* means "reddish." This bird gets its name from the red color of some of its feathers and from the shape of its beak.**
- Ask: **When does the roseate spoonbill build its nest?** (in November at the beginning of Florida's dry season) **How is that different from most other birds?** (Most birds build their nests in early spring, when the weather is getting warmer.) **How does it help the spoonbill to wait until November?** (There is less rain, so the water level in lakes and rivers is lower. It is easier for them to find food to feed their young.)
- Explain: **In Florida, the temperature stays about the same all year long. But some times of the year are rainier than others.** Remind students that the nesting of the spoonbill is a response to the amount of rain, not temperature.

myNGconnect.com

Have students use the Digital Library to find images of alligators.

Digital Booklet Students can use the information to make a digital booklet with captions.

The birds shown here are roseate spoonbills. Like other birds, they build a nest to get ready to lay eggs. Most birds build their nests in early spring because the weather is getting warmer. Warm weather helps young birds survive. However, spoonbills start building nests in November. This is the beginning of the dry season in Florida, where most spoonbills live. With less rain, the water level in lakes and rivers goes down. That makes it easier for the birds to gather fish and insects to feed their young.
190

Roseate spoonbill chicks peck at food in their parent's bill.

Differentiated Instruction

Extra Support

Have students fold a blank sheet of paper in half. On the left side of the paper, have them draw and label an animal in winter in a northern state. On the right side of the paper, have them draw and label an animal in winter in Florida.

Challenge

Have students research more about their own state's seasons and the changes in plants and animals that accompany them. Have them write a two-page report on their findings. Encourage students to share their findings with the class.

The alligator in the photo copes with dry weather in a different way—by digging gator holes. These holes collect water from the surrounding swamps and marshes. So the holes almost always have water, even during dry weather. Alligators hunt the animals that come to drink from these pools of water.

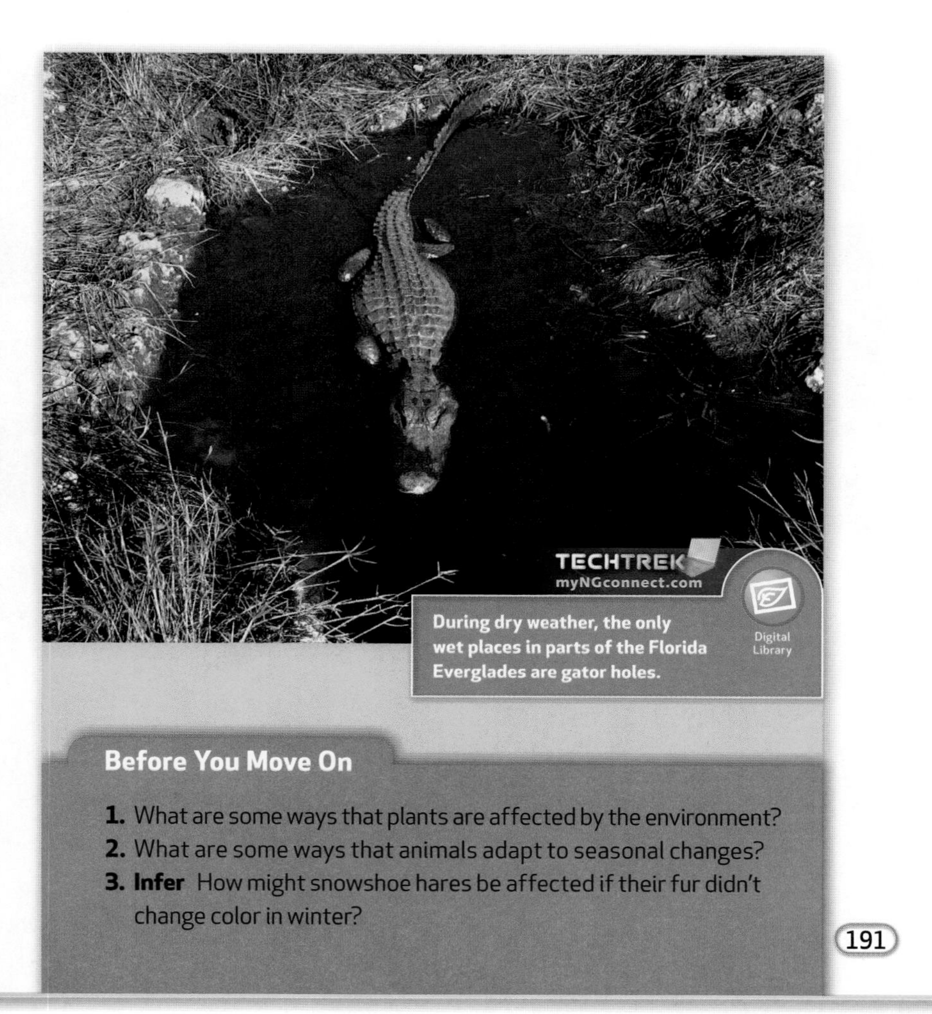

During dry weather, the only wet places in parts of the Florida Everglades are gator holes.

Before You Move On

1. What are some ways that plants are affected by the environment?
2. What are some ways that animals adapt to seasonal changes?
3. **Infer** How might snowshoe hares be affected if their fur didn't change color in winter?

NATIONAL GEOGRAPHIC **Raise Your SciQ!**

Alligators in Cold Weather In the Everglades, temperatures rarely become very cold, and alligators typically remain active all year long. But in places farther north, the bodies of alligators slow down when the weather turns cold. Alligators do not eat when temperatures drop below 23°C (73°F), and their bodies live off reserves of stored food. Alligators can even survive freezing temperatures. They stick their nostrils above the water and let the ice freeze around the rest of their bodies.

Teach, continued

Text Feature: Photos and Captions

- If a map of Florida is available, point out the Everglades. Ask: **How does the weather change with the seasons in the Everglades?** (Some times of the year are rainy, and others are dry.)
- Ask: **What is a gator hole?** (a hole that alligators dig in dry weather and that fills up with water) Ask: **How does digging gator holes help alligators survive?** (Other animals visit gator holes to drink the water, and the alligators can catch and eat them.) Discuss how digging a gator hole would not be useful in the rainy season, when the Everglades animals have plenty of water.

Describe The Effects of Fires and Floods

- Have students describe a forest and the plants and animals that live there. Say: **Fires sometimes occur in dry seasons.** Ask: **What would happen if a fire burned down the forest?** (Many or all of the trees would die, as would many animals that live in the forest.) Discuss how fires can be destructive, yet they also are natural events. Over time, the trees and animals of a forest would return.
- Explain that floods may form when heavy rains fall during wet seasons or when rivers overflow their banks. Ask: **What kind of damage can a flood cause?** (Floods can drown or wash away the plants and animals of a region.) Explain that like fires, floods are natural events. Over time, the environment recovers from a flood.

❸ Assess

» Before You Move On

1. **Recall What are some ways that plants are affected by the environment?** (Possible answers: Flowers bloom in the spring. The leaves of some trees change color during the fall.)
2. **Compare and Contrast What are some ways that animals adapt to seasonal changes?** (Some animals grow thick fur in the winter and shed some fur in the summer. Some animals hibernate in the winter. Some animals migrate to warmer places.)
3. **Infer How might snowshoe hares be affected if their fur didn't change color in winter?** (They would be in greater danger from enemies because they would be easier to see.)

CHAPTER 5
LESSON 4 ▫ Plants Change the Environment

Objectives
Students will be able to:
- Explain how plants can change the environment in many ways.

❶ Introduce

Tap Prior Knowledge
- Ask students to describe the roots of trees, grasses, or other plants that they have observed. Ask if any have seen plants growing out of sidewalks or other pavement.

Set a Purpose and Read
- Read the heading. Tell students that they will read about ways that plants can change the environment.
- Have students read pages 192–195.

❷ Teach

Describe How Plants Impact the Environment
- Ask: **How do plants cause changes to the environment?** (Possible answers: Plants add gas to the air; they keep the soil in place; they help build up the soil; they break apart rocks.)
- Ask: **Why are the gases that plants give off important?** (We need these gases to breathe.) Identify oxygen as the gas that plants add to the air. Explain that all animals, including humans, depend on plants and related organisms to renew the air's supply of oxygen. Plants also take in carbon dioxide from the air.

Plants Change the Environment

Changing the Land The changing seasons show how the environment affects plants and animals. But this cause-and-effect relationship works both ways. Plants and animals also affect the environment. Even something as simple as grass affects the world around you. The grass and other green plants give off gases that we use to breathe. The grass roots also hold the soil in place. The tiny roots form a net that keeps the soil from blowing away or washing away.

Yellow sea oats help keep this sand dune in place.

192

NATIONAL GEOGRAPHIC Raise Your SciQ!

Dune Grasses Hardy grasses often grow on sand dunes. The grasses help anchor the shifting sands from the effects of gravity and strong coastal winds. This allows other plants to grow on the dunes, and they provide shelter for animals.

While grasses help stabilize sand dunes, sand dunes typically are very fragile. On many beaches, laws protect dunes from human activities.

Plant roots not only keep soil in place—they help build it up. The roots trap tiny pieces of rock and decaying plant and animal matter. These materials slowly add to the soil. Roots even break apart rocks. Roots often grow into the cracks of rock. As the roots grow, they push against the sides of the cracks. The cracks widen and split the rock.

Tree roots are splitting this boulder in Rock High Falls Gorge, New York.

193

Teach, continued

Text Feature: Photos and Captions

- Have students observe the photo of the beach on pages 192–193 and then compare the left side of the beach, where the sea oats are growing, to the area of the beach at the water's edge.
- Ask: **How are the sea oats affecting the sand dunes?** (Their roots help hold the dunes in place.) **Why do you think sea oats aren't growing near the water's edge?** (Possible answer: The pounding waves would wash them away.) Discuss how you can sift sand grains through your fingers, but plant roots form a net that trap and hold sand or soil.
- Have students study the photo of the tree and rock on page 193 and then read the caption. Ask: **How is the tree changing the boulder?** (The tree's roots are growing around and inside the boulder, gradually cracking the boulder apart.) Point out that this is similar to what happens when a plant grows in a sidewalk.
- Ask: **How do you predict the boulder will change in the future?** (Possible answer: The tree's roots will continue to grow longer and wider and will continue to crack the boulder apart.)
- Explain that the cracking apart of a boulder may take hundreds or thousands of years. When the tree in the photo dies, new trees or other plants may begin growing in its place. Eventually, the boulder in the photo could be completely broken apart and changed into forest soil.

Make a Cause-and-Effect Chart

Have students make a cause-and-effect chart in their science notebook to show how plants change the environment.

Science Misconceptions

Soil Formation Some students may believe that soil forms independently of plants and other organisms. Explain that the actions of many living things help soil form. Soil is made of tiny pieces of rock and the remains of dead organisms. Plant roots help break apart rocks and other materials in the soil, and decaying plant matter enriches the soil with nutrients that new plants use to grow. Healthy soil is also home to a huge number of bacteria and other microorganisms.

Objectives

Students will be able to:

- Describe how plants can harm an environment.

Science Academic Vocabulary

invasive organism

Teach, continued

Academic Vocabulary: *invasive organism*

- Point out the term **invasive organism.** Reread the sentence to define it.
- Have students describe the **invasive organisms** water hyacinth and kudzu by observing the photos on pages 194–195.

Describe How Plants Can Harm the Environment

- Ask: **Why did people bring water hyacinths to the United States?** (The plants have pretty flowers, and people liked the way they float on water.) **What was the effect of this new plant?** (The plants grew quickly and have completely covered some water surfaces, killing fish and other plants.)
- Say: **Sometimes people's actions can have effects they did not intend. When people brought water hyacinths to the United States, they likely did not imagine the harm the plants would cause.**

Identify Invasive Plants

- Have students observe the photo of kudzu on page 195 and read the caption. Ask: **How is kudzu harming the environment shown here?** (It has grown so much that it is covering nearly all of the ground and the tree trunks, and it seems to be growing without limit.) Discuss how scenes like this one are common throughout the southeast United States.
- Ask: **Why did people first bring kudzu to the United States?** (They thought it would cover bare hills and stop soil from blowing and washing away.) Explain that people did not imagine that kudzu would cause this kind of damage.

Invasive Plants The water hyacinth plants shown below have beautiful flowers. That's why people first brought these plants from South America to the United States in the 1880s. The plants float on water, and people thought they looked nice on ponds. But water hyacinth grows quickly. It can cover a pond or marsh and kill fish and other plants. Water hyacinth is an **invasive organism** —a living thing that doesn't belong in a place and can harm it.

Water hyacinth covers part of the Myakka River in Florida.

The purple-pink flowers of water hyacinth are beautiful, but the plant can harm the environment.

194

NATIONAL GEOGRAPHIC Raise Your SciQ!

Invasive Plants Invasive plants cause environmental problems across the United States. An underwater plant called Eurasian watermilfoil, for example, has spread to many lakes, rivers, and streams. It grows rapidly into large mats that deplete the oxygen supply, choking fish and other aquatic life. Invasive plants on land include Japanese knotweed, purple loosestrife, garlic mustard, and some species of foxtail, all of which replace and threaten native plants.

Another invasive plant is kudzu. People brought kudzu to the United States from Asia. They wanted to use it to cover bare hills and stop soil from blowing and washing away. It worked. But kudzu grows so quickly that it crowds out and covers other plants. Kudzu has spread all over the southeastern United States.

What do you observe about kudzu in this picture?

Before You Move On

1. What is an invasive organism?
2. How do plants help build and protect soil?
3. **Generalize** How do you think kudzu harms other plants when it covers them?

195

Social Studies in Science

The Introduction of Kudzu Kudzu was first introduced to the United States in 1876 at the Philadelphia Centennial Exposition. From the 1930s to the 1950s, the U.S. encouraged farmers in the south to grow kudzu as a way to control soil erosion. By 1953, however, the Department of Agriculture recognized kudzu as an invasive species. Although kudzu was no longer permitted to grow as a cover crop, the plant had already spread widely. Today, kudzu covers millions of acres of land in the southeastern United States.

Teach, continued

Identify Pests

- Ask: **What is a pest?** (something that is annoying or causes harm or damage) **Do you think kudzu is a pest plant? Discuss.** (Possible answer: Yes. It harms the environment, and it is difficult to get rid of.)
- Discuss other pest plants, many of which are discussed in the feature at the bottom of page T194. Have students identify these plants from photos and look for them in the local environment. Students may also look for damage from molds, which typically grow on damp wood or damp leaves.

Monitor and Fix Up

Ask students if anything they read was confusing. Students may be confused by the term **invasive organism**. Discuss possible fix-up strategies. For example, students may rewrite the definition in their own words, study the pictures on pages 194–195, or reread the paragraphs in the text. Looking ahead to the text and photos on pages 200–201 may also help students figure out what an **invasive organism** is.

3 Assess

Before You Move On

1. **Define** **What is an invasive organism?** (An invasive organism is a plant or animal that does not belong in a place and that causes harm to a place.)
2. **Explain** **How do plants help build and protect soil?** (Plants help build soil by breaking down rock and trapping tiny pieces of rock. Plants help protect soil by helping to slow erosion.)
3. **Generalize** **How do you think kudzu harms other plants when it covers them?** (The kudzu blocks sunlight from reaching the other plant. Kudzu also takes water and nutrients from the soil that would go to the other plant.)

LESSON 5 ▫ Animals Change the Environment

Objectives

Students will be able to:

- Explain how animals can change the environment by making homes and changing soil.

PROGRAM RESOURCES

- Enrichment Activities at myNGconnect.com

❶ Introduce

Tap Prior Knowledge

- Read aloud the first sentence on page 196 and ask students to answer. Discuss the equipment used at a construction site and what that equipment does. Ask: **How does a construction project change the land?** (Possible answer: Trees are torn down; soil is paved over; the land is flattened.)

Set a Purpose and Read

- Read the heading. Remind students that they have just read about how plants change the environment. Now they will read about the changes that animals can cause.
- Have students read pages 196–201.

❷ Teach

Describe How Animals Change the Environment

- Have students observe the photo of the beaver lodge on page 196. Emphasize that the photo shows the beaver lodge, not the dam that the beavers built first. Ask: **How did beavers change the environment of this stream?** (By building a dam, they blocked the stream and formed a body of deep water. They also cut down and moved many trees and branches to build both the dam and the lodge.)

Animals Change the Environment
Making Homes Have you ever been to a construction site? You can see how people change the environment when they build homes. Animals change the environment when they build homes too. For example, beavers use their large teeth to cut down trees. They use the trees to build dams across streams. These dams block the flowing water. The water backs up and forms ponds and wetlands. Beavers build their lodges, or homes, in this deeper water.
TECHTREK
myNGconnect.com
Enrichment Activities
This beaver pushes a branch through the water to build its lodge.
Large amounts of branches and mud went into making this beaver lodge in Colorado.
196

Differentiated Instruction

Extra Support

Have students draw and label a beaver building a dam across a stream. Then have them write a caption to describe how the beavers changed their environment.

Challenge

Have students research facts about beavers and their dams and lodges. Students may present their findings in an oral report to the class.

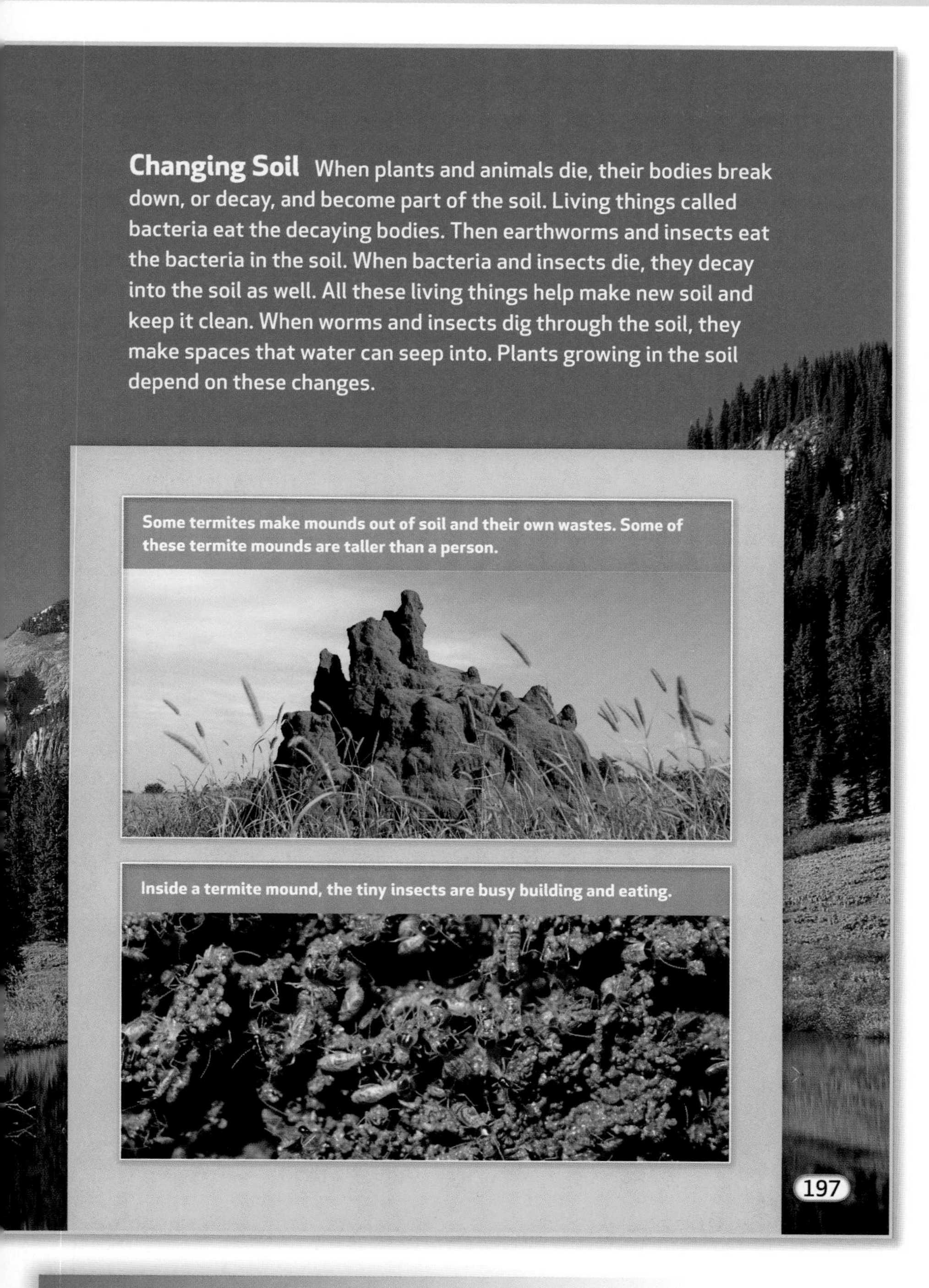

Changing Soil When plants and animals die, their bodies break down, or decay, and become part of the soil. Living things called bacteria eat the decaying bodies. Then earthworms and insects eat the bacteria in the soil. When bacteria and insects die, they decay into the soil as well. All these living things help make new soil and keep it clean. When worms and insects dig through the soil, they make spaces that water can seep into. Plants growing in the soil depend on these changes.

Some termites make mounds out of soil and their own wastes. Some of these termite mounds are taller than a person.

Inside a termite mound, the tiny insects are busy building and eating.

197

Teach, continued

Text Features: Photo and Caption

- Have students observe the photos of termite mounds on page 197. Ask: **How do termites change the soil?** (Termites eat the bacteria in the soil. They also create large mounds out of soil.) Discuss how termite mounds can provide shelter for plants and animals during rainy seasons when the ground becomes flooded.

myNGconnect.com

Have students use the Enrichment Activities to find out more about beavers.

Computer Presentation Students can use the information to make a computer presentation about beavers.

Make Inferences

- Say: **When you *infer*, you make an educated guess about something based on what you know.**
- Ask: **What can you infer would happen if dead organisms did not decay?** (Their bodies would remain in the environment. Nutrients would not return to the soil, and new living things would eventually stop growing.) Discuss how plants need healthy soil to grow in and how animals need plants for food.

Differentiated Instruction

ELL Language Support for How Animals Change Soil

BEGINNING

Have students use the verbs *decay, dig,* and *seep* to complete Academic Language Frames.

- *The remains of plants and animals ____ in the soil.*
- *Earthworms and insects ____ in the soil.*
- *Water ____ into spaces in the soil.*

INTERMEDIATE

Have students write new captions for the photos on page 197 using the verbs *decay, dig,* and *seep.* Remind students that their captions should discuss both the soil and the actions of living things.

ADVANCED

Have students write several sentences that explain in their own words how animals change soil. Have students use the words *first, next,* and *last* to sequence the order of events.

Objectives

Students will be able to:

- Explain how animals can change the environment by grazing and feeding.

Teach, continued

Describe How Animals Change the Environment by Feeding

- Make a two-column chart on the board, and label the headings *Animal* and *Food*. Write *cattle* and *grass* for the first entry, and then have students suggest other examples. Ask: **Which of these animals affect their environments by feeding?** (All of them do.) Discuss how animals and the organisms they eat share the same environment and how animals change that environment by eating other organisms.
- Point to the example of cattle in the chart. Ask: **What if cattle ate all of the grass in a field? What harm would this cause?** (Grass helps protect the soil, and new grass might take a long time to grow again.) Discuss how ranchers try to choose the ideal number of cows that can graze in a field and how they often move cattle from one field to the next to keep the grass and soil healthy.

Feeding Animals affect the environment just by eating. For example, cattle eat grass. Sometimes cattle graze too much in one place. They might dig up the roots or kill the grass and other plants. Plant roots help hold bits of soil together. If plants die, there is nothing to hold the soil in place. Wind and water can easily wear it away.

The picture shows another animal that changes the environment when it eats. Caterpillars eat huge amounts of leaves. They can kill trees and other plants, including crops that people grow for food.

Caterpillars interact with the environment when they feast on tree leaves. How does this interaction change the environment?

198

Differentiated Instruction

Extra Support

Have students draw a picture of an animal eating food in its environment. Suggest cattle or sheep eating grass, a robin catching or eating a worm, a bear eating berries, or examples of students' choice. Then have students write a brief caption that explains how the animal changes its environment by feeding.

Challenge

Encourage students to research examples of animals and the food they eat, and the effects of their feeding on the environment. Interesting choices for research include grasshoppers, boll weevils, and other agricultural pests. Have students write a short paragraph that presents their findings.

If tiny caterpillars can change the environment, imagine what elephants can do! Elephants eat only plants. These huge animals often tear bark and branches off trees or knock them over. But elephants help the environment too. They spread the seeds of trees when they drop their wastes in new places. Elephants use their tusks to dig for water under the ground. Other animals drink from the water holes they make.

These African elephants feed on bark from an acacia tree.

199

Teach, continued

Text Features: Photos and Captions

- Have students observe and compare the photos on pages 198 and 199 and read the captions. Ask: **How are the two photos alike?** (Both photos show animals that are feeding on plants.) **How are they different?** (The elephants are much larger than the caterpillars, and they are tearing the bark off a tree. The caterpillars are small but numerous, and they are eating the leaves of a tree)
- Point to the caterpillars. Ask: **What problem could these caterpillars cause?** (If too many caterpillars eat too many leaves, they could harm or kill the trees or other plants they are feeding on.) Discuss how a tree can live well after losing a few of its leaves but not all of them. Point out that caterpillars can also devastate garden plants when they eat too many leaves.
- Point to the elephants. Ask: **Although these elephants are harming the tree in the photo, how do their actions help other organisms?** (They help spread seeds to new places; they dig water holes that other animals use.)

myNGconnect.com

Have students use the **Digital Library** to find images of elephants and other animals eating.

Computer Presentation Students can use the images to make a computer presentation about how animals affect the environment by eating. You can help them show the presentation on a whiteboard.

NATIONAL GEOGRAPHIC Raise Your SciQ!

Gypsy Moths The gypsy moth, *Lymantria dispar*, is one of the most devastating invasive organisms in North America, especially in the northeast United States. The gypsy moth was introduced from Asia to Massachusetts in the late 1860s and then quickly spread. It commonly feeds on the leaves of aspen and oak trees but will readily eat the leaves of many other plants. A dense population of gypsy moths can quickly strip a tree of all its leaves, which may weaken the tree or kill it directly.

Objectives

Students will be able to:

- Describe how animals can hurt the environment.

Teach, continued

Recognize Problems Caused by Invasive Animals

- Review the definition of *invasive organism* from page 194. Ask: **What is an invasive animal?** (an animal that does not belong in a place and can harm it)
- Have students observe the photo of the nutria on page 200 and read the caption. Ask: **Why is the nutria an example of an invasive animal?** (The nutria is harming wetlands in the United States, where it had not lived until recently.) **How are nutria harming wetlands?** (They eat roots that hold the soil in place. Then the soil washes away quickly.)
- Ask: **Why were nutria brought to the United States?** (People wanted to raise them for fur.) Discuss how invasive animals, like other invasive organisms, may be introduced either on purpose or by accident. In either case, people do not typically expect the damage and problems that the organisms cause.

Identify Animal Pests

- Ask: **Why do you think people describe nutria as pests?** (Possible answer: because nutria destroy wetland plants, which are valuable resources) Discuss how animal pests may use up food, water, or other resources that humans or other animals need. Or pests may be parasites that live off the bodies of valuable plants or animals.
- Discuss examples of animal pests, such as aphids, ticks, zebra mussels, starlings, and mice. Have students observe photos of animal pests and look for them in environments in your community.

Invasive Animals You know that invasive plants can harm the environment. Invasive animals also cause harm. For example, people brought nutria from South America to the United States to raise them for their furs. Nutria have been released into the wild in many places. They eat plants that grow in wetlands. When nutria eat too many plants, no roots are left to hold soil in place. Then the soil washes away quickly.
Nutria can destroy wetlands by eating wetland plants.
200

Social Studies in Science

Damage from Nutria Nutria are voracious eaters of marsh grasses and tree seedlings, and they have caused serious harm to wetlands across the United States. In a cypress swamp in Louisiana, scientists found that nutria destroyed between 86 to 100 percent of cypress seedlings that had been planted by hand, even when protective devices had been used. Nutria also carve out large underground tunnels. Their burrowing has weakened the sides of drainage canals and levees, sometimes causing them to collapse.

The picture shows an invasive animal in Florida that is harmful in many ways—a Burmese python. This huge snake comes from the rainforests of southeast Asia. Some people keep pythons as pets. But when the snakes get too large, people sometimes release them into the wild. This is a problem. Like most invasive species, pythons have no natural enemy in their new home. So they live to reproduce and grow in large numbers. Hundreds of pythons have been captured. Scientists are looking for ways to control this and other invasive organisms.

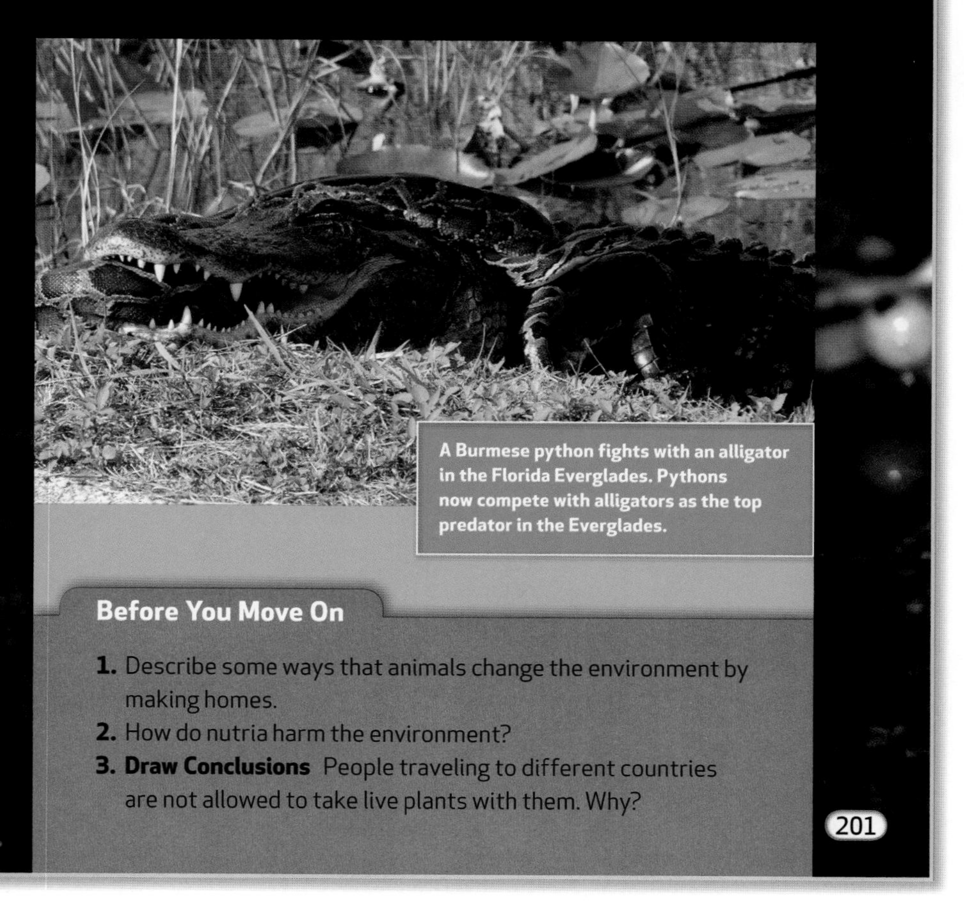

A Burmese python fights with an alligator in the Florida Everglades. Pythons now compete with alligators as the top predator in the Everglades.

Before You Move On

1. Describe some ways that animals change the environment by making homes.
2. How do nutria harm the environment?
3. **Draw Conclusions** People traveling to different countries are not allowed to take live plants with them. Why?

201

Raise Your SciQ!

NATIONAL GEOGRAPHIC

Zebra Mussels Zebra mussels, which are native to Russia, have caused great harm in lakes and rivers throughout the world. They became established in North America during the 1980s, when cargo ships accidentally dumped them into the Great Lakes with their ballast water. Zebra mussels eat huge amounts of plankton, which is the base of the food chain for many fish and other marine life. As the population of zebra mussels increases, the native animals lose more and more of their food supply. The rapid and dense growth of zebra mussels also fouls boat bottoms, docks, and water-intake pipes.

Teach, continued

Describe How Animals Can Harm the Environment

- Have students study the photo of the python and alligator on page 201 and then read the caption. Ask: **What is a predator?** (an animal that hunts and kills other animals for food) Discuss how the top predator is the predator that no other animals hunt. In the Everglades, alligators were definitely the top predators until Burmese pythons arrived.
- Ask: **How are Burmese pythons affecting the alligators of the Florida Everglades?** (The pythons are competing with the alligators, and they fight.) Explain that the pythons are likely eating food that alligators might eat otherwise.
- Ask: **Why do you think scientists have not been able to remove all the Burmese pythons from the Everglades?** (Possible answers: The Everglades are large and swampy, and pythons can hide easily in the tall grasses. Python eggs and young snakes may be especially difficult to find.)
- Ask: **Why are Burmese pythons an example of an invasive animal in the Everglades?** (They do not belong there, and they are harming the animals that do belong there.)

3 Assess

» Before You Move On

1. **Describe** **Describe some ways that animals change the environment by making homes.** (Beavers change the environment when they cut down trees and build dams and lodges. They block or change the flow of water in waterways. Termites build huge mounds.)
2. **Cause and Effect** **How do nutria harm the environment?** (Nutria eat too many plants, so there are not enough plants to hold the soil in place. The soil erodes quickly.)
3. **Draw Conclusions** **People traveling to different countries are not allowed to take live plants with them. Why?** (If people brought live plants into a country, the plants might become invasive organisms there.)

CHAPTER 5

LESSON 6 ▫ People Change the Environment

Objectives

Students will be able to:

- Explain that humans can change the environment in beneficial or harmful ways.
- Recognize that humans depend on their environments to meet their needs.

PROGRAM RESOURCES

- Science Inquiry and Writing Book: *Life Science*
- Science Inquiry and Writing Book eEdition at myNGconnect.com

❶ Introduce

Tap Prior Knowledge

- Ask: **What is an engineer?** (a person who designs or builds something) **What might engineers design or build?** (Possible answers: highways, buildings, and dams) Discuss how these structures are often longer-lasting than structures that other living things build or make.

Set a Purpose and Read

- Read the heading. Say: **Plants and animals can change the environment, but people can change it much more.** Have students read pages 202–209.

Discuss How People Can Change the Environment

- Explain that plants and animals can both cause and be affected by changes to the environment. Ask students if they have ever fed birds, squirrels, or other wild animals. Ask: **How do you think feeding a wild animal changes the way it lives?** (Possible answer: The animal gets food, but it may come to rely on people instead of natural sources for its food.) Discuss how people's actions can affect wild animals in many ways, both helpfully and harmfully. Sometimes it's not easy to predict whether an action is helpful or harmful.
- Ask: **Do you think people help or harm animals when they hunt and fish?** (Possible answer: They harm the animals that are killed.) Explain that people definitely hurt the environment when they hunt or fish too many of one kind of animal. However, limited hunting and fishing can help keep animal populations from becoming larger than the environment can support.

People Change the Environment

Using Natural Resources Plants and animals change the environment in some major ways. One kind of living thing, however, causes the most change—people.

People change the environment when they use natural resources. These include air, water, land, plants, and animals. People used lots of rocks and other land resources to make the concrete for this dam. How do you think building the dam changed the environment?

This part of the Columbia River became flooded when the Grand Coulee Dam was built.
The Grand Coulee Dam blocked the Columbia River, which backed up to form Lake Roosevelt.
202

Social Studies in Science

Grand Coulee Dam The Grand Coulee Dam opened in 1942 and has been renovated and expanded several times since then. It is one of the world's largest producers of hydroelectricity, or electricity from the flow of water. However, despite fish ladders and other measures, its construction dramatically reduced the population of salmon in the upper Columbia River. This greatly altered the culture and livelihood of the Native Americans of the region. The United States Government eventually compensated the Colville Indians for millions of dollars of lost income.

People use water behind the dam to make electricity. Water rushes through tunnels in the dam. The force of the water spins large magnets near wires to make electricity.

The dam has been helpful to people but harmful to other living things. A dam causes the river behind it to back up and flood the land. Plants and animals that lived near the river lost their homes. Dams in the northwestern United States sometimes block salmon from swimming upstream to lay their eggs. So people build fish ladders that go around the dams. The fish ladders help make sure plenty of salmon can live and grow in the wild.

A fish ladder is a chain of pools built like steps so fish can bypass a dam. The fish leap out of one pool and into the next.

203

NATIONAL GEOGRAPHIC **Raise Your SciQ!**

Changes from Dams Dams can change the environment far away from their location on a river. One reason is because dams block the flow of sediment down a river. In rivers without dams, sediment often forms sandy beaches along the river's shore and the coastline of the ocean where the river empties. The sediment also enriches the river valley when the river floods. Because of dams on the Nile River in Egypt, beaches of the Mediterranean Sea have become seriously eroded. In the valley around the Nile, farmers now depend on fertilizer to grow crops.

❷ Teach

Analyze Environmental Impact

Science Inquiry and Writing Book: *Life Science*, page 15

Read and follow the instructions together with students. Guide students to understand the way living things in their natural habitat interact and how humans impact the environment when they build structures. Students should record their observations and analysis in their science notebook.

In this environment, the trees help anchor the soil on the hill and prevent it from falling into the lake. The big horn sheep graze on plants in the environment. Humans built a highway in this environment, and the sheep strayed onto it. The highway reduced the food available to the sheep, and the cars may endanger them.

Compare Changes to the Environment

- Have students compare the photos of the beaver lodge on page 196 with the photo on pages 202–203. Ask: **How are the dams and lodges that beavers build like the Grand Coulee Dam shown?** (Both change the flow of water, and both affect the plants and animals of the environment.) **How are they different?** (The Grand Coulee Dam is much larger, will last longer, and has a much greater impact on the environment.)
- Point out that the inset photo shows the river before the dam was built. Ask: **How did the Grand Coulee Dam change the Columbia River?** (Before, the Columbia River flowed across the countryside. The dam blocked the river's flow, and now there is a lake instead of a river valley.)

Describe The Effects of Floods on the Environment

- Explain that a river causes a flood when it overflows its banks. Dams and levees on a river may prevent floods along some sections but make them worse elsewhere. Ask: **What kind of damage can a flood cause?** (Floods can drown or wash away the plants and animals of a region.)
- Ask: **After the flood ends, how do you predict the land will change?** (Possible answer: Over time, the plants and animals will return.)

Objectives

Students will be able to:

- Explain that humans can change the environment in beneficial or harmful ways.
- Recognize that humans depend on their environments to meet their needs.

Science Academic Vocabulary

recycle, conserve

Teach, continued

Academic Vocabulary: *recycle, conserve*

- Point out the term **recycle** and write it on the board. Have a student read the definition. As a class, list and record materials that can be **recycled**, such as glass, plastic, metal, and paper.
- Point out the term **conserve** and write it on the board. Have a student read the definition.
- Ask: **What are some ways we can conserve resources?** (Possible answers: To conserve gasoline, we can walk, ride a bicycle, or use public transportation instead of an automobile. To conserve electricity, we can turn off lights and computers when they are not in use.) List students' responses on the board and discuss them.

Identify Major Land Uses

- Point to the photos of the two housing developments. Say: **These photos each show a suburban community.** Describe a suburban community as near a dense city, or urban community.
- Ask: **How do people use the land in suburban communities?** (for houses with yards, as well as businesses and roads)
- Ask: **How is land used in a city or urban community?** (It is packed tightly with houses, apartment buildings, office buildings, and factories.) **How is land used in a rural, or farming, community?** (Much of the land is used for farms and ranches.) Discuss how in all three kinds of communities, people use the land for their own purposes, and they change the communities of native plants and animals.

Building Homes and Roads People often change the environment to meet their needs. Think about building homes and roads. This construction usually harms the environment in some way. People cut down forests to make room for buildings and roads. They use wood to make furniture and paper. Many animals lose their homes when people cut down forests. Trees can grow back, but not as quickly as they are cut down.

How did building these homes change the environment?

204

NATIONAL GEOGRAPHIC Raise Your SciQ!

Uses for Recycled Materials Some materials, including metals and glass, can be recycled over and over again in the same products. The aluminum in a soft drink can, for example, can be readily cut, melted, and used to make new cans. Other recycled materials are put to new uses. Recycled plastic is often used in building materials. Recycled rubber from automobile tires is used to pave roads and playgrounds. Have interested students research how the materials that the school recycles are collected, processed, and used.

Some people are trying to cause less harm to the environment by building "green homes." These homes are made with **recycled** materials. To recycle is to use something again. Green homes are often built using recycled tires or wood taken from old buildings. This helps **conserve** natural resources. To conserve is to use something carefully and not waste it. Conserving saves trees and other resources. Conserving makes the environment healthier for all living things.

This green housing development in London, England, uses a lot of recycled materials.

205

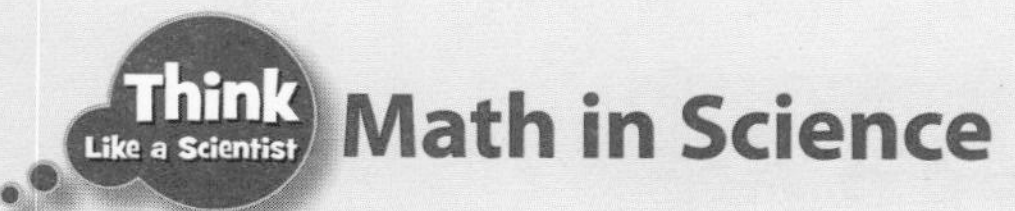

Math in Science

Recycling Survey With the permission of other teachers or school officials, have students conduct a survey around the school about recycling habits. Students may ask survey participants whether they regularly recycle paper, glass, plastic, and metal objects they use at school. Have students pool their survey results, tabulate them, and present them in a chart or graph. As a class, discuss what the results show about the students' recycling habits.

Teach, continued

Describe the Effects of Fires

- Point to the forest on page 204. Explain that both human actions and natural events, such as fires, affect forests. Ask: **What would happen if a fire burned down the forest?** (Many or all of the trees would die, as would many animals that live in the forest. The houses might burn down, too.)
- Ask: **What do you predict would happen to the land after the fire?** (Possible answer: Over time, the trees and animals would return.)

Describe the Effects of Construction

- Have students observe the photo on page 204 and suggest answers to the caption question. Discuss how the houses and road replaced land that was home to trees and forest animals. Grasses and other plants not found in the forest will cover the bare soil.
- Ask: **What do you think happens when old, unwanted houses or roads are torn apart and hauled away?** (Possible answer: Plants and animals gradually return to the land.) Explain that people sometimes do change developed lands into "green spaces" for plants and animals.

Describe Ways to Conserve Resources

- Ask: **What are "green homes?"** (houses made of recycled materials instead of new materials)
- Ask: **How does the green home differ from other houses?** (The green home is larger, and most of its windows are on one side.) Explain that in the green home, many families share the same building; windows form nearly all of one side of the building and are absent from other sides. Discuss how the design of the windows helps **conserve** the energy needed to keep the house warm or cool.

Discuss the Need for Resources

- Have students list activities that they do every day, such as washing and bathing, eating food, traveling to and from school, and using school supplies, sports equipment, and other manufactured products. List their responses on the board.
- Ask: **Which of these activities depends on resources from the environment?** (all of the activities) Discuss how fuel, water, food, and manufactured products all depend on Earth's resources.

Objectives

Students will be able to:

- Explain that humans can change the environment in beneficial or harmful ways.
- Recognize that humans depend on their environments to meet their needs.

Teach, continued

Explain How Farming Practices Can Change the Environment

- Say: **Farming means growing crops and raising animals.** Ask: **How can farming practices harm the environment?** (They can cause soil erosion, use too much water, or harm soil and water with chemicals.)
- Ask: **Why are chemicals used on crops?** (to kill insects and weeds; to help the crops grow better) Ask: **Why can the chemicals cause problems?** (The chemicals stay in the soil and spread into the water supply, which can carry them to lakes and streams.) Discuss how water dissolves or carries many chemicals and can carry them far from the places where they were introduced into the environment.

Farming People also change the environment when they grow crops and raise animals. For example, some farming methods cause more soil erosion than others. Some methods use more water than others. When a lot of water is taken from rivers and lakes to grow crops, there is less for other plants and animals. Many farmers use chemicals on their crops to kill insects and weeds. These chemicals get into the soil. Rain washes the chemicals, as well as animal wastes, into streams and lakes. The chemicals and wastes can harm plants and animals.

Water that washes off this farm erodes the soil.

Contour farming slows soil erosion on this farm in Missouri.

206

Raise Your SciQ!

Environmentally Friendly Farming Practices Both farmers and consumers can make choices that help the environment. For example, some farmers are choosing to raise coffee plants in the shade of forests, the environment where coffee plants grow in nature. This is better for the environment than clearing forests to raise coffee in open fields. Consumers can support environmentally friendly practices with their food purchases. Consumers can also buy foods that are raised or produced locally. This conserves the fuel needed to transport foods across long distances.

Some people farm in ways that do less harm to the environment. Some farmers use contour farming to reduce soil erosion. In contour farming, farmers grow crops in rows that run across a hill instead of up and down it. You may have seen signs for organic foods in grocery stores. Organic farmers do not use chemicals to kill pests or grow their crops. Instead, they use natural substances, plants, animals, or other living things to keep their crops healthy.

People at this organic farmers' market are buying fresh tomatoes grown on nearby farms.

207

Teach, continued

Text Feature: Photos and Caption

- Have students observe the photo of contour farming on pages 206–207 and read the caption. Ask: **How are the rows of plants arranged in this field?** (They are curved and overlap, and they follow the general shape of the land.) Discuss how this field is an example of contour farming.
- Ask: **Why is contour farming a wiser practice than plowing in straight, parallel rows?** (Contour farming reduces soil erosion.) Explain that straight, parallel rows of crops are easier to plant and harvest with machines, but contour plowing is often better for the land.
- Have students observe the photo of the organic farmers' market on page 207. Ask: **How do the vegetables at this farmers' market compare to vegetables you might buy elsewhere?** (The vegetables appear fresh and good to eat but in other ways may be much like vegetables from other sources.) Discuss how many people prefer organic foods because they believe they taste better and are healthier to eat, as well as for their benefits to the environment.

Monitor and Fix Up

Ask students if everything they have read so far has made sense. Students may be confused by the terms *contour farming* or *organic farming*. Discuss possible fix-up strategies. For example, students may reread the paragraph in the text, study the picture on pages 206–207, draw an example of each situation, or find out more about these methods in the library or on the Internet.

Differentiated Instruction

ELL Language Support for Describing Farming Practices

BEGINNING

Provide Academic Language Frames to help students describe farming practices.

- *____ farming reduces soil erosion.*
- *Some farmers use ____ to kill pests.*
- *____ farming does not use chemicals to kill pests.*

INTERMEDIATE

Have students use these Academic Language Frames to write sentences describing farming practices.

- *Some farmers use ____ to ____.*
- *____ farming does not use ____.*

ADVANCED

Have students use these Academic Language Stems to describe farming practices.

- *Some farming methods cause . . .*
- *Some farmers use . . .*

Objectives

Students will be able to:

- Observe that humans can change the environment in beneficial or harmful ways.
- Explain that humans depend on their environments to meet their needs.

Science Academic Vocabulary

pollution

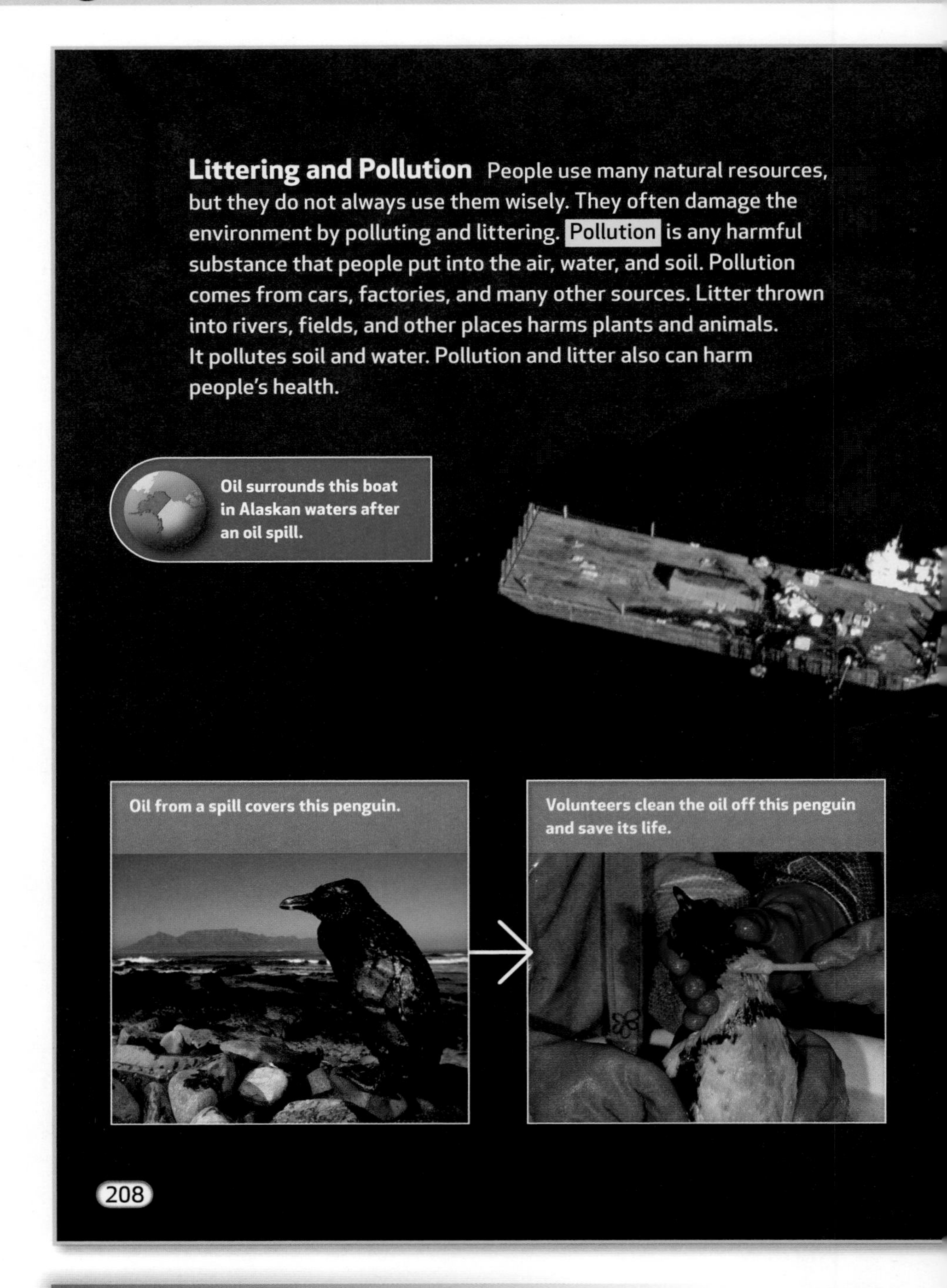

Littering and Pollution People use many natural resources, but they do not always use them wisely. They often damage the environment by polluting and littering. Pollution is any harmful substance that people put into the air, water, and soil. Pollution comes from cars, factories, and many other sources. Litter thrown into rivers, fields, and other places harms plants and animals. It pollutes soil and water. Pollution and litter also can harm people's health.
Oil surrounds this boat in Alaskan waters after an oil spill.
Oil from a spill covers this penguin.
Volunteers clean the oil off this penguin and save its life.
208

Teach, continued

Academic Vocabulary: *pollution*

- Point out the term **pollution.** Reread the sentence to define it.
- Have students observe the photos on pages 208–209 and then describe oil **pollution** in their own words.

Describe How Littering and Pollution Can Harm the Environment

- Ask: **What are people doing to reduce littering and pollution?** (finding new ways to power cars and factories; increasing recycling; forming groups to pick up litter from the land and water)
- Have students read the question that concludes the text on page 209, and then invite them to answer. (Possible answers: avoid littering, pick up litter, use less electricity and gasoline)

Differentiated Instruction

Extra Support

Have students make a three-column chart for air, water, and soil pollution. Have students draw and label pictures for each column.

Challenge

Have students write a paragraph that describes one kind of pollution, how it happens, and how they think people can prevent it. Students may also illustrate their paragraphs and read them to the class.

Today, more people understand the ways in which they harm the environment. People are working to stop pollution. Burning coal, oil, and natural gas causes pollution. So scientists are trying to find new ways to run cars and factories. Many cities have set up recycling programs. People get together in groups to pick up litter safely from beaches, rivers, and roads. What can you do to help the environment?

Science in a Snap! What Can You Recycle?

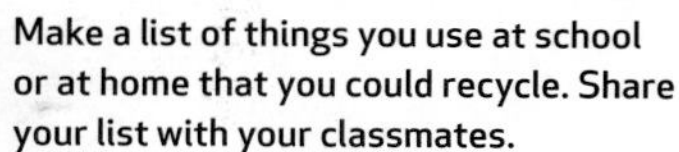

Make a list of things you use at school or at home that you could recycle. Share your list with your classmates.

Choose one of the items. Make a plan to recycle it.

How will recycling this item help the environment?

Before You Move On

1. What are some ways that people change the environment?
2. How is organic farming different from other kinds of farming?
3. **Apply** Suppose you want to help the environment by changing how you get your food. Describe what you might do.

Teach, continued

Describe Ways to Conserve Resources

- Ask: **Why do people want and need oil?** (Oil is used to make gasoline, which powers cars and trucks.) Discuss how Earth's oil is often located in remote regions, such as the Arctic Ocean.
- Ask: **Why do you think it is important to conserve oil?** (Possible answer: Oil supplies are limited, and once they are used they cannot be used again. Oil also causes pollution.)
- Ask: **How are scientists working to conserve oil?** (They are researching other resources to power cars and factories.) Explain that scientists are studying new ways to get energy from the sun, winds, tides, and Earth's hot interior.

3 Assess

» Before You Move On

1. **List What are some ways that people change the environment?** (when they use natural resources, build homes and roads, farm, and litter and pollute)
2. **Compare and Contrast How is organic farming different from other kinds of farming?** (Organic farmers use natural substances instead of chemicals.)
3. **Apply Suppose you want to help the environment by changing how you get your food. Describe what you might do.** (I could grow some of my own food or buy food from local growers. I could buy organic foods.)

Science in a Snap!

What Can You Recycle?

Materials paper, pencil or pen

Students' plans might involve picking up litter outdoors or setting up bins for recycling around the school or community. Students may present their plans on an illustrated poster.

What to Expect Students' plans for recycling should include ways to involve classmates or community members in recycling metal, glass, plastic, or paper products that would otherwise be discarded.

Quick Questions Ask students the following questions:

- **Why is it important to recycle?** Recycling helps conserve the raw materials and energy that are used to make many products. It also helps reduce the amount of trash that people throw away.
- **Why is it important for everyone in the community to recycle, not just a few people?** Recycling a few things won't make much of a difference. But recycling the huge number of things used in a community can help improve the environment.

Objectives

Students will be able to:

- Explain that humans can change the environment in beneficial or harmful ways.

❶ Introduce

Tap Prior Knowledge

- Ask students to define the term *invasive organisms* in their own words. Have them name some invasive organisms that they studied earlier in this chapter. Discuss how the concept of invasive organisms may seem bad or threatening, but that doesn't mean that invasive organisms can't be put to good use.

Preview and Read

- Read the heading. Then do a picture walk of pages 210–213. Tell students that they will read about how invasive organisms are being used in helpful ways.
- Have students read pages 210–213.

❷ Teach

Explain How Invasive Plants Can Be Controlled

- Say: **Invasive organisms can be harmful to the environment.** Ask: **How are people in Virginia using animals to remove invasive plants?** (They are having goats eat the invasive plants.)

NATIONAL GEOGRAPHIC
USING INVASIVE ORGANISMS TO HELP THE ENVIRONMENT
Invasive plants and animals often damage the environment. Invasive organisms usually spread very quickly, so they are difficult to remove. One way to solve the problem is to put invasive organisms to work! For example, people in Virginia are using invasive plants to feed goats. The goats graze on invasive plants such as Asiatic bittersweet, multiflora rose, and some kinds of honeysuckle.
Asiatic bittersweet is native to Asia. It is a vine that can pull down trees and crowd out other plants.
210

NATIONAL GEOGRAPHIC Raise Your SciQ!

Controlling Invasive Plants Scientists have investigated many methods for controlling invasive plants. Using goats is a form of biological control, a management method that uses organisms to remove plant material, kill plants, or limit plant growth. Another method is mechanical control, such as physically uprooting unwanted plants from an environment. In chemical control, herbicides or other chemicals are used to kill or slow the growth of unwanted plants.

The goats may have to graze different parts of the land a few times before they kill the invasive plants. It might take several years before the plants are gone. In the meantime, the goats have a new source of food. An added bonus: wastes from the goat help improve the soil.

Goats are known for eating almost anything. Here, they help get rid of invasive plants by grazing on them.

211

Teach, continued

Social Studies Link

- Show students a map of the United States and have them locate Virginia. Then have them locate your state and compare the two locations. Ask: **Could an invasive plant in Virginia affect our state someday? Why or why not?** (Possible answer: Yes. Seeds can travel easily from place to place.) Explain that different plants grow best in different kinds of environments. If your state and Virginia have similar climate and soil, then the same invasive plants could affect them both.

Recognize the Relationship Between Goats and the Environment

- Ask: **Why are goats a good choice for removing invasive plants?** (because goats eat almost everything) Discuss how goats do not necessarily prefer invasive plants to other plants but that they will eat almost all plant matter that is available to them.
- Ask: **How else are goats helpful to the environment when they graze on invasive plants?** (The wastes from the goats help improve the soil.) Explain: **Wastes from the goats add nutrients back to the soil.**

Make a Cause-and-Effect Chart

Have students make a cause-and-effect chart in their science notebook for how goats can help improve the environment.

Science Misconceptions

Harmful Plants? Some students may think that invasive plants are inherently more dangerous or harmful than other plants. Remind them that invasive plants, like other invasive organisms, are natural parts of ecosystems in other parts of the world. They become harmful only when they arrive in a new ecosystem, such as Asiatic bittersweet arriving in Virginia, where they have no natural checks on their growth.

Objectives

Students will be able to:

- Explain that humans can change the environment in beneficial or harmful ways.

PROGRAM RESOURCES

- Learning Masters Book, page 60, or at myNGconnect.com

Teach, continued

Describe How Invasive Organisms Can Be Helpful

- Say: **"To harvest" means to collect.** Ask: **Why would people want to harvest invasive organisms?** (Possible answer: to find out if the organisms could be helpful in some way)
- Ask: **How might saltcedar be useful?** (It might be useful for building homes.) **How might the red lionfish be used?** (It might be used for food.) Remind students that the term *invasive* does not necessarily mean bad or dangerous, at least not in all ways.

Harvesting Invasive Organisms In the southwestern United States, saltcedar is an invasive plant that crowds out native trees and bushes. Scientists are testing saltcedar to see if it would be useful in building homes. If it is, people could use less native wood to build homes and get rid of an invasive plant all at once!
Saltcedar is native to Europe, Asia, and Africa. People brought it to the United States to add beauty to the land.
212

Think Like a Scientist Math in Science

Doubling in Number Invasive organisms cause damage because they typically reproduce quickly and few or no organisms harm them in the environment. To model one example, describe an invasive plant that makes seeds every month. If 2 seeds grow into new adult plants that make seeds, then the population changes from 1 plant in the first month, to 2 plants in the second month, to 4 plants in the third month, and so on. After 2 years of unchecked growth, the plant population would be over 16 million plants!

Invasive organisms damage underwater environments too. Some invasive fish crowd out native fish by eating them or their food. But what if you could turn invasive fish into food for people?

That is what some people are trying to do with the red lionfish. This fish comes from the western Pacific Ocean. It is an invasive animal in the Florida Keys and the Caribbean Sea. Lionfish damage coral reefs and the organisms that live there. People have started catching red lionfish and serving them at restaurants. It is hoped that as this new source of food becomes more popular, the numbers of lionfish in the Caribbean will start going down.

These poisonous spines protect the red lionfish. Few other fish can eat them.

213

Teach, continued

Text Feature: Photos and Captions

- Have students observe the photo of saltcedar on page 212 and read the caption. Ask: **What does saltcedar look like?** (a spindly plant with flowers) Discuss how the plant is attractive, which is why people planted it. Unfortunately, the plant proved to be invasive.
- Ask: **Why has saltcedar become harmful?** (It grows and spreads rapidly, taking over much of the land and replacing or harming native plants.) Explain that while many people might consider the saltcedar to be a beautiful addition to the land, it is also causing damage.
- Have students observe the photo of the red lionfish on page 213 and read the caption. Ask: **What does the red lionfish look like?** (It is a colorful, attractive fish that has many spines; it may look a little like a lion.) Describe how the lionfish's attractive appearance makes it popular for aquariums.
- Ask: **Why will humans eating the red lionfish help the environment of the Caribbean?** (Lionfish are invasive animals of the Caribbean. By eating them, humans can reduce their number and help the native fish.)

3 Assess

1. **Describe How does the red lionfish affect its environment?** (It damages coral reefs and the organisms that live there.)
2. **Cause and Effect How can goats help improve the environment?** (Possible response: Goats eat invasive plants. This helps native plants. The goats' waste improves the soil, which also helps native plants grow.)
3. **Analyze How can invasive organisms be helpful?** (Possible response: People can use the invasive organisms instead of native organisms. This helps conserve native organisms.)

Share and Compare

my SCIENCE notebook

Examples of Environmental Change Give students the Learning Master. Have them find examples of how plants and animals change with the seasons and how they change the environment. Have students record these examples in the chart. Check their work, and then have them write a summary statement about the different ways that the environment can change. Have partners discuss their work.

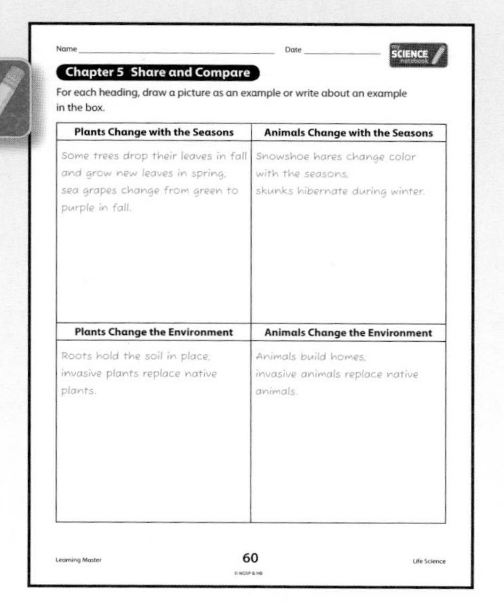
Name ______ Date ______ my SCIENCE notebook

Chapter 5 Share and Compare

For each heading, draw a picture as an example or write about an example in the box.

Plants Change with the Seasons	Animals Change with the Seasons
Some trees drop their leaves in fall and grow new leaves in spring; sea grapes change from green to purple in fall.	Snowshoe hares change color with the seasons; skunks hibernate during winter.
Plants Change the Environment	**Animals Change the Environment**
Roots hold the soil in place; invasive plants replace native plants.	Animals build homes; invasive animals replace native animals.

Learning Master 60 Life Science

© NGSP & HB

Learning Master 60 or at myNGconnect.com

Objectives

Students will be able to:

- Investigate through Guided Inquiry (answer a question; make and compare observations; collect and record data and observations; generate explanations and conclusions based on evidence; share findings; ask questions based on observations to increase understanding; adjust explanations based on findings and new ideas).
- Recognize the characteristics of a fair and unbiased test and the importance of keeping conditions the same.
- State, orally and in writing, any inferences or generalizations indicated by the data collected.

Science Process Vocabulary

investigate, infer

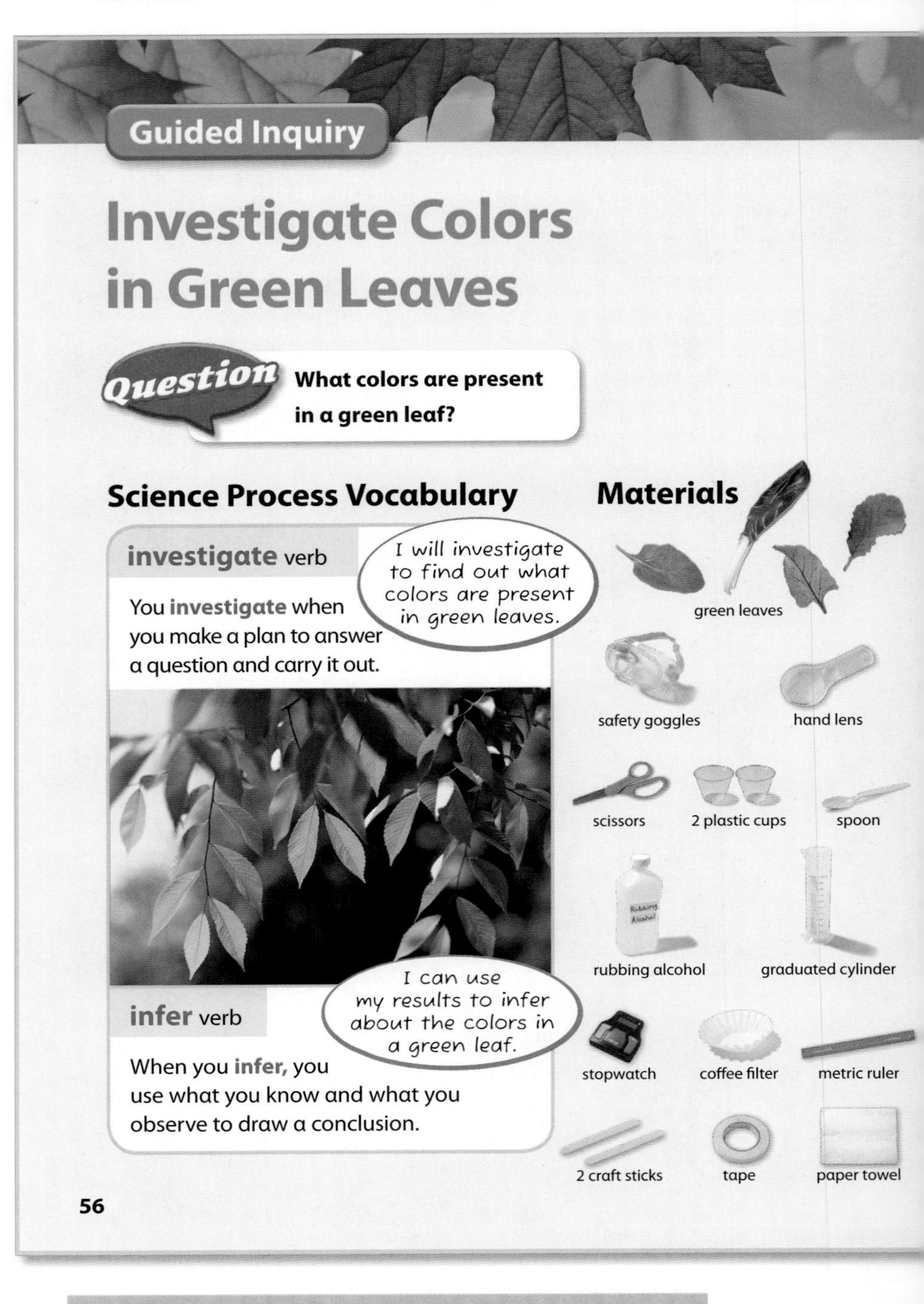

Guided Inquiry

Investigate Colors in Green Leaves

Question What colors are present in a green leaf?

Science Process Vocabulary

investigate verb

You **investigate** when you make a plan to answer a question and carry it out.

I will investigate to find out what colors are present in green leaves.

infer verb

When you **infer,** you use what you know and what you observe to draw a conclusion.

I can use my results to infer about the colors in a green leaf.

Materials

green leaves
safety goggles
hand lens
scissors
2 plastic cups
spoon
rubbing alcohol
graduated cylinder
stopwatch
coffee filter
metric ruler
2 craft sticks
tape
paper towel

56

PROGRAM RESOURCES

- Science Inquiry and Writing Book: *Life Science*
- Science Inquiry and Writing Book eEdition at myNGconnect.com
- Inquiry eHelp at myNGconnect.com
- Science Inquiry Kit: *Life Science*
- Learning Masters Book, pages 61–64, or at myNGconnect.com
- Inquiry Rubric: Assessment Handbook, page 186, or at myNGconnect.com
- Inquiry Self-Reflection: Assessment Handbook, page 198, or at myNGconnect.com

MATERIALS

Kit materials are listed in italics.

green leaves; safety goggles; *hand lens;* scissors; *2 plastic cups (9 oz); plastic spoon;* rubbing alcohol; *graduated cylinder (100 mL); stopwatch; coffee filter;* metric ruler; *2 craft sticks;* masking tape; paper towel; colored pencils (optional)

MANAGING THE INVESTIGATION

Time

40 minutes and time for observations after 1 hour

Groups

Small groups of 4

Advance Preparation

- Provide and label a container of rubbing alcohol. Each group will need about 100 mL.
- Collect leaves from several types of deciduous trees, such as oak, elm, hickory, maple, and sycamore. You may also use green leaves from produce, such as spinach or chard.
- Have paper towels on hand to wipe up spills.

❶ Introduce

Tap Prior Knowledge

- Ask: **Have you seen leaves change color?** Some areas of the country experience dramatic seasonal changes. Students may have observed leaves changing color in the fall or leaves that turn yellow as they wilt.

Do an Experiment

Write your plan in your science notebook.

Make a Hypothesis

In this investigation, you will separate the colors in a green leaf. What colors will you observe? Write your **hypothesis.**

Identify, Manipulate, and Control Variables

Which variable will you change?
Which variable will you observe or measure?
Which variables will you keep the same?

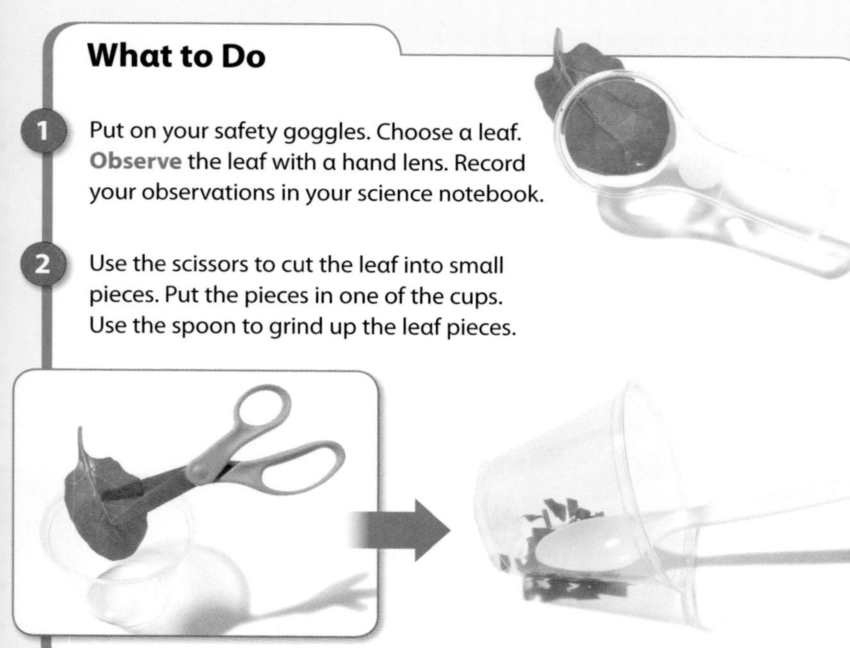

What to Do

1. Put on your safety goggles. Choose a leaf. **Observe** the leaf with a hand lens. Record your observations in your science notebook.

2. Use the scissors to cut the leaf into small pieces. Put the pieces in one of the cups. Use the spoon to grind up the leaf pieces.

57

Teaching Tips

- In step 2, instruct students to twist or roll up the leaf before they cut it. If students cut the pieces to be very small and grind them up well, they will get clearer results.
- Dispose of the rubbing alcohol and the plastic cup in accordance with federal, state, and local requirements.

What to Expect

The rubbing alcohol and leaf solution will turn green, and the plain rubbing alcohol will not change color. Different leaves will yield different colors on the strips, but most leaves contain green, yellow, and orange.

Introduce, continued

Connect to the Big Idea

- Review the Big Idea Question, *How do living things interact with their environment?* Explain to students that this inquiry will help them understand that leaves change color in certain conditions.
- Have students open their Science Inquiry and Writing Books to page 56. Read the Question and invite students to share ideas about the colors present in green leaves.

2 Build Vocabulary

Science Process Words: investigate, infer

Use this routine to introduce the words.

1. **Pronounce the Word** Say **investigate.** Have students repeat it in syllables.
2. **Explain Its Meaning** Choral read the sentence. Ask students for another word or phrase that means the same as **investigate.** (perform tests or experiments to find an answer)
3. **Encourage Elaboration** Ask: **What are some parts of an investigation?** (making a plan, making a hypothesis or prediction, making and recording observations, and drawing conclusions)

Repeat for the word **infer.** To encourage elaboration, ask: **Can you tell the class about an inference you made after doing an investigation?** (Answers will vary, but make sure students understand the meaning of an inference.)

3 Guide the Investigation

- Ask students to describe some of the colors they may have seen in leaves as they wilt or change in the fall. Students may use this prior experience to support their hypotheses.
- Ask students to brainstorm a list of variables in this investigation, such as type of leaf and time for leaves to remain in the alcohol. Discuss which variables to change, observe, and keep the same.
- Discuss the importance of keeping conditions the same so that comparisons can be fair and unbiased. Ask: **What do you think would happen if the conditions of each cup were not the same?** (There might be different results.)

Guide the Investigation, continued

- Distribute materials. Read the inquiry steps on pages 57–58 together with students. Move from group to group and clarify steps if necessary.
- In step 1, encourage different groups to choose different leaves to test.
- After step 5, ask: **How do you think the leaf cup will look after one hour? Will the plain cup look different?** (Predictions will vary.)
- In step 6, provide colored pencils or markers so students can record the colors on the coffee filters. Ask: **Which color shows the most on your coffee filter? Which shows the least?**

Students should not touch, taste, or smell rubbing alcohol. Wipe up spills immediately.

❹ Explain and Conclude

- Display the coffee filters for several different leaves side by side. Ask: **Did different leaves produce different results? How were they alike and different?** (Answers will vary.) Explain that different leaves have different amounts of colored substances, or pigments.
- Students should give evidence as they answer scientific questions orally. To encourage this, ask: **How do you know that green is not the only color in the leaf?** (because there were other colors on the coffee filter)
- Encourage students to share their inferences from question 2 on page 59 with the class. Ask: **Where do you think the colors came from and why?**
- Help students make the connection that the other colors present in a green leaf become visible when the seasons and the weather change.

Answers

1. Possible answer: The leaf solution was green. The rubbing alcohol with no leaves had no color.

What to Do, continued

3. Use the graduated cylinder to add 25 mL of alcohol to the cup. Let the leaves and the rubbing alcohol sit for 20 minutes. Observe and record the color of the solution in the cup.
4. Pour 25 mL of rubbing alcohol into the second cup. Do not add anything else to this cup. Observe and record the color of the alcohol.
5. Cut 2 strips from a coffee filter. Each should be about 3 cm wide and 9 cm long. Cut a point at one end of each strip. Tape each strip around a craft stick. Place each craft stick across the top of one of the cups. The points of the strips should just touch the alcohol in each cup. Observe the coffee filter strip in each cup and record your observations.
6. After 1 hour, remove the strips from the cups and place them on a paper towel. Observe the strips after they have dried. Record your observations.

58

NATIONAL GEOGRAPHIC Raise Your SciQ!

Plant Pigments Leaves get their color from molecules called pigments. The green color of leaves is a result of a green pigment called *chlorophyll.* Chlorophyll absorbs energy from sunlight and begins the process of photosynthesis. Plants must constantly produce chlorophyll because it breaks down easily in sunlight.

Many plants also contain a pigment called *carotene,* which looks yellow or orange. When plants go dormant because of colder weather and reduced amount of daylight, they stop producing chlorophyll and the yellow carotene becomes visible. Another type of pigment called *anthocyanin* appears red and forms when sugars react with the protein in plant sap.

Record

Write or draw in your science notebook.
Use a table like this one.

my SCIENCE notebook

Observations of Cups

	Color of Liquid	Coffee Filter Strip: Start	Coffee Filter Strip: Dried
Leaf solution			
Alcohol			

Explain and Conclude

1. What color was the solution in the cup with the leaf pieces? **Compare** your **observation** of the leaf solution and the rubbing alcohol with no leaves.
2. What colors did you observe on each coffee filter strip after 1 hour? **Infer** where the colors on the strip came from. Explain your reasoning.
3. Shorter daylight hours and cooler temperatures cause some trees to stop making the material that gives leaves their green color. **Infer** how the leaves change color.

Think of Another Question

What else would you like to find out about colors in leaves? How could you find an answer to this new question?

What colors are present in the leaves of these trees?

59

Explain and Conclude, continued

2. I observed green, yellow, and orange on the strip. I infer that the colors came from the leaf because the strip in the cup with just alcohol did not change color.
3. Possible answer: Green leaves contain different colors. When the trees stop making the material that gives leaves their green color, the other colors become visible in the leaves.

5 Find Out More

Think of Another Question

- Have students use the following stems to form testable questions: *What happens when…? What would happen if…? How can I/we…?* Discuss what students could do to find answers to the questions.

6 Reflect and Assess

- To assess student work with the Inquiry Rubric shown below, see Assessment Handbook, page 186, or go online at myNGconnect.com
- Have students use the Inquiry Self-Reflection on Assessment Handbook, page 198, or at myNGconnect.com

Inquiry Rubric at myNGconnect.com	Scale			
The student made a **hypothesis** about the colors in a green leaf.	4	3	2	1
The student identified **variables** to change, to control, and to **observe.**	4	3	2	1
The student collected and recorded **data** and **compared** observations with others.	4	3	2	1
The student made **inferences** about why the filters showed several colors and why leaves of some trees change color.	4	3	2	1
The student **shared** results and compared ideas based on the evidence gathered in the **investigation.**	4	3	2	1
Overall Score	4	3	2	1

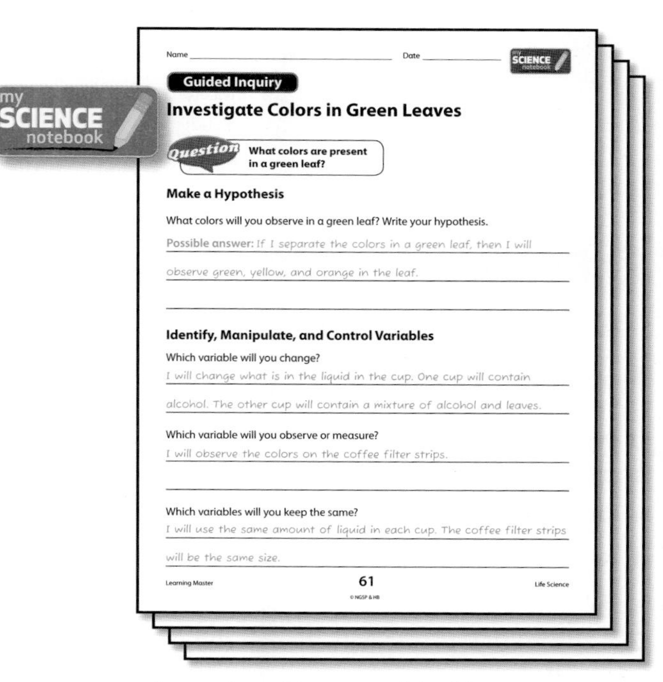
my SCIENCE notebook

Name ______ Date ______

Guided Inquiry

Investigate Colors in Green Leaves

Question: What colors are present in a green leaf?

Make a Hypothesis

What colors will you observe in a green leaf? Write your hypothesis.

Possible answer: If I separate the colors in a green leaf, then I will observe green, yellow, and orange in the leaf.

Identify, Manipulate, and Control Variables

Which variable will you change?

I will change what is in the liquid in the cup. One cup will contain alcohol. The other cup will contain a mixture of alcohol and leaves.

Which variable will you observe or measure?

I will observe the colors on the coffee filter strips.

Which variables will you keep the same?

I will use the same amount of liquid in each cup. The coffee filter strips will be the same size.

Learning Master 61 Life Science

Learning Masters 61–64 or at myNGconnect.com

LESSON 9 ▫ Conclusion and Review

PROGRAM RESOURCES

- Chapter 5 Test, Assessment Handbook, pages 27–30, or at myNGconnect.com
- NGSP ExamView CD-ROM

❶ Sum Up the Big Idea

- Display the chart from page T186–T187. Read the Big Idea Question. Then add new information to the chart.
- Ask: **What did you find out in each section? Is this what you expected to find?**

How Do Living Things Interact with Their Environment?

Changes in Seasons

Alejandro: Plants and animals adapt to seasonal changes in many ways, such as by dropping leaves or changing their fur.

Plants Change the Environment

Kurt: Plants can change the environment by keeping soil in place and adding gas to the air.

Animals Change the Environment

Kathy: Animals can change the environment by making homes, changing the soil, and feeding.

People Change the Environment

Grace: People can change the environment in ways that are helpful and ways that are harmful.

Conclusion

Living things interact with their environment in many ways. The different seasons cause changes in plants and animals. Plants change the environment by helping to make soil, slow erosion, clean the air, and cool Earth. Animals change the environment by building homes, helping to make soil, and feeding. Invasive plants and animals damage the environment by using up resources. People change the environment by using natural resources. They harm it by polluting and littering. People also can do things to improve the environment by conserving natural resources, recycling, and stopping pollution.

Big Idea Living things interact with their environment by changing the environment and changing in response to it.

Vocabulary Review

Match each of the following terms with the correct definition.

A. invasive organism
B. conserve
C. recycle
D. pollution

1. To use resources in a careful way
2. Harmful substances that people put into the air, water, and soil
3. To use something again
4. A plant or animal that does not belong in a place and can harm it

214

❷ Discuss the Big Idea

Notetaking

Have students write in their science notebook to show what they know about the Big Idea. Have them:

1. Write about how seasonal changes affect plants and animals.
2. Explain how plants can change the environment.

Review Academic Vocabulary

Academic Vocabulary Tell students that scientists use special words when describing their work to others. For example:

invasive organism **conserve** **recycle** **pollution**

Have students write the list in their science notebook and use each word in a sentence that tells about how organisms change the environment. They should then share their sentences with a partner.

Big Idea Review

1. **Describe** Describe how seasonal changes in plants are different in different parts of the United States.
2. **Recall** What do some animals do to survive cold weather?
3. **Recall** Explain what building "green homes" has to do with conserving resources.
4. **Compare** How are invasive plants and invasive animals the same?
5. **Predict** Suppose you see a sign that says "New Homes Coming Soon" outside of a forest. What do you think might happen to the forest and the living things in it?
6. **Infer** A river backs up behind a dam and forms a lake. What do you think happens to the river on the other side of the dam? How does this affect the plants and animals downstream from the dam?

Write About Helping the Environment

Evaluate Look at the picture. How does choosing organic foods help the environment? What can you do to help the environment?

Name ____________ Date ____________ Chapter 5

Chapter Test How Do Living Things Interact with Their Environment?

Directions: Read each question. Then choose the correct answer.

1. Which of these adaptations would help an animal that lives in an environment that is snowy in winter?
   - Ⓐ It sheds its fur.
   - Ⓑ It changes its color.
   - Ⓒ It becomes more active.
   - Ⓓ It moves to a colder area.

2. How does a tree **most likely** change in response to the beginning of fall?
   - Ⓐ Its fruit will grow.
   - Ⓑ Its flowers will bloom.
   - Ⓒ Its branches will fall off.
   - Ⓓ Its leaves will change color.

3. A certain kind of tree grows in a northern state and a southern state. How will the trees in the two states **most likely** change when winter comes?
   - Ⓐ The trees will both have flowers.
   - Ⓑ The leaves of both trees will turn red.
   - Ⓒ The northern tree will have no leaves, but the southern tree will.
   - Ⓓ The southern tree will have flowers and the northern tree will have fruit.

GO ON

27 Grade 4 Life Science

Assess Student Progress Chapter 5 Test

Have students complete their Chapter 5 Test to assess their progress in this chapter.

Chapter 5 Test, Assessment Handbook, pages 27–30, or at myNGconnect.com

or NGSP ExamView CD-ROM

Discuss the Big Idea, continued

3. Explain how animals can change the environment.
4. Explain how humans can change the environment, using the vocabulary words.

❸ Assess the Big Idea

Vocabulary Review

1. B **2.** D **3.** C **4.** A

Big Idea Review

1. Plants usually change more during the seasons in the northern part of the United States than in the southern part.
2. When the weather turns cold, some animals hibernate, or go into a sleep-like state. They eat a lot of food in the fall, and that keeps them alive during the winter. Other animals migrate, or move, to warmer places. They come back when the weather turns warm again.
3. "Green" homes are often built with recycled materials. That conserves natural resources.
4. Both invasive plants and animals do not belong in their new environment. They crowd out and kill native organisms. They can do great damage to the environment.
5. Possible answer: I think the forest will be cut down to make room for the new houses. People might use the wood from the forest to build the houses. Many of the plants and animals that live in the forest will die. Some of the animals might go somewhere else to live.
6. The level of the river might drop on the other side of the dam. There may be less water for the plants and animals that live downstream from the dam.

Write About Helping the Environment

Choosing organic foods supports farmers who don't use chemicals to kill pests or to grow crops. Students can also help by recycling, by not wasting water or electricity, and by not littering.

Objectives

Students will be able to:

- Recognize that scientists from many different gender and ethnic backgrounds make contributions to science.

PROGRAM RESOURCES

- Big Ideas Book: *Life Science*
- Big Ideas Book: *Life Science* eEdition at myNGconnect.com
- Digital Library at myNGconnect.com

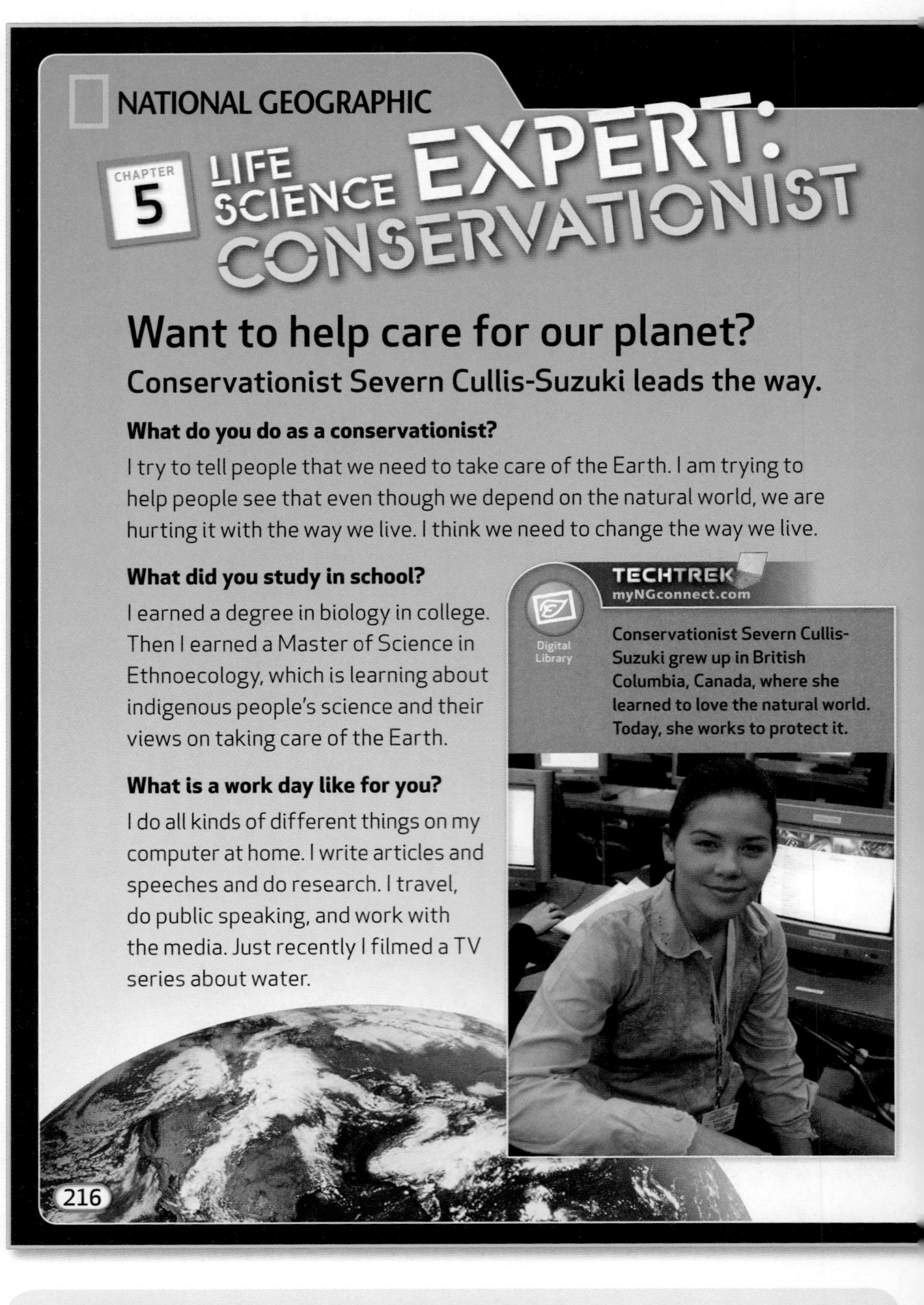

NATIONAL GEOGRAPHIC

CHAPTER 5 LIFE SCIENCE EXPERT: CONSERVATIONIST

Want to help care for our planet?

Conservationist Severn Cullis-Suzuki leads the way.

What do you do as a conservationist?

I try to tell people that we need to take care of the Earth. I am trying to help people see that even though we depend on the natural world, we are hurting it with the way we live. I think we need to change the way we live.

What did you study in school?

I earned a degree in biology in college. Then I earned a Master of Science in Ethnoecology, which is learning about indigenous people's science and their views on taking care of the Earth.

What is a work day like for you?

I do all kinds of different things on my computer at home. I write articles and speeches and do research. I travel, do public speaking, and work with the media. Just recently I filmed a TV series about water.

TECHTREK myNGconnect.com

Digital Library

Conservationist Severn Cullis-Suzuki grew up in British Columbia, Canada, where she learned to love the natural world. Today, she works to protect it.

216

❶ Introduce

Tap Prior Knowledge

- Ask: **What does the word *conserve* mean?** (to use resources in a wise way) Have students list resources that can be conserved.

Preview and Read

- Read the heading. Have students preview the photos on pages 216–217. Ask them to think about the kind of work that a conservationist might do.
- Have students read pages 216–217.

❷ Teach

Analyze Language

- Write *conserve* and *conservationist* on the board. Ask: **What do these two words have in common?** (They both contain the letters "conserv.") Explain: ***Conserve* is a verb. *Conservationist* is a noun that describes a person involved in conservation.**

myNGconnect.com

Have students use the Digital Library to find images of conservation efforts.

Computer Presentation Students can use the information to make a computer presentation about conservation efforts.

NATIONAL GEOGRAPHIC **Raise Your SciQ!**

Ethnoecology Ethnoecology is the study of indigenous ecological knowledge. Ethnoecologists study the way groups of people understand the interactions among the plants, animals, and environment around them. These understandings include the gathering of food and medicines, raising farm crops and animals, aquaculture, and astronomy. By comparing data from many different cultures, ethnoecologists gain a better understanding of human nature. They also find new solutions to various environmental problems.

Cullis-Suzuki speaks to groups of people to spread the word about caring for our planet.

What's been your greatest accomplishment so far?

When I was a teenager, I was invited by the United Nations to work with people from all over the world to put together the Earth Charter. The charter is a code of conduct for the way people should act toward the Earth. I am very proud of the Earth Charter.

What advice would you give young people who are interested in doing what you do?

I think a love of the Earth is the most important thing you need if you intend to fight for it. I always give the same advice to any young person who is interested in working for change in the world: Follow your heart. You will be most successful if you are doing what you are excited about and are good at.

Where do you hope your job has made the biggest impact?

I like to think that what I have done has helped youth voices be heard and that they have helped people take action in their communities. Now, more than ever, we need the voices of young people to remind the world of what is truly important.

217

Differentiated Instruction

ELL Language Support for Describing the Work of Conservationists

BEGINNING

Help students make a word web to describe conservationists. Have them write words that relate to this career, such as *protect*, *conserve*, and *environment*. Help students define and learn unfamiliar words.

INTERMEDIATE

Help students describe conservationists using Academic Language Frames.

- *A ____ works to protect the environment.*
- *To ____ resources is to use them wisely.*
- *The surroundings of living things are their ____.*

ADVANCED

Have students write several sentences describing the work of conservationists and why it is important. Have them include the words *conserve*, *protect*, *conservationist*, and *environment*.

Teach, continued

Describe the Work of Conservationists

- Have students identify Severn Cullis-Suzuki in her photo on page 216. Ask: **According to Severn Cullis-Suzuki, what is the most important thing you need if you want to become a conservationist?** (a love of the Earth and the desire to fight for it)
- Ask: **How does Cullis-Suzuki communicate with the public?** (Answers include writing articles, giving speeches, traveling and meeting people, and working with media.)
- Ask: **Why did Cullis-Suzuki want a career as a conservationist?** (She was and remains concerned about Earth and how its resources are being used.)
- Ask: **How much education does Cullis-Suzuki have?** (a college degree in biology and a Masters of Science degree in Ethnoecology) Discuss how professional scientists must study in colleges and universities for many years. However, people in a wide variety of careers and with a variety of backgrounds work to protect Earth's environments.

Find Out More

- Ask: **Is this a career you would find interesting? Why or why not?** (Accept all answers, positive or negative.)
- Encourage interested students to research what other conservationists do.

3 Assess

1. **Explain** **What does a conservationist do?** (helps people take better care of Earth)
2. **Infer** **Why is it important for young people to remind the world to conserve resources?** (Possible answer: The future belongs to young people, and resources must be conserved for when they are adults.)
3. **Draw Conclusions** **Who should pursue a career as a conservationist?** (anyone who is very interested in and excited about protecting Earth's resources)

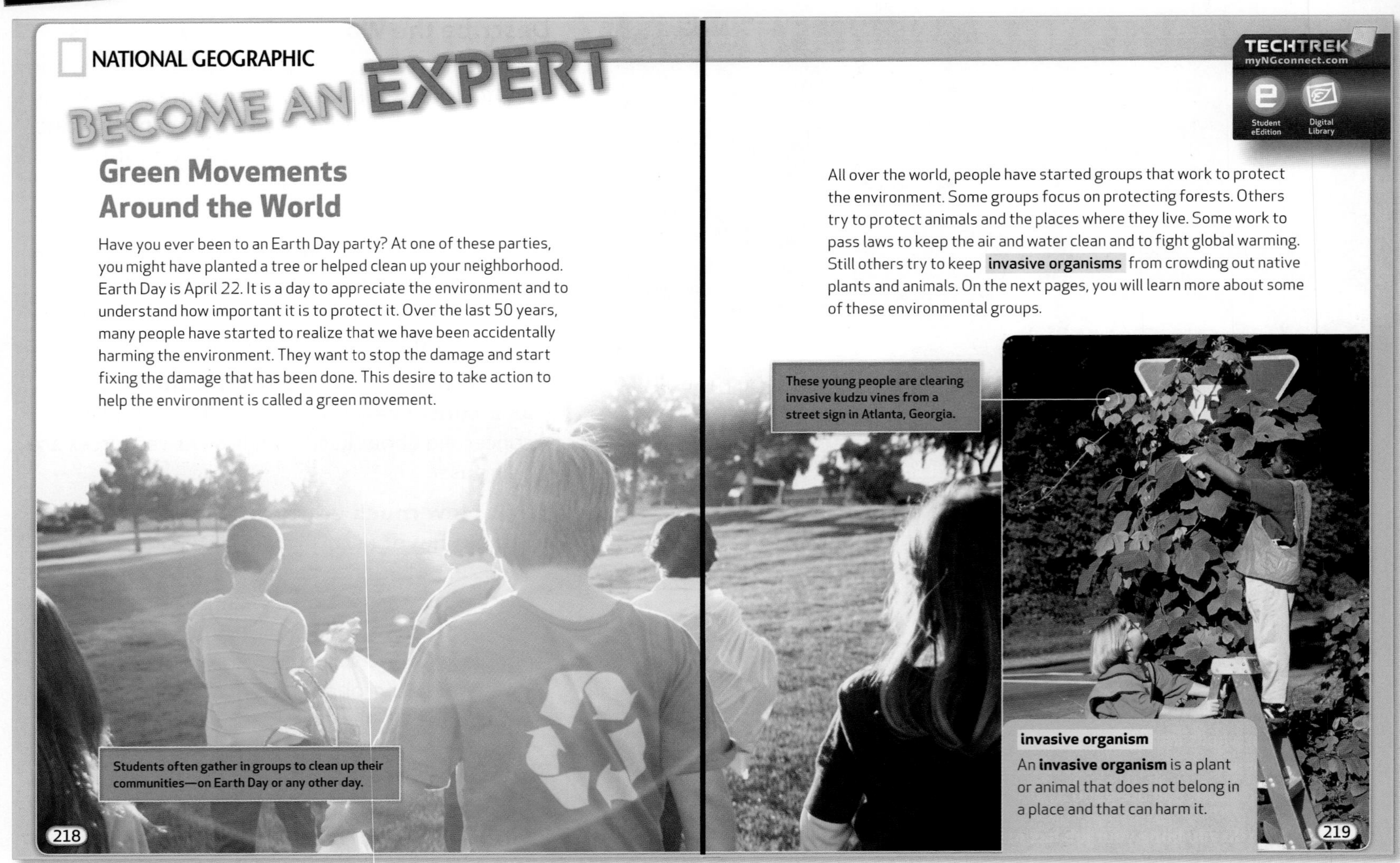

NATIONAL GEOGRAPHIC
BECOME AN EXPERT
Green Movements Around the World
Have you ever been to an Earth Day party? At one of these parties, you might have planted a tree or helped clean up your neighborhood. Earth Day is April 22. It is a day to appreciate the environment and to understand how important it is to protect it. Over the last 50 years, many people have started to realize that we have been accidentally harming the environment. They want to stop the damage and start fixing the damage that has been done. This desire to take action to help the environment is called a green movement.
TECHTREK
myNGconnect.com
Student eEdition
Digital Library
All over the world, people have started groups that work to protect the environment. Some groups focus on protecting forests. Others try to protect animals and the places where they live. Some work to pass laws to keep the air and water clean and to fight global warming. Still others try to keep invasive organisms from crowding out native plants and animals. On the next pages, you will learn more about some of these environmental groups.
These young people are clearing invasive kudzu vines from a street sign in Atlanta, Georgia.
Students often gather in groups to clean up their communities—on Earth Day or any other day.
invasive organism
An invasive organism is a plant or animal that does not belong in a place and that can harm it.
218
219

PROGRAM RESOURCES

- Big Ideas Book: *Life Science*
- Big Ideas Book: *Life Science* eEdition at myNGconnect.com
- Digital Library at myNGconnect.com

Access Science Content

Describe the Green Movement

- Have students read the text and observe the pictures on pages 218–219. Point to the larger picture and ask: **What are these students doing?** (cleaning up trash from the park) Ask: **What are green movements?** (people working together to help the environment) Discuss how students can take part in green movements, just like those shown in the photos.
- Point out the term **invasive organisms** and have students read the definition at the bottom of the page. Ask: **How can invasive organisms harm the environment?** (They can crowd out native plants and animals.) **How else are people harming the environment?** (Sample answers: cutting down forests, polluting the air and water, burning fossil fuels that increase global warming)

Notetaking

As students read pages 218–219, ask them to write down new terms they encounter in their science notebook. Then have students reread the pages to find the definitions of these words. For each word, ask students to write a descriptive sentence. Encourage them to draw a picture for each word to help them understand it better.

BECOME AN EXPERT

NATIONAL GEOGRAPHIC

Green Belt Movement "She thinks globally and acts locally," said the chair of the committee who awarded Dr. Wangari Maathai the 2004 Nobel Peace Prize. Maathai won the prize for her work with the Green Belt Movement (GBM). She started the GBM in 1977. The GBM works to **conserve** Africa's trees. Trees slow soil erosion. They also provide places for animals and plants to live. Because of the work of the GBM, more than 40 million trees have been planted in Africa!

Wangari Maathai of Kenya, Africa, is known all over the world for getting people together for good causes, such as planting trees.

conserve
To **conserve** is to use resources in a careful way.

220

The GBM teaches local women how to plant tree seeds and care for young trees. When the trees are big enough, the women plant them on people's farms and near their homes. Then their neighbors care for the trees. The trees give the people fruit to eat, food for their farm animals, and wood for fires and for building houses. The trees also help make new, healthier soil, which helps farmers. The GBM has shown the people of Africa that trees are good for the environment and for their own lives.

Women gather for a meeting of the Green Belt Movement.

Each of these containers holds a young tree that members of the GBM will plant.

221

Access Science Content

Describe the Green Belt Movement

- Have students read pages 220–221 and look at the photos. Ask: **What does it mean to "think globally and act locally?"** (Most of the world's environmental problems are on a global scale, but if everyone takes care of their local area it helps the whole world.) If a globe or world map is available, have a volunteer find the continent of Africa. Say: **The Green Belt Movement is helping protect trees in Africa.**
- Point out the word **conserve** and have students read the definition. Ask: **Why is it important to conserve trees?** (Trees help slow soil erosion, provide homes for animals and plants, and provide food.) Explain that trees are important in Africa and around the world.

Make Inferences

Have students reread the last sentence on page 221. Ask: **Why should trees be conserved in forests all over the world, not just in Africa?** (The benefits that trees provide African forests are also provided in forests everywhere.)

Assess

1. **Define What is the meaning of the word conserve?** (to use resources in a careful way)
2. **Cause and Effect Why does Wangari Maathi want the people of Africa to know that trees are good for the environment?** (If people understand how helpful trees are, they will be more likely to protect trees from being cut down or harmed.)

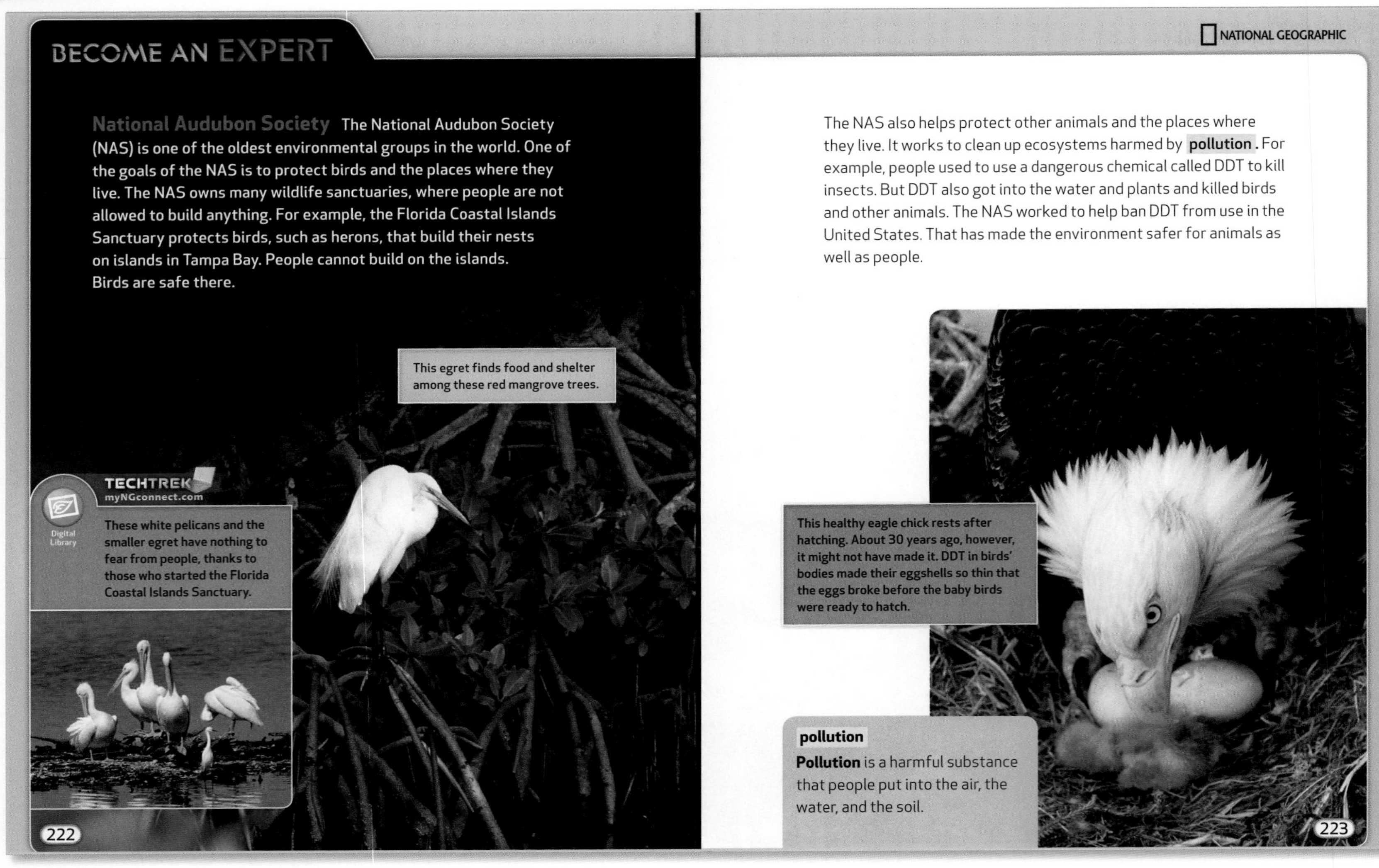

BECOME AN EXPERT
NATIONAL GEOGRAPHIC
National Audubon Society The National Audubon Society (NAS) is one of the oldest environmental groups in the world. One of the goals of the NAS is to protect birds and the places where they live. The NAS owns many wildlife sanctuaries, where people are not allowed to build anything. For example, the Florida Coastal Islands Sanctuary protects birds, such as herons, that build their nests on islands in Tampa Bay. People cannot build on the islands. Birds are safe there.
This egret finds food and shelter among these red mangrove trees.
TECHTREK
myNGconnect.com
Digital Library
These white pelicans and the smaller egret have nothing to fear from people, thanks to those who started the Florida Coastal Islands Sanctuary.
222
The NAS also helps protect other animals and the places where they live. It works to clean up ecosystems harmed by pollution. For example, people used to use a dangerous chemical called DDT to kill insects. But DDT also got into the water and plants and killed birds and other animals. The NAS worked to help ban DDT from use in the United States. That has made the environment safer for animals as well as people.
This healthy eagle chick rests after hatching. About 30 years ago, however, it might not have made it. DDT in birds' bodies made their eggshells so thin that the eggs broke before the baby birds were ready to hatch.
pollution
Pollution is a harmful substance that people put into the air, the water, and the soil.
223

PROGRAM RESOURCES

- Digital Library at myNGconnect.com

Access Science Content

Describe the National Audubon Society

- Have students read pages 222–223, observe the photos, and read the captions. Ask: **What does the National Audubon Society do?** (protects birds and other animals and the places where birds live) **How does establishing wildlife sanctuaries protect birds and other animals?** (People are not allowed to build structures on wildlife sanctuaries, allowing animals to live without human harm.) Ask: **How did DDT cause pollution?** (DDT was a dangerous chemical that people used to kill insects, but it harmed the environment.)

myNGconnect.com

Have students use the Digital Library to find images of other birds.

Computer Presentation Help students share the information in a computer presentation.

Notetaking

Have students prepare an outline of the Become an Expert section in their science notebook. Tell them first to write the headings on the pages and then add details for each.

Wildlife Protection Society of India The Wildlife Protection Society of India (WPSI) helps Indian state governments train forest and police officers to catch people who hurt wild animals. Some people catch and sell wild animals—including monkeys, birds, and reptiles—for pets or to circuses. This is against the law. Ending poaching is another goal of WPSI. Poaching is killing animals in order to sell their hides, fur, or other body parts. Tigers are often poached in India. Poaching is against the law.

The Wildlife Protection Society of India helps protect animals, such as Bengal tigers, from poachers.

The WPSI also works to help other animals in India, such as otters, sea turtles, and elephants. Pollution and construction have damaged or destroyed many of the places these animals used to live. For example, people have built buildings and iron mines on many elephant paths. Elephants used these paths to move around the country. The WPSI has studied ways to make new paths or bring back the old ones. Businesses and state governments are using their ideas to help elephants.

It's not a parade. It's a herd of elephants heading along a wide path to a river for a drink. The WPSI helps to keep these paths open, or make new ones.

224 225

Access Science Content

Describe the Wildlife Protection Society of India

- Have students read pages 224–225, observe the photos, and read the captions. If a world map or globe is available, show India's location. Ask: **What is the goal of the Wildlife Protection Society of India?** (to stop people from illegally catching or killing animals) Discuss how laws around the world protect many wild animals, yet those laws sometimes are difficult to enforce.
- Ask: **What is poaching?** (killing animals in order to sell their hides, fur, or other body parts) **Why do you think tigers are threatened by poaching?** (Possible answer: Tigers have beautiful fur.)
- Point to the photo of elephants on page 225. Ask: **How is the WPSI protecting elephants?** (They are protecting the paths that elephants use to travel.)

Assess

1. **Define What is poaching?** (killing animals illegally to sell their body parts)
2. **Explain How can people's actions harm wildlife in unintended ways? Use DDT as an example.** (Harm can occur when people do not understand all the effects of their actions. People used DDT to kill insects, but DDT spread into the environment and killed other animals, too.)

BECOME AN EXPERT

Movement of Small Farmers In Brazil, the Movement of Small Farmers (Movimiento de Pequeños Agricultores), or the MPA, is trying to stop the spread of very large farms owned by companies. Large areas of rainforest are often cleared to make room for large company-owned farms. When rainforests are cut down, soil erodes. Plants and animals lose their homes. The MPA is working for a future in which Brazilian land is farmed by local farmers, not by large companies.

Huge areas of rainforest in Brazil have been cleared to make room for large farms and cattle grazing.

226

NATIONAL GEOGRAPHIC

Like the MPA, people in all countries can do things to help the environment. They can do something small, such as convince their family to **recycle**. They can do something large, such as start a group like the MPA. Either way, they are making the world a better place.

School and community gardens have had a positive impact on the environment. Here, a teacher and her students pull weeds from a school garden.

recycle
To **recycle** is to use something again.

227

Access Science Content

Describe the Movement of Small Farmers

- Have students read pages 226–227. Display a globe or world map and have a volunteer point to South America and then the country of Brazil. Discuss how Brazil has huge tracts of rainforests.
- Ask: **Why are rainforests valuable?** (They are home to many plants and animals, and the trees prevent soil erosion.) Discuss how trees also release oxygen, the gas that we breathe, and take up carbon dioxide. **What is happening to the rainforests in Brazil?** (They are being cut down in part to make way for large farms.) Discuss the goal of the Movement of Small Farmers, which is to preserve rainforests and promote small, locally owned farms.
- Point out the word **recycle** on page 227. Ask: **How does recycling help the environment?** (It helps people conserve natural resources.)

Assess

1. **Define** **What does it mean to recycle something?** (to use something again instead of throwing it away)
2. **Evaluate** **What do the Green Belt Movement, the National Audubon Society, the Wildlife Protection Society of India, and the Movement of Small Farmers all have in common?** (They are all helping people to become aware of environmental problems and to conserve natural resources.)

BECOME AN EXPERT

NATIONAL GEOGRAPHIC

CHAPTER 5

SHARE AND COMPARE

Turn and Talk How do different groups around the world help protect the environment? Form a complete answer to this question together with a partner.

Read Select two pages in this section that are the most interesting to you. Practice reading the pages so that you can read them smoothly. Then read them aloud to a partner or small group. Talk about why the pages are interesting.

my SCIENCE notebook **Write** Write a conclusion that summarizes what you have learned about groups that help the environment. State what you think is the Big Idea of this section. Share what you wrote with a classmate. Compare your conclusions. Did you recall some of the specific things that different groups in the green movement do?

my SCIENCE notebook **Draw** Draw a picture of what you think a National Audubon Society sanctuary for birds might look like. Combine ideas from your drawing with those of your classmates to make a new drawing that includes the best ideas.

228

Sum Up

Tell students that to sum up a text helps them pull together the ideas of the text. Remind students that they summed up the Become an Expert lesson in the Write section of Share and Compare. Have students take turns reading the conclusions they wrote to a partner. Invite them to compare and contrast their conclusions.

Share and Compare

Turn and Talk

Ask students to turn to partners and talk about what they learned about protecting the environment. Prompt students by asking:

1. **Describe What is a natural resource? Give an example.** (something in the environment that people can use, such as trees, wildlife, air, land, and water)
2. **Explain How do organizations like the Green Belt Movement help people conserve natural resources?** (First, they make people aware of the problem. Then they show people how to protect the environment.)
3. **Apply How do different groups around the world help protect the environment?** (They teach people about environmental problems, help people solve problems, and work with governments and law-enforcement officials to enact and enforce wise laws.)

Read

Ask students to choose the two pages that are most interesting to them. Have them take turns reading those pages with a partner or small group. Ask students leading questions to help them express why they found those pages interesting.

Have students write a conclusion that summarizes what they learned about green movements. Have partners compare their answers and ask each other questions about green movements and their effect on the environment.

Have students draw a picture of a bird sanctuary. Ask students to share their drawing with a partner and describe the features of the sanctuary they drew. Then have partners work together to make a new drawing that combines the best of their ideas.

Notes

How Do the Parts of an Organism Work Together?

LIFE SCIENCE

CHAPTER
6
HOW DO THE PARTS OF AN ORGANISM WORK TOGETHER?
TECHTREK
myNGconnect.com
You can whirl across a stage, climb a mountain, or walk a dog. Have you ever wondered how your body works? Like the bodies of all living things, your body needs certain things not only to survive but also to thrive. Some parts of your body help you get oxygen, food, and water. Other parts help you respond to changes.
230
231

After reading Chapter 6, you will be able to:

- Explain that humans and other living things have different body parts that help them get what they need to survive. **CIRCULATORY AND RESPIRATORY SYSTEMS, DIGESTIVE SYSTEM, SKELETAL, MUSCULAR, AND NERVOUS SYSTEMS**
- Recognize that animals are composed of different parts performing different functions but working together for the well-being of the organism. **CIRCULATORY AND RESPIRATORY SYSTEMS, DIGESTIVE SYSTEM, SKELETAL, MUSCULAR, AND NERVOUS SYSTEMS**
- Describe the major organs and function of different body systems, including the circulatory, respiratory, digestive, skeletal, muscular, and nervous systems. **CIRCULATORY AND RESPIRATORY SYSTEMS, DIGESTIVE SYSTEM, SKELETAL, MUSCULAR, AND NERVOUS SYSTEMS**
- Recognize that healthful habits include eating a balanced diet, getting regular exercise, using proper hygiene, and avoiding drugs, alcohol, and tobacco. **KEEPING BODY SYSTEMS HEALTHY**
- Science in a Snap! Describe the major organs and function of different body systems, including the circulatory, respiratory, digestive, skeletal, muscular, and nervous systems. **CIRCULATORY AND RESPIRATORY SYSTEMS**

229

Chapter 6 Contents

Preview and Predict

Tell students that they can preview a text by looking over it to get an idea of what the text is about and how it is organized. Tell students that predicting is forming ideas about what they will read. Students should confirm their ideas with others.

Have students preview and make predictions about the chapter. Remind them to use section heads, vocabulary words, pictures, and their background knowledge to make predictions.

CHAPTER 6 ▫ How Do the Parts of an Organism Work Together?

LESSON	PACING	OBJECTIVES
1 **Directed Inquiry** *Investigate How Your Brain Can Work* pages T229g–T229j **Science Inquiry and Writing Book** pages 60–63	**30** minutes	Investigate through Directed Inquiry (answer a question; make and compare observations; collect and record data and observations; generate explanations and conclusions based on evidence; share findings; ask questions based on observations to increase understanding). Investigate and describe basic functions of the brain. Recognize that the human body is composed of different parts that perform different functions and work together. Conduct investigations in which distinctions are made among observations, conclusions, and inferences. Analyze and organize data in tables and recognize simple patterns.
2 **Big Idea Question and Vocabulary** pages T230–T231—T232–T233	**25** minutes	Explain that humans and other living things have different body parts that help them get what they need to survive.
3 **Circulatory and Respiratory Systems** pages T234–T239	**30** minutes	Recognize that animals are composed of different parts performing different functions and working together for the well-being of the organism. Describe the major organs of the circulatory system and how they function. Describe the major organs of the respiratory system and how they function.
4 **Digestive System** pages T240–T241	**20** minutes	Describe the major organs of the digestive system and how they function.
5 **Skeletal, Muscular, and Nervous Systems** pages T242–T247	**30** minutes	Describe the major organs of the skeletal system and how they function. Describe the major organs of the muscular system and how they function. Describe the major organs of the nervous system and how they function.

VOCABULARY	RESOURCES	ASSESSMENT
data **compare**	Science Inquiry and Writing Book: *Life Science* Science Inquiry Kit: *Life Science* Directed Inquiry: Learning Masters 65–67	Inquiry Rubric: Assessment Handbook, page 187 Inquiry Self-Reflection: Assessment Handbook, page 199 Reflect and Assess, page T229j
	Vocabulary: Learning Master 68 *Life Science* Big Ideas Book	
organ **system** **circulatory system** **respiratory system**		Assess, page T239
digestive system		Assess, page T241
nervous system	Science Inquiry and Writing Book: *Life Science*	Assess, page T247

TECHNOLOGY RESOURCES

STUDENT RESOURCES

myNGconnect.com

- Student eEdition
- Big Ideas Book
- Science Inquiry and Writing Book
- Read with Me
- Vocabulary Games
- Enrichment Activities
- Digital Library

National Geographic Kids

National Geographic Explorer!

TEACHER RESOURCES

myNGconnect.com

- Teacher eEdition
- Teacher's Edition
- Science Inquiry and Writing Book
- Online Lesson Planner
- National Geographic Unit Launch Videos
- Assessment Handbook
- Presentation Tool
- Digital Library

NGSP ExamView CD-ROM

IF TIME IS SHORT...
FAST FORWARD.

CHAPTER 6 ▫ How Do the Parts of an Organism Work Together?

LESSON	PACING	OBJECTIVES
6 **Keeping Body Systems Healthy** pages T248–T251	**25** minutes	Recognize that eating a balanced diet and getting regular exercise are healthful habits. Recognize that washing your hands frequently and avoiding drugs, alcohol, and tobacco are healthful habits.
7 NATIONAL GEOGRAPHIC **The Body Systems of Animals** pages T252–T255	**25** minutes	Recognize that animals are composed of different parts performing different functions and working together for the well-being of the organism.
8 **Guided Inquiry** *Investigate Exercise and Heart Rate* pages T255a–T255d **Science Inquiry and Writing Book** pages 64–67	**30** minutes	Investigate through Guided Inquiry (answer a question; make and compare observations; collect and record data and observations; generate explanations and conclusions based on evidence; share findings; ask questions based on observations to increase understanding; adjust explanations based on findings and new ideas). Recognize that humans need healthful food, cleanliness, exercise, and rest to grow and maintain good health. Use a stopwatch to aid observations and data collection. Judge whether measurements and computations are reasonable.
9 **Conclusion and Review** pages T256–T257	**15** minutes	

VOCABULARY	RESOURCES	ASSESSMENT
		Assess, page T251
	Share and Compare: Learning Master 69	Assess, page T255
count **conclude**	Science Inquiry and Writing Book: *Life Science* Science Inquiry Kit: *Life Science* Guided Inquiry: Learning Masters 70–73	Inquiry Rubric: Assessment Handbook, page 187 Inquiry Self-Reflection: Assessment Handbook, page 200 Reflect and Assess, page T255d
		Assess the Big Idea, page T257 Chapter 6 Test: Assessment Handbook, pages 32–35 NGSP ExamView CD-ROM

TECHNOLOGY RESOURCES

STUDENT RESOURCES

myNGconnect.com

- Student eEdition
- Big Ideas Book
- Science Inquiry and Writing Book
- Read with Me
- Vocabulary Games
- Enrichment Activities
- Digital Library

National Geographic Kids

National Geographic Explorer!

TEACHER RESOURCES

myNGconnect.com

- Teacher eEdition
- Teacher's Edition
- Science Inquiry and Writing Book
- Online Lesson Planner
- National Geographic Unit Launch Videos
- Assessment Handbook
- Presentation Tool
- Digital Library

NGSP ExamView CD-ROM

IF TIME IS SHORT...

FAST FORWARD.

CHAPTER 6 ▫ How Do the Parts of an Organism Work Together?

LESSON	PACING	OBJECTIVES
10 NATIONAL GEOGRAPHIC **LIFE SCIENCE EXPERT** *Pharmacologist* pages T258–T259 NATIONAL GEOGRAPHIC **BECOME AN EXPERT** *Technology and Body Systems* pages T260–T261—T268	**35** minutes	Recognize how different scientists' work has contributed to general scientific understanding. Recognize that humans need body parts and systems that work together.
11 **Open Inquiry** **Do Your Own Investigation** pages T268b–T268c **Science Inquiry and Writing Book** pages 68–69 **Write Like a Scientist** pages T268d–T268e **Science Inquiry and Writing Book** pages 70–77 **Investigation Model** pages T268f–T268i	**50** minutes	Investigate through Open Inquiry (generate questions to investigate; plan, make, and compare investigations; collect and record data and observations; generate explanations and conclusions based on evidence or observations; share findings; ask questions based on observations to increase understanding; adjust interpretations based on findings and new ideas). Investigate through Write Like a Scientist (choose a question; gather materials to use; make a hypothesis or prediction; identify, manipulate, and control variables if needed; make and carry out a plan for an investigation; collect and record data in charts, tables, and graphs; analyze data; explain and share results; tell a conclusion; think of another question). Indicate materials to be used and steps to follow to conduct an investigation, and describe how data will be recorded. Explore and solve problems generated from school, home, and community situations, using concrete objects and materials. Recognize ways animals such as earthworms can impact plants in the environment.
12 **Think Like a Scientist** *How Scientists Work: Designing Investigations and Using Evidence* pages T268j–T268m **Science Inquiry and Writing Book** pages 78–81	**30** minutes	Identify a question that was asked, or a problem that needed to be solved when given a brief scenario. Differentiate fact from opinion and explain that scientists do not rely on conclusions unless they are backed by observations that can be confirmed. Recognize and explain that scientists base their explanations on evidence.

FAST FORWARD
ACCELERATED PACING GUIDE

DAY 1 30 minutes

Directed Inquiry

Investigate How Your Brain Can Work, page T229g

DAY 2 35 minutes

NATIONAL GEOGRAPHIC **LIFE SCIENCE EXPERT** *Pharmacologist,* page T258

NATIONAL GEOGRAPHIC **BECOME AN EXPERT** *Technology and Body Systems,* page T260–T261

VOCABULARY	RESOURCES	ASSESSMENT
		Assess, pages T259, T262–T263, T266–T267
measure	Science Inquiry and Writing Book: *Life Science* Science Inquiry Kit: *Life Science* Open Inquiry: Learning Masters 74–77 Investigation Model: Learning Masters 78–79	Inquiry Rubric: Assessment Handbook, page 188 Inquiry Self-Reflection: Assessment Handbook, page 201 Reflect and Assess, page T268c, T268i
	Science Inquiry and Writing Book: *Life Science* Think Like a Scientist: Learning Masters 80–81	Assess, page T268m

TECHNOLOGY RESOURCES

STUDENT RESOURCES

myNGconnect.com

- Student eEdition
- Big Ideas Book
- Science Inquiry and Writing Book
- Read with Me
- Vocabulary Games
- Enrichment Activities
- Digital Library

National Geographic Kids

National Geographic Explorer!

TEACHER RESOURCES

myNGconnect.com

- Teacher eEdition
- Teacher's Edition
- Science Inquiry and Writing Book
- Online Lesson Planner
- National Geographic Unit Launch Videos
- Assessment Handbook
- Presentation Tool
- Digital Library

NGSP ExamView CD-ROM

DAY 3 30 minutes

Guided Inquiry

Investigate Exercise and Heart Rate, page T255a

DAY 4 50 minutes

Open Inquiry

Do Your Own Investigation, page T268b

SCIENCE

Objectives

Students will be able to:

- Investigate through Directed Inquiry (answer a question; make and compare observations; collect and record data and observations; generate explanations and conclusions based on evidence; share findings; ask questions based on observations to increase understanding).
- Investigate and describe basic functions of the brain.
- Recognize that the human body is composed of different parts that perform different functions and work together.
- Conduct investigations in which distinctions are made among observations, conclusions, and inferences.
- Analyze and organize data in tables and recognize simple patterns.

Science Process Vocabulary

data, compare

PROGRAM RESOURCES

- Science Inquiry and Writing Book: *Life Science*
- Science Inquiry and Writing Book eEdition at myNGconnect.com
- Inquiry eHelp at myNGconnect.com
- Science Inquiry Kit: *Life Science*
- Learning Masters Book, pages 65–67, or at myNGconnect.com
- Inquiry Rubric: Assessment Handbook, page 187, or at myNGconnect.com
- Inquiry Self-Reflection: Assessment Handbook, page 199, or at myNGconnect.com

MATERIALS

Kit materials are listed in italics.

2 sheets of white paper; markers; *stopwatch*

❶ Introduce

Tap Prior Knowledge

- Ask: **What parts of your body do you use to read?** (the eyes and the brain) Explain to students that reading is a complex activity that involves many different parts of the brain.

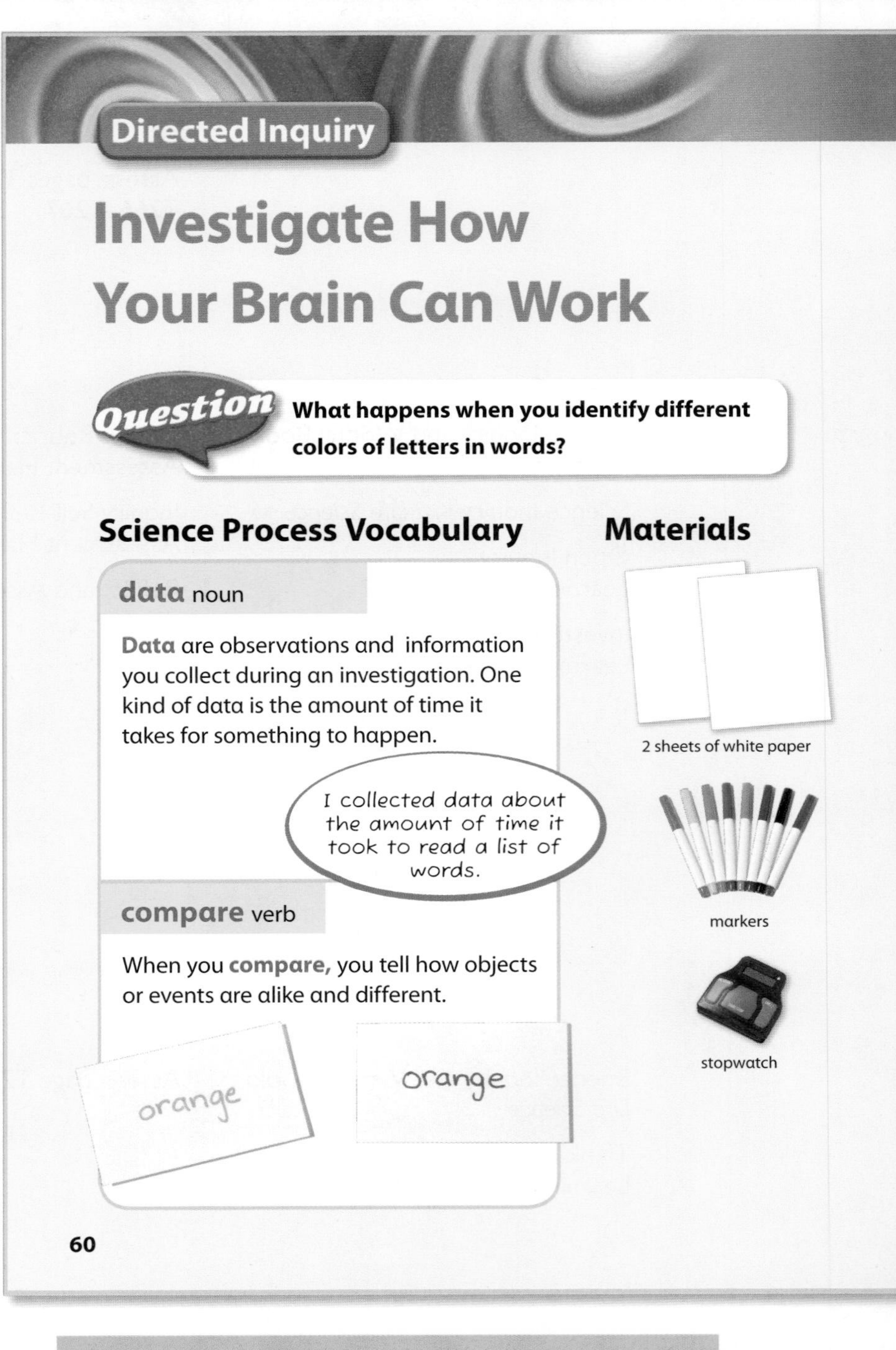

Directed Inquiry

Investigate How Your Brain Can Work

Question **What happens when you identify different colors of letters in words?**

Science Process Vocabulary

data noun

Data are observations and information you collect during an investigation. One kind of data is the amount of time it takes for something to happen.

I collected data about the amount of time it took to read a list of words.

compare verb

When you **compare,** you tell how objects or events are alike and different.

orange
orange

Materials

2 sheets of white paper

markers

stopwatch

60

MANAGING THE INVESTIGATION

Time

30 minutes

Groups

Small groups of 4

Advance Preparation

- Supply markers for each group. Include red, orange, yellow, green, blue, purple, black, and brown markers.

What to Expect

- Most groups should observe that it takes longer to read Test B than Test A because in Test B the colors of the words did not match the names of the colors.

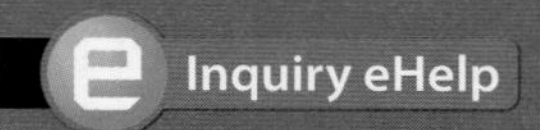

What to Do

1 On a sheet of paper, write the following words in 3 rows of 4 words, as shown in the photo: **red, yellow, orange, green, blue, purple, black, brown.** You can put the words in any order and repeat words to get a total of 12 words. Use the markers to write each word in the color that matches it. For example, the word **red** should be written using the red marker. Label the paper **Test A.**

red blue brown orange
green blue purple red
black brown yellow green

Test A

61

Teaching Tips

- Students in each group should work in pairs, with two pairs sharing a stopwatch.
- Students should read clearly and accurately. Encourage partners to check for accuracy as well as speed.
- Remind students that this is not a competition and there are no right or wrong times for these tests. Point out that the times for each test should vary.

Introduce, continued

Connect to the Big Idea

- Review the Big Idea Question, *How do the parts of an organism work together?* Explain to students that this inquiry will help them understand the way parts of the body work together to process information.
- Have students open their Science Inquiry and Writing Books to page 60. Read the Question and invite students to share ideas about how the brain processes information to read.

❷ Build Vocabulary

Science Process Words: data, compare

Use this routine to introduce the words.

1. **Pronounce the Word** Say **data.** Have students repeat it in syllables.
2. **Explain Its Meaning** Choral read the sentence. Ask students for other words or phrases that mean the same as **data.** (observations, measurements, and information collected during an investigation)
3. **Encourage Elaboration** Ask: **What type of data can you collect using a stopwatch?** (the length of time something takes)

ELL Use Cognates

Spanish speakers may know the noun **los datos** in their home language. Use it to help them access the English word **data.**

Repeat for the word **compare.** To encourage elaboration, ask: **How could you compare the heights of the students in this class?** (You could measure each student's height with a measuring tape and list the heights in a table or graph.)

❸ Guide the Investigation

- Distribute materials. Read the inquiry steps on pages 61–62 together with students. Move from group to group and clarify steps if necessary.
- In step 1, ask students to write as clearly as possible so that the words are legible and will not affect the test results.

Guide the Investigation, continued

- Ask: **How are the two tests alike and different?** (The words and the order of the words are the same in both tests. The colors in which the words are written are different in the tests.)
- Guide students in a discussion about observations, conclusions, and inferences. Ask students to explain the difference between each and how each can be used in this investigation. Discuss how students can distinguish between the three as they keep records.

❹ Explain and Conclude

- Guide students as they record their observations in their tables. If students have difficulty seeing patterns, display one group's data that show clear results.
- Ask groups to compare times with each other. Ask students to find similarities and differences in their data and explain any differences.
- Say: **Compare the times from Test A and Test B. Which test took longer? Why do you think that happened?** Students should find that in Test B they were trying to identify the color while ignoring the word. This took more time.
- Explain to students that when they read, their eyes read the words and then send messages through the nervous system and to the brain. The brain then interprets those messages and sends a message to the vocal cords and the mouth, leading to speech. Discuss with students how in Test B, their brains were getting messages about the color of the word and the word itself.
- Ask: **Which senses did you use to do this test and collect your data? How did each sense help you?** Students may mention they used sight to read the words and hearing to listen to their partner identify the colors.

Answers

1. Possible answer: Test B took longer. I think it is because the brain is trying to read a word and identify the color it is written in at the same time. In Test A, the word and the color are the same. In Test B, the word and the color are different. The brain receives conflicting information and it takes longer to respond.

What to Do, continued

2 On the second sheet of paper, write the words again, in the same order you wrote them the first time. This time use a different color for each word. For example, you may choose to write the word **brown** using the blue marker. Label the paper **Test B.**

red blue brown orange
green blue purple red
black brown yellow green

Test B

3 Use the stopwatch to time your partner as he or she correctly names the COLOR of the letters of each word on Test A. Record the **data.**

4 Then time your partner as he or she correctly names the COLOR of the letters of each word on Test B. For example, if the word black is written in purple marker, say "Purple." Record the data.

5 Switch roles with your partner and repeat steps 3 and 4.

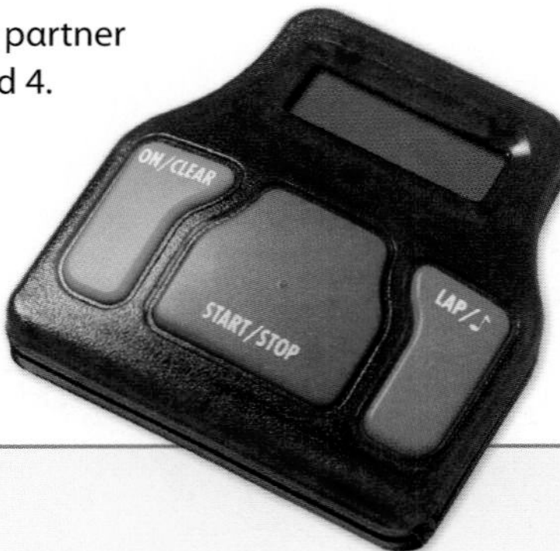

62

Raise Your SciQ!

The Stroop Effect In this activity students are testing the Stroop Effect, which was discovered by Ridley Stroop in the 1930s. It takes longer to read Test B than Test A. This occurs because the brain is processing two conflicting pieces of information at the same time: reading the word itself and identifying the color in which the word is written. The brain reads the word faster and more automatically than it interprets the ink color, making it difficult to say the ink color and not read the word.

Record

Write in your science notebook.
Use a table like this one.

my SCIENCE notebook

Identifying Colors

Who?	Time to Read Test A (s)	Time to Read Test B (s)
Me		
Partner		

Explain and Conclude

1. **Compare** the times needed to complete Test A and Test B. Which test did you take longer to finish? Why do you think that is so?
2. Compare your times with your partner's times. Explain any differences.
3. What body parts did you use as you took these tests? How did each part help you with the task?

Think of Another Question

What else would you like to find out about how the brain processes information? How could you find an answer to this new question?

Tests like this can help determine how a person sees different colors.

63

Inquiry Rubric at myNGconnect.com	Scale			
The student **observed** what happened as the brain processed information.	4	3	2	1
The student **measured** how long it took his or her partner to read each test.	4	3	2	1
The student recorded **data** in a table.	4	3	2	1
The student formed a **conclusion** about how the brain processes information based on observations and data.	4	3	2	1
The student **shared** and **compared** data and conclusions with others.	4	3	2	1
Overall Score	4	3	2	1

Explain and Conclude, continued

2. Possible answer: I read the words a little faster than my partner. I think the reason is that I had already seen the words when my partner read them.
3. Possible answer: I used my eyes to see the words and their colors. Nerves sent information about what I saw to my brain. My brain interpreted the information and sent messages to use my mouth, throat, and body to speak. Muscles moved my eyes and mouth as I read and said the words.

5 Find Out More

Think of Another Question

- Students should use their observations to generate new questions that arise from their investigations. Record student questions and discuss what students could do to find answers.

6 Reflect and Assess

- To assess student work with the Inquiry Rubric shown below, see Assessment Handbook, page 187, or go online at myNGconnect.com
- Have students use the Inquiry Self-Reflection on Assessment Handbook, page 199, or at myNGconnect.com

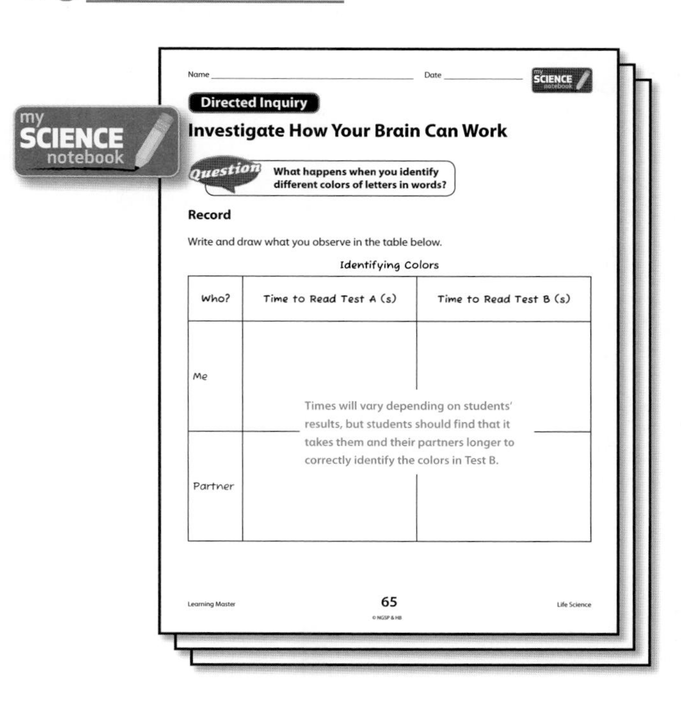

my SCIENCE notebook

Name ____ Date ____

Directed Inquiry

Investigate How Your Brain Can Work

Question: What happens when you identify different colors of letters in words?

Record

Write and draw what you observe in the table below.

Identifying Colors

Who?	Time to Read Test A (s)	Time to Read Test B (s)
Me	Times will vary depending on students' results, but students should find that it takes them and their partners longer to correctly identify the colors in Test B.	
Partner		

Learning Master 65 Life Science

Learning Masters 65–67
or at myNGconnect.com

LESSON 2 ▫ Big Idea Question and Vocabulary

Objectives

Students will be able to:

- Explain that humans and other living things have different body parts that help them get what they need to survive.

Science Academic Vocabulary

organ, system, circulatory system, respiratory system, digestive system, nervous system

CHAPTER
6
HOW DO THE PARTS OF AN ORGANISM WORK TOGETHER?
You can whirl across a stage, climb a mountain, or walk a dog. Have you ever wondered how your body works? Like the bodies of all living things, your body needs certain things not only to survive but also to thrive. Some parts of your body help you get oxygen, food, and water. Other parts help you respond to changes.
TECHTREK
myNGconnect.com
All the parts of this young woman's body are working hard to help her perform at her peak.
230
231

PROGRAM RESOURCES

- Big Ideas Book: *Life Science*
- Big Ideas Book: *Life Science* eEdition at myNGconnect.com
- Vocabulary Games at myNGconnect.com
- Digital Library at myNGconnect.com
- Enrichment Activities at myNGconnect.com
- Read with Me at myNGconnect.com
- Learning Masters Book, page 68, or at myNGconnect.com

❶ Introduce

Tap Prior Knowledge

- Have students observe the photo of the skater on pages 230–231. Then have them list different parts of the body that work together to help her skate. Encourage students to recognize that athletes use a wide variety of body parts and systems, not just bones and muscles.

❷ Focus on the Big Idea

Big Idea Question

- Read the Big Idea Question aloud and have students echo it.
- Preview pages 234–239, 240–241, 242–247, and 248–251, linking the headings with the Big Idea Question.

Differentiated Instruction

ELL Vocabulary Support

BEGINNING

Have students match the organ system with its purpose. For example:

Carrying blood *(circulatory system)*

Breathing *(respiratory system)*

Sending and receiving messages *(nervous system)*

INTERMEDIATE

Provide Academic Language Frames to help students describe organ systems.

- *Blood is carried by the ____ system.*
- *The lungs are part of the ____ system.*
- *The ____ system acts as a communication network.*

ADVANCED

Provide Academic Language Stems to help students describe organ systems.

- *The circulatory system works to . . .*
- *The respiratory system acts to . . .*
- *The nervous system acts to . . .*

Focus on the Big Idea, continued

- With student input, post a chart for the chapter headings. Have students share what they expect to find in each section. Then read page 230 aloud.

How Do the Parts of an Organism Work Together?

Circulatory and Respiratory Systems

Digestive System

Skeletal, Muscular, and Nervous Systems

Keeping Body Systems Healthy

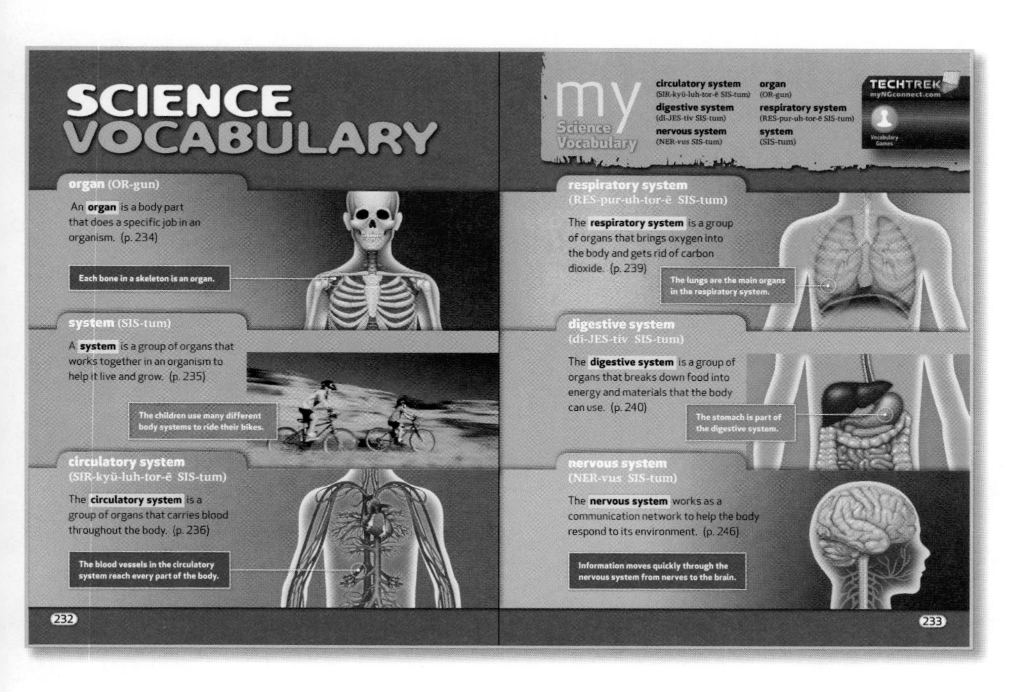

SCIENCE VOCABULARY

organ (OR-gun)

An **organ** is a body part that does a specific job in an organism. (p. 234)

Each bone in a skeleton is an organ.

system (SIS-tum)

A **system** is a group of organs that works together in an organism to help it live and grow. (p. 235)

The children use many different body systems to ride their bikes.

circulatory system (SIR-kyū-luh-tor-ē SIS-tum)

The **circulatory system** is a group of organs that carries blood throughout the body. (p. 236)

The blood vessels in the circulatory system reach every part of the body.

232

my Science Vocabulary

circulatory system (SIR-kyū-luh-tor-ē SIS-tum)
digestive system (di-JES-tiv SIS-tum)
nervous system (NER-vus SIS-tum)
organ (OR-gun)
respiratory system (RES-pur-uh-tor-ē SIS-tum)
system (SIS-tum)

TECHTREK myNGconnect.com Vocabulary Games

respiratory system (RES-pur-uh-tor-ē SIS-tum)

The **respiratory system** is a group of organs that brings oxygen into the body and gets rid of carbon dioxide. (p. 239)

The lungs are the main organs in the respiratory system.

digestive system (di-JES-tiv SIS-tum)

The **digestive system** is a group of organs that breaks down food into energy and materials that the body can use. (p. 240)

The stomach is part of the digestive system.

nervous system (NER-vus SIS-tum)

The **nervous system** works as a communication network to help the body respond to its environment. (p. 246)

Information moves quickly through the nervous system from nerves to the brain.

233

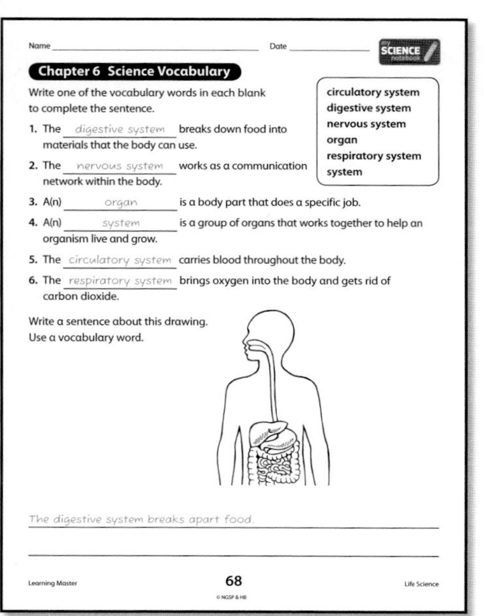

Name __________ Date __________ my SCIENCE notebook

Chapter 6 Science Vocabulary

Write one of the vocabulary words in each blank to complete the sentence.

circulatory system
digestive system
nervous system
organ
respiratory system
system

1. The _digestive system_ breaks down food into materials that the body can use.
2. The _nervous system_ works as a communication network within the body.
3. A(n) _organ_ is a body part that does a specific job.
4. A(n) _system_ is a group of organs that works together to help an organism live and grow.
5. The _circulatory system_ carries blood throughout the body.
6. The _respiratory system_ brings oxygen into the body and gets rid of carbon dioxide.

Write a sentence about this drawing. Use a vocabulary word.

The digestive system breaks apart food.

Learning Master 68 Life Science

Learning Master 68 or at myNGconnect.com

❸ Teach Vocabulary

Have students look at pages 232–233. Use this routine to teach each word. For example:

1. **Pronounce the Word** Say **organ** and have students repeat it.
2. **Explain Its Meaning** Read the word, definition, and sample sentence, and use the photo to explain the word's meaning.
3. **Encourage Elaboration** Say: **Name some other organs that are part of the human body.** (Examples include the heart, eyes, ears, kidneys, lungs, and skin.)

Repeat for the words **system**, **circulatory system**, **respiratory system**, **digestive system**, and **nervous system** using the following Elaboration Prompts:

- **Which system of the body does the stomach belong to?** (the digestive system)
- **What organs make up the circulatory system?** (heart, blood vessels)
- **Where are the organs of the respiratory system?** (The lungs are in the chest. Tubes carry air between the mouth and lungs.)
- **Where in the digestive system are foods first broken apart?** (inside the mouth, where teeth and saliva act on food)
- **What kinds of messages does the nervous system carry?** (Answers include messages about pain, hunger, thirst, and position and messages that move body parts.)

LESSON 3 ▫ Circulatory and Respiratory Systems

Objectives

Students will be able to:

- Recognize that animals are composed of different parts performing different functions and working together for the well-being of the organism.

Science Academic Vocabulary

organ, system

❶ Introduce

Tap Prior Knowledge

- Have students observe the photos of bicycles on pages 234–235 and identify the parts of the bikes they recognize. Have them discuss why it's important for all the parts to connect properly and to work together.

Set a Purpose

- Read the heading. Tell students that in this section they will learn about two of the different **systems,** or groups of body parts, that make up the human body.

❷ Teach

Academic Vocabulary: *organ, system*

- Point out the word **organs** on page 234. Have a volunteer read the sentence that defines it.
- Ask: **Do you think the organs of the human body are located in just a few places, in most of the body, or throughout the entire body?** (Possible answer: throughout the entire body) Explain that the body is filled with bones, muscles, nerves, and other organs, each working when called upon to complete a specific task.
- Point out the word **system** on page 235. Have a volunteer read the sentence that defines it.
- Ask: **Aside from systems of the body, what other systems are you familiar with?** (Possible answers: coding systems for organizing books in libraries; computer systems, such as the Internet; a system of streets and avenues in a city) Discuss how these systems, much like body systems, have different parts that work together for a useful purpose.

Circulatory and Respiratory Systems

When you ride a bicycle, you turn the pedals. The pedals turn a gear, which moves a chain. As the chain turns, it moves another gear at the back wheel. Each part has a special job to do. When all of the parts work together, the bike moves forward.

You have body parts that work together too. These body parts are called **organs**. An organ is a structure that does a specific job in an organism. The human body has many organs.

Bicycles have parts that work together just like your body does.

234

Differentiated Instruction

Extra Support

Have students scan the illustrations in this chapter and identify the organs that they recognize. Then have them define organ and system in their own words.

Challenge

Have students choose a specific organ and find out more about it. Work with students so that a variety of organs are chosen. Have students draw the organ on a sheet of paper, then add labels and a caption. Students may display their drawings around the classroom.

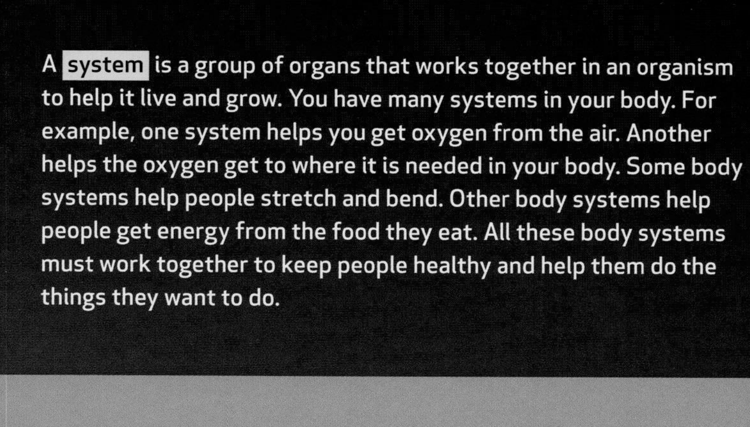

A system is a group of organs that works together in an organism to help it live and grow. You have many systems in your body. For example, one system helps you get oxygen from the air. Another helps the oxygen get to where it is needed in your body. Some body systems help people stretch and bend. Other body systems help people get energy from the food they eat. All these body systems must work together to keep people healthy and help them do the things they want to do.

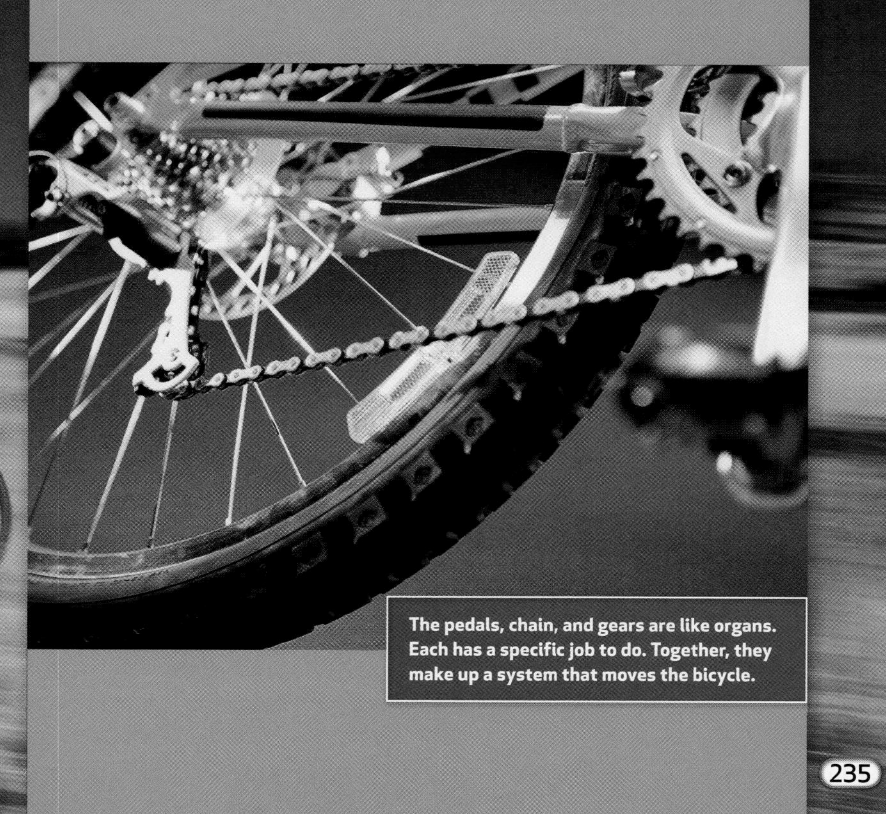

The pedals, chain, and gears are like organs. Each has a specific job to do. Together, they make up a system that moves the bicycle.

235

Science Misconceptions

Working Together Some students may believe that they use only one or two systems at a time. For example, they might believe that bicycling involves only the muscular and skeletal system. Although body systems may work harder at some times than at others, many systems must work together to complete even very simple tasks, such as picking up a penny from the floor. Athletic activities, such as riding a bicycle, involve responses and cooperation from many systems, including the circulatory, respiratory, nervous, muscular, skeletal, and endocrine systems.

Teach, continued

Explain How Organs Work Together

- Have students read pages 234–235 and read the captions to the photos.
- Ask: **How are the parts of a bicycle like the organs of the body?** (Like the parts of the bicycle, the body's organs are separate parts that work together to accomplish complex tasks.)
- Ask: **What would happen if one part of a bicycle, such as a tire or chain, broke or stopped working?** (The bicycle would no longer work or it would work poorly.) Explain that if an organ became injured or stopped working, the entire body could be affected.

Describe How Systems Function

- Ask: **When you walk or run, do you think you use just a few muscles in your legs, muscles in only your arms and legs, or muscles throughout the body?** (Possible answer: muscles throughout the body) Explain that even simple motions involve bones and muscles throughout the body. To demonstrate, invite students to walk around the room. Encourage them to pay attention to which body parts are in motion.
- Ask: **Why does walking depend on many muscles throughout the body, not just in the legs?** (because walking involves movement of many body parts, not just the legs) Explain that many muscles move to keep the body in balance.
- Ask: **Along with bones and muscles, what other organs are at work when you walk?** (Answers include the heart, lungs, and brain, which work all the time.) Discuss how the heart, lung, and brain are part of different **systems** that work together. When you run, for example, the **systems** work together to increase the rates of your heartbeat and breathing.

Notetaking

Have students take notes about each body **system** as they encounter it in this lesson. Encourage students to make a data table, such as the one below, in their science notebook.

Body System	Organs	Function
Circulatory		
Respiratory		

LESSON 3 ▫ Circulatory and Respiratory Systems

Objectives

Students will be able to:

- Describe the major organs of the circulatory system and how they function.

Science Academic Vocabulary

circulatory system

PROGRAM RESOURCES

- Digital Library at myNGconnect.com

Teach, continued

Academic Vocabulary: *circulatory system*

- Point out the term **circulatory system**. Have students repeat it.
- Say: **The name of a familiar shape is almost a part of the word *circulatory*.** Ask: **What shape is it?** (a circle) Explain that the **circulatory system** moves blood through the body in a never-ending path like a circle's path.

Explain the Purpose of the Circulatory System and Identify Its Parts

- Have students read pages 236–237, and then study the diagram on page 236 and its caption.
- Ask: **What is the job of the circulatory system?** (moving blood through the body) **Why do body parts need blood?** (Possible answer: Blood carries oxygen, a gas that body parts need to do their jobs.) Explain that blood carries other things that body parts need, including nutrients that provide them with energy.
- Ask: **What is the main organ of the circulatory system?** (the heart) Have students point to the heart in the diagram. Ask: **What does the heart do?** (pumps blood throughout the circulatory system) Explain that when the heart is beating, it is pumping blood throughout the body.
- Ask: **What are two other types of organs of the circulatory system?** (veins and arteries) Discuss how arteries carry blood away from the heart. Most of this blood is rich in oxygen. Veins carry blood back to the heart.

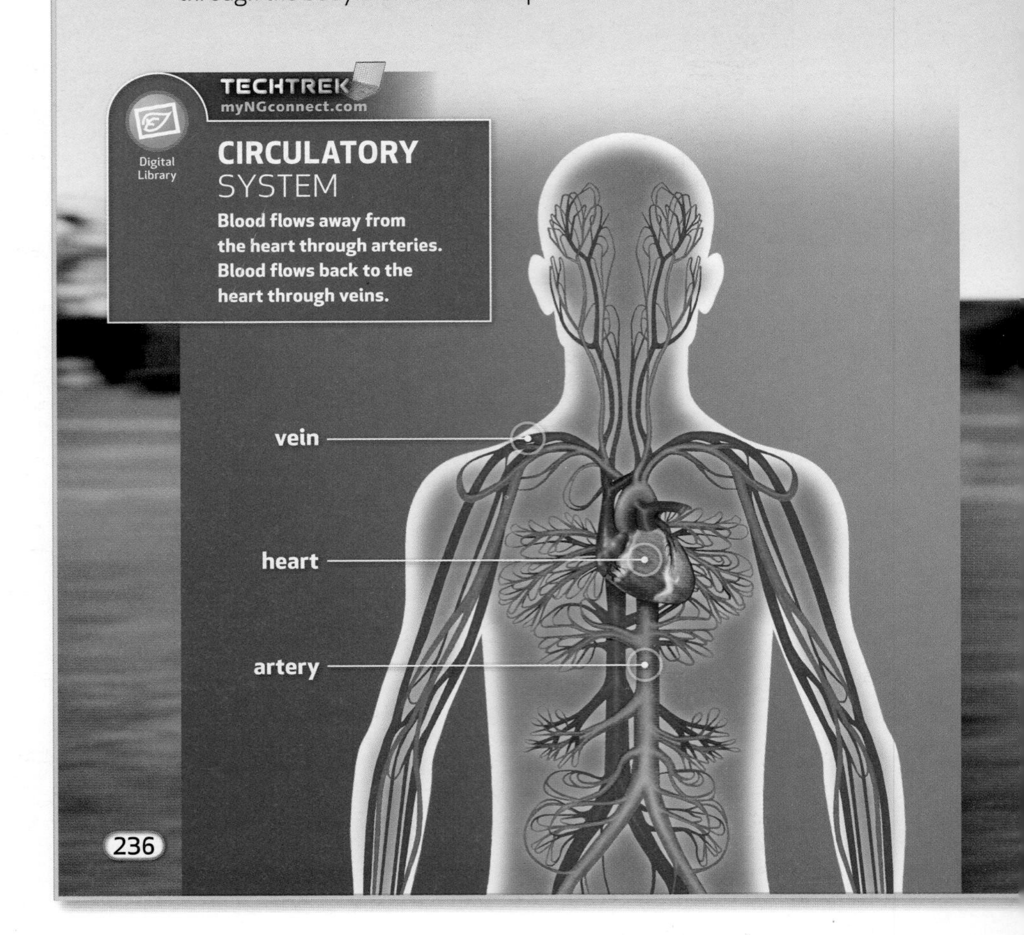

Circulatory System Have you ever felt your heart pounding while running a race? Your heart is the main organ in your circulatory system. This system carries blood throughout your body. Each time your heart beats, it pumps blood. The blood travels around the body through blood vessels.
Each part of your body needs oxygen to do its job. Your body takes the oxygen it needs from the blood. Blood, rich with oxygen, circles through the body in an endless loop.
TECHTREK
myNGconnect.com
Digital Library
CIRCULATORY SYSTEM
Blood flows away from the heart through arteries. Blood flows back to the heart through veins.
vein
heart
artery
236

Taking a Pulse

my SCIENCE notebook

Materials clay, straw, stopwatch

Have students work in pairs. Distribute a straw and a piece of modeling clay to each pair of students. Read the instructions on page 237 together. Then have partners follow the instructions. Have one partner act as observer to measure the pulse of the other partner, and then have students switch roles and repeat the activity. You may wish to act as timekeeper for the whole class. Students should record their observations and results in their science notebook.

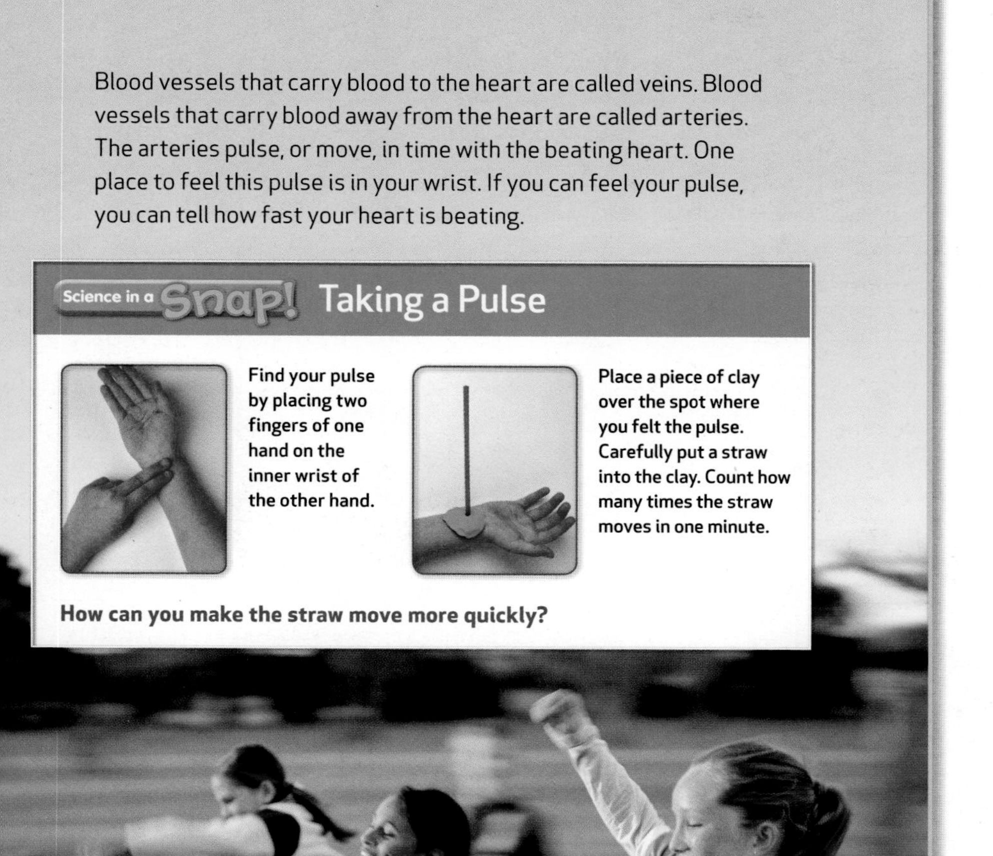
Blood vessels that carry blood to the heart are called veins. Blood vessels that carry blood away from the heart are called arteries. The arteries pulse, or move, in time with the beating heart. One place to feel this pulse is in your wrist. If you can feel your pulse, you can tell how fast your heart is beating.

Science in a Snap! Taking a Pulse

Find your pulse by placing two fingers of one hand on the inner wrist of the other hand.

Place a piece of clay over the spot where you felt the pulse. Carefully put a straw into the clay. Count how many times the straw moves in one minute.

How can you make the straw move more quickly?

The harder the body works, the faster the heart pumps. This helps get more oxygen to all the parts of the body.

237

What to Expect Pulse rates will vary among students. A typical pulse for a 10-year old is 60 to 100 beats per minute.

Quick Questions Ask students the following questions:

- **What are you feeling when you take your pulse?** (the blood moving through an artery) Explain that arteries in the wrist and neck are located just beneath the skin, and thus easy to feel.
- **How can you make the straw move more quickly?** (Exercise before you measure your pulse.) **Why does exercise increase the pulse rate?** (During exercise, the body needs more oxygen-rich blood, so the heart beats faster.)

Teach, continued

- Say: **Arteries and veins are both examples of blood vessels, or tubes that carry blood. Capillaries are the third kind of blood vessel. They are thin, tiny vessels that connect arteries and veins.** Point to examples of capillaries in the illustration. Capillaries are found throughout the body.

Describe How the Circulatory System Functions

- Point again to the diagram on page 236. Have students use their fingers to trace the flow of blood away from the heart through a set of arteries (colored in red), then through a capillary, then through veins (colored in purple) back to the heart.
- Ask: **Which organs of the body get blood from the circulatory system?** (all organs, from the top of the head to the tips of the toes) **To reach these organs, does blood travel through one pathway in and out of the heart, two or three pathways, or a very large number of pathways?** (a very large number) Point out that a large, wide artery leaves the heart and then branches into many smaller arteries. The branching continues to supply blood to the entire body. Blood travels through each branch.
- Ask: **Why do you think the circulatory system must work closely with other systems to do its job?** (Possible answer: Body parts use up oxygen that the blood brings them. More oxygen must be constantly added to the blood.) Discuss how bringing oxygen to the blood is a job for the respiratory system, which students will read about on the next page.

myNGconnect.com

Have students use the Digital Library to find other images that illustrate the **circulatory system**.

Whiteboard Presentation Have students use the information to make a computer presentation. Help them show the presentation on a whiteboard.

Objectives

Students will be able to:

- Describe the major organs of the respiratory system and how they function.

Science Academic Vocabulary

respiratory system

PROGRAM RESOURCES

- Digital Library at myNGconnect.com

Teach, continued

Academic Vocabulary: *respiratory system*

- Point out the term **respiratory system** on page 239. Ask a volunteer to define it.
- Point out that the adjective *respiratory* is similar to the verb *respire*. Write both words on the board. Say: **To *respire* is to breathe. The respiratory system is the system that you use to breathe.**

Describe the Respiratory System

- Have students read pages 238–239, observe the photo, and read the diagram.
- Ask students to take in a deep breath, hold it for a moment, then let it out. Ask: **Which body system have you just used?** (the respiratory system) **What is the job of the respiratory system?** (to bring oxygen into the body and remove carbon dioxide, a waste gas) **Why is oxygen important?** (The body needs oxygen to get energy from food.)
- Ask: **What are the main organs of the respiratory system?** (the lungs) Have students identify the lungs in the diagram on page 239. **What other organs are part of the respiratory system?** (the mouth, nose, and the tubes that lead from the mouth and nose to the lungs)
- Have students use the diagram to trace the flow of air. When you breathe in, air rushes through the nose or mouth, down air tubes, and into the lungs. When you breathe out, air reverses direction.
- Ask: **How is the air that enters the lungs different from the air that leaves them?** (The air that enters the lungs is rich in oxygen, while the air that leaves has less oxygen and more carbon dioxide.)

Respiratory System Humans need oxygen to live. Your body uses oxygen to get energy from the food you eat. Oxygen helps your body do the work it needs to do.
These soccer players' respiratory systems help them get the oxygen they need.
238

NATIONAL GEOGRAPHIC Raise Your SciQ!

The Respiratory System Air enters the respiratory system through the mouth and nose. Then it passes through the larynx, or voice box, and into a long, wide tube called the trachea. Near the lungs, the trachea splits into two tubes called bronchi (singular: bronchus). The bronchi enter the lungs and divide into bronchioles, which lead to tiny air sacs called alveoli. Alveoli have thin membranes that border capillaries. Gases are exchanged across these capillaries. When air is breathed out, it travels in the same path in the reverse direction. Air moves in and out of the lungs because of the diaphragm, a large muscle that shrinks and expands the chest cavity.

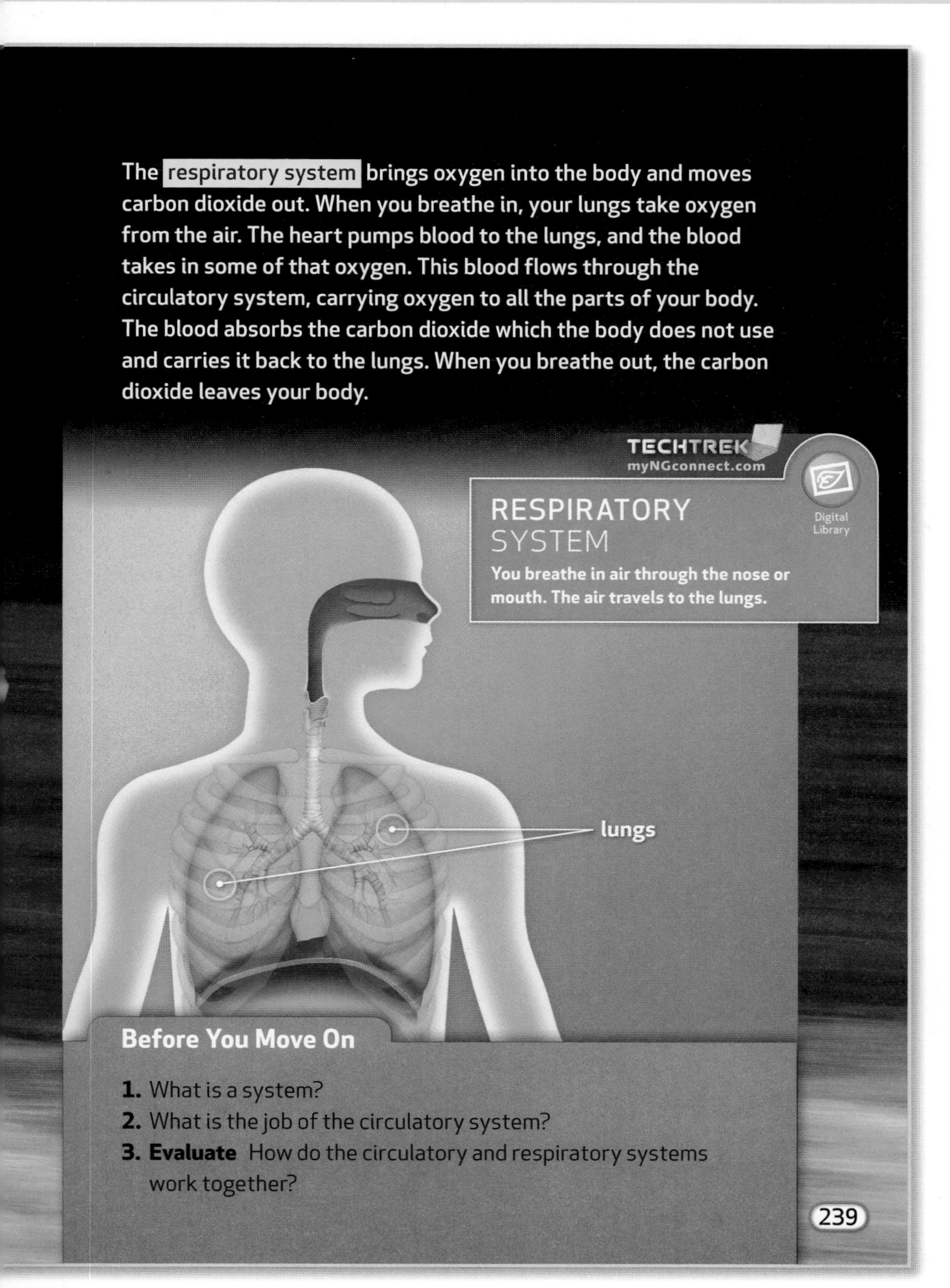

The respiratory system brings oxygen into the body and moves carbon dioxide out. When you breathe in, your lungs take oxygen from the air. The heart pumps blood to the lungs, and the blood takes in some of that oxygen. This blood flows through the circulatory system, carrying oxygen to all the parts of your body. The blood absorbs the carbon dioxide which the body does not use and carries it back to the lungs. When you breathe out, the carbon dioxide leaves your body.
TECHTREK
myNGconnect.com
Digital Library
RESPIRATORY SYSTEM
You breathe in air through the nose or mouth. The air travels to the lungs.
lungs
Before You Move On
1. What is a system?
2. What is the job of the circulatory system?
3. Evaluate How do the circulatory and respiratory systems work together?
239

Differentiated Instruction

ELL Language Support for Describing the Circulatory and Respiratory Systems

BEGINNING

Have students match organs with their body systems.

lungs *(respiratory system)*

heart *(circulatory system)*

arteries and veins *(circulatory system)*

INTERMEDIATE

Help students use Academic Language Frames to describe the circulatory and respiratory systems. For example:

- *Breathing is the job of the ____ system.*
- *Moving blood is the job of the ____ system.*

ADVANCED

Provide Academic Language Stems to help students describe the circulatory and respiratory systems. For example:

- *Breathing is the job of . . .*
- *Moving blood is the job of . . .*
- *The heart is the main organ of . . .*

Teach, continued

Explain How the Respiratory System and Circulatory System Work Together

- Have students review the diagram of the circulatory system on page 236. Ask: **If we added the lungs to this diagram, where do you think they should be drawn?** (Possible answer: to the left and right of the heart, around the network of arteries, veins, and capillaries already shown) Explain that the heart and lungs are located close together in the chest.
- Ask: **What happens to blood when it reaches the lungs?** (It loses the carbon dioxide it carries and it gains oxygen.) Explain that after leaving the lungs, the blood travels back to the heart and then to the rest of the body.
- Ask: **Why must the circulatory and respiratory systems work closely together?** (Both have the job of bringing oxygen into the body and removing carbon dioxide.)

myNGconnect.com

Have students use the Digital Library to find more images that illustrate the **respiratory system**.

Digital Booklet Have students write captions for the images they find. Students can arrange the images into a digital booklet.

3 Assess

» Before You Move On

1. **Define What is a system?** (a group of parts that work together to do a specific job)
2. **Recall What is the job of the circulatory system?** (to carry blood throughout the body)
3. **Evaluate How do the circulatory and respiratory systems work together?** (The respiratory system brings oxygen from the air into the body. It also expels carbon dioxide. The circulatory system helps move oxygen and carbon dioxide between the lungs and the rest of the body.)

LESSON 4 ▫ Digestive System

Objectives

Students will be able to:

- Describe the major organs of the digestive system and how they function.

Science Academic Vocabulary

digestive system

PROGRAM RESOURCES

- Enrichment Activities at myNGconnect.com

❶ Introduce

Tap Prior Knowledge

- Ask students what they had for breakfast or lunch. Have them describe the actions of body parts, such as the mouth, teeth, and stomach, that they used while they ate.

Set a Purpose and Read

- Read the heading. Tell students that they will read about the organs of the **digestive system,** the system that breaks apart the food they eat. Have students read pages 240–241.

❷ Teach

Academic Vocabulary: *digestive system*

- Point out the term **digestive system.** Have students repeat it.
- Point out that the root of the adjective *digestive* is the verb *digest*. Write both words on the board. Say: **To *digest* is to break apart the food you eat. The digestive system is the set of organs that do this.**

Explain the Purpose of the Digestive System

- Have students read page 240, observe the photo, and read the caption. Ask: **Why does the body need the digestive system?** (The foods we eat are too large for the body to use. The foods must be broken down into smaller pieces.)
- Ask: **Why does the body need food?** (Food provides the energy that the body needs to live, grow, and move.) Explain that every organ in the body needs energy to survive.

Digestive System

What kinds of foods do you like to eat? Whether you are eating a salad or a pizza, your body is able to change that food into energy. The digestive system breaks down food into energy and materials that the body can use.

The girl in the picture is eating. First, she chews the food, breaking it up into smaller pieces. Her saliva breaks down the pieces even further. When she swallows, the pieces go down a tube called the esophagus, and then into the stomach.

The stomach churns the pieces, breaking them down even more. From the stomach, the digested food passes into the intestines. Now the intestines can absorb materials from the food that the whole body will use.

Eating is just the first step in getting energy from food.

240

Science Misconceptions

Very Tiny Pieces of Food Some students may not understand the extent that food particles are broken apart by the digestive system. Explain that digestion breaks apart food into individual molecules that are too small to be observed with the unaided eye. Only pieces this small can move across the lining of the small intestine into the blood. Larger pieces of food, many of which the body cannot break apart, are excreted as waste.

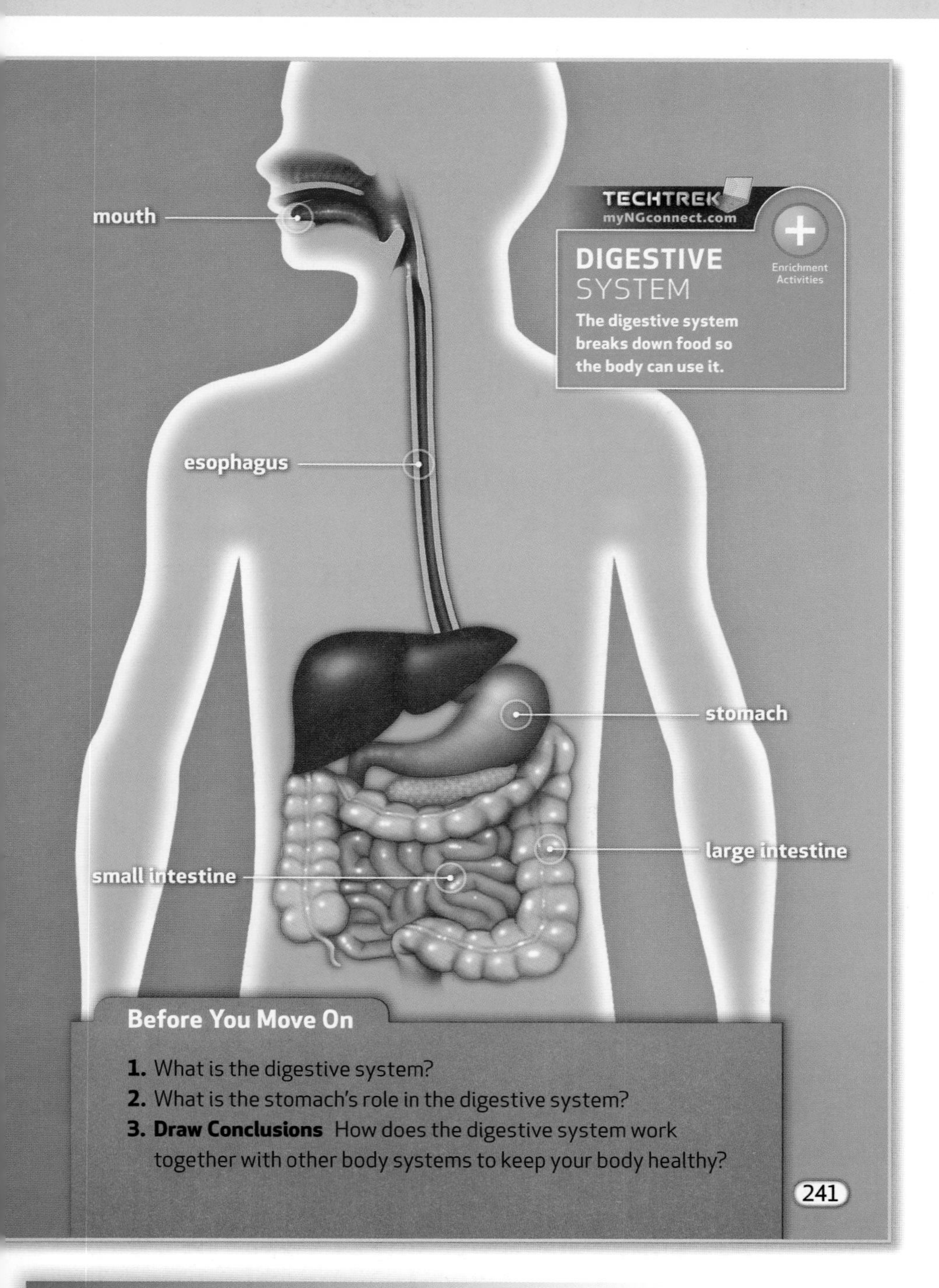
mouth
esophagus
stomach
large intestine
small intestine
TECHTREK
myNGconnect.com
DIGESTIVE SYSTEM
The digestive system breaks down food so the body can use it.
Enrichment Activities
Before You Move On
1. What is the digestive system?
2. What is the stomach's role in the digestive system?
3. Draw Conclusions How does the digestive system work together with other body systems to keep your body healthy?
241

Teach, continued

Describe the Digestive System

- Have students study the diagram of the digestive system on page 241 and read the captions and labels. Ask: **Where does digestion begin?** (the mouth) Discuss how the mouth, teeth, and saliva act to wet, mash, and break apart food.
- Ask: **Where does food go after you swallow it?** (down the esophagus) Have students trace the path of food down the esophagus to the stomach.
- Ask: **What happens to the food inside the stomach?** (The stomach churns the pieces of food, breaking them down into smaller pieces.) Explain that the stomach is lined with strong muscles and makes acid. Both the churning action and the acid help break apart food.
- Ask: **Where does the food go next?** (to the intestines) Explain that food first travels to the small intestine, where most digestion is completed. Here the tiny pieces of food are absorbed into the blood. Then the remaining food enters the large intestine, where water is absorbed into the blood. Any remaining food leaves the body as waste.

Enrichment Activities

myNGconnect.com

Have students use the Enrichment Activities to learn more about the **digestive system**.

Integrated Technology

Computer Presentation Students may use the information to make a computer presentation.

Differentiated Instruction

ELL Language Support for Describing the Digestive System

BEGINNING

Have students place the organs of the digestive system into the correct sequence: **stomach, large intestine, mouth, small intestine, esophagus**

INTERMEDIATE

Provide Academic Language Frames to help students describe the sequence of digestion:

- *Digestion begins in the ____.*
- *Next, food travels through the ____ to the ____.*
- *From the stomach, food enters the ____.*

ADVANCED

Have students write one or two sentences to answer a question about the digestive system, such as:

- *What does the digestive system do?*
- *How does food travel through the digestive system?*

3 Assess

» Before You Move On

1. **Define** **What is the digestive system?** (the system that breaks down food)
2. **Explain** **What is the stomach's role in the digestive system?** (It churns the food so it breaks down into tiny pieces.)
3. **Draw Conclusions** **How does the digestive systems work together with other body systems to keep your body healthy?** (It uses oxygen from the respiratory and circulatory systems. It uses the circulatory system to send nutrients to other body systems.)

Objectives

Students will be able to:

- Describe the major organs of the skeletal system and how they function.

❶ Introduce

Tap Prior Knowledge

- Have students describe different ways their body can move and ways it cannot move. For example, the arm can bend at the elbow but not just above or below the elbow. Then have them discuss ways that their bodies are learning to move, such as for sports or for playing a musical instrument. Lead them to recognize that learning new movements takes thought and practice.

Preview

- Read the heading. Tell students that they will read about three more systems of the body and how they work.
- Ask students to preview the images on pages 242–247. Point out the diagrams that illustrate the three body systems.

❷ Teach

Describe the Function of the Skeletal System

- Have students read pages 242–243.
- Ask: **What make up the skeletal system?** (a little over 200 bones) **What is the purpose of the skeletal system?** (supporting and giving shape to the body and allowing it to move)
- Have students compare the shapes and positions of the children shown in the photo with the diagram of the skeleton shown on page 243. Ask: **How does the skeleton inside each of these children compare to the skeleton in the drawing?** (The same bones are in the same order and arrangement, but they are in different positions.) Point out that in two children, the skeleton is upside down!

Skeletal, Muscular, and Nervous Systems
Skeletal System Hundreds of bones make up your skeletal system. The skeletal system supports and gives shape to the body. In many places in your body, two bones come together and form a joint. Your knees, shoulders, and other joints allow you to bend and move in different ways.
242

NATIONAL GEOGRAPHIC Raise Your SciQ!

Structure of the Skeleton Vertebrate skeletons are divided into two components. The axial skeleton includes the skull, spinal column, and rib cage. It supports the central axis of the body. The appendicular skeleton includes the bones of the appendages, such as limbs, wings, or fins. Girdles are collections of bones that connect appendages to the rest of the skeleton. Humans have a pelvic girdle, which connects the axial skeleton to the legs, and a pectoral girdle, which connects the axial skeleton to the arms.

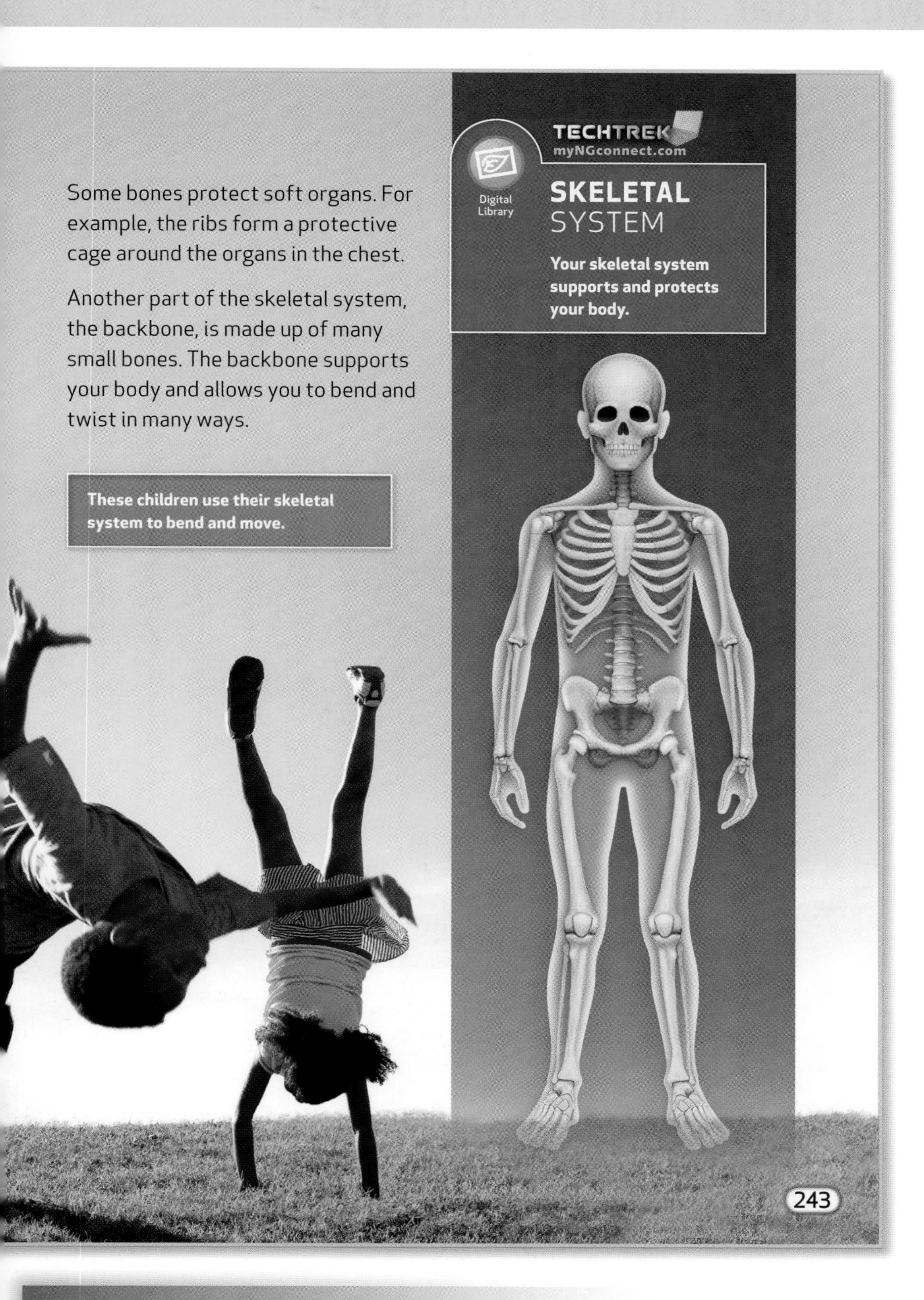

Some bones protect soft organs. For example, the ribs form a protective cage around the organs in the chest.
Another part of the skeletal system, the backbone, is made up of many small bones. The backbone supports your body and allows you to bend and twist in many ways.
TECHTREK
myNGconnect.com
Digital Library
SKELETAL SYSTEM
Your skeletal system supports and protects your body.
243

These children use their skeletal system to bend and move.

Differentiated Instruction

ELL Language Support for Describing Bones and Joints

BEGINNING

Have students match the word to its definition. For example:

The system made of bones *(skeletal system)*

Where two bones meet *(joint)*

Bones that form a cage in the chest *(ribs)*

INTERMEDIATE

Help students describe the skeletal system by using Academic Language Frames.

- *The ____ system is made of bones.*
- *Two bones meet at a ____.*
- *Two examples of joints are the ____ and ____.*

ADVANCED

Help students describe the skeletal system by using Academic Language Stems.

- *The skeletal system is . . .*
- *The function of the skeletal system is to . . .*
- *Joints form where . . .*

Teach, continued

Compare and Contrast Bones

- Point to the diagram of the skeleton. Say: **The adult skeleton has over 200 bones.** Ask: **How are all of the bones alike?** (All are relatively hard organs; all are connected to other bones at joints.)
- Ask: **How are the bones different from one another?** (The bones have different shapes and sizes.)
- Ask: **How are the bones in the hands and feet different from the bones of the legs and arms?** (Bones in the hands and feet are short, thin, and numerous. Bones in the arms and legs are longer and wider, and there are fewer of them.) Discuss how the bones in the hands and feet allow for fine, delicate movement, while the long bones of the arms and legs provide strength and support.

Identify and Describe Joints

- Ask: **What is a joint?** (a place where two bones meet) **Why are joints important?** (They allow the body to move.) Explain that bones cannot bend. But at joints, two or more bones move in relation to one another.
- Ask: **Where are some of the joints in the body?** (Answers include the shoulder, knee, elbow, ankle, within the hands and feet, and up and down the spine.)
- Ask: **Does each joint allow the same kind of movement, or do different joints allow different movements? Give examples.** (Different joints allow different movements. For example, the shoulder and hip joints allow limbs to move in a full circle, while elbow and knee joints bend, or glide, in one direction only.) Explain that some joints, such as those in the skull, allow no movement at all.

myNGconnect.com

Have students use the **Digital Library** to find images of people moving in different ways.

Digital Booklet Students may use the information to make a digital booklet about how the skeleton helps people move in different ways.

Objectives

Students will be able to:

- Describe the major organs of the muscular system and how they function.

Teach, continued

Describe the Function of the Muscular System

- Have students read pages 244–245, observe the photos, and read the captions.
- Ask: **What makes up the muscular system?** (more than 600 muscles of the body) Ask: **What does the muscular system do?** (It moves the body parts.)
- Point to the rowers on page 245. Ask: **How are these people using their muscles to move their bodies?** (They are using the muscles in their hands, arms, and shoulders to move the oars through the water.) Discuss how the rowers are using other muscles, too. Muscles in their legs and back help keep their bodies upright. Tiny muscles move the eyeball. Muscles in the face form smiles and other expressions.

Text Feature: Diagram and Labels

- Have students study the diagram of arm muscles on page 244. Then invite them to hold one arm above their head, and then slowly move it behind their head, mimicking the action in the diagram.
- Ask: **How many muscles can you see in the diagram that are involved when the elbow bends or straightens?** (two muscles: one on either side of the upper arm) **How do these muscles work together to bend or straighten the elbow?** (One muscle contracts, or gets shorter, while the other muscle extends, or gets longer.) Explain that muscles come in pairs. One muscle of the pair pulls a joint in one direction, while the other muscle pulls a joint in the opposite direction. Muscles can only pull on joints, they cannot push them.

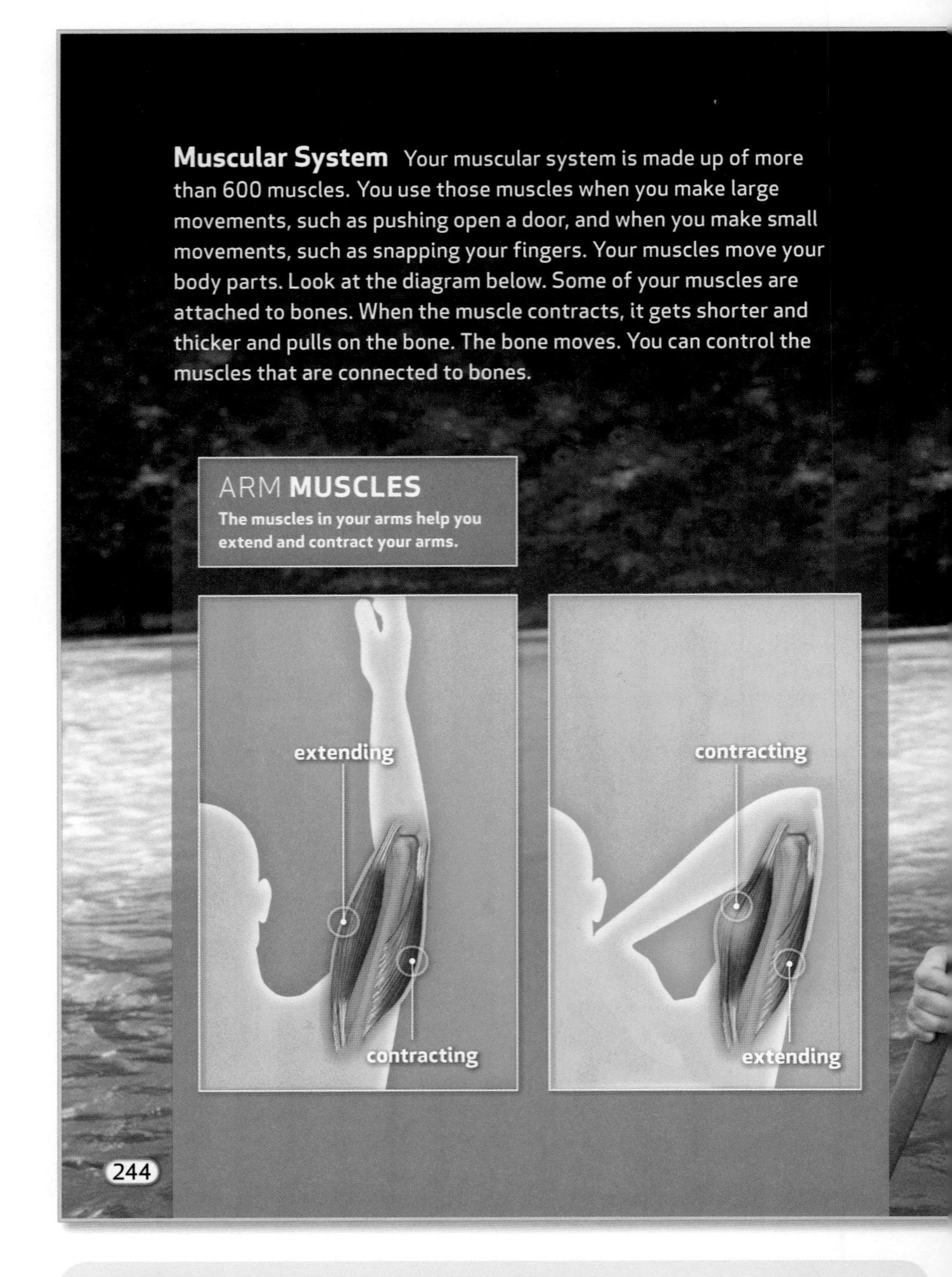

Muscular System Your muscular system is made up of more than 600 muscles. You use those muscles when you make large movements, such as pushing open a door, and when you make small movements, such as snapping your fingers. Your muscles move your body parts. Look at the diagram below. Some of your muscles are attached to bones. When the muscle contracts, it gets shorter and thicker and pulls on the bone. The bone moves. You can control the muscles that are connected to bones.
ARM MUSCLES
The muscles in your arms help you extend and contract your arms.
extending
contracting
contracting
extending
244

Raise Your SciQ!

Muscle Tissues Three types of muscle tissue are found in vertebrates. Skeletal muscles are usually attached to bones and are under voluntary control, meaning we can choose when and how to use them. Skeletal muscles are described as striated, a word that means *striped*, because of their striped appearance under a microscope. Smooth muscles are under involuntary control and are not striated. They form the walls of hollow organs such as the stomach, arteries, and intestines, where their actions help move materials through the organ. Cardiac muscle is only found in the heart. It is striated like skeletal muscle but is involuntarily controlled.

There are other muscles in the body that you do not control. For example, heart muscles keep pumping blood throughout your body. You don't have to think about moving these muscles. They move automatically.
Muscles let you make many facial expressions.
This boy's muscles help him move the paddle through the water.
245

Teach, continued

Compare the Parts of the Muscular System

- Point to the inset photo of the girl on page 245. Ask: **How is this girl using her muscles?** (She is using muscles in her face to close her eyes and purse her lips.) Explain that many muscles move the skin that covers the face.
- Ask: **Are bones and skin the only body parts that muscles move? Give an example.** (No. The heart is a muscle that pumps blood.) Explain that muscles also move food through the digestive system and help push blood through arteries.
- Ask: **Can you control all of the body's muscles? Explain.** (No. You can control the muscles that move your bones and face but not the muscles of the heart and digestive organs, which work automatically.)
- Ask: **Why is it helpful that muscles in many organs work automatically, without us thinking about them or taking control of them?** (Organs such as the heart and digestive organs are very important. Our thoughts might not be able to control them properly or constantly.)

Monitor and Fix Up

Ask students if anything they read was puzzling. The diagram on page 244 may confuse students. Discuss possible fix-up strategies. For example, students may write a summary of the diagram in their own words, compare the diagram to the photo of boaters on pages 244–245, or complete the Differentiated Instruction activity at the bottom of this page.

Differentiated Instruction

Extra Support

Have students list any new or unfamiliar words they encounter on pages 244–245. Examples include *muscular, extending, contracting,* and *automatically.* Have students work with a partner to define these words using context clues, or by consulting the glossary or a dictionary if necessary.

Challenge

Have students prepare a poster that illustrates the muscular system or both the muscular and skeletal systems. Encourage them to research the names and locations of several muscles and bones in the body, and to use drawings or clipped photographs from old magazines to show how these organs move the body.

Objectives

Students will be able to:

- Describe the major organs of the nervous system and how they function.

Science Academic Vocabulary

nervous system

PROGRAM RESOURCES

- Science Inquiry and Writing Book: *Life Science*
- Science Inquiry and Writing Book eEdition at myNGconnect.com
- Digital Library at myNGconnect.com

Teach, continued

Academic Vocabulary: *nervous system*

- Point out the term **nervous system** on page 246. Have a volunteer read the sentence that defines it. Say: **A *nerve* is an organ that carries messages. The nervous system is the system that includes nerves.**

Identify the Major Organs of the Nervous System

- Have students read pages 246–247 and then study the diagram of the **nervous system**.
- Ask: **Why is the brain the most important organ of the nervous system?** (The brain receives messages from the body and controls most of the actions of the body.) Explain that nearly all of the messages of the nervous system either begin or end in the brain.
- Ask: **Along with the brain, what are the other organs of the nervous system?** (many nerves, some of which travel through the spinal cord) Discuss how nerves are like telephone wires for the body. They carry messages very quickly.

myNGconnect.com

Have students use the Digital Library to find other images that illustrate the **nervous system**.

Computer Presentation Have students organize the information for a computer presentation.

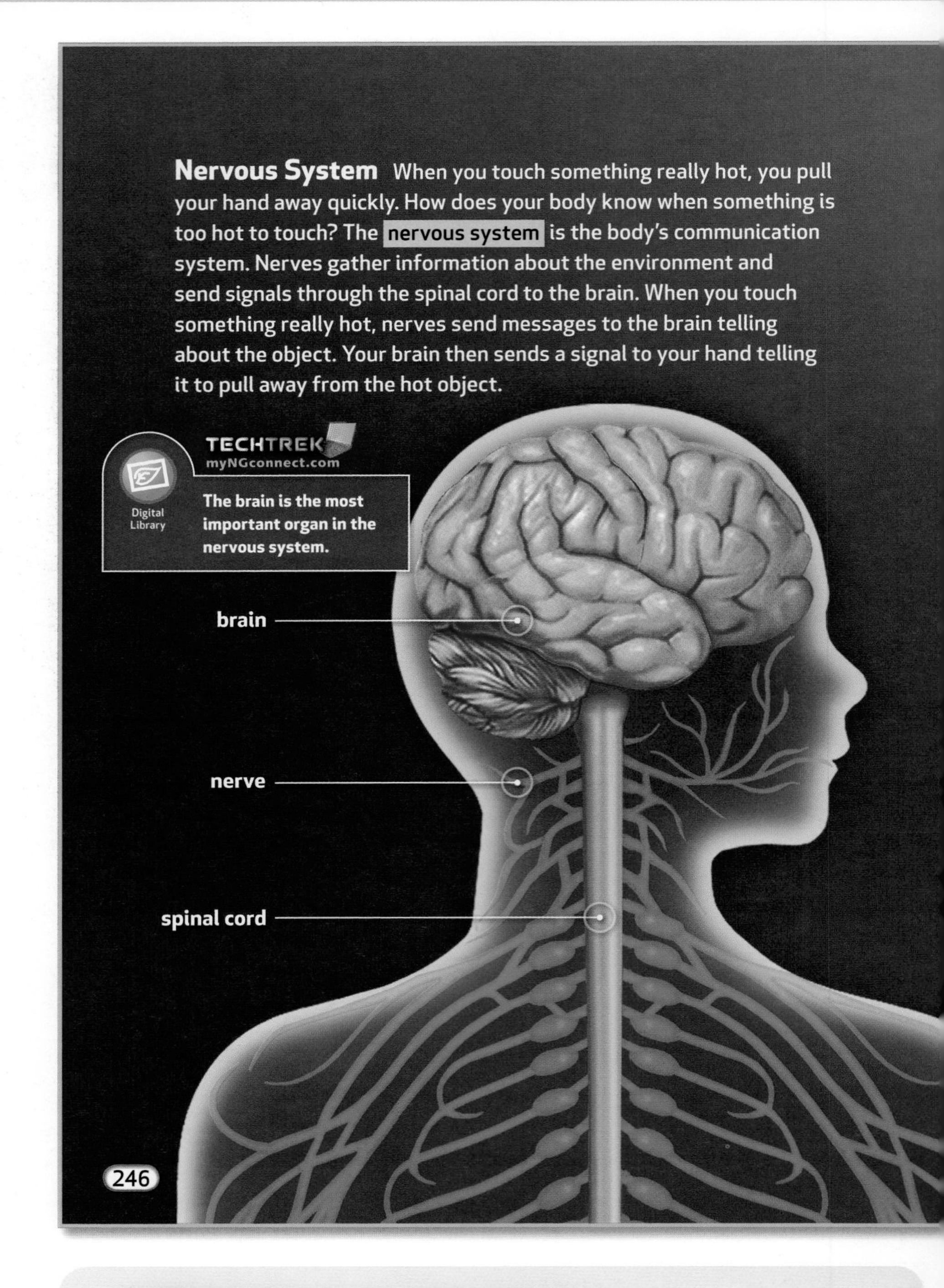

Nervous System When you touch something really hot, you pull your hand away quickly. How does your body know when something is too hot to touch? The nervous system is the body's communication system. Nerves gather information about the environment and send signals through the spinal cord to the brain. When you touch something really hot, nerves send messages to the brain telling about the object. Your brain then sends a signal to your hand telling it to pull away from the hot object.
TECHTREK
myNGconnect.com
Digital Library
The brain is the most important organ in the nervous system.
brain
nerve
spinal cord
246

Raise Your SciQ!

NATIONAL GEOGRAPHIC

Organization of the Nervous System The nervous system is divided into two main divisions. The central nervous system (CNS) includes the brain and spinal cord. The peripheral nervous system (PNS) consists of all the nerves that communicate with the brain and spinal cord. The PNS is further divided into sensory nerves and motor nerves. Sensory nerves carry messages toward the brain. These nerves lead from special sense organs, such as the eyes, ears, and nose, as well as from the skin, where they may carry information about touch, temperature, or pain. Motor nerves carry messages away from the brain. Many lead to the muscles that move the body. Others lead to organs that release hormones, which are chemical messages in the body.

The brain is like a control center. It gets messages about all your senses—sight, hearing, touch, taste, and smell. The brain uses the messages to tell your body how to respond to changes in your environment. For example, when you get hot, your brain tells your body to perspire. The perspiration cools your body.

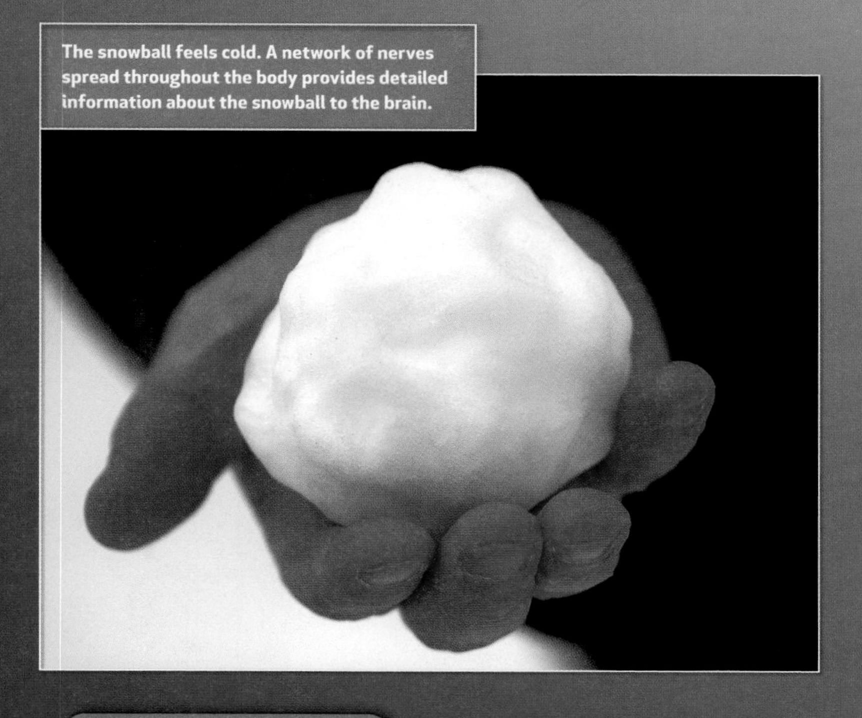

The snowball feels cold. A network of nerves spread throughout the body provides detailed information about the snowball to the brain.

Before You Move On

1. What is the skeletal system?
2. What is the brain's role in the nervous system?
3. **Infer** Suppose you go to a hot beach. How might your nervous system help your body respond?

247

Differentiated Instruction

ELL Language Support for the Nervous System

BEGINNING

Ask questions with choices to help students understand the parts of the nervous system and their role:

Does the nervous system send messages, receive messages, or send and receive them? Is the control center the brain or spinal cord?

INTERMEDIATE

Help students describe the nervous system using Academic Language Frames.

- *The ____ system allows body parts to communicate.*
- *The ____ is the control center of the nervous system.*
- *The ____ connects the brain to many nerves.*

ADVANCED

Help students describe the nervous system using Academic Language Stems.

- *The function of the nervous system is . . .*
- *The parts of the nervous system include . . .*
- *The brain is important because it . . .*

Teach, continued

Describe the Nervous System

- Have students observe the photo on page 247. Ask: **How does the brain know that the snowball is cold?** (Nerves in the skin sense the temperature of the snowball, and then other nerves carry this message to the brain.) Explain that nerves carry messages to the brain very quickly. When you touch a snowball, you know almost instantly that it is cold.
- Ask: **How does it help the body to have nerves that work so quickly?** (Quick nerves allow the body to respond quickly, which can help it accomplish tasks and stay safe.) Explain that the quick speed of nerves allows people to catch a baseball, dance ballet, and sing songs together. Fast nerves also help jerk a hand away from a hot stove.

Science in a Snap!

Observe How Eyes React to Light

Science Inquiry and Writing Book: *Life Science,* page 15

Read the instructions together with students. Have students work with a partner. Students should record their observations in their science notebook.

What to Expect When the lights are dimmed, students will observe that their partner's pupils are large. When the lights become brighter, their partner's pupils become smaller. Pupils become larger or smaller as they react to the amount of light.

3 Assess

» Before You Move On

1. **Define** **What is the skeletal system?** (the framework of bones that gives a body structure and protects soft organs.)
2. **Explain** **What is the brain's role in the nervous system?** (The brain interprets the messages about the environment. It controls body responses and allows us to think and react.)
3. **Infer** **Suppose you go to a hot beach. How might your nervous system help your body respond?** (Possible response: My nerves would tell my brain that it was hot. Then I would start to perspire. I might decide to drink water, look for shade, or cool off in the ocean.)

Objectives

Students will be able to:

- Recognize that eating a balanced diet and getting regular exercise are healthful habits.

❶ Introduce

Tap Prior Knowledge

- Ask students to discuss the last time they were sick or injured. Lead them to recognize that illnesses and injuries are not pleasant and should be avoided.

Preview and Read

- Read the heading. Have students preview the photos on pages 248–251. Tell students they will read about ways to keep body systems healthy.
- Have students read pages 248–251.

❷ Teach

Discuss the Importance of Staying Healthy

- Have students observe the photos on pages 248–249 and read the captions. Ask: **How do these photos show people staying healthy?** (The basketball players are getting exercise, and the family is eating a balanced meal.)
- Point out that the people in the photos are smiling and appear happy. Say: **Good health is important, because you feel better when you are healthy. You are able to do things you want to do, and you may live a longer life.**

Recognize Exercise as a Healthful Habit

- Ask: **What sports do you enjoy playing?** (Accept all answers.) **Why is playing sports a good form of exercise?** (It gets body systems to work hard and keeps them in top condition.) Discuss how many activities can provide useful exercise, including team sports, individual sports, playing on the playground, going for a ride on your bike, or even walking around your neighborhood.

Keeping Body Systems Healthy

Diet, Exercise, and Rest What can you do to keep your body systems working the way they should? One of the best things to do is to eat many different kinds of food. Eating a balanced diet can help you get all the nutrients your body needs. Nutrients are materials that help organisms live and grow. Eating healthful foods helps your digestive system work well. It also helps keep your circulatory system free of unhealthful materials that can clog blood vessels.

Most experts say young people should get at least an hour of exercise a day. You might walk, ride a bike, or play basketball.

248

Science Misconceptions

The Value of Exercise Students may have many misconceptions about the value of exercise and ways to get it. Exercise does more than just strengthen muscles. It helps all body systems stay healthy, improves sleep, and keeps the mind sharp and active. While team sports can provide useful exercise, so can individual sports and other activities, such as jogging, bicycling, and swimming. Fitness experts also recommend that people vary the kinds of exercise they do to build both strength and endurance in their muscles.

What are your favorite activities? Getting regular exercise is a health habit that helps all your body systems stay strong. It keeps your lungs healthy. That helps your respiratory system. It keeps your heart healthy. That helps your heart pump blood through your body. Regular exercise can make your bones and muscles stronger too.

In addition to exercise, your body needs rest to stay healthy and grow. Most ten-year-olds need about ten hours of sleep a night. Getting enough rest helps all your body systems work well.

A balanced meal contains many nutrients that people need to stay healthy.

249

Teach, continued

Recognize Eating a Balanced Diet as a Healthful Habit

- Ask: **What is a balanced diet?** (a diet made of many different kinds of foods) Explain that no single food contains all the nutrients that the body needs. A balanced diet is made of a variety of foods, including breads, meats, dairy products, fruits, and vegetables. Together they provide all the nutrients the body needs.
- Point to the photo on page 249. Ask: **What is the family eating?** (spaghetti, a salad, and bread) **Why is this a healthy meal?** (It contains a variety of foods, all of which are healthful because the foods provide nutrients the body needs.) **What other foods might the family include in this meal to promote a balanced diet?** (Possible answers: milk, cheese, or other dairy products; more vegetables; and meat or fish)
- Ask: **What foods do you think are not part of a balanced diet and should be eaten only in small amounts?** (Answers include foods that are high in sugar, such as candy, soda, and other sweet foods.) Discuss how these foods are often called *junk foods*. They provide sugar but not the other nutrients that the body needs.

Make a Chart

Have students make a chart in their science notebook for the different practices and habits that promote good health that are presented in the lesson. Encourage students to record their ideas about how they can develop these habits.

NATIONAL GEOGRAPHIC Raise Your SciQ!

The Food Guide Pyramid The U.S. Department of Agriculture publishes the Food Guide Pyramid, a general guide to a balanced diet and proper nutrition. The pyramid describes the six food groups—grains, vegetables, fruits, oils, milk, and meat & beans—that make up a balanced diet. Because dietary needs vary from person to person, the new Food Guide Pyramid does not prescribe specific servings for each food group. Instead, the pyramid shows how eating a variety of foods that represent each food group is the key to a balanced diet and proper nutrition.

Objectives

Students will be able to:

- Recognize that washing your hands frequently and avoiding drugs, alcohol, and tobacco are healthful habits.

Teach, continued

Recognize Cleanliness as a Healthful Habit

- As instructed in the first paragraph on page 250, have students look at their hands. Ask: **Why might your hands not be as clean as they may look?** (Many germs may be on the hands. Germs are too small to see without a microscope.)
- Have students observe the photos on page 250 and read the captions. Ask: **Why is washing your hands a healthful habit?** (Hand washing kills the germs on your hands, and some germs can make you sick.) Discuss how hands can pick up germs from any object they grasp. Then the germs can easily transfer from the hands to the food we eat.
- Ask: **When should you wash your hands?** (Answers include before eating meals, after using the restroom, and after playing outdoors.)
- Ask: **Why do you think doctors and nurses wear gloves?** (to protect themselves and their patients from germs) Discuss other jobs where workers wear gloves, such as kitchen staff in restaurants and sanitation workers.

Text Feature: Photo and Caption

- Point to the photo of bacteria on human skin. Ask: **Along with a camera, what tools do you think were used to take this photo?** (Possible answer: a microscope and a computer) Explain that a computer added the colors to this image to make the bacteria easier to observe. This kind of bacteria are oval shaped, but they are not actually purple.
- Ask: **What does the photo show about your skin?** (It shows that while you cannot ordinarily see bacteria, many of them may be living on your skin.)

Healthful Habits Look at your hands. Do they look clean? You might not think so if you could see them under a microscope!

When people touch something, such as a book or a doorknob, they often leave some germs behind. When you touch that book or doorknob, you pick up the germs the other people left on it. Germs can cause you to become ill. To keep from getting sick, wash your hands before you eat and after using the restroom. Washing your hands often is the best way to protect yourself from harmful germs.

Your hands can be loaded with germs and bacteria. The bacteria shown below are magnified many times.

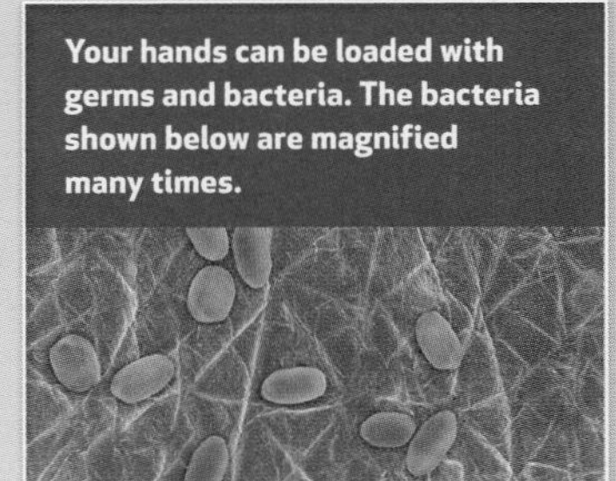

Washing your hands can keep you from getting sick and from passing your germs to other people.

250

NATIONAL GEOGRAPHIC Raise Your SciQ!

More Healthful Habits Along with regular hand washing, good health habits include daily baths or showers, keeping nails and hair clean and neat, and brushing and flossing teeth regularly. All of these habits help keep germs from spreading. Germs can cause acne, bad breath, dental caries (cavities), and skin rashes, as well as more serious infections and illnesses. However, germs are a natural part of the environment, and many are beneficial to humans. For example, bacteria in the intestines help break down food and produce important vitamins. Bacteria also help break apart dead skin cells. Excessive vigilance against germs is not usually healthful.

Staying away from things that can harm your body also keeps your body healthy. These things include drugs, alcohol, and tobacco. They could cause diseases in many of your body systems, and may temporarily alter your thinking and behavior.

Healthful habits keep these kids happy and growing strong.

Before You Move On

1. Why is eating a balanced diet important?
2. How can washing your hands help protect you from getting a cold?
3. **Draw Conclusions** A doctor advises a patient who does not exercise to find a physical activity she likes to do. Why might the doctor have asked her to choose her own activity?

Differentiated Instruction

ELL Language Support for Creating Healthful Habits

BEGINNING

Ask students either/or questions about healthful habits: ***Should you wash your hands with soap or without soap? Should you drink alcohol or stay away from alcohol?***

INTERMEDIATE

Have students use Academic Language Frames to write sentences about healthful habits.

- *____ can cause illness.*
- *One way to stop germs from spreading is to ____.*
- *You should stay away from ____, ____, and ____.*

ADVANCED

Provide Academic Language Stems to help students describe healthful habits.

- *Germs can be harmful because . . .*
- *Washing hands is healthful because . . .*
- *You should stay away from . . .*

Teach, continued

Recognize the Dangers of Harmful Substances

- Ask: **Why should you avoid using drugs, alcohol, and tobacco?** (These things can harm the body, often by changing how you think or act.) Discuss how drugs, alcohol, and tobacco can each be addictive. This means that once you begin using them, it can be very hard to quit.
- Ask: **What should you do if someone offers you a product that contains alcohol, tobacco, or drugs?** (Tell them no.) Describe how alcohol and tobacco are part of many products, such as alcoholic beverages and cigarettes, that are illegal for children to buy or use. Some drugs can only be used in medicines, while other drugs are illegal in all circumstances.
- Say: **By avoiding these products, you can keep your body healthy and live a longer, happier life.**

3 Assess

» Before You Move On

1. **Identify Why is eating a balanced diet important?** (Your body needs a variety of nutrients to stay healthy. You can get all these nutrients only by eating a variety of foods, and these foods make up a balanced diet.)
2. **Cause and Effect How can washing your hands help protect you from getting a cold?** (People touch things and leave behind germs, especially when they are ill with a cold, flu, or other disease caused by germs. If you touch the same thing, you may pick up those germs. When you wash your hands, you can wash off the germs.)
3. **Draw Conclusions A doctor advises a patient that is new to exercise to find a physical activity she likes to do. Why might the doctor have asked her to choose her activity?** (The patient will be more likely to keep exercising if she enjoys the activity.)

LESSON 7 ▫ The Body Systems of Animals

Objectives

Students will be able to:

- Recognize that animals are composed of different parts performing different functions and working together for the well being of the organism.

❶ Introduce

Tap Prior Knowledge

- Have students name an animal and describe a special way it can move or use its body. Record their responses. Examples include kangaroos (hopping), tigers (leaping and pouncing), dolphins (swimming), and most birds (flying). Discuss how different animals have different abilities—and differences in their organ systems—that help them survive.

Preview

- Read the heading. Have students preview the photos on pages 252–255. Remind students that they have learned about the systems inside the human body. Tell students they will now apply what they have learned to the bodies of different animals.

❷ Teach

Compare the Circulatory System in Different Animals

- Have students read pages 252–253.
- Point to the photo of the grasshopper on page 252. Have a volunteer read the caption. Ask: **What organ system does the caption discuss?** (the circulatory system) Ask: **Why does a grasshopper need a circulatory system?** (for the same reason that humans do, to bring nutrients and oxygen to its body and to remove waste) Explain that many kinds of animals have a system to circulate blood or a fluid that acts like blood.
- Ask: **How is the circulatory system similar in a grasshopper and human?** (In both, a heart pumps blood throughout the body.) **How is it different?** (A grasshopper's blood flows through a tube and into open spaces in the body. Human blood flows only through blood vessels.)

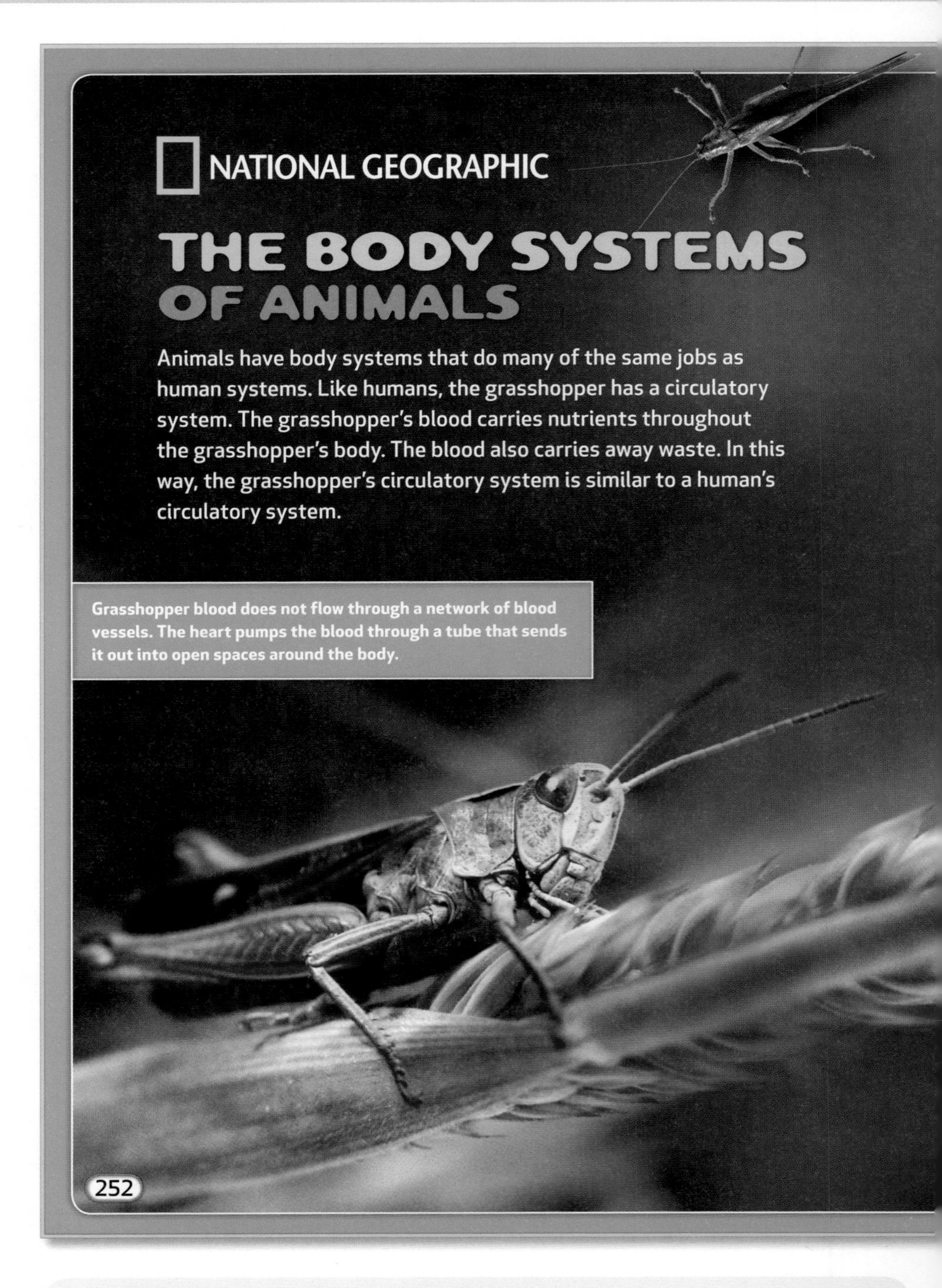

NATIONAL GEOGRAPHIC
THE BODY SYSTEMS OF ANIMALS
Animals have body systems that do many of the same jobs as human systems. Like humans, the grasshopper has a circulatory system. The grasshopper's blood carries nutrients throughout the grasshopper's body. The blood also carries away waste. In this way, the grasshopper's circulatory system is similar to a human's circulatory system.
Grasshopper blood does not flow through a network of blood vessels. The heart pumps the blood through a tube that sends it out into open spaces around the body.
252

NATIONAL GEOGRAPHIC Raise Your SciQ!

Open and Closed Circulatory Systems All vertebrates and a few invertebrates have a closed circulatory system, in which blood flows only through blood vessels. Many invertebrates, including insects, have an open circulatory system, in which the heart pumps blood into one or more sinuses, or body cavities. Inside the sinuses, blood bathes exposed tissues and gases and other materials are exchanged. Blood diffuses between cells back to the heart. Other invertebrates do not have a heart or blood, but they still move fluids through their bodies. Sea stars and other echinoderms have a water vascular system, which is a set of canals and tube feet that carry water and nutrients. Sponges do not have organs or organ systems, but special cells help move water through hollow openings in their bodies.

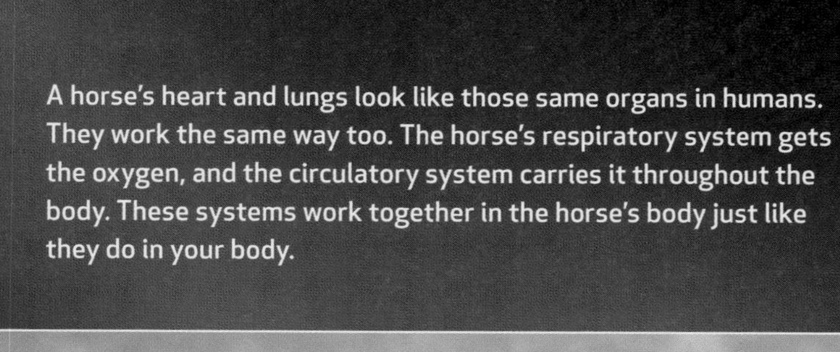

A horse's heart and lungs look like those same organs in humans. They work the same way too. The horse's respiratory system gets the oxygen, and the circulatory system carries it throughout the body. These systems work together in the horse's body just like they do in your body.

Galloping horses need energy and oxygen just like human runners do.

253

Teach, continued

Text Feature: Photos and Captions

- Have students observe the photo of horses on page 253 and read the caption.
- Ask: **How can you tell that these horses have healthy respiratory and circulatory systems?** (The horses are running, which means their muscles are working hard. The circulatory and respiratory systems are providing the nutrients and oxygen that muscles need.)
- Point to the grasshopper. Ask: **Do you think a grasshopper could move as powerfully as a horse? Why or why not?** (Possible answer: No. The grasshopper does not have the muscles and bones needed to run like a horse.) Explain that the grasshopper's circulatory and respiratory systems also limit its size and speed. While the systems are efficient enough for an animal the size of a grasshopper, they would not allow a large horse to run quickly.

Make Inferences

Ask students to name other animals that are fast, powerful runners. Answers include cheetahs, ostriches, dogs, wolves, gazelles, and coyotes. Ask: **Do you think that the respiratory and circulatory systems of fast-running animals are more like those of a human or a grasshopper? Explain your inference.** (Possible answer: A human. Horses are fast runners that have similar circulatory and respiratory systems to humans, so the other fast runners likely are similar to humans, too.)

Social Studies in Science

Mustangs The photo on page 253 shows a group of wild horses, also called mustangs. The term *mustang* comes from the Spanish word *mustengo*, which means "ownerless animal." American mustangs originally came from a Spanish stock of horses that were brought here in the early sixteenth century. At one time, an estimated 2 million mustangs lived throughout the United States. Today, only about 25,000 mustangs remain. Have students research the populations of wild horses that still roam North America.

Objectives

Students will be able to:

- Recognize that animals are composed of different parts performing different functions and working together for the well-being of the organism.

PROGRAM RESOURCES

- Learning Masters Book, page 69, or at myNGconnect.com

Teach, continued

Compare the Skeletal Systems of Animals

- Have students read page 254.
- Ask: **What is the human skeleton made of?** (bones) **Where is the skeleton located?** (inside the body, spanning from head to foot) Discuss how horses, cats, dogs, and many other animals have internal skeletons like the human skeleton. These animals are called the *vertebrates*.
- Have students observe the photo on page 254 and read the caption. Ask: **Where is the lobster's skeleton?** (the shell on the outside of its body) **How does the lobster's skeleton compare to a human skeleton?** (The two types of skeletons look different from each other and are located in different places. Both skeletons protect and give shape to the body.)

Compare the Nervous Systems of Animals

- Have students read page 255.
- Ask: **What is the function of the nervous system?** (to detect and respond to changes in the environment; to control all body activities) Remind students that the nervous system gets information from sense organs, which in humans include the eyes, ears, and nose.
- Point to the photo of the lobster on page 254. Ask: **What senses do lobsters use to detect changes in their environments?** (eyes, antenna, and hair-like parts on their shells) **Why do you think the lobster needs to use its senses?** (Possible answers: to find food, to avoid enemies, to find a mate) Explain that all animals use their senses for many of the same reasons.

Many animals have a skeleton to support their body, protect the soft parts, and allow the body to move. Some animals have a backbone. Their skeleton is inside their body. Horses, cats, and dogs have a backbone. Humans do too. Some animals, such as lobsters, have a skeleton on the outside of the body. A lobster's tough shell is different from your skeletal system, but it does the same job. It gives the lobster its structure and protects it.
Unlike human skeletons, a lobster's skeleton does not grow as the lobster grows. Instead, it splits open and the lobster forms a new, bigger skeleton around itself.
254

NATIONAL GEOGRAPHIC Raise Your SciQ!

Hydrostatic Skeletons Not all invertebrates have an exoskeleton. Jellies and many kinds of worms have a hydrostatic skeleton. This type of skeleton uses water for support instead of bones or hard shells. A hydrostatic skeleton does not provide hard parts for muscles to attach to. Instead the muscles surround a fluid-filled body cavity. By contracting certain muscles in the body wall, the animal moves by shortening or elongating different parts of its body. The wriggling of a worm and the contractions of the bell of a jelly are examples of this type of motion.

Nervous systems help animals survive in their environment. Lobsters use their eyes, antennae, and hair-like parts on their shells to gather information underwater. This information is sent along the nerve cord to the brain.

In animals with backbones, information passes from nerves to the spinal cord to the brain. In this way, birds, dogs, and humans are alike. They all see through their eyes, hear through their ears, and smell through their nose.

Hawks and other birds of prey use their sharp eyesight to gather information. They can detect small animals moving along the ground, even when the birds are soaring high above.

255

Share and Compare

Comparing Body Systems

my SCIENCE notebook

Give students the Learning Master. Have them review Chapter 6 to compare the different body systems. Then have them complete the graphic organizer. Have partners discuss their graphic organizers and compare the different body systems.

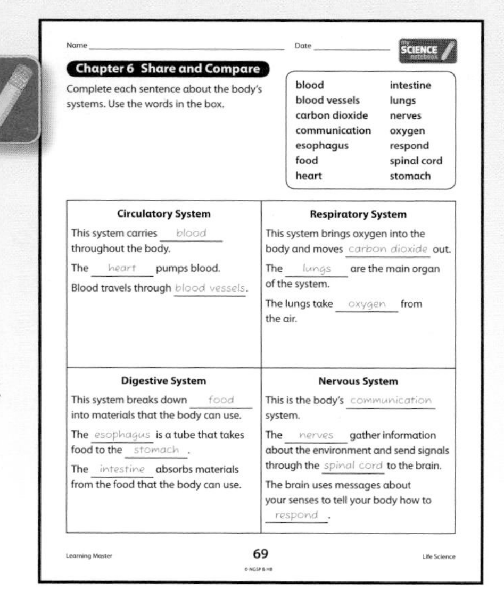

Name ______ Date ______

Chapter 6 Share and Compare

Complete each sentence about the body's systems. Use the words in the box.

blood	intestine
blood vessels	lungs
carbon dioxide	nerves
communication	oxygen
esophagus	respond
food	spinal cord
heart	stomach

Circulatory System	Respiratory System
This system carries blood throughout the body. The heart pumps blood. Blood travels through blood vessels.	This system brings oxygen into the body and moves carbon dioxide out. The lungs are the main organ of the system. The lungs take oxygen from the air.

Digestive System	Nervous System
This system breaks down food into materials that the body can use. The esophagus is a tube that takes food to the stomach. The intestine absorbs materials from the food that the body can use.	This is the body's communication system. The nerves gather information about the environment and send signals through the spinal cord to the brain. The brain uses messages about your senses to tell your body how to respond.

Learning Master 69 Life Science

Learning Master 69 at myNGconnect.com

Teach, continued

- Have students observe the photo of the hawk on pages 254–255 and read the caption. Ask: **Do you think a lobster or a hawk has a more powerful nervous system? Explain.** (Possible answer: The hawk. Its nervous system allows it to fly and dive in the air, which requires precise coordination. It also has keen eyesight, as well as senses of hearing and smell.)

3 Assess

1. **Recall How is the circulatory system of a grasshopper different from the human circulatory system?** (Blood vessels do not carry blood to all parts of a grasshopper's body. Instead the blood collects in open spaces.)
2. **Compare and Contrast How do skeletal systems compare between animals with backbones and animals without backbones?** (In animals with backbones, such as humans and horses, the skeleton includes a backbone and is inside the animal's body. Animals without backbones do not have this type of skeleton. Many of these animals, including lobsters, have a hard outer skeleton that acts to protect and give shape to their bodies.)
3. **Generalize Why must many organ systems work together to keep an animal alive? Describe a specific example.** (Possible answer: All organ systems depend on other organ systems to bring them oxygen and nutrients, remove wastes, and provide coordination and control. For example, for a hawk to get the food it needs, it must use its nervous system to find prey and coordinate its muscles, its circulatory and respiratory system to bring oxygen to its body, and its digestive system to eat and digest the prey.)

Objectives

Students will be able to:

- Investigate through Guided Inquiry (answer a question; make and compare observations; collect and record data and observations; generate explanations and conclusions based on evidence; share findings; ask questions based on observations to increase understanding; adjust explanations based on findings and new ideas).
- Recognize that humans need healthy food, cleanliness, exercise, and rest to grow and maintain good health.
- Use a stopwatch to aid observations and data collection.
- Judge whether measurements and computations are reasonable.

Science Process Vocabulary

count, conclude

PROGRAM RESOURCES

- Science Inquiry and Writing Book: *Life Science*
- Science Inquiry and Writing Book eEdition at myNGconnect.com
- Inquiry eHelp at myNGconnect.com
- Science Inquiry Kit: *Life Science*
- Learning Masters Book, pages 70–73, or at myNGconnect.com
- Inquiry Rubric: Assessment Handbook, page 187, or at myNGconnect.com
- Inquiry Self-Reflection: Assessment Handbook, page 200, or at myNGconnect.com

MATERIALS

Kit materials are listed in italics.

stopwatch

❶ Introduce

Tap Prior Knowledge

- Ask: **What changes do you notice in your body when you run?** (I have to breathe more rapidly, my heart beats faster, and I get tired.)

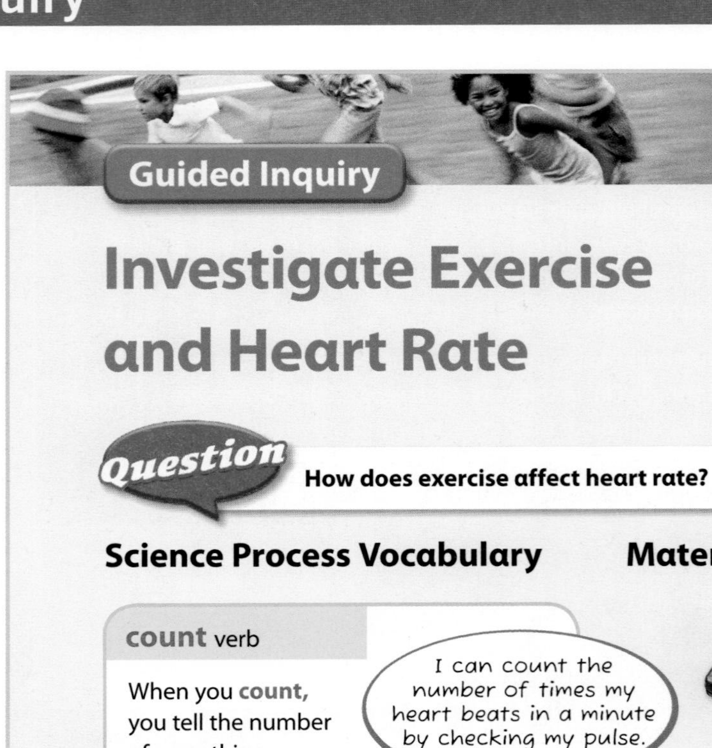

Guided Inquiry

Investigate Exercise and Heart Rate

Question How does exercise affect heart rate?

Science Process Vocabulary

count verb

When you **count,** you tell the number of something.

I can count the number of times my heart beats in a minute by checking my pulse.

conclude verb

When you **conclude,** you decide something based on evidence.

I conclude that a change in activity can change how fast my heart beats.

Materials

stopwatch

64

MANAGING THE INVESTIGATION

Time

30 minutes

Groups

Small groups of 4

Advance Preparation

- Clear an area where each student can exercise.

What To Expect

- Students should observe that their heart rate increases after they exercise. Students should also conclude that more strenuous exercise causes their heart rate to increase more.

Do an Experiment

Write your plan in your science notebook.

Make a Hypothesis

In this investigation, you will choose an exercise and measure your partner's heart rate before and after he or she does the exercise. How will exercise affect heart rate? Write your **hypothesis.**

Identify, Manipulate, and Control Variables

Which variable will you change?
Which variable will you observe or measure?
Which variables will you keep the same?

What to Do

1. Work with a partner. Choose an exercise you will do in this **investigation.** You may choose running in place, jumping jacks, or sit-ups.

2. Have your partner sit on a chair. Find your partner's pulse by placing your first 2 fingers on the thumb-side of his or her wrist, below the bottom of the thumb. Press firmly with flat fingers. Slowly move your fingers around until you feel the beats. Practice **counting** the beats until you feel comfortable doing it.

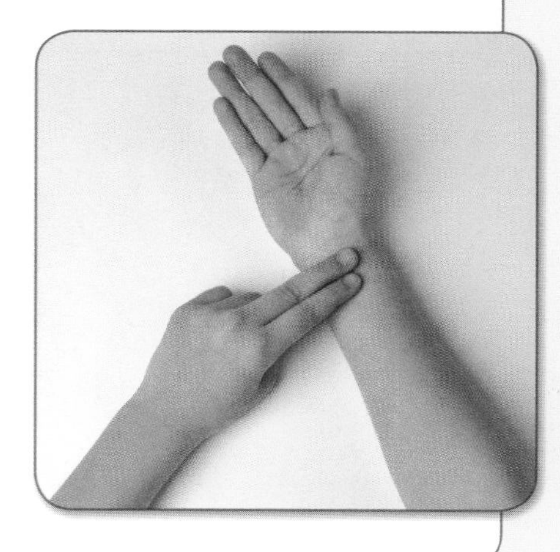

65

Teaching Tips

- Other exercises may include dancing, jumping jacks, or jogging in place. If possible, students could also walk or march in the hallway.
- Explain that our bodies respond to changes in the environment. Ask students to give examples of this, such as changes that cause eye blinking, shivering, sweating, increased or decreased heart rate, and salivating.

Introduce, continued

Connect to the Big Idea

- Review the Big Idea Question, *How do the parts of an organism work together?* Explain that in this inquiry, students will explore how heart rate is affected by exercise.
- Have students open their Science Inquiry and Writing Books to page 64. Read the Question and invite students to share ideas about how exercise affects heart rate.

2 Build Vocabulary

Science Process Words: count, conclude

Use this routine to introduce the words.

1. **Pronounce the Word** Say **count**. Have students repeat the word.
2. **Explain Its Meaning** Choral read the sentence. Ask students for another word or phrase that means the same as **count**. (tell the number of something)
3. **Encourage Elaboration** Ask: **How many students can you count in this class?** (Answers will vary.)

ELL Use Language Frames

Write *I* ***count*** _____. Ask: **How many books are on your desk?** Have students use the language frame to respond. Repeat with other items in the classroom.

Repeat for the word **conclude**. To encourage elaboration, ask: **What information do you need to analyze in order to form a conclusion?** (data and observations)

3 Guide the Investigation

- Distribute materials. Read the inquiry steps on pages 64–65 together with students.
- Guide students in writing their hypotheses. They may use "If..., then..." statements.
- Guide students in planning an experiment. Students should list what they will change, what they will observe or measure, and what they will keep the same.
- In step 2, explain that each pulse beat represents a heartbeat. Heart rate is measured by taking a pulse, usually in beats per minute.

Guide the Investigation, continued

- Discuss that experiments are an important way, but not the only way, to carry out scientific investigations. Remind students that in experiments, scientists change one variable at a time and control other variables to test a hypothesis. Scientists can also make models or conduct systematic observations without manipulating variables.
- Before step 4, remind students not to overexert themselves. Have each student within a group do the same exercise so that they can compare data.

❹ Explain and Conclude

- Guide students as they record observations in their tables and explain patterns in the data. Have students share observations with other groups.
- Say: **After comparing observations with other groups, does your data seem reasonable?** If not, discuss possible sources of error.
- Have students compare their results with their hypotheses. Tell students not to alter their hypotheses if they do not match their results. Instead, they should use the data to explain what they learned.
- Say: **Compare your resting pulse to your pulse after exercising. How did it change?** Students should mention that their resting pulse is slower than their pulse after exercising.
- Tell students to use the data from other groups to infer why one exercise may increase their heart rate more than another. Students should conclude that the more strenuous the exercise, the more it will cause their heart rate to increase.
- Say: **Exercise is a healthy habit. Why? Can you name some other healthy habits?** Guide students in a discussion about the benefits of exercise, eating a variety of healthy foods, and getting rest. They may also mention personal cleanliness and avoiding harmful substances.

Answers

1. Possible answer: My heart rate was higher after exercising. I can infer that exercise increases heart rate.

What to Do, continued

3. **Measure** your partner's resting pulse by counting the beats for 1 minute. Use the stopwatch to time yourself. Record the **data** in your science notebook.
4. Have your partner do the exercise you chose for 1 minute. As soon as your partner stops, measure his or her pulse for 1 minute. Record the data.
5. Wait 3 minutes. Then check your partner's pulse to make sure it has returned to the resting rate. Repeat step 4 two more times.
6. Switch roles with your partner and repeat steps 2–5.

66

NATIONAL GEOGRAPHIC Raise Your SciQ!

A Workout for Your Heart The heart is a muscle that pumps blood throughout the body. Exercise causes the heart to work harder and the heart rate to increase. This work helps the heart become stronger. As it gets stronger, the heart becomes more efficient in circulating blood. It can then pump the extra blood needed during exercise with less strain. A resting heart may pump about 5 liters per minute throughout the body, while a heart during exercise may pump about 20–25 liters per minute.

Record

Write in your science notebook.
Use a table like this one.

Effects of Exercise on Heart Rate

Who?	Kind of Exercise	Resting Heart Rate (beats/min)	Heart Rate After Exercising (beats/min)
Me, trial 1			
Me, trial 2			

Explain and Conclude

1. **Compare** your heart rate before and after exercising. What can you **conclude** about how exercise affects heart rate?
2. Compare your results with groups that did the same exercise. Did they get similar results? Explain any differences.
3. Compare your results with groups that did different exercises. Did one kind of exercise affect heart rate more than other kinds? Why do you think that is so?

Think of Another Question

What else would you like to find out about how exercise affects heart rate? How could you find an answer to this new question?

How does running affect heart rate?

67

Inquiry Rubric at myNGconnect.com	Scale			
The student made a **hypothesis** about how exercise would affect heart rate.	4	3	2	1
The student **planned** and conducted an **experiment.**	4	3	2	1
The student **measured** their heart rate before and after exercising.	4	3	2	1
The student formed a **conclusion** about how exercise affects heart rate.	4	3	2	1
The student **shared** and **compared data** with other students.	4	3	2	1
Overall Score	**4**	**3**	**2**	**1**

Explain and Conclude, continued

2. Answers will vary, but all groups should observe an increase in heart rate with exercise. However, the amount of increase may vary. Factors that can affect the rate of increase include the type of exercise and how strenuously students exercise.
3. Answers will vary. Students might respond that a particular kind of exercise increased heart rate more because the students doing that exercise had to work harder.

5 Find Out More

Think of Another Question

- Students should use observations to generate other questions that could be studied using readily available materials. Discuss how students could find answers to the new questions.

6 Reflect and Assess

- To assess student work with the Inquiry Rubric below, see Assessment Handbook, page 187, or go online to myNGconnect.com
- Have students use the Inquiry Self-Reflection on Assessment Handbook, page 200, or at myNGconnect.com

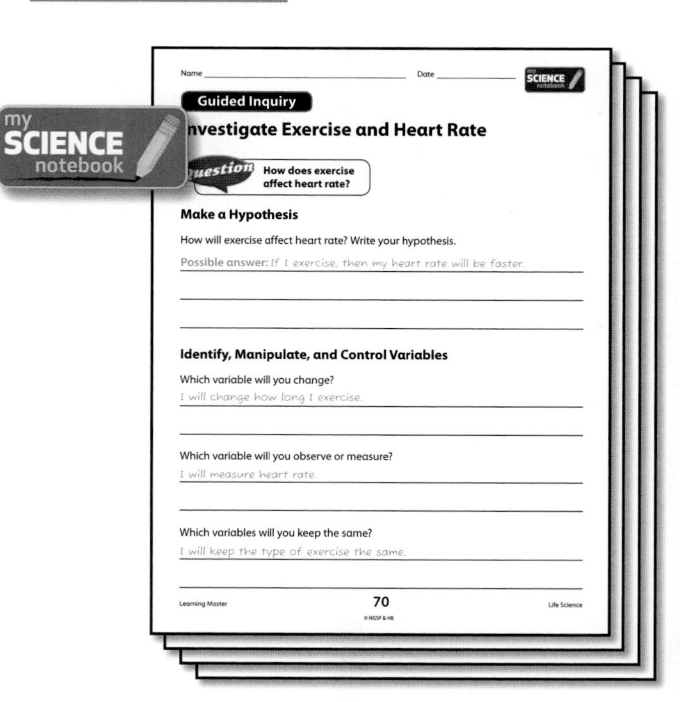

Name ______ Date ______

Guided Inquiry

Investigate Exercise and Heart Rate

Question: How does exercise affect heart rate?

Make a Hypothesis

How will exercise affect heart rate? Write your hypothesis.

Possible answer: If I exercise, then my heart rate will be faster.

Identify, Manipulate, and Control Variables

Which variable will you change?

I will change how long I exercise.

Which variable will you observe or measure?

I will measure heart rate.

Which variables will you keep the same?

I will keep the type of exercise the same.

Learning Master 70 Life Science

© NGSP & HB

Learning Masters 70–73
or at myNGconnect.com

CHAPTER 6
LESSON 9 ▫ Conclusion and Review

PROGRAM RESOURCES
- Chapter 6 Test, Assessment Handbook, pages 32–35, or at myNGconnect.com
- NGSP ExamView CD-ROM

1 Sum Up the Big Idea
- Display the chart from page T232–T233. Read the Big Idea Question. Then add new information to the chart.
- Ask: **What did you find out in each section? Is this what you expected to find?**

How Do the Parts of an Organism Work Together?

Circulatory and Respiratory Systems

Sarah Jane: The circulatory and respiratory systems work together to bring oxygen and remove wastes from all body parts.

Digestive System

Harry: The digestive system breaks down food into energy and materials that the body can use.

Skeletal, Muscular, and Nervous Systems

Leela: The nervous system controls how the muscular and skeletal systems move the body.

Keeping Body Systems Healthy

Tegan: Diet, exercise, and rest can help keep the body healthy and all organ systems working at their best.

Conclusion
People have many body systems that do certain jobs. The respiratory system takes in oxygen. Then the circulatory system moves the oxygen-rich blood throughout the body. The digestive system breaks down food. The skeletal system gives the body structure and protects it, while the muscular system helps the body to move. The nervous system helps the body respond to its environment. Healthful habits help people keep their body systems healthy.

Big Idea The parts of an organism work together in systems. These systems help the body get oxygen and nutrients it needs. Systems also help organisms respond to their environment.

Vocabulary Review
Match each of the following terms with the correct definition.

A. system
B. nervous system
C. organ
D. digestive system
E. respiratory system
F. circulatory system

1. A group of organs that brings oxygen into the body and gets rid of carbon dioxide
2. A group of organs that carries blood throughout the body
3. A group of organs that works as a communication network to help the body respond to its environment
4. A group of organs that works together in an organism to help it live and grow
5. A group of organs that breaks down food into energy and materials that the body can use
6. A body part that does a specific job in an organism

256

Review Academic Vocabulary
Academic Vocabulary Tell students that scientists use special words when describing their work to others. For example:

organ	system	circulatory system
respiratory system	digestive system	nervous system

Have students write the list in their science notebook and use each word in a sentence that tells about how the parts of an organism work together. They should then share their sentences with a partner.

Big Idea Review

1. **Identify** What is the main job of the muscular system?
2. **Name** Which two body systems help your body parts get the oxygen they need?
3. **Explain** Why is it a healthful habit to wash your hands often?
4. **Relate** How do the skeletal and muscular systems work together?
5. **Evaluate** Suppose your brain could not understand nerve signals for pain. Why would that be dangerous?
6. **Apply** Zoos often remind visitors not to feed the animals any popcorn, peanuts, or other food. Why might zookeepers give this warning?

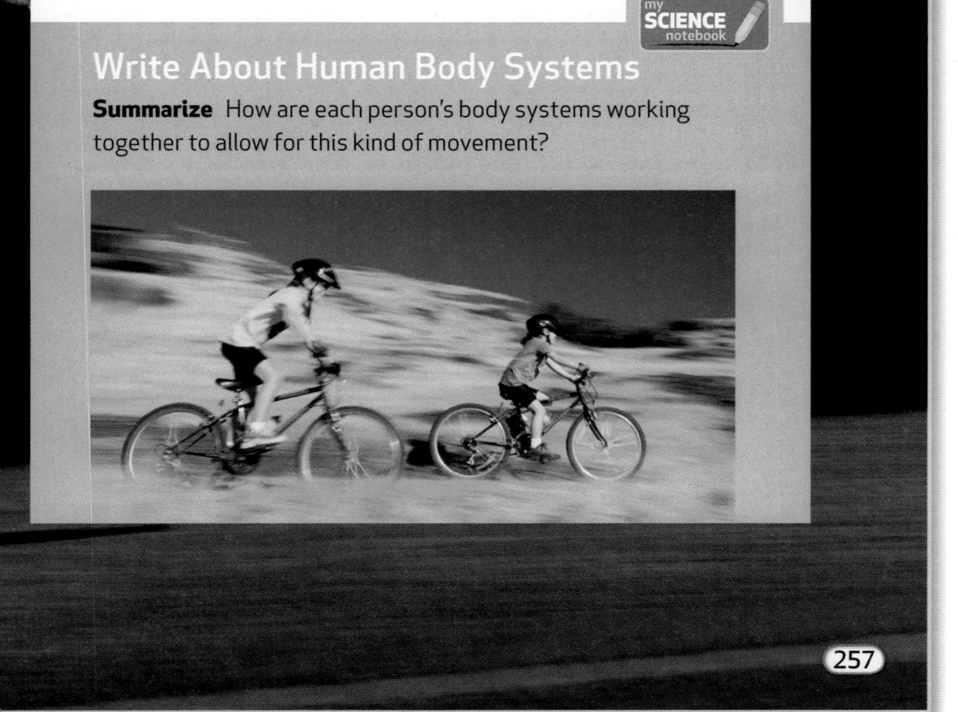
my SCIENCE notebook
Write About Human Body Systems
Summarize How are each person's body systems working together to allow for this kind of movement?
257

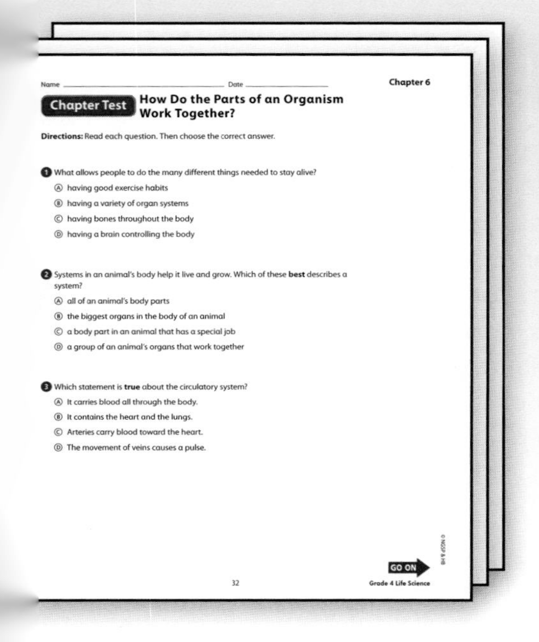
Name ______ Date ______ Chapter 6

Chapter Test **How Do the Parts of an Organism Work Together?**

Directions: Read each question. Then choose the correct answer.

1. What allows people to do the many different things needed to stay alive?
   - Ⓐ having good exercise habits
   - Ⓑ having a variety of organ systems
   - Ⓒ having bones throughout the body
   - Ⓓ having a brain controlling the body
2. Systems in an animal's body help it live and grow. Which of these **best** describes a system?
   - Ⓐ all of an animal's body parts
   - Ⓑ the biggest organs in the body of an animal
   - Ⓒ a body part in an animal that has a special job
   - Ⓓ a group of an animal's organs that work together
3. Which statement is **true** about the circulatory system?
   - Ⓐ It carries blood all through the body.
   - Ⓑ It contains the heart and the lungs.
   - Ⓒ Arteries carry blood toward the heart.
   - Ⓓ The movement of veins causes a pulse.

GO ON

32 Grade 4 Life Science

Assess Student Progress Chapter 6 Test

Have students complete their Chapter 6 Test to assess their progress in this chapter.

Chapter 6 Test, Assessment Handbook, pages 32–35, or at myNGconnect.com

or NGSP ExamView CD-ROM

❷ Discuss the Big Idea

Notetaking

Have students write in their science notebook to show what they know about the Big Idea. Have them:

1. Explain what a system is.
2. Describe the function of different systems of the human body.
3. Compare systems in humans and other animals.

❸ Assess the Big Idea

Vocabulary Review

1. E **2.** F **3.** B **4.** A **5.** D **6.** C

Big Idea Review

1. The muscular system moves the body parts.
2. The respiratory system takes oxygen into the body, and the circulatory system carries that oxygen throughout the body.
3. Washing your hands can keep you from becoming sick and passing germs to others.
4. Muscles contract to pull and move the bones they are connected to.
5. If your brain could not interpret pain, it would not be able to protect you from some harmful situations. For example, it would not make you pull your hand out of a fire.
6. Human foods could harm an animal's digestive system because those kinds of foods probably do not occur in that animal's natural environment. The animals' digestive system might not be able to break down that food.

Write About Human Body Systems

The muscular and skeletal systems work together to allow the bikers to push the pedals and grasp the handlebars. The respiratory and circulatory systems work together to provide oxygen to the body and remove wastes. The nervous system controls the other systems to let them work together efficiently.

Objectives

Students will be able to:

- Recognize how different scientists' work has contributed to general scientific understanding.

PROGRAM RESOURCES

- Big Ideas Book: *Life Science*
- Big Ideas Book: *Life Science* eEdition at myNGconnect.com
- Digital Library at myNGconnect.com

❶ Introduce

Tap Prior Knowledge

- Have students describe their experiences taking medicine. Ask if they know which medicine they used, how long they used it, and whether or it was prescribed by a doctor. Remind students that they should only take medicines given to them by their parents, guardians, or medical staff.

Preview and Read

- Read the heading on page 258 and have students preview the photos. Ask students to think about what it might be like to work with medicines.
- Have students read pages 258–259.

❷ Teach

Describe Pharmacologists

- Write the word *pharmacologist* on the board and have students pronounce it. Say: **This word comes in part from the Greek word *pharmakon*, meaning "medicine or poison."**
- Ask: **Why might medicine also be considered a poison?** (If you take too much medicine or the wrong kind of medicine, it can be harmful to your body.) Say: **Pharmacologists know how to use the right amounts and right kinds of medicine for their customers.**
- Ask: **According to Dr. Snead, what is a pharmacologist's job?** (to use medicines and chemicals to change how body parts work) **What is the goal of a pharmacologist?** (to gain a better understanding of how living things function)

NATIONAL GEOGRAPHIC
CHAPTER 6
LIFE SCIENCE EXPERT: PHARMACOLOGIST
Dr. Aaron Snead, Pharmacologist
What is a pharmacologist's job?
A pharmacologist's job is to use known medicines and chemicals to change how the body parts of different organisms work. The goal is to gain a better understanding of how living things function.
What do you remember liking about science when you were in elementary school?
The thing I liked most about science when I was young was the idea that there was more than one right way to be successful. Since the goal of science is to learn new things, I liked being challenged to come up with new ways to solve old problems.
Dr. Aaron Snead works in his laboratory to study how the body parts of organisms function.
258

Raise Your SciQ!

Careers in Pharmacology and Pharmacy Students who are interested in drugs and medicine can pursue a wide variety of careers. Pharmacologists may choose to conduct research at universities or private laboratories, or they may study and patients in hospitals. They may also work for the government to test new medicines and approve patents. Pharmacists typically work in drug stores. They package and distribute prescription medicines, advise customers about their medications, and recommend over-the-counter medications for minor illnesses and maladies.

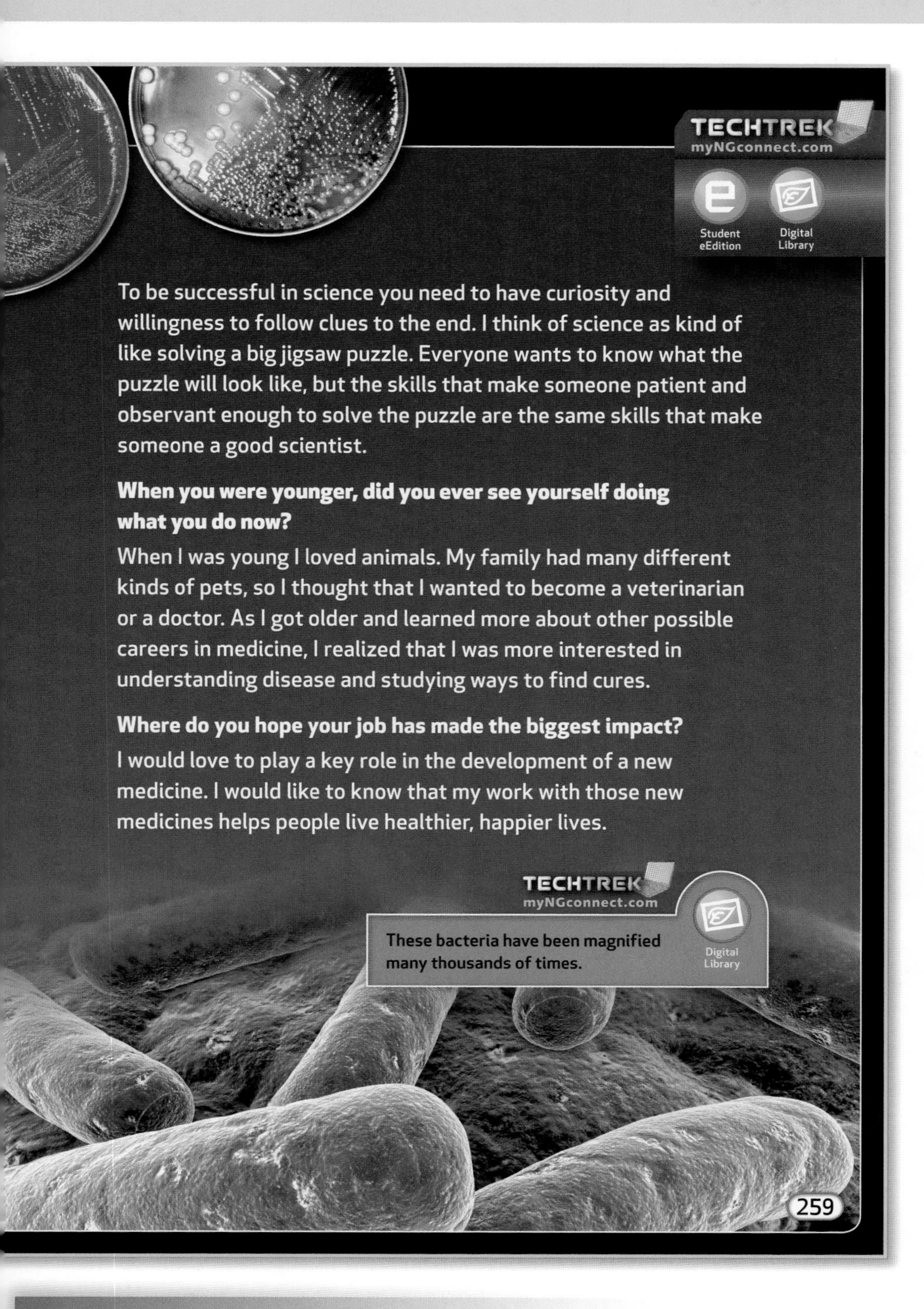

To be successful in science you need to have curiosity and willingness to follow clues to the end. I think of science as kind of like solving a big jigsaw puzzle. Everyone wants to know what the puzzle will look like, but the skills that make someone patient and observant enough to solve the puzzle are the same skills that make someone a good scientist.

When you were younger, did you ever see yourself doing what you do now?
When I was young I loved animals. My family had many different kinds of pets, so I thought that I wanted to become a veterinarian or a doctor. As I got older and learned more about other possible careers in medicine, I realized that I was more interested in understanding disease and studying ways to find cures.

Where do you hope your job has made the biggest impact?
I would love to play a key role in the development of a new medicine. I would like to know that my work with those new medicines helps people live healthier, happier lives.

These bacteria have been magnified many thousands of times.

259

Differentiated Instruction

ELL Language Support for Describing Pharmacologists

BEGINNING

Have students review pages 258–259 and identify any new or unfamiliar words, such as *pharmacology, bacteria,* and *observant.* Have students work with partners to define these words using context clues, the glossary, or a dictionary.

INTERMEDIATE

Help students describe the work of a pharmacologist using Academic Language Frames.

- *A ____ helps people use medicine to improve their lives.*
- *A pharmacologist needs skills such as ____ and ____.*

ADVANCED

Have students write sentences to answer basic questions about the work of pharmacologists. For example:

- *What do pharmacologists study?*
- *Where do pharmacologists work?*
- *How do pharmacologists help people?*

Teach, continued

Technology in Life Science

- Point to the photo of the bacteria on page 259. Have a volunteer to read the caption. Ask: **What technology was used to take this photo of bacteria?** (a microscope and a computer) Explain that a computer was used to add the purple color to the photo. The bacteria are not really purple.
- Ask: **Why do you think pharmacologists need to use microscopes?** (Possible answer: Bacteria and other germs are too small to see without microscopes.)

Find Out More

- Ask: **Is this a career you would find interesting? Why or why not?** (Accept all answers, positive or negative.) Encourage interested students to research a career in pharmacology.

myNGconnect.com

Have students use the Digital Library to find other photos that relate to careers in health care.

Whiteboard Presentation Have students use the information to make a computer presentation about careers in health care. Help them show the presentation on a whiteboard.

3 Assess

1. **Describe What does a pharmacologist do?** (A pharmacologist uses medicines and chemicals to change how the body parts of different organisms work.)
2. **Explain What traits and skills do scientists need to be successful? List three examples and explain why they are important.** (Answers include curiosity, observation, and patience. These traits and skills help scientists recognize and understand a problem, and then keep working until the problem is solved.)
3. **Generalize Why might being a pharmacologist be a rewarding career?** (You might gain a new understanding of how living things function. You might discover a new way of helping people live healthier lives.)

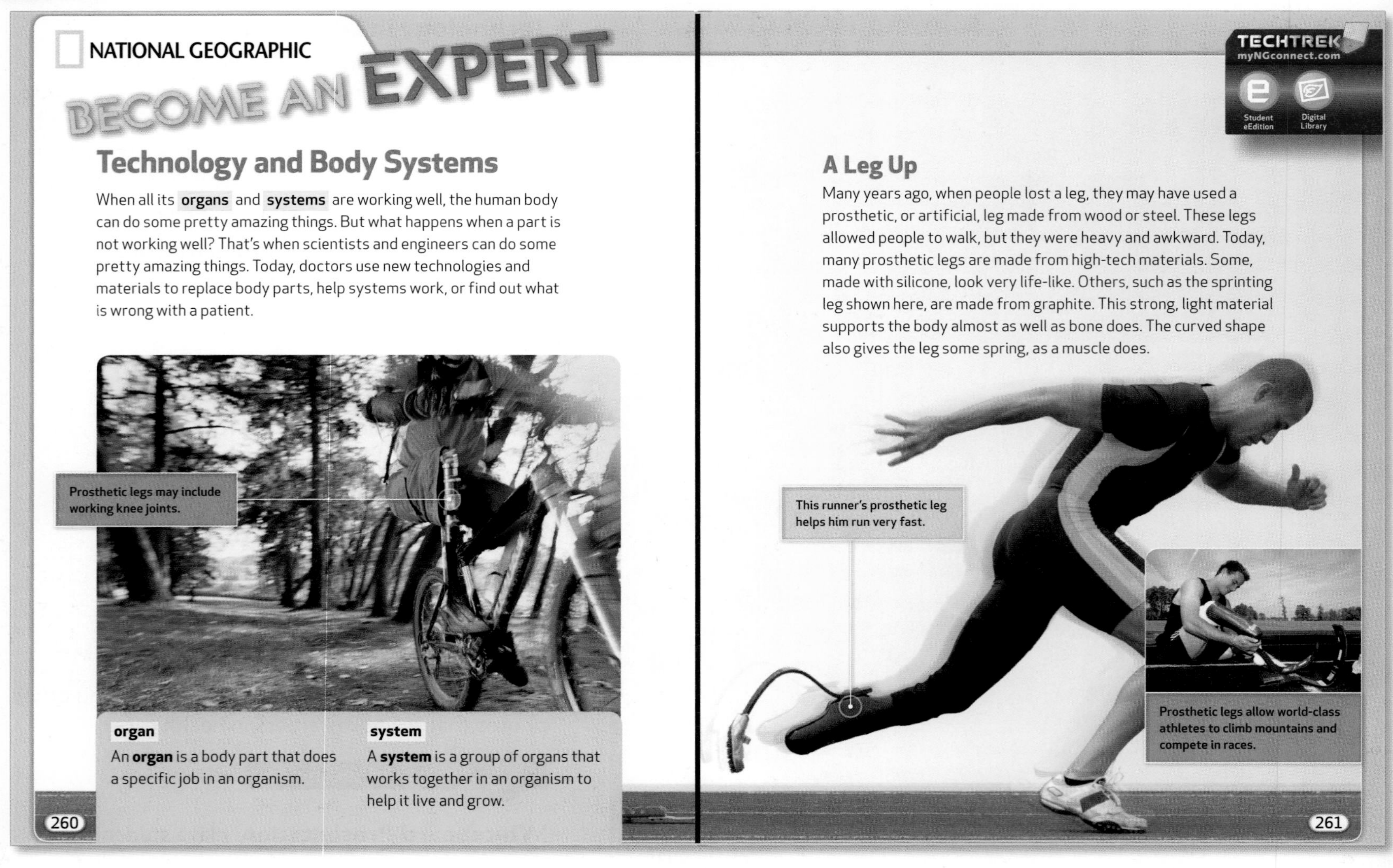

NATIONAL GEOGRAPHIC

BECOME AN EXPERT

Technology and Body Systems

When all its **organs** and **systems** are working well, the human body can do some pretty amazing things. But what happens when a part is not working well? That's when scientists and engineers can do some pretty amazing things. Today, doctors use new technologies and materials to replace body parts, help systems work, or find out what is wrong with a patient.

Prosthetic legs may include working knee joints.

organ
An **organ** is a body part that does a specific job in an organism.

system
A **system** is a group of organs that works together in an organism to help it live and grow.

260

TECHTREK
myNGconnect.com
Student eEdition
Digital Library

A Leg Up

Many years ago, when people lost a leg, they may have used a prosthetic, or artificial, leg made from wood or steel. These legs allowed people to walk, but they were heavy and awkward. Today, many prosthetic legs are made from high-tech materials. Some, made with silicone, look very life-like. Others, such as the sprinting leg shown here, are made from graphite. This strong, light material supports the body almost as well as bone does. The curved shape also gives the leg some spring, as a muscle does.

This runner's prosthetic leg helps him run very fast.

Prosthetic legs allow world-class athletes to climb mountains and compete in races.

261

PROGRAM RESOURCES

- Big Ideas Book: *Life Science*
- Big Ideas Book: *Life Science* eEdition at myNGconnect.com
- Digital Library at myNGconnect.com

Access Science Content

Describe How Technology Can Help Body Systems

- Have students read pages 260–261, observe the photos, and read the captions. Ask: **What do you think happened to the people shown in the photos?** (Possible answer: Each lost a leg or part of the leg, perhaps to injury. Doctors replaced the missing part.) Help students define the word *prosthetic,* which means an *artificial* replacement for a body part. Ask: **What organs work together to move the body?** (bones and muscles) Discuss how bones are part of the skeletal **system** and muscles are part of the muscular **system**. Have students review the definitions of these words at the bottom of page 260.

Notetaking

As students read about different kinds of medical technology, have them record their observations and thoughts in their science notebook.

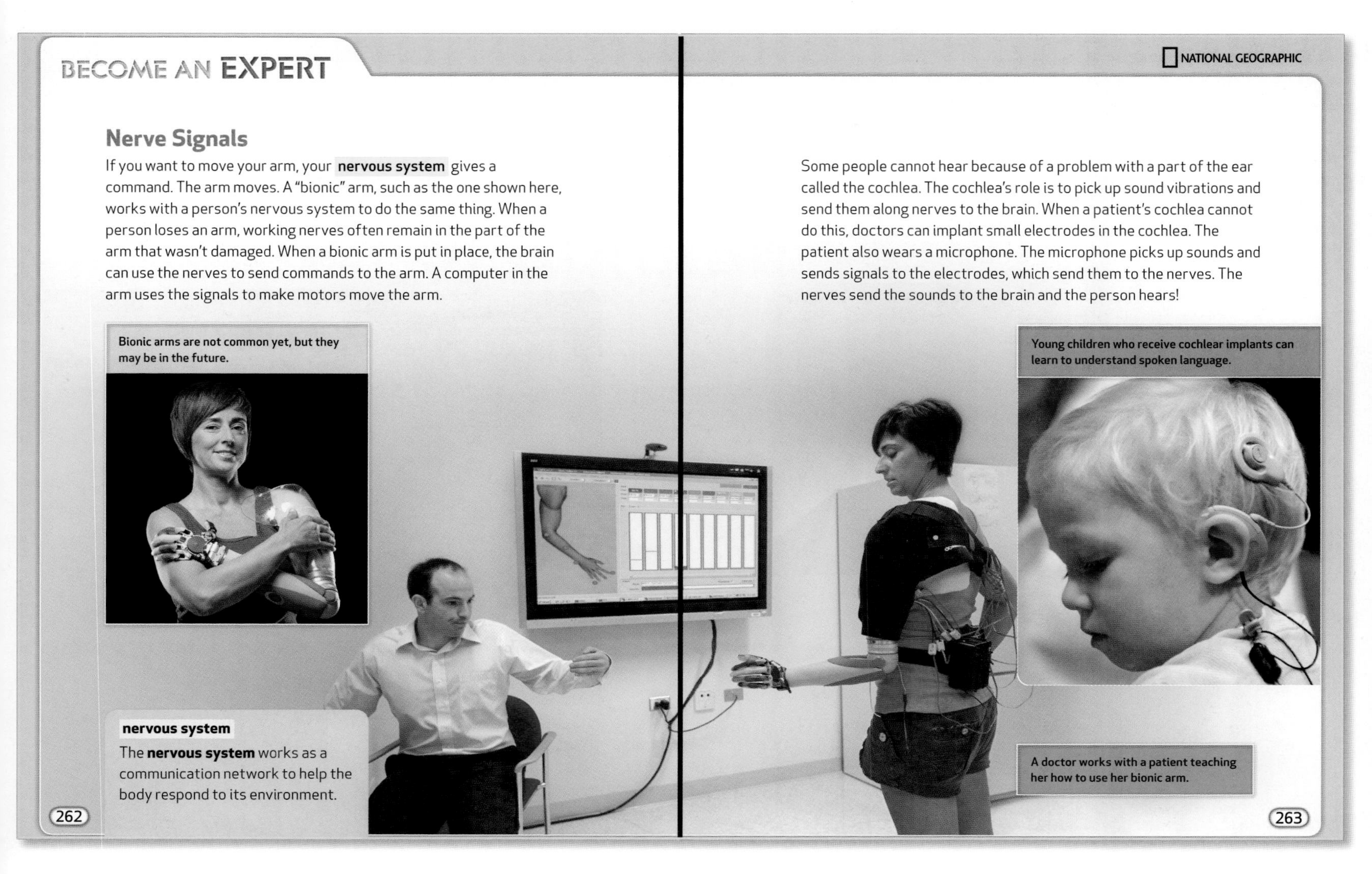

BECOME AN EXPERT

NATIONAL GEOGRAPHIC

Nerve Signals

If you want to move your arm, your **nervous system** gives a command. The arm moves. A "bionic" arm, such as the one shown here, works with a person's nervous system to do the same thing. When a person loses an arm, working nerves often remain in the part of the arm that wasn't damaged. When a bionic arm is put in place, the brain can use the nerves to send commands to the arm. A computer in the arm uses the signals to make motors move the arm.

Bionic arms are not common yet, but they may be in the future.

nervous system

The **nervous system** works as a communication network to help the body respond to its environment.

Some people cannot hear because of a problem with a part of the ear called the cochlea. The cochlea's role is to pick up sound vibrations and send them along nerves to the brain. When a patient's cochlea cannot do this, doctors can implant small electrodes in the cochlea. The patient also wears a microphone. The microphone picks up sounds and sends signals to the electrodes, which send them to the nerves. The nerves send the sounds to the brain and the person hears!

Young children who receive cochlear implants can learn to understand spoken language.

A doctor works with a patient teaching her how to use her bionic arm.

262 263

Access Science Content

Describe How Technology Helps the Nervous System

- Have students read pages 262–263, observe the photos, and read the captions. Ask: **What does the nervous system do?** (It communicates with the rest of the body and controls body functions.) Point to the photo of the woman on page 262. Ask: **How is this woman's bionic arm different from an ordinary prosthetic limb?** (The bionic arm connects to the nervous system and allows the brain to control it; ordinary prosthetic limbs do not have nerves.) Remind students that the brain controls all muscle movements, which often can be very complex.
- Point to the photo of the child on page 263. Ask: **What is the cochlea?** (a part of the inner ear) **What does a cochlear implant do?** (It takes over the job of a damaged cochlea and allows the person to hear sounds.)

Make Inferences

Discuss how all of the technology shown in this feature is expensive to develop and use. Ask: **Why might a bionic hand be more useful than a bionic foot?** (Possible answer: Because grasping, writing, and many other actions of hands require delicate muscle movements, which the brain must control. Feet can be useful even without such movements)

Assess

1. **Define** **What is a body system?** (a group of organs that work together in an organism to help it live and grow)
2. **Infer** **How does the nervous system work with all the other systems of the body?** (The nervous system is the body's communication network. It sends and receives signals from all of the other body systems to move the body, control body functions, and help the body respond to the environment.)

BECOME AN EXPERT

Helping the Heart

The heart has a key role in the **circulatory system**. A patient whose heart is not working well may need a heart transplant. Often, a real heart cannot be found right away. Doctors may give patients a machine to help the heart pump blood or an artificial heart that does the job of the real heart. These machines can't take the place of a heart transplant, but they keep many patients healthy while they wait for a heart to become available. In the future, scientists hope to create an artificial heart that will work as well as a transplanted heart.

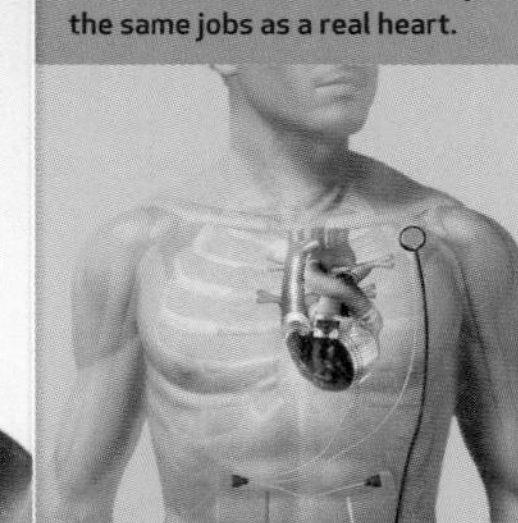
An artificial heart can do many of the same jobs as a real heart.

circulatory system
The **circulatory system** is a group of organs that carries blood throughout the body.

Help is also available for the **respiratory system**. In a working respiratory system, the lungs take carbon dioxide out of the blood and put oxygen into the blood. For people with lung problems, the blood circles out of a patient's body and through an artificial lung. The artificial lung adds oxygen and removes carbon dioxide from the body just like real lungs do.

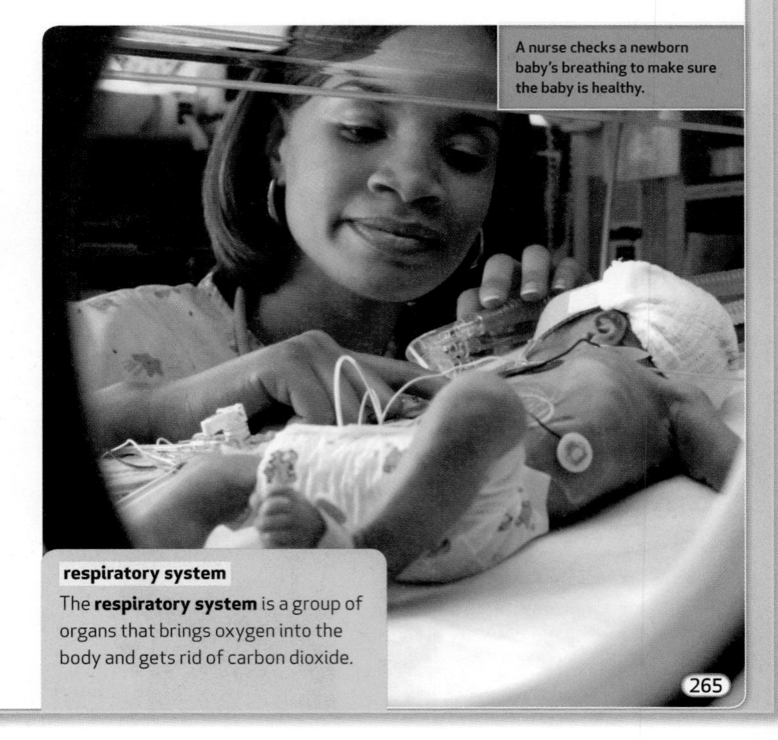
A nurse checks a newborn baby's breathing to make sure the baby is healthy.

respiratory system
The **respiratory system** is a group of organs that brings oxygen into the body and gets rid of carbon dioxide.

Access Science Content

Describe How Technology Helps Circulation and Breathing

- Have students read pages 264–265, observe the photos, and read the captions. Ask: **What is the role of the heart in the circulatory system?** (pumping blood) Discuss how the body would quickly die if the heart stopped beating. Ask: **Why are artificial hearts valuable?** (They can take over temporarily for a diseased heart.) Discuss how artificial lungs are useful in similar ways in the **respiratory system.**

Differentiated Instruction

ELL Language Support for Body Systems

BEGINNING

Have students match the organ with the body system it is a part of. For example:

Brain *(nervous system)*
Heart *(circulatory system)*
Lungs *(respiratory system)*

INTERMEDIATE

Provide Academic Language Frames. For example:

- *The ____ system controls body functions.*
- *The ____ system moves blood through the body.*

ADVANCED

Provide Academic Language Stems. For example:

- *The nervous system has the job of . . .*
- *The respiratory system has the job of . . .*

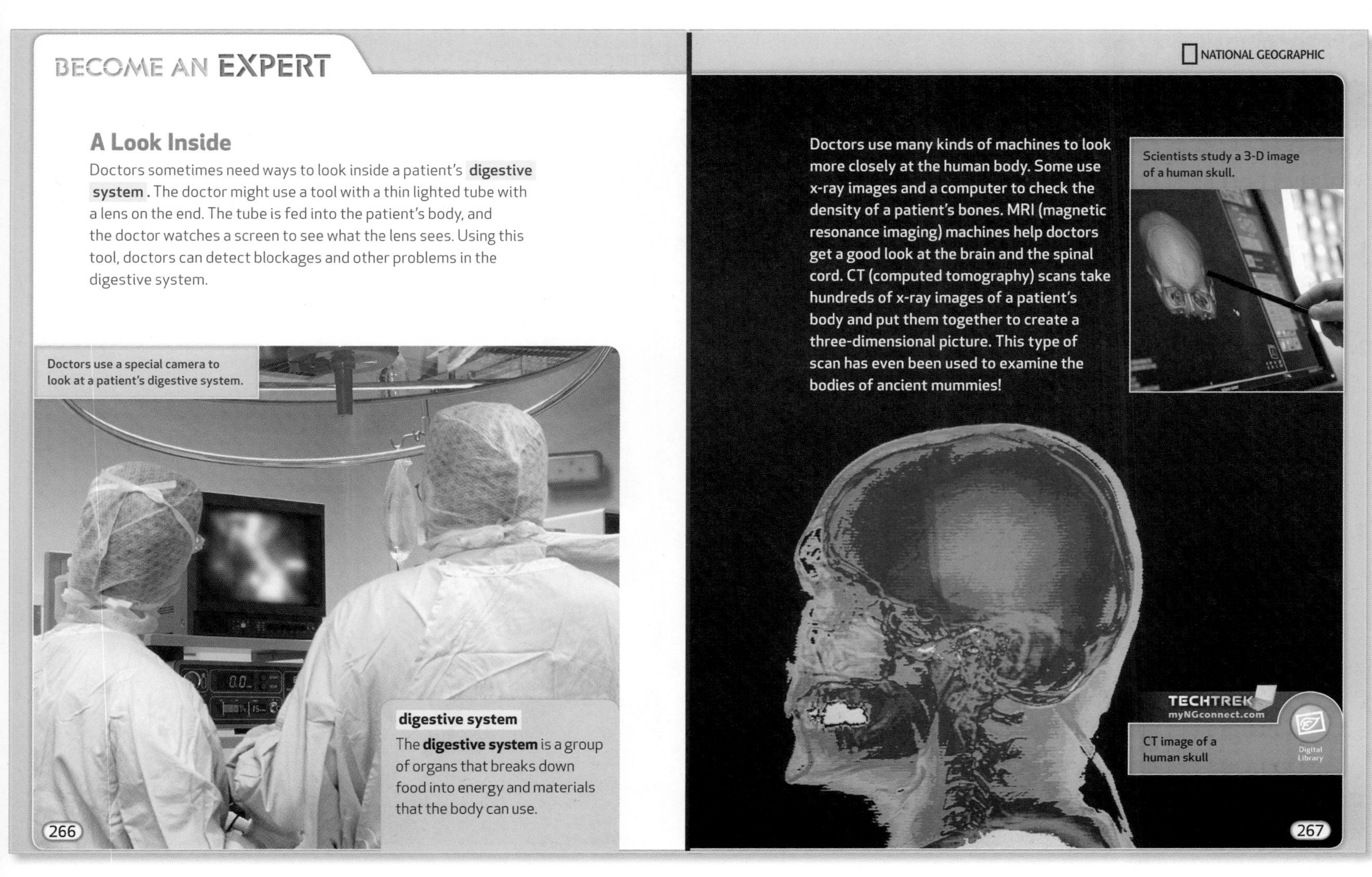
BECOME AN EXPERT

NATIONAL GEOGRAPHIC

A Look Inside

Doctors sometimes need ways to look inside a patient's **digestive system**. The doctor might use a tool with a thin lighted tube with a lens on the end. The tube is fed into the patient's body, and the doctor watches a screen to see what the lens sees. Using this tool, doctors can detect blockages and other problems in the digestive system.

Doctors use a special camera to look at a patient's digestive system.

digestive system

The **digestive system** is a group of organs that breaks down food into energy and materials that the body can use.

266

Doctors use many kinds of machines to look more closely at the human body. Some use x-ray images and a computer to check the density of a patient's bones. MRI (magnetic resonance imaging) machines help doctors get a good look at the brain and the spinal cord. CT (computed tomography) scans take hundreds of x-ray images of a patient's body and put them together to create a three-dimensional picture. This type of scan has even been used to examine the bodies of ancient mummies!

Scientists study a 3-D image of a human skull.

TECHTREK myNGconnect.com

Digital Library

CT image of a human skull

267

PROGRAM RESOURCES

- Digital Library at myNGconnect.com

Access Science Content

Describe Technology for Imaging the Body

- Have students read pages 266–267, observe the photos, and read the captions. Point to the photo on page 266. Ask: **How are doctors able to take this picture of the digestive system?** (They fed a tube down the patient's mouth and down the digestive system. The tube has a camera at its end.) **Why might they want to see inside the digestive system?** (Possible answers: to find if it is blocked or injured; to remove an object lodged inside it)
- Point to the photo of the CT image on page 267. Ask: **What body parts can you identify from this image?** (Answers include the skull, several round bones [vertebrae] in the neck, and the jaw, mouth, and teeth.) **Why do you think it is useful to make a three-dimensional image of the head, rather than a flat image?** (Possible answer: because the head contains many organs that are layered on top of one another, including the skull and the brain inside it, as well as different parts of the brain)

myNGconnect.com

Have students use the Digital Library to find more media of medical imaging and body systems.

Computer Presentation Help students use the information to make a computer presentation.

Assess

1. **Explain Why might doctors want to look inside the human body?** (to see problems, such as a blocked digestive organ)
2. **Infer Why is it challenging to replace a diseased heart or lung?** (Possible answer: because the heart and lungs must work all the time, including during an operation)

Share and Compare

Turn and Talk

Ask students to turn to partners and talk about what they learned about human body systems. Prompt students by asking:

1. **Recall Name six systems of the human body.** (skeletal, muscular, nervous, circulatory, respiratory, and digestive systems)
2. **Compare and Contrast How do artificial organs compare and contrast with the organs they replace?** (Artificial organs are made of metal or plastic instead of living tissue. Many artificial organs work well, but not as well as a real organ.)
3. **Draw Conclusions How can technology be used to help keep a patient's body parts and systems working well?** (When body parts fail due to injury or disease, artificial parts may be used to replace them, either temporarily or permanently. Many artificial parts replace vital organs, such as the heart and lungs, which are essential to body systems.)

Read

As partners read and discuss their selections, encourage them to ask questions of one another about details that interest them, points they find confusing or unclear, or subjects they would like to investigate further.

Students may conclude that a variety of prosthetic devices and artificial organs now help people who have suffered from diseases and injuries, and that technology also helps doctors take images of the body and identify problems and treatments. Have partners share and compare their conclusions, and then revise their work to incorporate new ideas.

Work with students so that the class selects a variety of prosthetic devices. Assemble students' drawings into a catalog in book form.

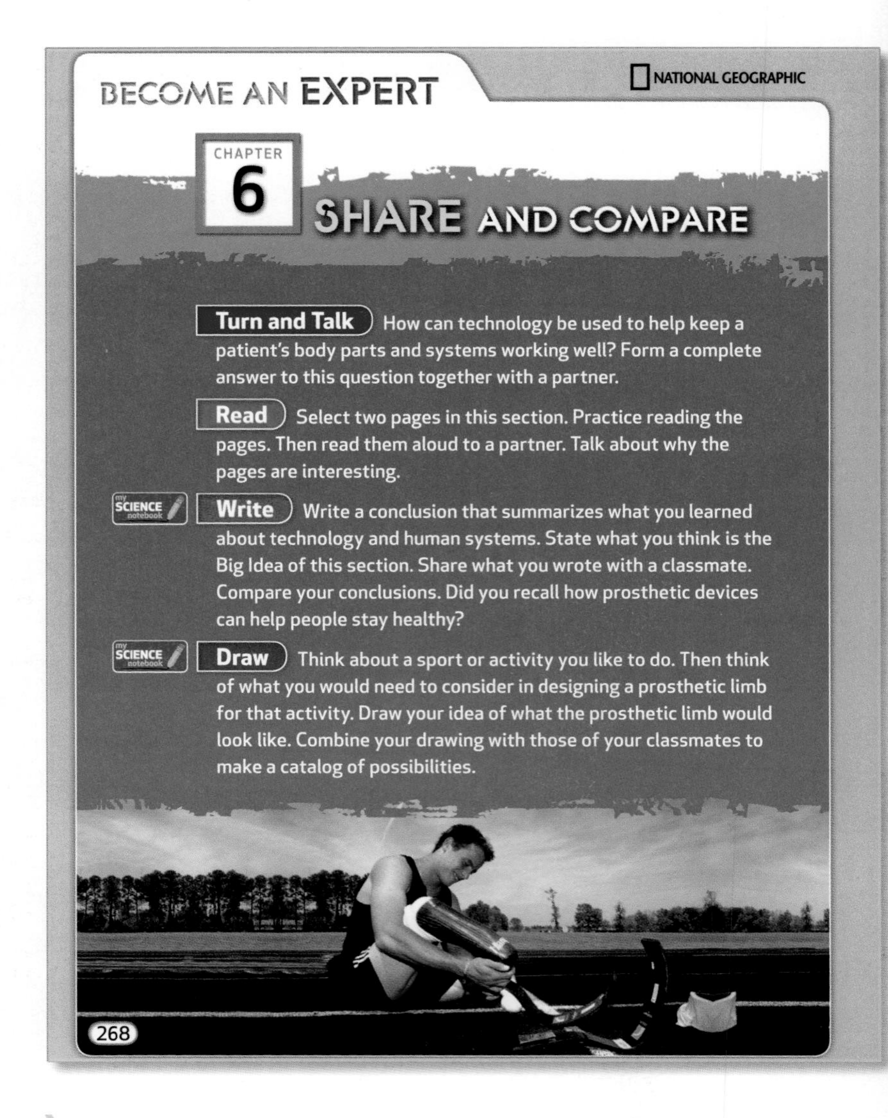

BECOME AN EXPERT

NATIONAL GEOGRAPHIC

CHAPTER 6 SHARE AND COMPARE

Turn and Talk How can technology be used to help keep a patient's body parts and systems working well? Form a complete answer to this question together with a partner.

Read Select two pages in this section. Practice reading the pages. Then read them aloud to a partner. Talk about why the pages are interesting.

Write Write a conclusion that summarizes what you learned about technology and human systems. State what you think is the Big Idea of this section. Share what you wrote with a classmate. Compare your conclusions. Did you recall how prosthetic devices can help people stay healthy?

Draw Think about a sport or activity you like to do. Then think of what you would need to consider in designing a prosthetic limb for that activity. Draw your idea of what the prosthetic limb would look like. Combine your drawing with those of your classmates to make a catalog of possibilities.

268

Sum Up

Tell students that to sum up a text helps them pull together the ideas of the text. Remind students that they summed up the Become an Expert lesson in the Write section of Share and Compare. Have students take turns reading the conclusions they wrote to a partner. Invite them to compare and contrast their thoughts and conclusions.

Notes

Objectives

Students will be able to:

- Investigate through Open Inquiry (generate questions to investigate; plan, make, and compare investigations; collect and record data and observations; generate explanations and conclusions based on evidence or observations; share findings; ask questions based on observations to increase understanding; adjust interpretations based on findings and new ideas).

Science Process Vocabulary

measure

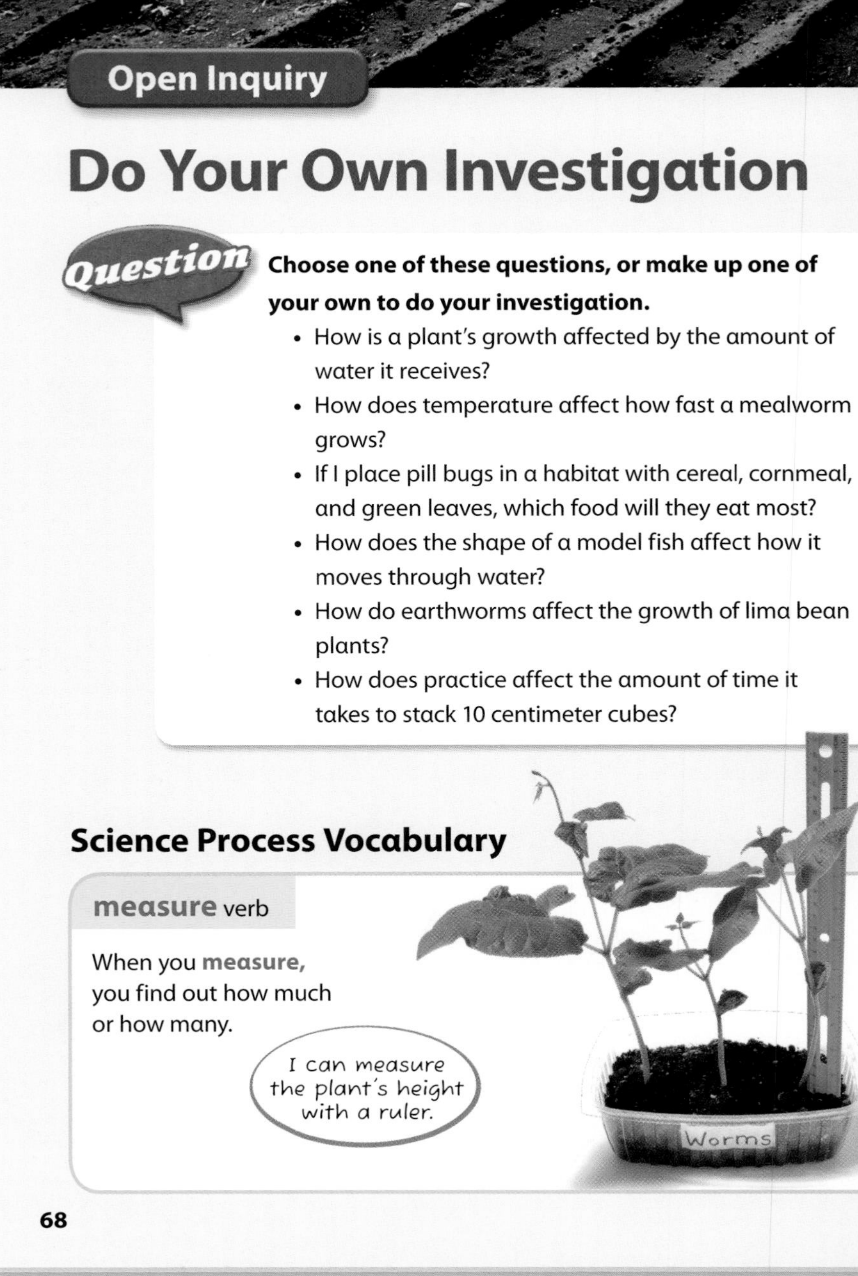

Open Inquiry

Do Your Own Investigation

Question **Choose one of these questions, or make up one of your own to do your investigation.**

- How is a plant's growth affected by the amount of water it receives?
- How does temperature affect how fast a mealworm grows?
- If I place pill bugs in a habitat with cereal, cornmeal, and green leaves, which food will they eat most?
- How does the shape of a model fish affect how it moves through water?
- How do earthworms affect the growth of lima bean plants?
- How does practice affect the amount of time it takes to stack 10 centimeter cubes?

Science Process Vocabulary

measure verb

When you **measure,** you find out how much or how many.

I can measure the plant's height with a ruler.
Worms

68

PROGRAM RESOURCES

- Science Inquiry and Writing Book: *Life Science*
- Science Inquiry and Writing Book eEdition at myNGconnect.com
- Inquiry eHelp at myNGconnect.com
- Learning Masters Book, pages 74–77, or at myNGconnect.com
- Inquiry Rubric: Assessment Handbook, page 188, or at myNGconnect.com
- Inquiry Self-Reflection: Assessment Handbook, page 201, or at myNGconnect.com

MATERIALS

Materials will vary based on students' choices of investigations.

❶ Introduce

Tap Prior Knowledge

- In Open Inquiry, students ask questions, develop written plans, and then design and conduct investigations.
- With students, brainstorm questions and evaluate explanations about scientific topics in this unit or in the natural world around them. Ask students to choose a sample question on page 68 or make up one of their own. They should then make a prediction or hypothesis.
- Students may use the Open Inquiry Checklist as they plan and conduct their investigations.

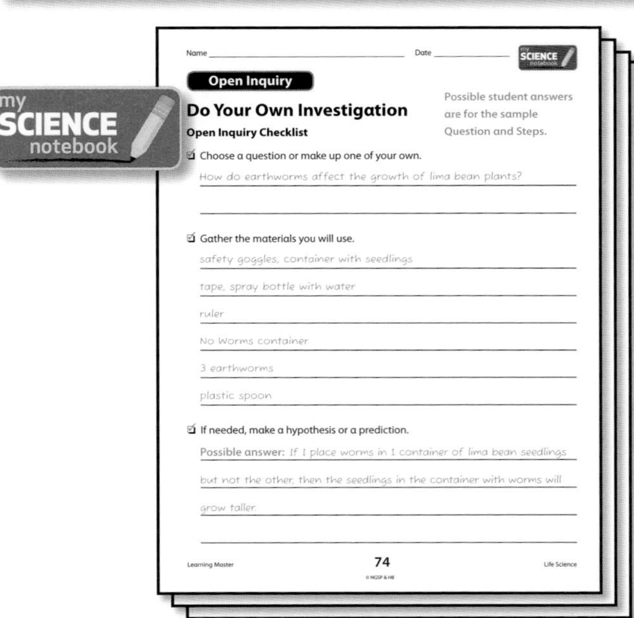

Name ______ Date ______

Open Inquiry

Do Your Own Investigation

Open Inquiry Checklist

Possible student answers are for the sample Question and Steps.

☑ Choose a question or make up one of your own.

How do earthworms affect the growth of lima bean plants?

☑ Gather the materials you will use.

safety goggles, container with seedlings

tape, spray bottle with water

ruler

No Worms container

3 earthworms

plastic spoon

☑ If needed, make a hypothesis or a prediction.

Possible answer: If I place worms in 1 container of lima bean seedlings but not the other, then the seedlings in the container with worms will grow taller.

Learning Master 74 Life Science

Learning Masters 74–77 or at myNGconnect.com

Open Inquiry Checklist

Here is a checklist you can use when you investigate.

- ☐ Choose a **question** or make up one of your own.
- ☐ Gather the materials you will use.
- ☐ If needed, make a **hypothesis** or a **prediction.**
- ☐ If needed, identify, manipulate, and control **variables.**
- ☐ Make a **plan** for your **investigation.**
- ☐ Carry out your **plan.**
- ☐ Collect and record **data. Analyze** your data.
- ☐ Explain and **share** your results.
- ☐ Tell what you **conclude.**
- ☐ Think of another question.

As they burrow, earthworms mix and enrich the soil.

69

Differentiated Instruction

Understand Science Process Vocabulary

Provide the following sentences for students to complete.

- *When you ______, you tell what you think will happen.* (predict)
- *When you ______, you tell how things are alike or different.* (compare)
- *When you ______, you conclude from what you observe and know.* (infer)

❷ Build Vocabulary

Science Process Word: measure

Use this routine to introduce the word.

1. **Pronounce the Word** Say **measure.** Have students repeat it in syllables.
2. **Explain Its Meaning** Choral read the sentence. Ask students to define **measure.** (to find out how much or how many)
3. **Encourage** Ask: **What parts of the plant could you measure?** (the height; the width of its leaves, etc.)

❸ Guide the Investigation

Have students choose steps to answer their questions. Encourage them to share their plans with others and revise them based on feedback. Then guide students as they seek answers to their questions through careful observation or experimentation. They should use appropriate tools and techniques to collect data. See pages T268f–T268i for a model of an Open Inquiry investigation.

❹ Explain Your Results

Have students gather data and organize it in charts or graphs. They should not change their data if it differs from someone else's work. Then students should share results, inferences, and conclusions. See pages T268h–T268i for sample student responses and guidance on explaining results.

❺ Find Out More

Encourage students to use observations made in their investigations to define other problems and generate more questions. See page T268i for more information on developing questions.

❻ Reflect and Assess

- Use the Inquiry Rubric on page T268i, Assessment Handbook, page 188, or online at myNGconnect.com
- Have students use the Inquiry Self-Reflection on Assessment Handbook, page 201, or at myNGconnect.com

Objectives

Students will be able to:

- Investigate through Write Like a Scientist (choose a question; gather materials to use; make a hypothesis or prediction; identify, manipulate, and control variables if needed; make and carry out a plan for an investigation; collect and record data in charts, tables, and graphs; analyze data; explain and share results; tell a conclusion; think of another question).
- Indicate materials to be used and steps to follow to conduct an investigation, and describe how data will be recorded.
- Explore and solve problems generated from school, home, and community situations, using concrete objects and materials.

Write About an Investigation

How Animals Get Energy

- Use this writing model to help students write about their investigations. The writing model uses the Open Inquiry checklist to guide writing.
- Page 70 briefly explains Taj's investigation. Ask: **Why do you think Taj chose pill bugs for his investigation?** (Pill bugs are easy and safe to keep.) Explain that when pill bugs eat food, the energy stored in the food is passed on to them.
- The writing model begins on page 71, with Taj's question and list of materials. Remind students that they, like scientists, can raise questions about the natural world to answer through investigations.
- Taj's hypothesis is on page 72. Help students notice the "If… then…" construction. Remind students that a hypothesis is a possible answer to a question tested in an experiment.
- Page 73 tells what variables Taj will change, observe or measure, and keep the same. Ask: **Which variable is Taj changing?** (the food) **Which variable will he observe or measure?** (the behavior of the pill bugs and how much food is left)
- Make sure students understand that scientists also use other types of investigations, such as systematic observations without manipulating variables, to find answers to questions about the natural world.

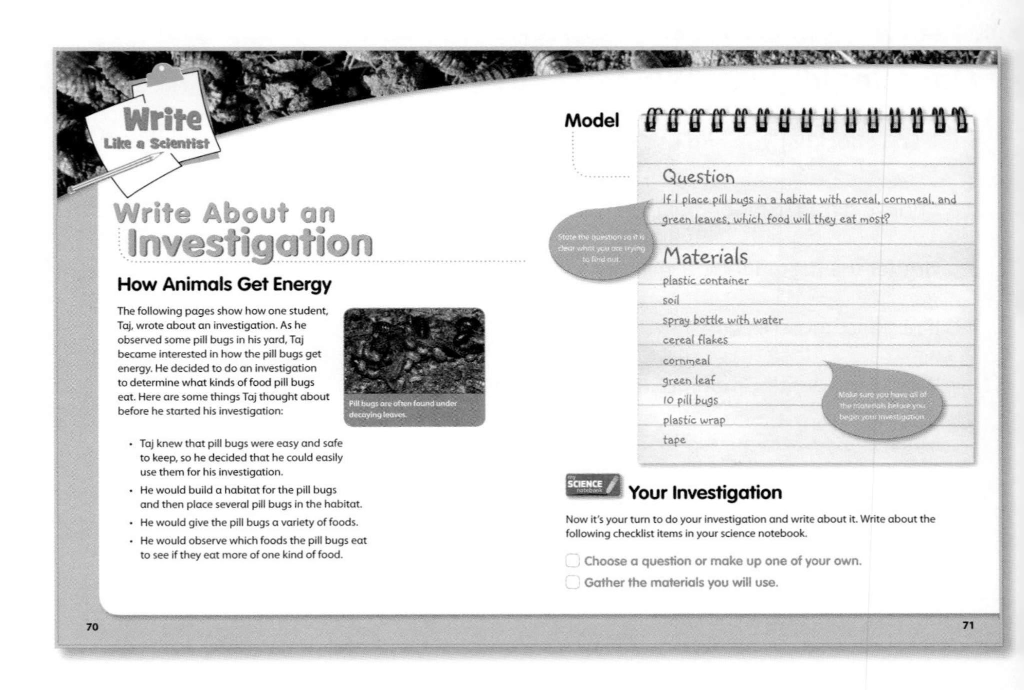

Write Like a Scientist

Write About an Investigation

How Animals Get Energy

The following pages show how one student, Taj, wrote about an investigation. As he observed some pill bugs in his yard, Taj became interested in how the pill bugs get energy. He decided to do an investigation to determine what kinds of food pill bugs eat. Here are some things Taj thought about before he started his investigation:

Pill bugs are often found under decaying leaves.

- Taj knew that pill bugs were easy and safe to keep, so he decided that he could easily use them for his investigation.
- He would build a habitat for the pill bugs and then place several pill bugs in the habitat.
- He would give the pill bugs a variety of foods.
- He would observe which foods the pill bugs eat to see if they eat more of one kind of food.

70

Model

Question

If I place pill bugs in a habitat with cereal, cornmeal, and green leaves, which food will they eat most?

State the question so it is clear what you are trying to find out.

Materials

plastic container
soil
spray bottle with water
cereal flakes
cornmeal
green leaf
10 pill bugs
plastic wrap
tape

Make sure you have all of the materials before you begin your investigation.

Your Investigation

Now it's your turn to do your investigation and write about it. Write about the following checklist items in your science notebook.

- [ ] Choose a question or make up one of your own.
- [ ] Gather the materials you will use.

71

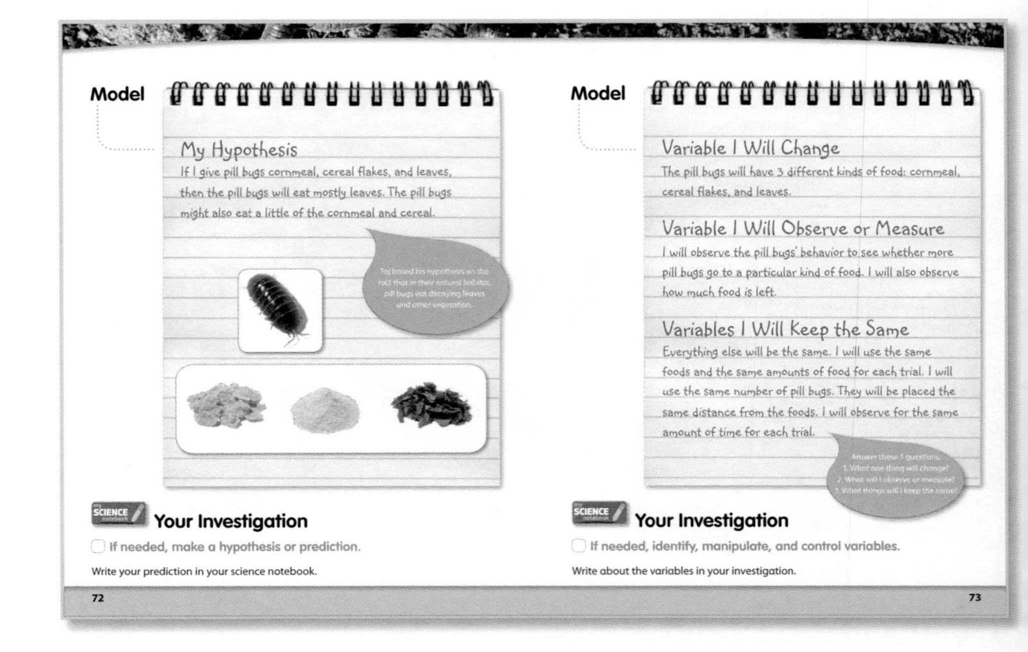

Model

My Hypothesis

If I give pill bugs cornmeal, cereal flakes, and leaves, then the pill bugs will eat mostly leaves. The pill bugs might also eat a little of the cornmeal and cereal.

Taj based his hypothesis on the fact that in their natural habitat, pill bugs eat decaying leaves and other vegetation.

Your Investigation

- [ ] If needed, make a hypothesis or prediction.

Write your prediction in your science notebook.

72

Model

Variable I Will Change

The pill bugs will have 3 different kinds of food: cornmeal, cereal flakes, and leaves.

Variable I Will Observe or Measure

I will observe the pill bugs' behavior to see whether more pill bugs go to a particular kind of food. I will also observe how much food is left.

Variables I Will Keep the Same

Everything else will be the same. I will use the same foods and the same amounts of food for each trial. I will use the same number of pill bugs. They will be placed the same distance from the foods. I will observe for the same amount of time for each trial.

Answer these 3 questions:
1. What one thing will change?
2. What will I observe or measure?
3. What things will I keep the same?

Your Investigation

- [ ] If needed, identify, manipulate, and control variables.

Write about the variables in your investigation.

73

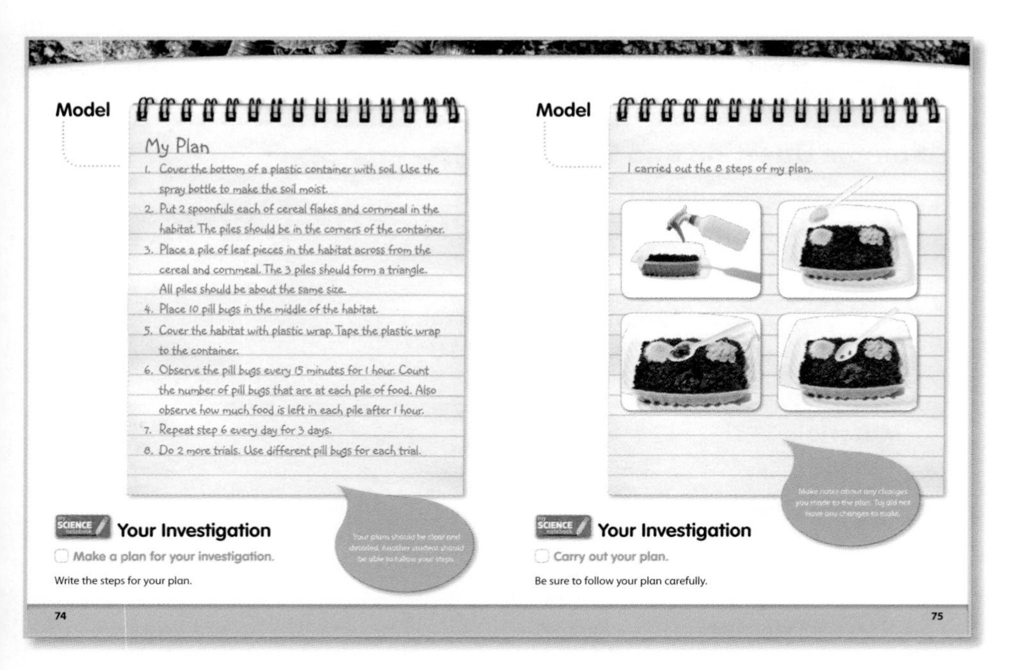

Model

My Plan

1. Cover the bottom of a plastic container with soil. Use the spray bottle to make the soil moist.
2. Put 2 spoonfuls each of cereal flakes and cornmeal in the habitat. The piles should be in the corners of the container.
3. Place a pile of leaf pieces in the habitat across from the cereal and cornmeal. The 3 piles should form a triangle. All piles should be about the same size.
4. Place 10 pill bugs in the middle of the habitat.
5. Cover the habitat with plastic wrap. Tape the plastic wrap to the container.
6. Observe the pill bugs every 15 minutes for 1 hour. Count the number of pill bugs that are at each pile of food. Also observe how much food is left in each pile after 1 hour.
7. Repeat step 6 every day for 3 days.
8. Do 2 more trials. Use different pill bugs for each trial.

Your Investigation

- Make a plan for your investigation.

Write the steps for your plan.

Model

I carried out the 8 steps of my plan.

Your Investigation

- Carry out your plan.

Be sure to follow your plan carefully.

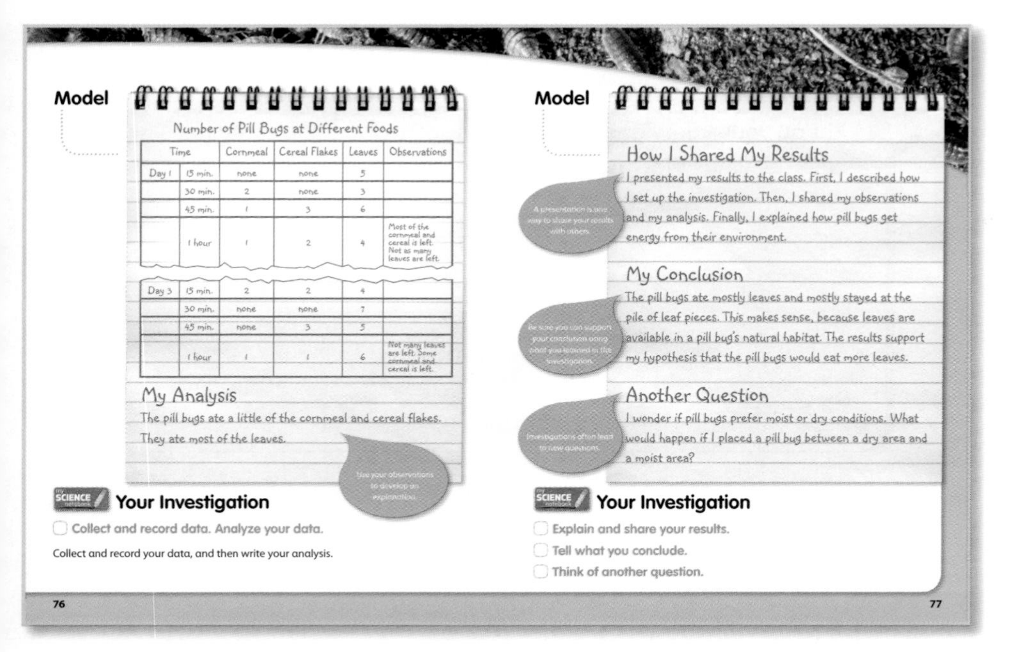

Model

Number of Pill Bugs at Different Foods

Time		Cornmeal	Cereal Flakes	Leaves	Observations
Day 1	15 min.	none	none	5	
	30 min.	2	none	3	
	45 min.	1	3	6	
	1 hour	1	2	4	Most of the cornmeal and cereal is left. Not as many leaves are left.
Day 3	15 min.	2	2	4	
	30 min.	none	none	7	
	45 min.	none	3	5	
	1 hour	1	1	6	Not many leaves are left. Some cornmeal and cereal is left.

My Analysis

The pill bugs ate a little of the cornmeal and cereal flakes. They ate most of the leaves.

Your Investigation

- Collect and record data. Analyze your data.

Collect and record your data, and then write your analysis.

Model

How I Shared My Results

I presented my results to the class. First, I described how I set up the investigation. Then, I shared my observations and my analysis. Finally, I explained how pill bugs get energy from their environment.

My Conclusion

The pill bugs ate mostly leaves and mostly stayed at the pile of leaf pieces. This makes sense, because leaves are available in a pill bug's natural habitat. The results support my hypothesis that the pill bugs would eat more leaves.

Another Question

I wonder if pill bugs prefer moist or dry conditions. What would happen if I placed a pill bug between a dry area and a moist area?

Your Investigation

- Explain and share your results.
- Tell what you conclude.
- Think of another question.

Write About an Investigation, continued

- Page 74 outlines Taj's plan for the investigation. Ask: **What specific details did Taj include in his plan?** (the amount of cornmeal, cereal flakes, and leaf pieces he used; the location of the piles; the number of pill bugs; the type of habitat; how often he observed the pill bugs)
- Page 75 shows a simple statement Taj wrote. Remind students that if they adjust their plans, they should keep notes about the changes they make.
- Taj's data is on page 76. Point to each column and review the organization of the data in the table. Remind students that the data was recorded over three days. Let students know that there are many ways they can record their data.
- Show students how Taj kept his observations separate from any ideas or inferences about those observations. Ask: **Why is it important to distinguish between observations and inferences?** (You observe when you use your senses to find out about something. An inference is made by using observations and prior knowledge to make a decision or conclusion. Someone could look at Taj's observations and make a different set of inferences than Taj did.) As students keep records, make sure they distinguish their observations from their ideas and inferences.
- Guide students in analyzing Taj's data. Ask: **What did the most pill bugs like to eat?** (leaves) **What inferences can you make about the eating habits of pill bugs?** (I can infer that in their natural habitat, pill bugs are more likely to eat leaves than other types of food.)
- After students collect and record data, they should generate explanations for what they observed. Read Taj's analysis aloud. Ask: **Do you have anything to add to Taj's conclusions?** Remind students to look at the data table or think about what they know about pill bugs. They should cite evidence to support their ideas.
- Page 77 shows how results were shared, Taj's conclusion, and another question that he wondered about after completing the investigation. Ask: **How could Taj answer his question about how pill bugs survive in wet and dry conditions?** Write ideas on the board.

LESSON 11 ▫ Investigation Model

Objectives

Students will be able to:

- Investigate through Open Inquiry (generate questions to investigate; plan, make, and compare investigations; collect and record data and observations; generate explanations and conclusions based on evidence or observations; share findings; ask questions based on observations to increase understanding; adjust interpretations based on findings and new ideas).
- Recognize ways animals such as earthworms can impact plants in the environment.

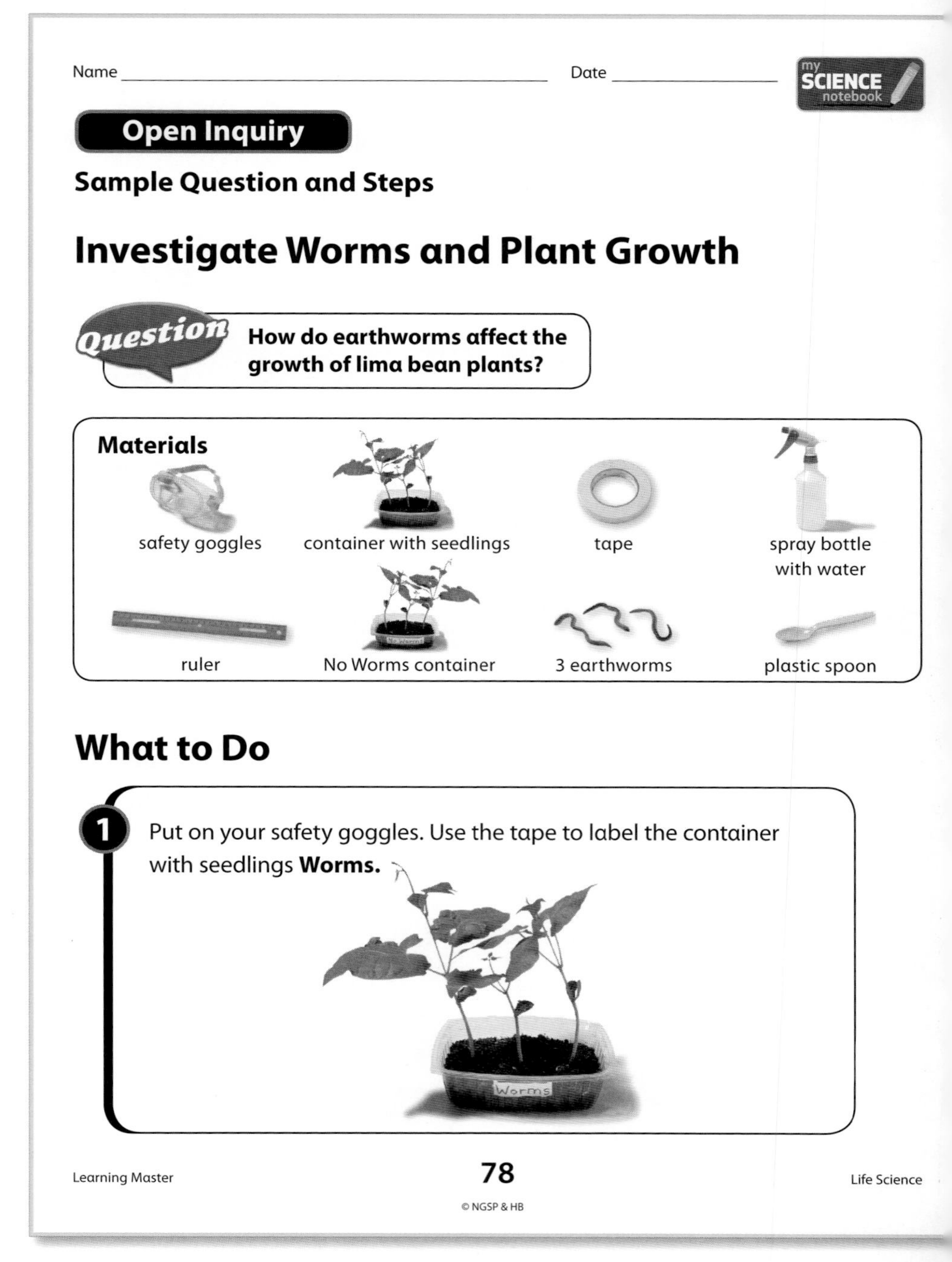

Name ______________________ Date ____________

Open Inquiry

Sample Question and Steps

Investigate Worms and Plant Growth

Question How do earthworms affect the growth of lima bean plants?

Materials

safety goggles, container with seedlings, tape, spray bottle with water, ruler, No Worms container, 3 earthworms, plastic spoon

What to Do

1. Put on your safety goggles. Use the tape to label the container with seedlings **Worms.**

PROGRAM RESOURCES

- Science Inquiry and Writing Book: *Life Science*
- Science Inquiry and Writing Book **eEdition** at myNGconnect.com
- **Inquiry eHelp** at myNGconnect.com
- Science Inquiry Kit: *Life Science*
- Learning Masters Book, pages 74–79, or at myNGconnect.com
- Inquiry Rubric: Assessment Handbook, page 188, or at myNGconnect.com
- Inquiry Self-Reflection: Assessment Handbook, page 201, or at myNGconnect.com

MATERIALS

Kit materials are listed in italics.

safety goggles; *lima beans; 2 clear plastic containers; loamy soil;* masking tape; marker; *spray bottle;* water; ruler; *3 earthworms; plastic spoon*

❶ Introduce

Tap Prior Knowledge

- Ask students if they have ever observed a worm in the soil. Some students may have seen worms while gardening, and they may have noticed that worms are often found in soil that is moist and fertile. Explain to students that worms help mix the soil. Ask: **Do you think worms would help plants to grow better or make them wilt?** Accept all answers. Encourage students to support their ideas.

MANAGING THE INVESTIGATION

Time

30 minutes

Groups

Small groups of 4

Advance Preparation

- Begin growing the lima bean seedlings a few weeks before the investigation. Use fertilizer-free, loamy soil and mix a few dry leaves in the soil for best results. Plant 3–5 seeds in each container to ensure that enough will grow. Each group should get two containers of seedlings, one labeled No Worms.
- Use the earthworms from the Chapter 2 Directed Inquiry.

Name ______________________ Date ____________ my SCIENCE notebook

What to Do, continued

2 Moisten the soil with the spray bottle.

3 **Measure** the height of each seedling with the ruler. Also measure the height of the seedlings in the teacher-made **No Worms** container. Record your measurements in your science notebook.

4 Use the spoon to add 3 earthworms to the container. Place the plants in a sunny area. Use the spray bottle to keep the soil moist.

5 Measure the height of the plants every day for 2 weeks. Also measure the height of the plants in the teacher-made No Worms container. Observe and compare the surface of the soil in each container. Record your measurements and observations.

Learning Master 79 Life Science

Teaching Tips

- Give each group seedlings that are approximately the same height.
- Each group should have the same number of seedlings. If one container grows more seedlings, transfer some to another container.

What to Expect

- Students may find that when worms have been placed in the soil, the plants grow better and the soil looks more mixed. Often, those plants will become taller or look healthier. If there are no worms in the soil, the soil may appear more compacted and not as mixed as the soil with worms.

Introduce, continued

Sample Question and Steps

- Students can use the Sample Question and Steps to practice the steps of an Open Inquiry investigation. Provide guidance as needed. Keep in mind that the goal is for students to learn how to make their own plans to answer questions through scientific investigations.
- Have students use the Open Inquiry Checklist on Science Inquiry and Writing Book, page 69. This checklist also appears on Learning Masters 74–77.

Connect to the Big Idea

- Review the Big Idea Question, *How do living things interact with their environment?* Distribute Learning Masters 74–79. Read the questions and invite students to talk about things that might affect the growth of bean plants. Remind students that plants grow differently depending on the conditions of the environment.

❷ Guide the Investigation

- Assist students as necessary to develop a list of steps for their investigations. Stress that the steps should be able to be repeated by anyone without additional instruction. See the sample student answers on pages T268h–T268i.
- Have groups share the steps to their plans and revise them based on each other's suggestions.
- Distribute materials. Read the steps on Learning Masters 78–79 together with students. Move from group to group and clarify steps if necessary.
- Remind students to handle the earthworms and seedlings gently.
- When moistening the soil in step 2, students should spray each pan with the same amount of water. Have them count the number of sprays.
- In step 3, students should start at the top of the soil when they measure the height of the plant. They should measure the highest point of the plant and also draw a sketch of how the plant looks.
- When students observe the containers each day, ask them to observe the soil in each pan. Ask: **Does the soil look mixed or packed down?** Encourage groups to share any differences between the two containers. Eventually, students may notice that the Worms container has soil that looks more mixed.

❸ Explain and Conclude

- Have groups tell about the steps they followed and the data they gathered. Have groups compare results and discuss similarities and differences.
- Ask the students to compare the plants and soil in the two containers. Ask: **How are the plants and soil in the Worms container different?** (The plants look healthier and they are taller. The soil looks mixed and not packed down.)
- Guide students to see patterns in their results. If some groups did not see differences in plant height, encourage them to look for other signs of plant health. For example, ask: **How do the plants look in the different containers? Do some plants have more leaves than others?**
- Have students push down on the soil or feel it with their fingers. Ask: **Do you see a difference in the soil in the two containers? How do you think the worms changed the soil?** (The worms mixed up the soil.)
- Ask students to make inferences about how worms affect soil and plant growth. Ask: **How do you think the worms help the plants grow?** (They loosen the soil and they provide nutrients.) Remind students to back up their ideas with evidence from the investigation.
- Explain to students that earthworms plow the soil by tunneling through it. When an earthworm moves through soil, it leaves behind an empty space. It is easier for air and water to get into these spaces and circulate around the soil. This means that plant roots can grow more easily and get more water and nutrients from the soil.
- Earthworm droppings, or castings, are also good for plants. They are rich in nitrogen, calcium, magnesium, and phosphorus, and all of these nutrients are good for plants.
- Ask students how they would improve or extend their investigations if they were to repeat them. Students may suggest using different plants or allowing the investigation to last longer.

Sample Student Responses

Name ______ Date ______

my SCIENCE notebook

Open Inquiry

Do Your Own Investigation

Open Inquiry Checklist

Possible student answers are for the sample Question and Steps.

☑ Choose a question or make up one of your own.

How do earthworms affect the growth of lima bean plants?

☑ Gather the materials you will use.

safety goggles, container with seedlings
tape, spray bottle with water
ruler
No Worms container
3 earthworms
plastic spoon

☑ If needed, make a hypothesis or a prediction.

Possible answer: If I place worms in 1 container of lima bean seedlings but not the other, then the seedlings in the container with worms will grow taller.

Learning Master 74 Life Science

© NGSP & HB

Name ______ Date ______

my SCIENCE notebook

Open Inquiry continued

☑ If needed, identify, manipulate, and control variables.

Variable I will change: I will only put worms in one container. **Variable I will measure or observe:** I will measure the growth of the lima bean seedlings and observe the soil. **Variables I will keep the same:** I will keep the amount of water and sunlight, the size and the location of the containers, the soil, and the kind and number of plants the same.

☑ Make a plan for your investigation.

1. Put on my safety goggles. Use the tape to label 1 container Worms.
2. Moisten the soil with the spray bottle.
3. Measure and record the height of each seedling in the Worms and No Worms containers.
4. Add 3 earthworms to the Worms container. Place both containers in a sunny spot. Keep the soil moist.
5. Measure and record the heights of the seedlings in both containers every day for 2 weeks. Observe and compare the surface of the soil in each container.

Learning Master 75 Life Science

© NGSP & HB

Name ________________ Date ________ my SCIENCE notebook

Open Inquiry continued

☑ Carry out your plan.

☑ Collect and record data. Analyze your data.

Observations of Lima Bean Growth

	Height of Seedlings (cm) Worms Container	Height of Seedlings (cm) No Worms Container	Soil Observations Worms Container	Soil Observations No Worms Container
Day 1	Seedling 1:	Seedling 1:		
	Seedling 2:	Seedling 2:		
	Seedling 3:	Seedling 3:		
Day 2	Seedling 1:	Seedling 1:		
	Seedling 2:	Seedling 2:		
	Seedling 3:	Seedling 3:		
Day 3	Seedling 1:	Seedling 1:		
	Seedling 2:	Seedling 2:		
	Seedling 3:	Seedling 3:		
Day 4	Seedling 1:	Seedling 1:		
	Seedling 2:	Seedling 2:		
	Seedling 3:	Seedling 3:		
Day 5	Seedling 1:	Seedling 1:		
	Seedling 2:	Seedling 2:		
	Seedling 3:	Seedling 3:		
Day 6	Seedling 1:	Seedling 1:		
	Seedling 2:	Seedling 2:		
	Seedling 3:	Seedling 3:		
Day 7	Seedling 1:	Seedling 1:		
	Seedling 2:	Seedling 2:		
	Seedling 3:	Seedling 3:		
Day 8	Seedling 1:	Seedling 1:		
	Seedling 2:	Seedling 2:		
	Seedling 3:	Seedling 3:		
Day 9	Seedling 1:	Seedling 1:		
	Seedling 2:	Seedling 2:		
	Seedling 3:	Seedling 3:		
Day 10	Seedling 1:	Seedling 1:		
	Seedling 2:	Seedling 2:		
	Seedling 3:	Seedling 3:		

Measurements and observations will vary depending on the initial heights of the seedlings. Students may find that the seedlings in the Worms container grew taller than the seedlings in the No Worms container. Students should observe that the soil is mixed in the Worms container and may appear compacted in the No Worms container.

Learning Master 76 Life Science

Name ________________ Date ________ my SCIENCE notebook

Open Inquiry continued

☑ Explain and share your results.

Possible answer: The plants in the Worms container grew taller than the plants in the No Worms container. The soil in the Worms container was mixed, and the soil in the No Worms container was not mixed.

☑ Tell what you conclude.

Possible answer: Earthworms help to mix the soil. Mixing the soil makes plants grow healthy and tall.

☑ Think of another question.

Possible answer: How do insects affect the growth of lima bean plants?

Learning Master 77 Life Science

❹ Find Out More

Think of Another Question

- Students should use observations made in this investigation to generate other questions for study. Have them use these question stems: *What happens when…? What would happen if…? How can I/we…?*
- Record student questions for possible future investigations. Discuss what students could do to find answers to the new questions.

❺ Reflect and Assess

- To assess student work with the Inquiry Rubric shown below, see Assessment Handbook, page 188, or go online at myNGconnect.com
- Score each item separately and then decide on one overall score.
- Have students use the Inquiry Self-Reflection on Assessment Handbook, page 201, or at myNGconnect.com

❻ Extend

- Continue this activity with any plants that are still healthy and growing well. Over a few weeks, students may see increasing differences in height as well as differences in which plants flower and produce beans.

Inquiry Rubric at myNGconnect.com	Scale			
The student generated or chose a **question** to **investigate.**	4	3	2	1
The student planned an **experiment** by identifying, manipulating, and controlling **variables.**	4	3	2	1
The student made and **compared observations** and collected and recorded **data.**	4	3	2	1
The student formed a **conclusion** and explained results based on evidence from the collected data and observations.	4	3	2	1
The student **shared** observations and conclusions with other students.	4	3	2	1
Overall Score	**4**	**3**	**2**	**1**

Objectives

Students will be able to:

- Identify a question that was asked, or a problem that needed to be solved when given a brief scenario.
- Differentiate fact from opinion and explain that scientists do not rely on conclusions unless they are backed by observations that can be confirmed.
- Recognize and explain that scientists base their explanations on evidence.

PROGRAM RESOURCES

- Learning Masters Book, pages 80–81, or at myNGconnect.com

1 Introduce

Tap Prior Knowledge

- Invite students to brainstorm questions they have about the natural world. Write questions on the board. Ask: **How would you try to find answers to these questions?** Students may suggest experimenting or making observations.
- If there are differences in the methods students suggest, emphasize these differences. Say: **Scientists come up with many ways to test and answer scientific questions.**

2 Teach

Designing Investigations and Using Evidence

- Read page 78 aloud with students. Ask: **What do all scientific investigations begin with?** (a question or a problem to solve)
- Ask: **Why do you think it is helpful to have a plan before you begin an investigation?** Guide students to understand that having a plan will help them investigate in a more logical way, think about any variables that need to be identified, manipulated, and controlled, and figure out exactly what they want to observe and how they will make those observations.

How Scientists Work

Designing Investigations and Using Evidence

Scientists do investigations to answer questions about the natural world. They make and follow inquiry plans to gather, organize, and analyze information. The organized plans that scientists use to investigate questions are called scientific methods.

All investigations begin with a question or a problem that scientists want to solve. Scientists then make a plan that they will follow to answer the question.

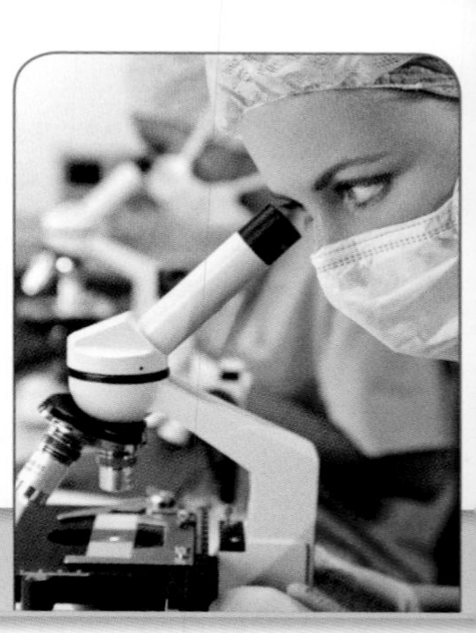

Each of these scientists will follow his or her own plan.

78

Scientists collect and record data. They analyze the data to explain what it shows. They make conclusions to answer their questions. Scientists use their data as evidence to support their explanations.

All investigations follow scientific methods, but there is no one set of steps that every investigation must follow. When planning an investigation, scientists follow the methods that will best help them to answer their question. Scientists have to be creative when they plan an investigation. A plan that helps to answer one question might not work as well to answer a different question.

79

Teach, continued

- Ask: **What are the scientists doing in the pictures on pages 78 and 79?** Write students' answers on the board.
- Read page 79 aloud with students. Ask: **What are some steps in a scientific investigation?** Write students' responses on the board. Guide students to mention asking a question, making observations, collecting and recording data, analyzing data, and drawing conclusions based on evidence.
- Ask: **Why is it important to be creative when designing a plan?** Discuss how different scientists come up with different ideas and different plans to answer scientific questions.
- Ask students to share examples of scientific investigations that they have carried out in school or elsewhere. Be sure to point out parts of these investigations that followed scientific methods. Students should understand that science does not always follow a rigidly defined method, but it does involve the use of observations and evidence. The order or number of scientific methods scientists use may vary.
- Ask: **Why do you think it is important for scientists to use evidence to support their explanations and conclusions?** Discuss how explanations and conclusions only are reliable when they are backed up with evidence. Also, evidence allows other scientists to build upon each other's work, which can lead to new understandings and discoveries.

Teach, continued

- Read page 80 aloud. Ask: **What question did Karl von Frisch investigate?** (Do honeybees see color?) Ask students to describe Frisch's investigation.
- Ask: **What did von Frisch conclude?** (Honeybees can see color.)
- Ask: **What evidence did he use to support his conclusion?** (The bees could be trained to seek food on a colored card. Bees continued to look for food on that colored card even when gray cards surrounded it.)
- Discuss the difference between fact and opinion. Ask: **Do you think von Frisch's conclusion that honeybees can see color is based on a fact or opinion? Why?** (Possible answer: Fact, because von Frisch gathered evidence to support his conclusion.) Point out that scientific methods are based on the use of facts and evidence, not opinion.
- Brainstorm a list of questions students have about honeybees. For several of the questions, invite students to suggest methods that could be used to answer those questions.

What Did You Find Out?

1. Scientists use different methods because some are more helpful in answering different types of questions.
2. Most of the bees visited the colored card. He concluded that the bees could see color.

Karl von Frisch

In the early 1900s, Austrian scientist Karl von Frisch wanted to find out whether honeybees could see color. He used scientific methods to investigate this question. His plan needed to be organized, but it also had to be creative.

In his investigation, von Frisch trained honeybees to feed from a dish that was set on a colored card. Then he placed the colored card in the middle of a group of gray cards. His hypothesis was that if most of the bees visited the colored card, then they could see color. During the investigation, most of the bees did visit the colored card. Von Frisch used this evidence to support his hypothesis. He concluded that honeybees see color.

Karl von Frisch won the Nobel Prize for his work with honeybees.

The honeybees were able to pick out the colored card from the gray cards.

SUMMARIZE

What Did You Find Out?

1. Why don't scientists use the same methods to investigate all scientific questions?
2. What evidence did Karl von Frisch use to support his hypothesis?

80

my SCIENCE notebook

Name ______ Date ______

Think Like a Scientist

How Scientists Work

Designing Investigations and Using Evidence

What Did You Find Out?

1. Why don't scientists use the same methods to investigate all scientific questions?

   Scientists use different methods because some are more helpful in answering different types of questions.

2. What evidence did Karl von Frisch use to support his hypothesis?

   Most of the bees visited the blue card. He concluded that the bees could see color.

Learning Master 80 Life Science

© NGSP & HB

Learning Masters 80–81 or at myNGconnect.com

Use Evidence to Answer a Question

Karl von Frisch also studied how bees find food. He observed that some worker bees left the hive to look for food. When these bees returned to the hive, they did a dance. Based on his observations, von Frisch concluded that the bees used different dances to communicate to other bees how far away the food was from the hive.

The graph shows data collected during an investigation of honeybee dances. Use the evidence in the graph to explain how the honeybee dance changes when food is farther away from the hive.

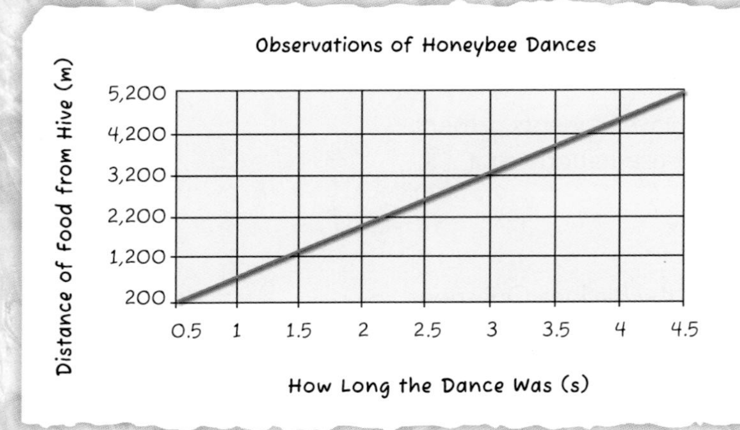

81

Raise Your SciQ!

NATIONAL GEOGRAPHIC

Bee Communication Karl von Frisch also observed that bees communicated through different types of dances, specifically through circling and wagging. Von Frisch found that the circling dance communicated that food was nearby, usually within about 75 m (250 ft) of the hive. The wagging dance, on the other hand, communicated that the food source was farther away.

Teach, continued

Use Evidence to Answer a Question

- Read page 81 aloud with students. Ask students to summarize Karl von Frisch's second bee investigation. Ask: **What did he conclude about the bee dances?** (The dances communicated to the other bees how far away the food was.)
- Help students interpret the graph and name the units used. Ask: **What does the *x*-axis show?** (the length of the bee dance in seconds) **What does the *y*-axis show?** (the distance of the food from the hive in meters)
- Ask: **When the food was about 1,200 m from the hive, how long was the dance?** (about 1.5 s) **When the food was about 3,200 m from the hive, how long was the dance?** (about 3 seconds)
- Ask: **What conclusion can you make from the data in this graph?** (As the distance of the food from the hive increases, the length of the dance increases.)
- Discuss the meaning of *scientific fact,* or observations that have been confirmed repeatedly and accepted as true. Ask: **Could von Frisch's conclusion that bees use different dances to communicate be considered scientific fact? Why or why not?** (It could be considered scientific fact only if the results of many different investigations supported it.) Point out that presenting an opinion as scientific fact is contrary to the scientific process.
- Have students identify examples of scientific facts from previous activities, such as how butterflies undergo complete metamorphosis or that owls eat small animals such as mice and voles.
- Encourage students to do research on Karl von Frisch or another scientist of their choice. Remind them to use appropriate references and identify any sources.

3 Assess

- Use Learning Masters Book, pages 80–81. Students should be able to answer questions about von Frisch's investigation and the graph on page 81.

Make a Life Cycle Poster

PAIRS

Help students make a poster to show the life cycle of a plant or animal, possibly native to your region.

Steps

1. Assign each pair a different plant or animal. Make sure that flowering plants, conifers, mosses, and ferns, as well as a variety of animals, are represented among the class.
2. Have pairs research the life cycle of their assigned plant or animal. Students may use reference books or the Internet, and they may consult with the school library or a media specialist.
3. Have pairs work together to plan their poster. The poster should illustrate the stages of the plant's or animal's life cycle in an attractive and engaging manner. Students may use drawings or photos, as well as text. Tell students to include a title, labels, and captions for their posters.
4. Designate a space for students to display their posters. Have students present their posters to each other. Then have them give their presentations on a family science night.

Discuss the Big Ideas

After students present their posters, lead a discussion about the Big Idea Questions:

- How Do Animals Grow and Change?
- How Do Plants Grow and Reproduce?

Help students see how their posters contribute to the understanding of the Big Idea Questions.

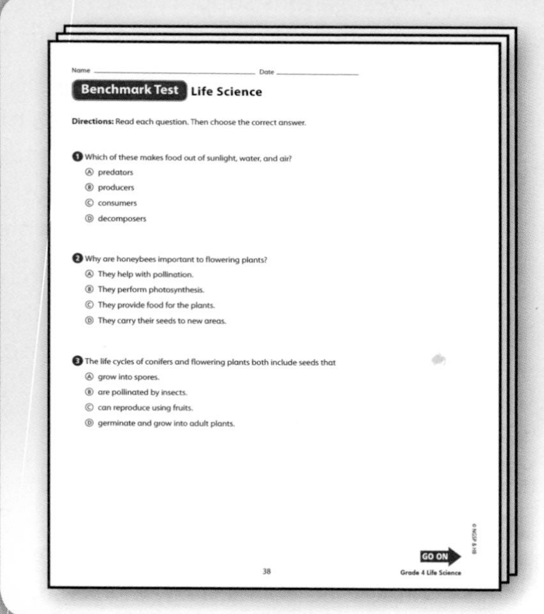

Name ______________________ Date ____________

Benchmark Test **Life Science**

Directions: Read each question. Then choose the correct answer.

1. Which of these makes food out of sunlight, water, and air?
   - Ⓐ predators
   - Ⓑ producers
   - Ⓒ consumers
   - Ⓓ decomposers

2. Why are honeybees important to flowering plants?
   - Ⓐ They help with pollination.
   - Ⓑ They perform photosynthesis.
   - Ⓒ They provide food for the plants.
   - Ⓓ They carry their seeds to new areas.

3. The life cycles of conifers and flowering plants both include seeds that
   - Ⓐ grow into spores.
   - Ⓑ are pollinated by insects.
   - Ⓒ can reproduce using fruits.
   - Ⓓ germinate and grow into adult plants.

GO ON

38 Grade 4 Life Science

© NGSP & HB

Assess Student Progress Benchmark Test

Have students complete their Life Science Benchmark Test to assess their progress in this unit.

Benchmark Test, Assessment Handbook, pages 38–44, or at myNGconnect.com

or NGSP ExamView CD-ROM

Make a Web Page About Living Things and Their Environment

SMALL GROUP

If students have a class Web page or school Web site, help them create a set of pages that show how the living things of their state depend on and adapt to the environment.

Steps

1. Work with groups so that each group chooses or is assigned a unique plant or animal. Students should then research the following information for their plant or animal:
   - role in environment (producer, carnivore, herbivore, omnivore, or decomposer)
   - type of environment where it lives (such as forest, prairie, tundra, or ocean)
   - adaptations that help the plant or animal survive in their environment
   - sample food chain or food web
2. If possible, students may use digital cameras to take photos of their plant or animal. Or students may draw or sketch their subject from observing it in nature or from observing photos. Scan in the drawings or upload photos to the Web page. If necessary, help students use the technology.
3. Have students write captions for each image, and then enter the captions onto the Web page. Remind students that the captions and illustrations should show how the plant or animal depends on and adapts to its environment.
4. Have groups present their Web pages to the class. You may also wish students to present their work to other fourth-grade classes.

Discuss the Big Ideas

After students present their Web pages, lead a discussion about the Big Idea Questions:

- How Do Living Things Depend on Their Environment?
- How Do Adaptations Help Living Things Survive?

Help students see how their Web pages contribute to the understanding of the Big Idea Questions.

Act Out an Interview with a Conservationist

PAIRS

Help students role-play an interview between a news reporter and a conservationist.

Steps

1. Have students investigate an environmental change or issue in their region or state, such as:
   - a new housing project or shopping center that is being proposed or will be built.
   - an invasive species, such as the Asian carp in the Illinois River, that is changing the lives of native species.
   - polluted land, air, or water and efforts to reduce the pollution.
2. Have pairs work together to research the environmental issue, write questions about it, and write answers to those questions. Encourage students to review the Big Ideas Book for examples of environmental issues and to learn about the job of conservationists.
3. When pairs are ready, have them perform their interviews for the class. You may want to use a video camera to film the interviews. Then have them perform their interviews on a family science night.

Discuss the Big Ideas

After students present their interviews, lead a discussion about the Big Idea Question:

- How Do Living Things Interact with Their Environment?

Help students see how their interviews contribute to the understanding of the Big Idea Question.

Make a Book About Body Systems

SMALL GROUP

Have students prepare a class book about the systems of the human body.

Steps

1. Assign each group one or two systems of the human body. The assignments should include the circulatory, respiratory, digestive, nervous, skeletal, and muscular systems.
2. Have groups research the parts and functions of their assigned system and how their system affects and depends on other body systems.
3. Choose a standard size and type of paper for the class book. Then distribute pages. Have each group prepare one or two book pages that illustrate and describe their assigned systems. The pages should include diagrams of the body systems with labels and captions.
4. Have each group present and discuss their page or pages. Then work with the class to create a cover and table of contents. Assemble the pages together into a class book.

Discuss the Big Ideas

After students present their diagrams, lead a discussion about the Big Idea Question:

- How Do the Parts of an Organism Work Together?

Help students see how their pages contribute to the understanding of the Big Idea Question.

NATIONAL GEOGRAPHIC
Science

Kindergarten

	Kindergarten
LIFE SCIENCE	Plants **How Are Plants Alike and Different?**
	Animals **How Are Animals Alike and Different?**
EARTH SCIENCE	Day and Night **How Are Day and Night Different?**
	Weather and Seasons **What Can You Observe About the Weather?**
PHYSICAL SCIENCE	Observing Objects **What Can You Observe About Objects?**
	How Things Move **How Do Things Move?**

K–2 Scope and Sequence

Grades 1–2

Living Things
Chapter 1 **How Are Living and Nonliving Things Different?**
Chapter 2 **What Are the Basic Needs of Plants?**
Chapter 3 **What Are the Basic Needs of Humans and Animals?**

Habitats
Chapter 1 **Where Do Plants and Animals Live?**
Chapter 2 **What Do Plants and Animals Need to Survive?**
Chapter 3 **How Do Plants and Animals Depend on Each Other?**

Plants and Animals
Chapter 1 **How Are Plants Alike and Different?**
Chapter 2 **How Are Animals Alike and Different?**
Chapter 3 **How Do Plants and Animals Change?**

Life Cycles
Chapter 1 **How Do Plants Grow and Change?**
Chapter 2 **How Do Humans and Animals Grow and Change?**
Chapter 3 **How Did Some Plants and Animals Become Extinct?**

Land and Water
Chapter 1 **How Do People Use Earth's Land?**
Chapter 2 **How Do People Use Earth's Water?**
Chapter 3 **How Does Earth's Land Change?**

Rocks and Soil
Chapter 1 **What Can You Observe About Rocks?**
Chapter 2 **How Do Rocks Change?**
Chapter 3 **What Can You Observe About Soil?**

Sun, Moon, and Stars
Chapter 1 **What Can You See in the Sky?**
Chapter 2 **What Can You Observe About the Sun?**
Chapter 3 **What Can You Observe About the Moon?**

Weather
Chapter 1 **How Does the Sun Affect Earth?**
Chapter 2 **How Does Weather Change?**
Chapter 3 **How Is Weather Measured?**

Properties
Chapter 1 **How Can You Describe Objects?**
Chapter 2 **How Can You Compare Objects?**
Chapter 3 **What Are Solids and Liquids?**

Solids, Liquids, and Gases
Chapter 1 **What Are Solids, Liquids, and Gases?**
Chapter 2 **How Can You Observe and Measure Properties?**
Chapter 3 **How Do Liquids and Solids Change?**

Pushes and Pulls
Chapter 1 **What Are Pushes and Pulls?**
Chapter 2 **What Ways Do Objects Move?**
Chapter 3 **How Do Magnets Pull Objects?**

Forces and Motion
Chapter 1 **What Is a Force?**
Chapter 2 **What Is Gravity?**
Chapter 3 **What Are Magnets?**

NATIONAL GEOGRAPHIC

Science

Grade 3

LIFE SCIENCE

Chapter 1 **How Do Plants Live and Grow?**

Chapter 2 **How Are Animals Alike and Different?**

Chapter 3 **How Do Plants and Animals Live in Their Environment?**

Chapter 4 **How Do Plants and Animals Survive?**

Chapter 5 **How Do Plants and Animals Respond to Seasons?**

EARTH SCIENCE

Chapter 1 **How Are the Sun, Earth, and Moon Connected?**

Chapter 2 **What Properties Can You Observe About the Sun?**

Chapter 3 **What Can You Observe About Stars?**

Chapter 4 **What Can You Observe About Earth's Materials?**

Chapter 5 **What Are Earth's Resources?**

Chapter 6 **How Do Weathering and Erosion Change the Land?**

Chapter 7 **How Does Earth's Surface Change Quickly?**

Chapter 8 **How Does Earth's Water Move and Change?**

PHYSICAL SCIENCE

Chapter 1 **How Can You Describe and Measure Matter?**

Chapter 2 **What Are States of Matter?**

Chapter 3 **How Does Force Change Motion?**

Chapter 4 **What Is Energy?**

Chapter 5 **What Is Light?**

3–5 Scope and Sequence

Grade 4

Chapter 1 **How Do Plants Grow and Reproduce?**
Chapter 2 **How Do Animals Grow and Change?**
Chapter 3 **How Do Living Things Depend on Their Environment?**
Chapter 4 **How Do Adaptations Help Living Things Survive?**
Chapter 5 **How Do Living Things Interact with Their Environment?**
Chapter 6 **How Do the Parts of an Organism Work Together?**

Chapter 1 **How Do Earth and Its Moon Move?**
Chapter 2 **How Are Rocks Alike and Different?**
Chapter 3 **What Are Renewable and Nonrenewable Resources?**
Chapter 4 **How Do Slow Processes Change Earth's Surface?**
Chapter 5 **What Changes Do Volcanoes and Earthquakes Cause?**
Chapter 6 **What Can We Observe About Weather?**

Chapter 1 **How Can You Describe and Measure Properties of Matter?**
Chapter 2 **What Are Physical and Chemical Changes?**
Chapter 3 **How Do Forces Act?**
Chapter 4 **What Is Magnetism?**
Chapter 5 **What Are Some Forms of Energy?**
Chapter 6 **What Is Sound?**
Chapter 7 **What Is Electricity?**

Grade 5

Chapter 1 **How Do Scientists Classify Living Things?**
Chapter 2 **What Are the Interactions in Ecosystems?**
Chapter 3 **How Does Energy Move in an Ecosystem?**
Chapter 4 **How Do Living Things Survive and Change?**
Chapter 5 **How Do Body Systems Work Together?**

Chapter 1 **How Do Earth and Its Moon Create Cycles?**
Chapter 2 **What Makes Up the Solar System?**
Chapter 3 **How Are Rocks and Minerals Identified?**
Chapter 4 **How Can We Protect Earth's Resources?**
Chapter 5 **How Are Weather and the Water Cycle Connected?**

Chapter 1 **How Can You Describe Matter, Mixtures, and Solutions?**
Chapter 2 **How Can Matter Change?**
Chapter 3 **How Do You Describe Force and the Laws of Motion?**
Chapter 4 **What Are Simple Machines?**
Chapter 5 **How Do You Describe Different Forms of Energy?**
Chapter 6 **How Does Electrical Energy Flow and Transform?**

Metric and SI Units

The metric system is also called the International System of Units, or SI. Scientists, as well as most countries throughout the world, use the metric system. The United States uses customary units. The chart below will help you convert units between metric measurements and customary measurements.

	METRIC/SI	CUSTOMARY	CONVERSION	
Distance	centimeter	inch	1 cm = .39 in	1 in = 2.54 cm
	meter	foot	1 m = 3.28 ft	1 ft = .30 m
	kilometer	mile	1 km = .62 mi	1 mi = 1.61 km
Volume	milliliter	teaspoon	1 mL = .20 tsp	1 tsp = 4.93 mL
	liter	pint	1 L = 2.12 pt	1 pt = .47 L
	liter	quart	1 L = 1.06 qt	1 qt = .95 L
	liter	gallon	1 L = .26 gal	1 gal = 3.79 L
Mass	gram	ounce	1 g = .035 oz	1 oz = 28.35 g
	kilogram	pound	1 kg = 2.2 lbs	1 lb = .45 kg
Temperature	Celsius	Fahrenheit		
	0° water freezes	32° water freezes	°C = 5/9 (°F – 32)	
	100° water boils	212° water boils	°F = 9/5 °C + 32	

CONVERTING AMONG CUSTOMARY UNITS	
Distance	12 inches = 1 foot; 3 feet = 1 yard; 5280 feet = 1 mile
Volume	1 pint = 16 ounces; 1 quart = 32 ounces; 1 gallon = 128 ounces
Mass	1 pound = 16 ounces; 1 ton = 2000 pounds

CONVERTING AMONG METRIC UNITS	
Distance	1 centimeter = .01 meter; 1 kilometer = 1000 meters
Volume	1 milliliter = .001 liter
Mass	1 milligram = .001 gram; 1 kilogram = 1000 grams

Materials List

	INQUIRY	KIT MATERIALS	SCHOOL-SUPPLIED MATERIALS
CHAPTER 1 **How Do Plants Grow and Reproduce?**	**Explore Activity** *Investigate a Plant Life Cycle* What are the stages in the life cycle of a radish plant? page T5e	spoon radish seeds plastic cup potting soil spray bottle hand lens brush	Purple Coneflower Life Cycle Learning Master water
	Directed Inquiry *Investigate Flowers* What are the parts of a flower? page T33e	hand lens microscope slide	flower Parts of a Flower Learning Master scissors metric ruler white paper
CHAPTER 2 **How Do Animals Grow and Change?**	**Directed Inquiry** *Investigate Earthworm Behavior* Will an earthworm move toward a light or dark environment? page T45e	plastic pan spray bottle black construction paper plastic spoon earthworms (live materials coupon) flashlight stopwatch medium-sized plastic container loamy soil cheesecloth rubber band	paper towels water masking tape food scraps
	Guided Inquiry *Investigate Insect Life Cycles* How does the life cycle of a butterfly compare with the life cycle of a grasshopper or dragonfly? page T69a	caterpillars (live materials coupon) hand lens butterfly habitat safety pin	Insect Life Cycle Learning Master facial tissue or carnation (optional) sugar water orange

Materials List, continued

	INQUIRY	KIT MATERIALS	SCHOOL-SUPPLIED MATERIALS
CHAPTER 3 **How Do Living Things Depend on Their Environment?**	**Directed Inquiry** *Investigate Owl Pellets* What objects can you find in an owl pellet? page T85e	protective gloves owl pellet paper plates hand lens craft stick forceps Bone Sorting Chart resealable plastic bags	safety goggles masking tape
	Guided Inquiry *Investigate Food Chains and Webs* How can you use an owl pellet to infer a food chain and food web? page T119a	protective gloves bones from owl pellet forceps index cards Bone Sorting Chart Skeleton diagrams Food Web diagram	glue white paper construction paper
CHAPTER 4 **How Do Adaptations Help Living Things Survive?**	**Directed Inquiry** *Investigate Plant Fossils* How can you use plant fossils to infer what some environments were like long ago? page T133g	fern, elm, or sycamore fossil hand lens tissue paper chenille stems clay	pencil Plant Information chart Learning Master construction paper markers glue scissors
	Guided Inquiry *Investigate Animal Survival* How can an animal's color help it survive in its environment? T169a	hole punch paper cup stopwatch white, patterned, and green cloth	construction paper

Materials List, continued

	INQUIRY	KIT MATERIALS	SCHOOL-SUPPLIED MATERIALS
CHAPTER 5 **How Do Living Things Interact with Their Environment?**	Directed Inquiry *Investigate Plants and Water* How can a plant affect the water in its environment? page T181e	hand lens index card waxed paper clay plastic cups	safety goggles plant cutting masking tape pencil water black marker
	Guided Inquiry *Investigate Colors in Green Leaves* What colors are present in a green leaf? page T213a	hand lens plastic cups plastic spoon graduated cylinder stopwatch coffee filter craft sticks	safety goggles green leaves scissors rubbing alcohol metric ruler masking tape paper towel colored pencils (optional)
CHAPTER 6 **How Do the Parts of an Organism Work Together?**	Directed Inquiry *Investigate How Your Brain Can Work* What happens when you identify different colors of letters in words? T229g	stopwatch	white paper markers
	Guided Inquiry *Investigate Exercise and Heart Rate* How does exercise affect heart rate? page T255a	stopwatch	
UNIT INQUIRY	Open Inquiry *Do Your Own Investigation* Sample Question and Steps: How do earthworms affect the growth of lima bean plants? page T268f	lima beans clear plastic containers loamy soil spray bottle earthworms (live materials coupon) plastic spoon	safety goggles masking tape marker water ruler

Glossary

A

adaptation
An adaptation is a feature that helps a living thing survive in its environment.

analyze
When you analyze, you look for patterns and relationships in the data.

C

camouflage
Camouflage is a color or shape that makes a living thing hard to see.

carnivore
A carnivore is an animal that eats other animals to survive.

circulatory system
The circulatory system is a group of organs that carries blood throughout the body.

classify
When you classify, you put things into groups according to their characteristics.

compare
When you compare, you tell how objects or events are alike and different.

conclude
You conclude when you use information, or data, from an investigation to come up with a decision or answer.

conifer
A conifer is a seed plant that reproduces with cones.

conserve
To conserve is to use resources in a careful way.

count
When you count, you tell the number of something.

D

data
Data are observations and information that you collect and record in an investigation.

digestive system
The digestive system is a group of organs that breaks down food into energy and materials that the body can use.

E

estimate
When you estimate, you tell what you think about how much or how many.

extinction
Extinction is the complete loss of one kind of living thing.

F

fertilization
In fertilization, an egg and a sperm cell join.

H

herbivore
An herbivore is an animal that eats plants to survive.

heredity
Heredity is the passing of traits from parents to their offspring.

hypothesis
When you make a hypothesis, you state a possible answer to a question that can be tested by an experiment.

I

infer
When you infer, you use what you know and what you observe to draw a conclusion.

inherited
An inherited characteristic is passed down from parents to offspring.

invasive organism
An invasive organism is a plant or animal that does not belong in a place and harms it.

investigate
You investigate when you make a plan to answer a question and carry it out.

L

larva
A larva is a young animal with a body form very different from the adult.

M

measure
When you measure, you find out how much or how many.

metamorphosis
Metamorphosis is a series of major changes in an animal's body form during its life cycle.

Glossary

model
You can make and use a model to show how something in real life works.

N

nervous system
The nervous system works as a communication network to help the body respond to its environment.

nymph
A nymph is the stage of incomplete metamorphosis in which the young adult looks like the adult.

O

observe
When you observe, you use your senses to learn about an object or event.

omnivore
An omnivore is an animal that eats plants or other animals to survive.

operational definition
When you make an operational definition, you use your own words to tell what something can do.

organ
An organ is a body part that does a specific job in an organism.

P

photosynthesis
In photosynthesis, plants use the energy of sunlight to make food.

plan
When you make a plan to answer a question, you list the materials and steps you need to take.

pollination
Pollination is the movement of pollen from a stamen to a pistil or from a male cone to a female cone.

pollution
Pollution is a harmful substance that people put into the air, water, and soil.

predator
A predator is an animal that hunts other animals for food.

predict
When you predict, you tell what you think will happen.

prey

Prey are animals that other animals hunt for food.

pupa

A pupa is the stage in metamorphosis in which the body form of a young animal changes from the larva to the adult.

R

recycle

To recycle is to use something again.

respiratory system

The respiratory system is a group of organs that brings oxygen into the body and gets rid of carbon dioxide.

S

seed dispersal

In seed dispersal, the seeds of a plant are carried to a new place.

share

When you share results, you tell or show what you have learned.

system

A system is a group of organs that work together in an organism to help it live and grow.

V

variable

A variable is a part of an experiment that you can change.

variation

A variation is a different form of a feature of the same kind of living thing.

Index

Note: EM refers to the End Matter and SN refers to the Nature of Science/ Science Notebook.

A

B

C

D

E

F

G

H

I

J

K

L

M

N

O

P

R

S

T

U

Notes

Notes

Notes

Notes

Notes

Notes

Notes

Credits

Acknowledgments

Grateful acknowledgment is given to the authors, artists, photographers, museums, publishers, and agents for permission to reprint copyrighted material. Every effort has been made to secure the appropriate permission. If any omissions have been made or if corrections are required, please contact the Publisher.

Photographic Credits

Front Matter

Front Cover, Tabs David Doubilet/National Geographic Image Collection. **Inside Front Cover** (tl) Albert Moldvay/National Geographic Image Collection. (tr) Paul Nicklen/National Geographic Image Collection. (cl) Bill Hatcher/National Geographic Image Collection. (cr) Michael S. Lewis/National Geographic Image Collection. (bl) Taylor S. Kennedy/National Geographic Image Collection. (br) Gordon Wiltsie/National Geographic Image Collection. **x–xi** (bg) PhotoDisc/Getty Images. **xii–xiii** Radius Images/Photolibrary. **xiv–xv** (bg) Ihoko Saito/Toshiyuki Tajima/Dex Image/Photolibrary. **xiv** (tl) NASA Human Space Flight Gallery. (tr) Harvey Lowe. (cl) Mark Thiessen/National Geographic Image Collection. (crt) Bruce Bennett. (crb) Renee Fadiman. (bl) John Livzey. (br) Mark Mallchok. **xv** (tl, tr) Brella Productions. (clt) Calit2. (clb) Mark Mallchok/Brella Productions. (crt) Robert Clark/National Geographic Image Collection. (crb) Brett Hobson. (bl) Mark Thiessen/National Geographic Society. (br) University of Alaska, Fairbanks – Marketing and Communications. **SN5** Digital Vision/Getty Images. **T4a** (bg) Edward Parker/Alamy Images. (inset) Jameson Weston/iStockphoto. **T4b** DigitalStock/Corbis.

Chapter 1

T5a–T5d Edward Parker/Alamy Images.

Chapter 2

T45a–T45d Edward Parker/Alamy Images.

Chapter 3

T85a–T85d Edward Parker/Alamy Images.

Chapter 4

T133a–T133f Edward Parker/Alamy Images.

Chapter 5

T181a–T181d Edward Parker/Alamy Images.

Chapter 6

T229a–T229f Edward Parker/Alamy Images.

End Matter

Back Cover (bg) David Doubilet/National Geographic Image Collection. **EM2–EM5** Tom Mead/Photolibrary.

Program Authors

Randy Bell, Ph.D., Associate Professor of Science Education, University of Virginia, Charlottesville, Virginia; Malcolm B. Butler, Ph.D., Associate Professor of Science Education, University of South Florida, St. Petersburg, Florida; Kathy Cabe Trundle, Ph.D., Associate Professor of Early Childhood Science Education, The Ohio State University, Columbus, Ohio; Judith Sweeney Lederman, Ph.D., Director of Teacher Education and Associate Professor of Science Education, Department of Mathematics and Science Education, Illinois Institute of Technology, Chicago, Illinois; David W. Moore, Ph.D., Professor of Education, College of Teacher Education and Leadership, Arizona State University, Tempe, Arizona

The National Geographic Society

National Geographic School Publishing
Hampton-Brown
www.NGSP.com

Printed in the USA.
RR Donnelley, Menasha, WI

ISBN: 978-0-7362-7781-5

11 12 13 14 15 16 17 18 19

10 9 8 7 6 5 4